Machine Learning in Earth, Environmental and Planetary Sciences

Theoretical and Practical Applications

Machine Learning in Earth, Environmental and Planetary Sciences

Theoretical and Practical Applications

Hossein Bonakdari
Department of Civil Engineering, Faculty of Engineering, University of Ottawa, Ottawa, Canada

Isa Ebtehaj
Soil and Environment Department, Faculty of Agriculture and Food Science, Laval University, Quebec City, Canada

Joseph D. Ladouceur
Department of Civil Engineering, Faculty of Engineering, University of Ottawa, Ottawa, Canada

Elsevier
Radarweg 29, PO Box 211, 1000 AE Amsterdam, Netherlands
The Boulevard, Langford Lane, Kidlington, Oxford OX5 1GB, United Kingdom
50 Hampshire Street, 5th Floor, Cambridge, MA 02139, United States

ISBN: 978-0-443-15284-9

For Information on all Elsevier publications
visit our website at https://www.elsevier.com/books-and-journals

Publisher: Candice Janco
Acquisitions Editor: Peter Llewellyn
Editorial Project Manager: Mason Malloy
Production Project Manager: Bharatwaj Varatharajan
Cover Designer: Miles Hitchen

Typeset by MPS Limited, Chennai, India

Dedication

Deepest thanks to our amazing families, who have shown endless support.

Contents

About the authors

Hossein Bonakdari, PhD, PEng., was recently appointed as an associate professor in the Department of Civil Engineering at the University of Ottawa, with advanced knowledge and organized research activities in water resources and artificial intelligence (AI) techniques for environmental issues under climate change. His research has primarily been focused on watershed-scale water quantity and quality models. Throughout his 15 years of professional experience in research, he has developed a body of work solving real engineering problems in the context of managing water-related risks in watersheds based on AI, satellite data and remote sensing, infrastructure design, optimizing current water resources, and infrastructure real-time modeling to protect vulnerable areas from water-related hazards in urban and agricultural sectors.

Results obtained from his research have been published in more than 270 papers in international journals (h-index = 48) with more than 7000 citations. He also has more than 150 presentations at national and international conferences. He is one of the most influential scientists in the field of developing novel algorithms based on AI and machine learning techniques for solving practical problems. Since 2019, he has been ranked in the list of the world's top 2% of scientists in various fields, which Stanford University publishes yearly. He is a Guest Editor for several special issues on the application of AI and machine learning techniques for solving applied science problems in *Informatics*, *Sustainability*, and *Earth* journals. He is in the editorial board of *Informatics*, Natural Resources Informatics, and *Earth* journals. He is an internationally recognized, prolific author in Smart and Sustainable Infrastructure Development and a high-impact productive researcher.

Isa Ebtehaj is a PhD candidate in soil and environments at the Laval University Department of Soils and Agri-Food Engineering, Québec, Canada. He holds a master degree in Civil Engineering-Water Engineering as well as a Bachelor of Civil Engineering diploma. Eight years of practical experience in two advanced research centers (i.e., Water and Wastewater Research Center and Environmental Research Center) led him to continue his postgraduate studies at Laval University. His research has primarily been focused on developing flood forecasting models using high-quality hybrid machine-learning techniques and climate change models. Through his PhD studies at Laval University, he has published 12 papers in international journals, six discussions in the ASCE, four conference papers, and two book chapters. Based on his experience in machine learning modeling, he has been invited to participate in a novel industrial project which uses data from the Quebec Pesticides Management Code to identify key practices resulting in pesticide use reduction on golf courses and to predict pesticide use evolution under different climate change scenarios.

Joseph D. Ladouceur, EIT, is a PhD student in civil engineering at the University of Ottawa, Canada. His research interests are primarily focused on applications of low-pressure membrane technology for drinking water production. Specifically, Joseph is focused on pretreatment strategies for membrane fouling reduction with an aim to make membrane processes more sustainable, economic, and better suited to changing global water quality. Joseph is a registered EIT with Professional Engineers Ontario.

Preface

Machine learning has garnered increased attention in recent years across most fields of science but especially in engineering. Several factors contribute to the popularity of machine learning techniques including their ability to be implemented using only simple coding, their capacity to overcome the existing challenges of traditional methodologies, and their use as a powerful tool for solving real-world problems. Despite the increased popularity, there is currently no book on this subject which provides readers with both the necessary theoretical and application-based background required when resolving complex problems. As an expert in developing and applying machine learning techniques in planetary, earth, and environmental sciences, it has always been a goal of mine to compose a comprehensive, yet user-friendly, professional reference for students, scholars, and practitioners alike.

The material in this book is presented over the course of 11 chapters and seeks to review fundamentals of machine learning approaches, develop novel approaches based on the extreme learning machine (ELM), and demonstrate the application of these approaches using practical real-world data. This text will provide the reader with the necessary background required to understand how to prepare data for use in machine learning, how to fine-tune the modeling process, and finally how to interpret the results. The first chapter introduces the machine learning–based modeling process in detail as well as defines the real-world example data that will be employed in the subsequent chapters. In the second chapter, four of the most common preprocessing techniques including normalization, standardization, data splitting, and cross-validation are presented in detail along with their associated MATLAB® coding. Moreover, each preprocessing technique is applied to the different real-world examples defined in the first chapter so that the outcomes may be compared and discussed. In the third chapter, postprocessing is introduced, with the techniques being categorized into two main groups: quantitative and qualitative. The quantitative tools include statistical indices, uncertainty analysis, and reliability analysis, while the qualitative tools include the Taylor diagram, scatter plot, barplot, histogram, and boxplot. For all quantitative and qualitative tools, the theoretical formulation and their detailed coding are presented. The fourth chapter comprised four main sections. In the first section, the fundamentals of the feed-forward neural network are reviewed. In the second section, the structure of the ELM is provided and its advantages and disadvantages are discussed. The third and fourth sections of the chapter present the mathematical definition of the ELM and its activation functions, respectively. The detailed coding of the ELM in MATLAB is provided in Chapter 5. Here, step-by-step coding of this model is provided including a detailed discussion of the two main functions used by the ELM for training and testing stages. Moreover, a calculator is introduced which can be used to apply the calibrated model to future unseen samples. Using the different real-world data sets introduced in Chapter 1, the effects of ELM parameters (i.e., iteration number, number of hidden neurons, activation function, and orthogonal effect) are also investigated. In Chapter 6, the performance of the ELM in the presence of outliers is discussed, and different approaches are introduced to overcome this limitation of the ELM. These approaches include the regularized extreme learning machine (RELM), weighted regularized extreme learning machine (WRELM), and outlier robust extreme learning machine (ORELM). The main concept of these techniques along with their mathematical formulations is provided in this chapter. In Chapter 7, a detailed description of how the developed ELM-based models improve the modeling performance of the ELM in MATLAB is presented. Besides, a calculator is introduced to apply the newly developed method for previously unseen samples. Moreover, the effects of user-defined parameters (i.e., regularization parameter, weight function in WRELM, and the maximum number of iterations in ORELM) on the modeling results of the ELM-based models are evaluated. In Chapter 8, the difference between online and batch learning is provided in detail. Using the online learning concept, the online sequential ELM (OSELM) is introduced and its mathematical formulation is presented in detail. In Chapter 9, the MATLAB coding process of the OSELM and its associated calculator tool is presented. Moreover, the effect of the OSELM parameters (i.e., the number of initial training samples and block size) on model performance for all defined problems in the first chapter is evaluated. In Chapter 10, an ELM-based method is optimized using an evolutionary technique known as self-adaptive evolutionary ELM (SaE-ELM). This method can efficiently overcome the limitation of the classical ELM in the random selection of the two of the three main matrices. The conceptual and mathematical definition of this method is described in this chapter. In Chapter 11, the last chapter of this book, a detailed description of how the SaE-ELM is implemented is offered. Moreover, a calculator is introduced to use the trained and calibrated SaE-ELM for practical studies. Besides, the sensitivity analysis of the SaE-ELM parameters to its parameters in different problems is evaluated.

Acknowledgments

I wish to extend my thanks to my coauthors, Isa and Joseph, without whom this book would not be possible. I especially wish to extent my gratitude to Isa, who, over the past 10 years, has supported me as an excellent, friendly, patient, kind, and well-established researcher. The authors acknowledge the financial support provided by the Natural Science and Engineering Research Council of Canada (NSERC) Discovery Grant (#RGPIN-2020-04583) and the "Fond de Recherche du Québec- Nature et Technologies", Québec Government (#B2X−315020). We would also like to acknowledge the Water Group of the Department of Civil Engineering, Ottawa University, for all the practical assistance and logistic support throughout this work. We particularly thank Dr. Jacques Beauvais, Dean of Faculty of Engineering, and Dr. Mamadou Fall, Chair of the Department of Civil Engineering, University of Ottawa, for their support and encouragement during the preparation of this book.

Hossein Bonakdari
Department of Civil Engineering, University of Ottawa, Ottawa, Canada

Ranked in the 2% list of the world's top scientists (h-index = 48)
Editorial board, *Earth*
Editorial board, *Informatics* journal
Editorial board, Natural Resources Informatics
Section Board Member, *Sustainability* journal

ORCID ID: https://orcid.org/0000-0001-6169-3654
Researcher ID: http://www.researcherid.com/rid/B-9305-2018
http://www.researchgate.net/profile/Hossein_Bonakdari
https://scholar.google.ca/citations?user = bBWIpAUAAAAJ&hl = en

Isa Ebtehaj
Department of Soils and Agri-Food Engineering, Laval University, Québec, Canada

Joseph D. Ladouceur
Department of Civil Engineering, University of Ottawa, Ottawa, Canada

About the cover image

The cover of this text, which happens to be the view from my office window, is the Rideau Canal, the oldest continuously operated canal system in North America and a UNESCO World Heritage Site. Located in the heart of the National Capital (Ottawa), the Rideau Canal hosts numerous attractions during the spring, summer, and fall seasons, although is perhaps best known for its transformation into the world's largest skating rink during the winter.

Hossein Bonakdari

Chapter 1

Dataset preparation

1.1 The modeling process

Prior to implementing any machine learning (ML) technique, the modeler must first have an understanding of the general approach to be followed when faced with any modeling problem. Often, in the analysis and resolution of any of these problems, a consistent modeling paradigm or methodology may be followed. Fig. 1.1 presents a schematic detailing all of the steps involved in the resolution of ML-based modeling problems. From this figure, the ML application process is composed of three main components: (1) data collection; (2) preprocessing; (3) modeling by ML techniques, and (4) postprocessing. During data collection, a set of data relating independent variables to one or more target variables is identified.

In Fig. 1.1, for example, three independent input variables are identified as "raw data." Data collection/ data generation can be done in many ways including through personal experimentation (laboratory experimental results), from open data sources (such as USGS or Statistics Canada), or from review documentation and/or existing published data sets (Fig. 1.2).

The second step is preprocessing. This is considered one of the most critical steps when modeling with ML techniques, as data transformation and cleaning can result in more meaningful modeling results and better model performance. In fact, without preprocessing, it may prove difficult in some circumstances to fit an adequate and generalizable model to the data set (Niu et al., 2020; Obaid et al., 2019). Consider the following example. Suppose the range of different input variables within a dataset is not the same and spans several orders of magnitude. In this case, the model will place more effort into optimizing adjustable parameters for the variables that have higher magnitudes compared to those variables which are small in magnitude. The changes obtained during the modeling process for the optimized

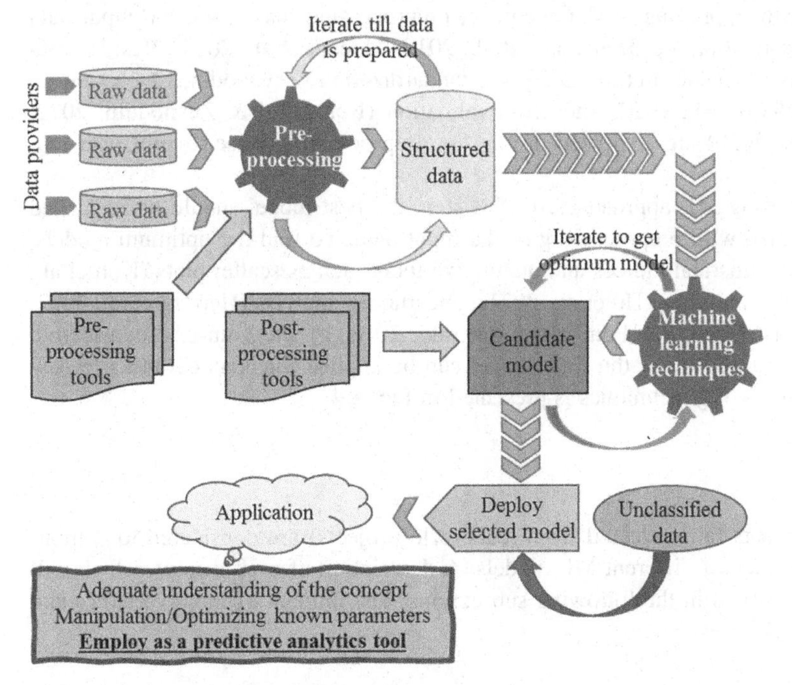

FIGURE 1.1 The modeling process/paradigm in machine learning.

Machine Learning in Earth, Environmental and Planetary Sciences. DOI: https://doi.org/10.1016/B978-0-443-15284-9.00002-1

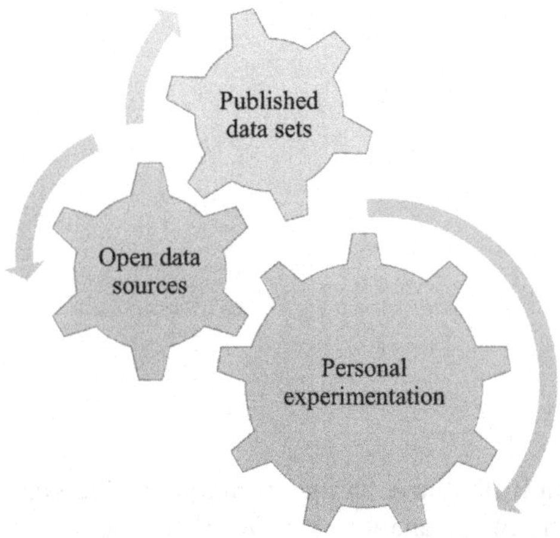

Published data sets
Open data sources
Personal experimentation

FIGURE 1.2 Different types of data collections.

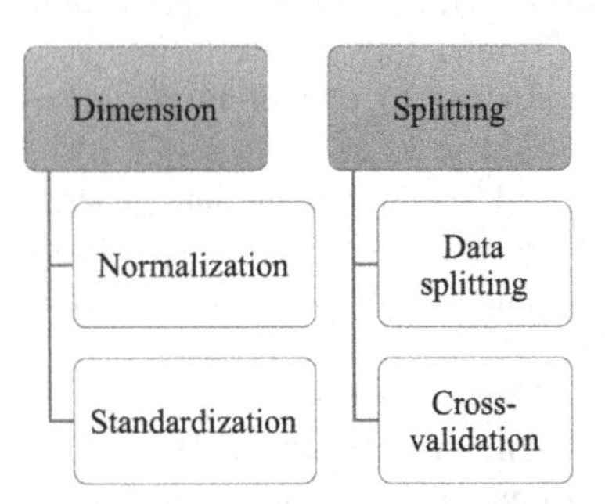

FIGURE 1.3 Summary of fundamental preprocessing techniques.

values of the variables with smaller values will be negligible, and the model may even choose to ignore the changes in the value of these parameters. One of the most well-known approaches to overcome this limitation of ML models is by applying preprocessing in the form of normalization (Ebtehaj & Bonakdari, 2016a; Ivanyuk & Soloviev, 2019; Qasem et al., 2017) or standardization (Ebtehaj et al., 2019; Gómez-Escalonilla et al., 2022; Zeynoddin et al., 2019) of the raw data. From this simple example, it can be readily appreciated that familiarity with preprocessing methods is fundamental to modeling with ML methods. In addition to improving model accuracy, preprocessing may make the input data simpler to understand for the modeler and easier to compare (Bonakdari et al., 2019; Moeeni et al., 2017). Besides data scaling in the form of normalization (Zeynoddin, Bonakdari et al., 2020) or standardization (Zeynoddin, Ebtehaj et al., 2020; Zhang et al., 2018), data splitting (Ebtehaj et al., 2020) and cross-validation (Bonakdari & Zeynoddin, 2022; Ebtehaj & Bonakdari, 2016b; Ferdinandy et al., 2020) are also used during the preprocessing phase to split data into training and testing samples (Fig. 1.3).

The third step in the paradigm is modeling using ML approaches. In this step, the best model should be identified through the use of optimization techniques coupled with preprocessing of the input data. To find the optimum models, a set of quantitative postprocessing tools such as statistical indices and qualitative tools such as scatter plots (Kim et al., 2019), box plots (Jato-Espino et al., 2019), Taylor diagrams (Hu et al., 2021), uncertainty analysis (Herrera et al., 2022; Sharafati et al., 2020); reliability analysis (Hariri-Ardebili & Pourkamali-Anaraki, 2018a,b), etc., must be considered. After selecting the best model through the use of these tools, the final model can be applied for practical tasks to new data sets. A summary of the fundamental postprocessing techniques is presented in Fig. 1.4.

1.2 Data description

Throughout this text, five different sample data sets (all collected from real-world projects) are considered to demonstrate the development, application, and performance of different ML models. A description of each dataset, which will be termed Examples 1–5 in Appendix 1A, is provided in the following subsections. The number of input variables and the number of all samples are provided in Fig. 1.5.

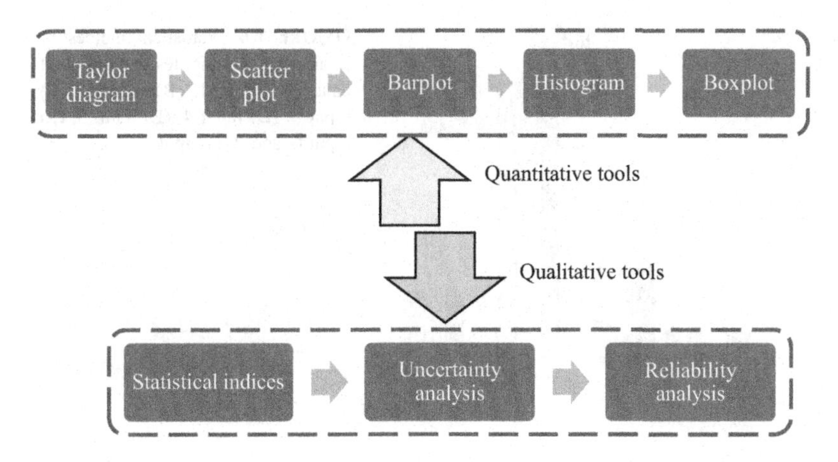

FIGURE 1.4 Summary of fundamental postprocessing techniques.

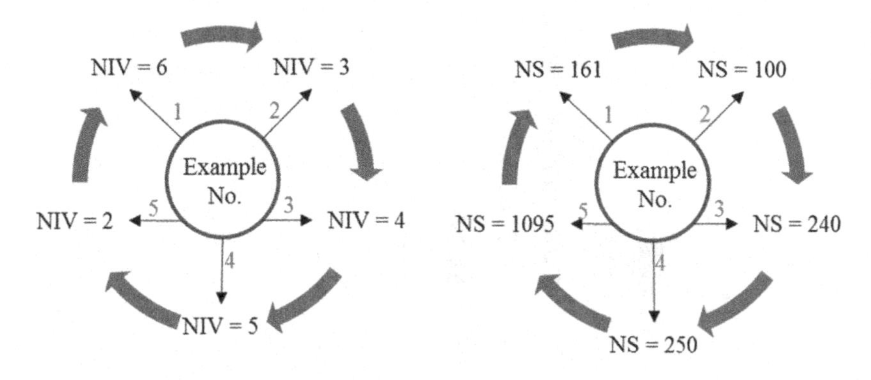

FIGURE 1.5 Characteristics of example data. *NIV*, Number of input variables; *NS*, number of all samples.

1.3 Different types of problems

1.3.1 Example 1: a problem with six input variables

The dataset considered in Example 1 is a composite set that was formed by the aggregation of data from two different studies (Bagheri et al., 2014; Cheong, 1991). Therefore the collected data for Example 1 is a combination of two published data sets. In order for a modeler to aggregate any number of datasets, the laboratory conditions of both datasets must be practically identical such that the data was collected in the same way. Given that it is time-consuming to perform experiments with different ranges of variables affecting the investigation, it is often the case that scholar(s) cannot examine all conditions in a single study. Therefore, by juxtaposing several studies in which the conditions of the experiments are consistent with each other, the limitations of each study can be overcome. In addition, the use of ML methods requires a wide range of independent input variables to train the desired model(s) and the typically large number of samples. High numbers of input variables are required to provide the model with the experience required to estimate the target variable with acceptable accuracy for unseen samples (i.e., testing samples). For example, if one dataset only covers a range of input values from 0 to 20, while another set ranges from 15 to 70, it may prove valuable to develop a model that spans the greater range of values defined by the juxtaposed set (i.e., from 0 to 70) so that it has a greater range of application in solving real-world problems. Considering more than one data set is a well-known approach in real-world practical applications of ML by scholars (Azimi et al., 2016; Ebtehaj & Bonakdari, 2014; Ebtehaj et al., 2015, 2016, 2017; Gholami et al., 2017).

In Example 1, the total number of samples is 161, with 113 samples being randomly selected to train the model (i.e., Training samples) while the rest of the samples (i.e., 48 samples) are used as a validation to check the performance of the developed ML-based model when faced with unseen samples (i.e., Testing samples). Modeling is performed considering 70% of all samples, while 30% of samples are reserved to verify the generalizability of the developed model. It should be noted that the modeling process for ML-based models should be controlled in such a way that the developed model performs well in both the testing phase and the training phase such that it is generalizable to application in a range of future tasks.

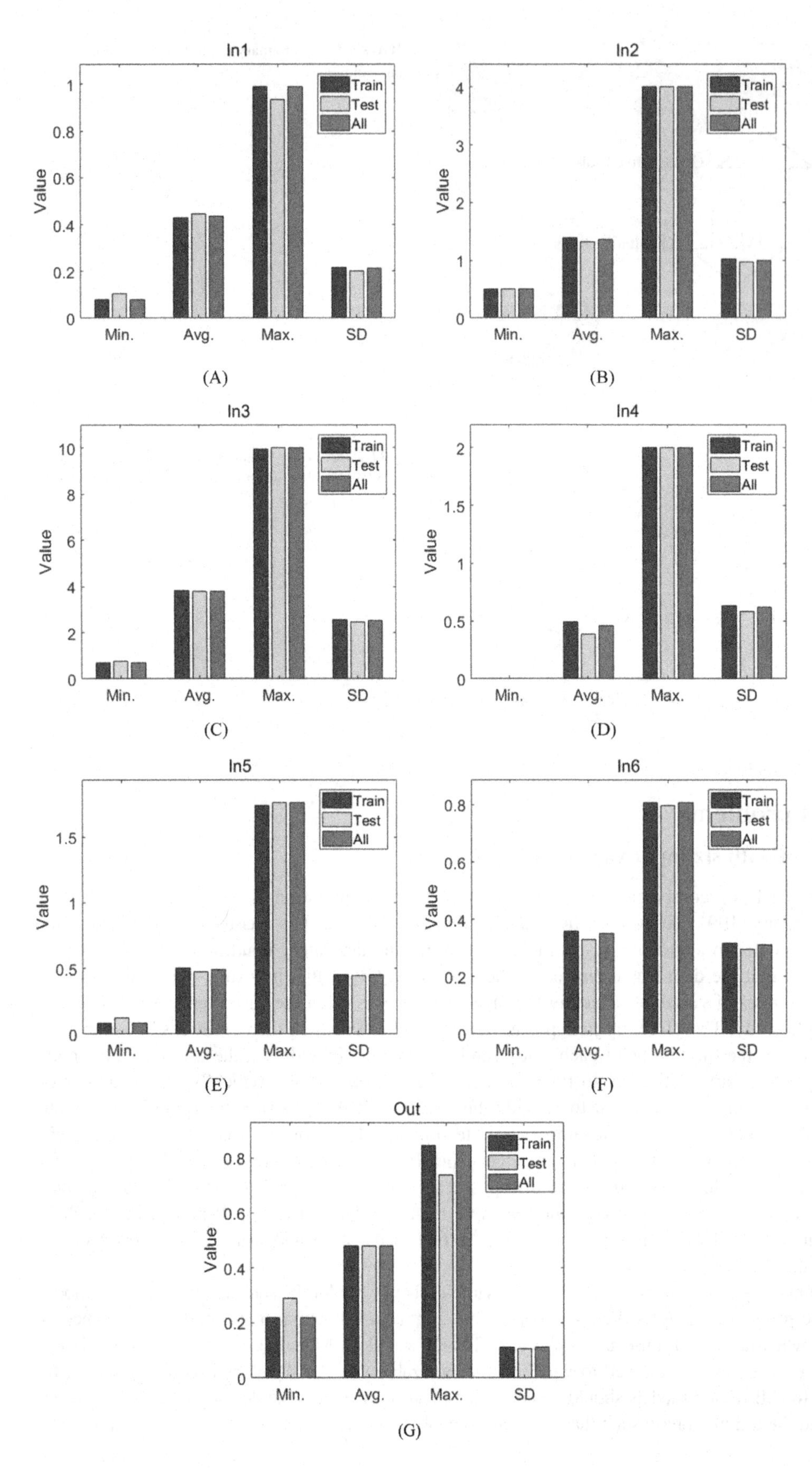

FIGURE 1.6 Statistical indices of Example 1 training, testing, and total data. (A) Input 1, (B) input 2, (C) input 3, (D) input 4, (E) input 5, (F) input 6, and (G) output.

Different splitting ratios may be considered for assigning training and testing data, where the maximum amount of test data is about 50% (i.e., 50% for the training stage and 50% for the testing stage) and the minimum is 10% (i.e., 90% for the training stage and 10% for the testing stage) (Ebtehaj et al., 2020). However, considering 30% of the total data as testing samples is a well-known distribution for the testing stage. Therefore this ratio is considered throughout this text. However, the reader should be aware that in some instances, the nature of the data set necessitates that different splitting ratios be studied to obtain the optimal distribution of training and testing samples. The optimal percentage is defined in such a way that the number of training and testing samples is not too small, but also the performance of the model in both training and testing modes must be very close to each other.

1.3.1.1 Statistical description of Example 1 data using barplot analysis

The minimum (Min.), average (Avg.), maximum (Max.), and standard deviation (SD) values for all independent inputs (In1, In2, In3, In4, In5, In6), as well as the dependent output (Out) for the training, testing, and total data, are provided in Fig. 1.6A–G for Example 1 data.

From this figure, it can be seen that the ranges of input values are significantly different from each other. This can also be said of the input variables compared to the output variable. For example, the maximum value of In3 is 10, while this same value is 4 (i.e., In3) or less than 4 (In1, In4, In5, In6) for all other inputs. If the range of values for the different inputs and output variables differs greatly from one another, the modeler may need to apply normalization during the preprocessing stage. Another consideration for the modeler is the similitude in the range of the data used for training and testing subsets. It is desirable to split the data such that the training subset will provide all the necessary experience to the developed model through exposure to the full range of input variables. If the data distribution spans different ranges in the training and testing subsets, then the model may yield poor results.

1.3.1.2 The barplot coding using MATLAB®

In Fig. 1.6, several statistical indices (minimum, maximum, standard deviation) were computed for each independent input variable as well as the output variable. In the following subsection, the detailed steps required to code and generate the barplots, including the aggregation of training and test data, the calculation of indices, and the plotting of figures, are presented.

The coding syntax for the generation of a barplot can be divided into several general categories: (1) load data; (2) Merge all samples; (3) Calculation of statistical indices; (4) Prepare data for plotting; (5) Plot results. First, the data must be read or loaded into the MATLAB environment. To do so, the data is first saved within a Microsoft Excel file that contains sheets (i.e., sheet1, sheet2, sheet3, sheet4), where sheets 1–4 contain the training input, training output, testing input, and testing output, respectively. From Fig. 1.7, For Example 1 data, the number of input variables is six, while the number of output variables is one. This was previously shown in the dataset description in Fig. 1.6. In addition, 70% of the data was considered as training data, while the other 30% was reserved as the testing subset. This results in 113 and 48 training and testing data samples, respectively.

Code 1.1 presents the required syntax for the "load data" and "Merge all samples" steps. Before providing the details of the code, its function is conceptualized in Fig. 1.8. According to this figure, using the **xlsread** command in the MATLAB environment, four different variables are loaded from the previously developed Excel spreadsheet (i.e., `TrainInputs; Testargets; TestInputs; TrainTargets`). In the next step, the train and test inputs, as well as their corresponding targets, are merged as Inputs and Targets.

Below, the coding details related to data loading (i.e., Code 1.1), calculating indicators (i.e., Code 1.2), as well as plotting figures (i.e., Code 1.3) are explained. In lines 1–3 of Code 1.1 presented next, some general MATLAB functions are used to prepare the MATLAB environment prior to the execution of any programming. These commands are used in almost all MATLAB syntaxes and are the real-life equivalent of wiping a whiteboard clean—a clean slate. In line 1, `clc` is used to clear the command window, which erases the text that was previously displayed. Once this command has been executed, the function history cannot be seen using the scroll bar, but the command history statements could be called using the up-arrow key ↑. In line 2, the second command is the `clear` command. This clears variables and functions from the memory of the program. Other times, the `clear all` command may be used, which is used to clear functions, variables, and other stored items from memory. Examples include cached memory, breakpoints, and persistent variables. It is often unnecessary to employ the `clear all` function, and the `clear` command is sufficient. The third command shown in line 3 is `close all` and is applied to remove all the figures whose handles are not hidden.

Code 1.1

```
1   clc
2   clear
3   close all
4
5   %% load data
6   TrainInputs  = xlsread ('Example1','sheet1'); % Train Inputs
7   TrainTargets = xlsread ('Example1','sheet2'); % Train Targets
8   TestInputs   = xlsread ('Example1','sheet3'); % Test Inputs
9   TestTargets  = xlsread ('Example1','sheet4'); % Test Targets
10
11  %% Merge all samples
12  Inputs  = [TrainInputs;TestInputs];                % Inputs
13  Targets = [TrainTargets;TestTargets];              % Targets
```

In lines 6–9 of Code 1.1, the `xlsread` command is used to load the data from the Excel spreadsheet developed previously (i.e., Fig. 1.7). The general format of this built-in MATLAB function is "`xlsread (filename, sheet)`," where `filename` is the name of the saved file and `sheet` is the name of the reference sheet where the data is contained.

In the case of the Example 1 data set, the data is saved under the Excel filename "Example1," which is used as the `filename` argument, while the `sheet` argument is specified as either sheet1, sheet2, sheet3, or sheet4, for the training

	A	B	C	D	E	F
1	In1	In2	In3	In4	In5	In6
2	0.988	0.814706	7.289474	1	0.111765	0
3	0.888	1	7.777778	0	0.128571	0
4	0.924	1.6	8.615385	0	0.185714	0
5	0.642	0.814706	9.892857	1	0.082353	0
⋮	⋮	⋮	⋮	⋮	⋮	⋮
111	0.275	0.51	0.931507	0	0.5475	0.242009
112	0.174	0.5075	0.821862	0	0.6175	0.408907
113	0.276	0.51	0.868085	0	0.5875	0.225532
114	0.222	0.5075	0.694017	0	0.73125	0.181197
115						

(A) Training Inputs

	A	B
1	Out	
2	0.219	
3	0.27	
4	0.297	
5	0.299	
⋮	⋮	
111	0.732551	
112	0.737137	
113	0.741814	
114	0.845961	
115		

(B) Training Target

	A	B	C	D	E	F
1	In1	In2	In3	In4	In5	In6
2	0.767	1	8.139535	0	0.122857	0
3	0.934	1.6	8.888889	0	0.18	0
4	0.847	1.6	9.180328	0	0.174286	0
5	0.784	1.6	10	0	0.16	0
⋮	⋮	⋮	⋮	⋮	⋮	⋮
46	0.239	0.5075	0.953052	0	0.5325	0.474178
47	0.118	4	2.366864	0.5	1.69	0.769
48	0.276	0.51	1.02	0	0.5	0.265
49	0.181	0.5075	0.751852	0	0.675	0.374074
50						

(C)Testing Inputs

	A	B
1	Out	
2	0.291	
3	0.306	
4	0.317	
5	0.334	
⋮	⋮	
46	0.629591	
47	0.653	
48	0.718385	
49	0.738453	
50		

(D)Testing Target

FIGURE 1.7 Data preparation in Microsoft Excel file. (A) Training inputs, (B) training targets, (C) testing inputs, and (D) testing target.

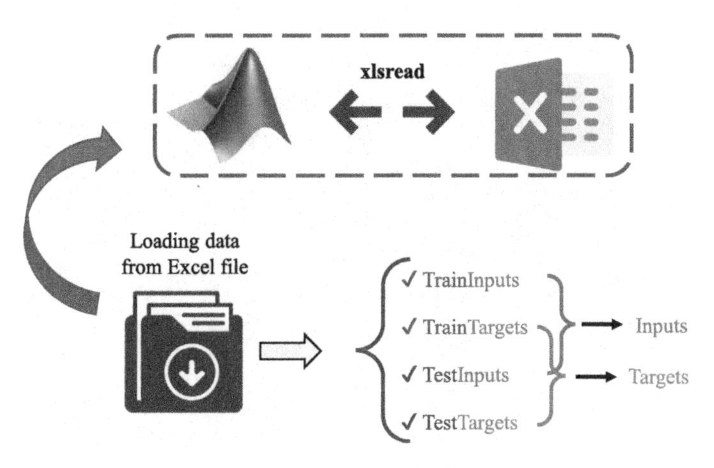

FIGURE 1.8 The conceptual coding process of Code 1.1.

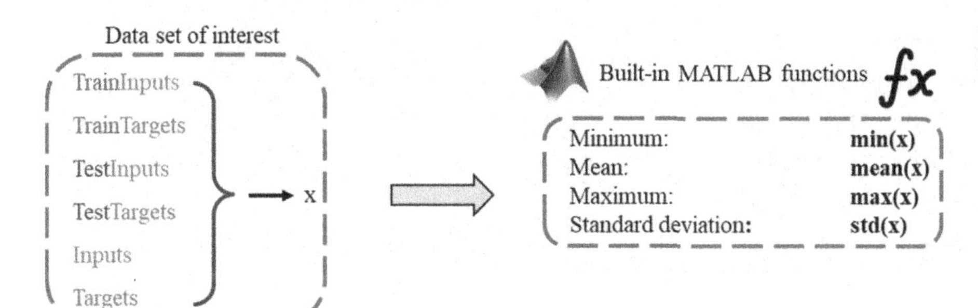

FIGURE 1.9 A graphical definition of Code 1.2.

input, training target, testing input, and testing target, respectively. Because the training and testing data are independently read into the MATLAB environment, it is necessary to define a variable capable of storing the data features for the entire set. To achieve this, the training inputs and testing inputs are merged and stored into the `Inputs` variable in line 12, while the training and testing output are merged and stored under the `Targets` variable in line 13.

Once the training, testing, and total data have been read into the MATLAB environment, the statistical indices, including the minimum, the average, the maximum, and the standard deviation, may be calculated (Code 1.2). Code 1.2 includes four different sections that are independently discussed as Code 1.2.A (finding the minimum), Code 1.2.B (finding the mean), Code 1.2.C (finding the maximum), and Code 1.2.D (finding the standard deviation). A simple graphical definition of Code 1.2 is provided in Fig. 1.9. The statistical indices are computed for each of the training input, training output, testing input, and testing output subsets, as well as the total input and total output. This results in 24 different parameters computed by the MATLAB code. The statistical indices are computed using the built-in MATLAB functions `min(x)`, `mean(x)`, `max(x)`, and `std(x)` where `x` contains the data set of interest. For example, `x` is `TrainInputs` for calculating the minimum, maximum, mean, and standard deviations of the training inputs.

Code 1.2.A

```
1  %% Calculation of statistical indices
2  % Minimum
3  Min_TrIn  = min(TrainInputs);    % Minimum of Train Inputs
4  Min_TrTa  = min(TrainTargets);   % Minimum of Train Targets
5  Min_TsIn  = min(TestInputs);     % Minimum of Test Inputs
6  Min_TsTa  = min(TestTargets);    % Minimum of Test Targets
7  Min_AllIn = min(Inputs);         % Minimum of All Inputs
8  Min_AllTa = min(Targets);        % Minimum of All Targets
```

Code 1.2.B

```
1  % Mean
2  Avg_TrIn  = mean(TrainInputs);     % Average of Train Inputs
3  Avg_TrTa  = mean(TrainTargets);    % Average of Train Targets
4  Avg_TsIn  = mean(TestInputs);      % Average of Test Inputs
5  Avg_TsTa  = mean(TestTargets);     % Average of Test Targets
6  Avg_AllIn = mean(Inputs);          % Average of All Inputs
7  Avg_AllTa = mean(Targets);         % Average of All Targets
```

Code 1.2.C

```
18 % Maximum
19 Max_TrIn  = max(TrainInputs);      % Maximum of Train Inputs
20 Max_TrTa  = max(TrainTargets);     % Maximum of Train Targets
21 Max_TsIn  = max(TestInputs);       % Maximum of Test Inputs
22 Max_TsTa  = max(TestTargets);      % Maximum of Test Targets
23 Max_AllIn = max(Inputs);           % Maximum of All Inputs
24 Max_AllTa = max(Targets);          % Maximum of All Targets
```

Code 1.2.D

```
1  % Standard deviation
2  SD_TrIn  = std(TrainInputs);       % Standard deviation of Train Inputs
3  SD_TrTa  = std(TrainTargets);      % Standard deviation of Train Targets
4  SD_TsIn  = std(TestInputs);        % Standard deviation of Test Inputs
5  SD_TsTa  = std(TestTargets);       % Standard deviation of Test Targets
6  SD_AllIn = std(Inputs);            % Standard deviation of All Inputs
7  SD_AllTa = std(Targets);           % Standard deviation of All Targets
```

During the fourth step, the data is prepared to be plotted by using Code 1.3, which is schematically represented in Fig. 1.10. For each input and output variable, the minimum, average, maximum, and standard deviation values computed for the testing and training subsets are merged into one parameter in Code 1.3.H. Considering line 3 of Code 1.3.A for example, the minimum values for the training and testing subsets (e.g., `Min_TrIn(1)` and `Min_TsIn(1)`) for `In(1)`, as well as the total data set (`Min_allIn(1)`) are merged into the variable `Min1`. Similarly, the mean, maximum, and standard deviation of In1 (i.e., input one) are also merged and saved into the variables `Avg1`, `Max1`, and `SD1` in lines 4 through 6, respectively. This process is repeated for all input variables (i.e., In2, In3, In4, In5, In6) in Codes 1.3.B to Code 1.3.G and for the output variable (i.e., Out) in Code 1.3.F. Following this, the information for each input and output variable is stored in an array format in Code 1.3.H. These final variables (i.e., `In1`, `In2`, `In3`, `In4`, `In5`, `In6`, and `Out`) are employed in the next step to plot all of the input and output characteristics. Indeed, each of the newly generated variables (i.e., `In1`, `In2`, `In3`, `In4`, `In5`, `In6`, and `Out`) is a matrix that contains four rows and three columns. As seen in Fig. 1.11, each row is associated with a given data statistic (i.e., minimum, maximum, mean, standard deviation), while the columns present the results for each subset of data (i.e., training, testing, total).

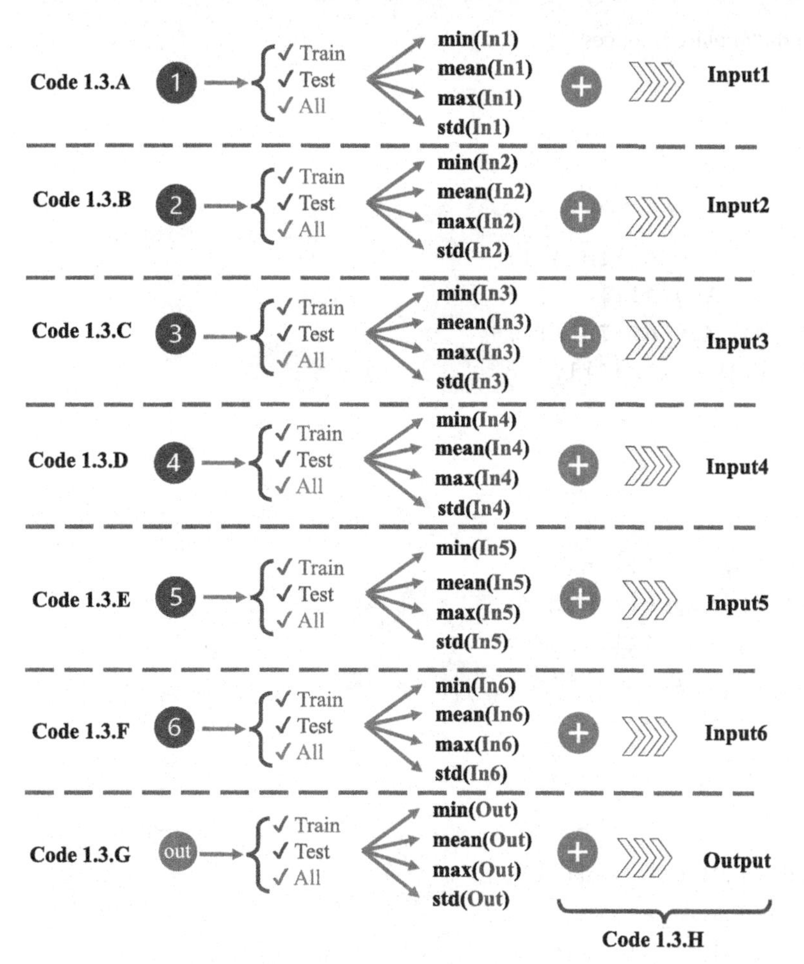

FIGURE 1.10 Schematic of Code 1.3.

Code 1.3.A

```
1  %% Prepare data for plotting
2  % Input 1
3  Min1 = [Min_TrIn(1),Min_TsIn(1),Min_AllIn(1)];
4  Avg1 = [Avg_TrIn(1),Avg_TsIn(1),Avg_AllIn(1)];
5  Max1 = [Max_TrIn(1),Max_TsIn(1),Max_AllIn(1)];
6  SD1  = [SD_TrIn(1),SD_TsIn(1),SD_AllIn(1)];
```

Code 1.3.B

```
1  % Input 2
2  Min2 = [Min_TrIn(2),Min_TsIn(2),Min_AllIn(2)];
3  Avg2 = [Avg_TrIn(2),Avg_TsIn(2),Avg_AllIn(2)];
4  Max2 = [Max_TrIn(2),Max_TsIn(2),Max_AllIn(2)];
5  SD2  = [SD_TrIn(2),SD_TsIn(2),SD_AllIn(2)];
```

Code 1.3.C

```
1  % Input 3
2  Min3 = [Min_TrIn(3),Min_TsIn(3),Min_AllIn(3)];
3  Avg3 = [Avg_TrIn(3),Avg_TsIn(3),Avg_AllIn(3)];
4  Max3 = [Max_TrIn(3),Max_TsIn(3),Max_AllIn(3)];
5  SD3  = [SD_TrIn(3),SD_TsIn(3),SD_AllIn(3)];
```

Code 1.3.D

```
1  % Input 4
2  Min4 = [Min_TrIn(4),Min_TsIn(4),Min_AllIn(4)];
3  Avg4 = [Avg_TrIn(4),Avg_TsIn(4),Avg_AllIn(4)];
4  Max4 = [Max_TrIn(4),Max_TsIn(4),Max_AllIn(4)];
5  SD4  = [SD_TrIn(4),SD_TsIn(4),SD_AllIn(4)];
```

Code 1.3.E

```
1  % Input 5
2  Min5 = [Min_TrIn(5),Min_TsIn(5),Min_AllIn(5)];
3  Avg5 = [Avg_TrIn(5),Avg_TsIn(5),Avg_AllIn(5)];
4  Max5 = [Max_TrIn(5),Max_TsIn(5),Max_AllIn(5)];
5  SD5  = [SD_TrIn(5),SD_TsIn(5),SD_AllIn(5)];
```

Code 1.3.F

```
1  % Input 6
2  Min6 = [Min_TrIn(6),Min_TsIn(6),Min_AllIn(6)];
3  Avg6 = [Avg_TrIn(6),Avg_TsIn(6),Avg_AllIn(6)];
4  Max6 = [Max_TrIn(6),Max_TsIn(6),Max_AllIn(6)];
5  SD6  = [SD_TrIn(6),SD_TsIn(6),SD_AllIn(6)];
```

Code 1.3.G

```
1  381  % Output
2  39   MinOut = [Min_TrTa,Min_TsTa,Min_AllTa];
3  40   AvgOut = [Avg_TrTa,Avg_TsTa,Avg_AllTa];
4  41   MaxOut = [Max_TrTa,Max_TsTa,Max_AllTa];
5  42   SDOut  = [SD_TrTa,SD_TsTa,SD_AllTa];
```

Code 1.3.H

```
1  % All
2  In1 = [Min1;Avg1;Max1;SD1];
3  In2 = [Min2;Avg2;Max2;SD2];
4  In3 = [Min3;Avg3;Max3;SD3];
5  In4 = [Min4;Avg4;Max4;SD4];
6  In5 = [Min5;Avg5;Max5;SD5];
7  In6 = [Min6;Avg6;Max6;SD6];
8  Out = [MinOut;AvgOut;MaxOut;SDOut];
```

Following data preparation for plotting, the compiled data from step four (i.e., In1, In2, In3, In4, In5, In6, and Out) is plotted in the fifth step (i.e., Plot results). From the column charts that were previously seen in Fig. 1.6, the characteristics of training and testing subsets as well as the total data set are provided within a single figure. Prior to plotting the data, it is necessary to define the title of each column in the charts. This can be performed by defining a categorical array which is saved under the variable name xx. As previously discussed, the first to the fourth row present the minimum, average, maximum, and standard deviation values, respectively, while the columns present the index values for each of the three data subsets. Therefore the xx array is initiated as a matrix with four rows and three columns. The output of variable xx in the command window is as shown in Fig. 1.11.

The built-in bar function is employed to create the column chart in MATLAB. The generic syntax used throughout Code 1.4 is bar(x,y), which draws the bars for data contained in the argument y at the locations specified by x. Direct use of the categorical variable xx without any other modifications in the bar function results in the order of the defined variables being incorrect. Hence, the xx array is redefined using reordercats in line 3 of Code 1.4.A so that the order of the plotted variables aligns with the stored data order. The built-in reordercats function follows the syntax reordercats(A,neworder), where A is the categorical array, and neworder is the order to be followed. The effect of reordercats on the bar plot for In1 can clearly be seen in Fig. 1.12. Prior to applying reordercats, the order of defined variables from the categorical array was not the same as the specified order of the data. Following the application of the reordercats function, the categories and the computed data are properly aligned.

To adjust the color of each bar in Code 1.4.A–Code 1.4.G next, the FaceColor function is used. This command uses a short name character input to change the color of each bar. In the case of Code 1.4.A, lines 8–10 set the colors of the training, testing, and total subsets of data to blue, green, and red, respectively. To achieve this, the short name character inputs b (blue), g (green), and r (red) are used. This distinction can clearly be seen in Fig. 1.6 and its accompanying legend. Other features of the chart may also be defined, such as font size, font weight, legend,

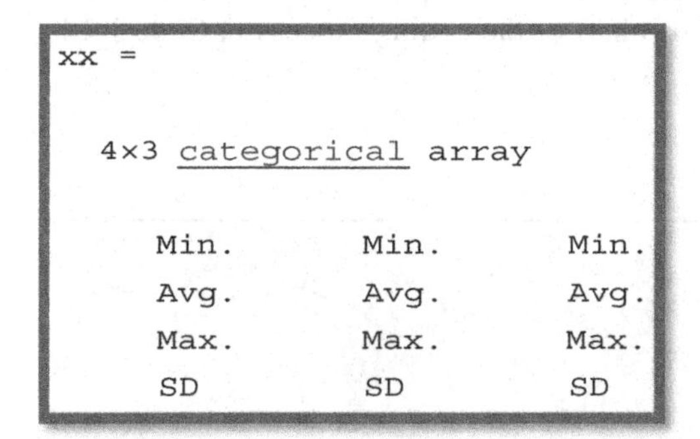

```
xx =

  4×3 categorical array

     Min.       Min.       Min.
     Avg.       Avg.       Avg.
     Max.       Max.       Max.
     SD         SD         SD
```

FIGURE 1.11 The size and type of the stored parameters in variable "xx."

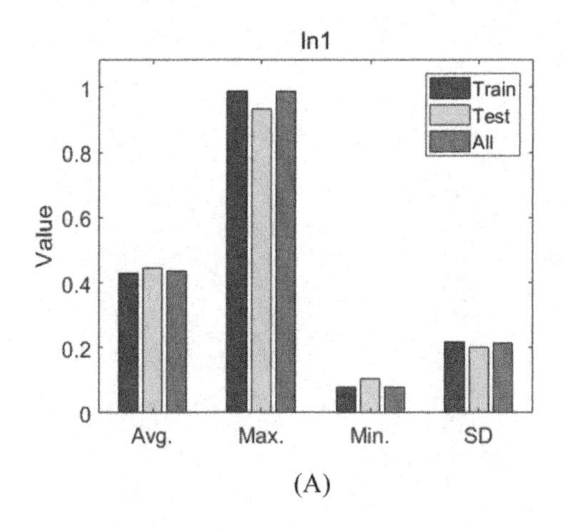

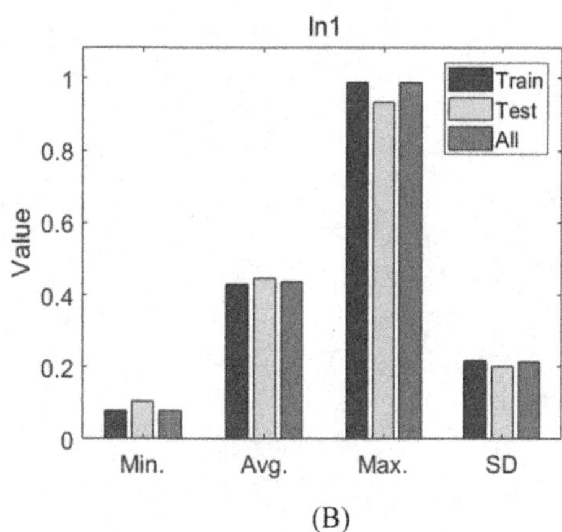

FIGURE 1.12 The effect of reordercats on the barplot (In1). (A) Before and (B) after.

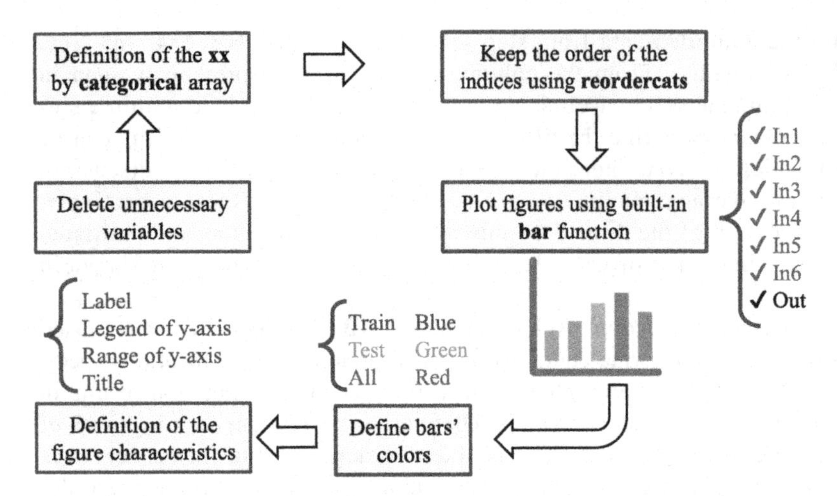

FIGURE 1.13 The conceptual coding process in Code 1.4.

axis titles, and axis ranges. The syntax `ax=gca` in line 11 of Code 1.4.A is used to create a cartesian axes object in the current figure. The font size and font weight may be adjusted by using the `FontSize` and `FontWeight` commands in lines 12 and 13, respectively. For font size, one "point" font is equivalent to 1/72″, while for the font weight, the options may be specified as the normal or boldened font. In this example, the `FontSize` is specified as 16 points, while the `FontWeight` is considered normal. The `ylabel`, specified in line 15, sets the *y*-axis title to "Value" (as seen in Fig. 1.6), while `ylim` sets the axis range from 0 to 1.1 times the maximum value. The `legend` function creates a legend with three labels, which are specified in the input arguments as Train, Test, and All. This plotting process is repeated for other inputs (i.e., In2, In3, In4, In5, in6) as well as the output (i.e., Out). Finally, in Code 1.4.G line 18, the "`clearvars -except`" function is applied to remove all variables except {In1, In2, In3, In4, In5, In6, Out}, which are specified in its input argument. The conceptual flowchart of the coding process in Code 1.4 is provided in Fig. 1.13.

Code 1.4.A

```
1   %% Plot results
    xx =
2   categorical({'Min.','Min.','Min.';'Avg.','Avg.','Avg.';'Max.','Max.','Max
    .';'SD','SD','SD'});
3   xx = reordercats(xx, {'Min.';'Avg.';'Max.';'SD'});
4
5   % Inputs #1
6   figure
7   B = bar(xx,In1);
8   B(1).FaceColor = 'b';
9   B(2).FaceColor = 'g';
10  B(3).FaceColor = 'r';
11  ax = gca;
12  ax.FontSize = 16;
13  ax.FontWeight='normal';
14
15  ylabel('Value')
16  legend('Train','Test','All')
17  ylim([0 1.1*max(Max1)])
18  title('In1','FontWeight','normal')
```

Code 1.4.B

```
1   % Inputs #2
2   figure
3   B = bar(xx,In2);
4   B(1).FaceColor = 'b';
5   B(2).FaceColor = 'g';
6   B(3).FaceColor = 'r';
7   ax = gca;
8   ax.FontSize = 16;
9   ax.FontWeight='normal';
10
11  ylabel('Value')
12  legend('Train','Test','All')
13  ylim([0 1.1*max(Max2)])
14  title('In2','FontWeight','normal')
```

Code 1.4.C

```
1   % Inputs #3
2   figure
3   B = bar(xx,In3);
4   B(1).FaceColor = 'b';
5   B(2).FaceColor = 'g';
6   B(3).FaceColor = 'r';
7   ax = gca;
8   ax.FontSize = 16;
9   ax.FontWeight='normal';
10
11  ylabel('Value')
12  legend('Train','Test','All')
13  ylim([0 1.1*max(Max3)])
14  title('In3','FontWeight','normal')
```

Code 1.4.D

```
1   % Inputs #4
2   figure
3   B = bar(xx,In4);
4   B(1).FaceColor = 'b';
5   B(2).FaceColor = 'g';
6   B(3).FaceColor = 'r';
7   ax = gca;
8   ax.FontSize = 16;
9   ax.FontWeight='normal';
10
11  ylabel('Value')
12  legend('Train','Test','All')
13  ylim([0 1.1*max(Max4)])
14  title('In4','FontWeight','normal')
```

Code 1.4.E

```
1   % Inputs #5
2   figure
3   B = bar(xx,In5);
4   B(1).FaceColor = 'b';
5   B(2).FaceColor = 'g';
6   B(3).FaceColor = 'r';
7   ax = gca;
8   ax.FontSize = 16;
9   ax.FontWeight='normal';
10
11  ylabel('Value')
12  legend('Train','Test','All')
13  ylim([0 1.1*max(Max5)])
14  title('In5','FontWeight','normal')
```

Code 1.4.F

```
1   % Inputs #6
2   figure
3   B = bar(xx,In6);
4   B(1).FaceColor = 'b';
5   B(2).FaceColor = 'g';
6   B(3).FaceColor = 'r';
7   ax = gca;
8   ax.FontSize = 16;
9   ax.FontWeight='normal';
10
11  ylabel('Value')
12  legend('Train','Test','All')
13  ylim([0 1.1*max(Max6)])
14  title('In6','FontWeight','normal')
```

Code 1.4.G

```
1   % Output
2   figure
3   B = bar(xx,Out);
4   B(1).FaceColor = 'b';
5   B(2).FaceColor = 'g';
6   B(3).FaceColor = 'r';
7   ax = gca;
8
9   ax.FontSize = 16;
10  ax.FontWeight='normal';
11
12  ylabel('Value')
13  legend('Train','Test','All')
14  ylim([0 1.1*max(MaxOut)])
15  title('Out','FontWeight','normal')
16
17  %% Clear all except mandatory information
18  clearvars -except In1 In2 In3 In4 In5 In6 Out
```

1.3.2 Example 2: A problem with three input variables

The Example 2 dataset consists of three input variables and one output variable. The total number of samples in this example is 100 so that 70 samples (i.e., 70% of all samples) are randomly selected to calibrate the model (i.e., Training samples) while the remaining 30 samples were employed to test the performance of the developed model for unseen samples (i.e., Testing samples). Similar to the previous example, the percentage of training and testing samples are 70% and 30%, respectively. This is standard practice for each category.

1.3.2.1 The histogram analysis for Example 2 training, testing, and total data

Fig. 1.14 provides the histograms of all inputs and output variables for the training, testing, and total sample subsets in Example 2. It can be seen from this figure that unique histograms are plotted for each input/output variable and subset of the data (Train, Test, All). The reason that all subsets are not plotted on the same histogram is that the number of samples in each set is not the same. An equal number of samples is required to plot all samples in a histogram at the same time.

From the figure, the number of samples in all bins for In1 and a given subset of data is approximately equal. When considering In2, the number of samples in each bin for a given subset of data (Train, Test, All) was also found to be the same. For In3, the samples' distribution in all bins is not similar to the first two input variables—there is much greater variability in this input variable compared to the previous two. It can be seen that the third bin, which is in the range [92, 102], contains the highest number of samples in the training and total sets of data, while for the testing subset, this bin contains almost no entries. A modeler must be aware that, should the test conditions shown in the histogram have represented the training data, then the developed model would have no prior experience in dealing with situations with such a data distribution. In that case, it could affect the modeling results because the model was not trained for the range of In3 variables being evaluated. For the conditions presented in this Example, however, since the training subset includes entries within [92,102], it would not affect the model training.

1.3.2.2 The coding of histogram in MATLAB

In Codes 1.5–1.7, the required MATLAB syntax to generate the data histograms seen in Fig. 1.14 is presented in detail. This coding includes four main steps: (1) loading of data; (2) merging of data; (3) preparation of data for plotting; and (4) histogram generation. The initial three lines of Code 1.5 specify the general commands `clc`, `clear`, and `close all`, which were introduced earlier in this chapter. Following this, the first step is initiated (i.e., loading the data), which includes reading four subsets of data into the MATLAB environment. These four subsets of data include the training inputs, training targets, testing inputs, and testing targets and are called from sheet1, sheet2, sheet3, and sheet 4 (respectively) of a Microsoft Excel file saved under the filename Example 2. In order to generate the histograms for all the data, two variables are created in lines 12–13 of Code 1.5 that store all of the input and target data. The data is read into MATLAB using the `xlsread` function, whose syntax was previously described in Code 1.1. The schematic of Code 1.5 is presented in Fig. 1.15.

Code 1.5

```
1   clc
2   clear
3   close all
4
5   %% load data
6   TrainInputs  = xlsread ('Example2','sheet1'); % Train Inputs
7   TrainTargets = xlsread ('Example2','sheet2'); % Train Targets
8   TestInputs   = xlsread ('Example2','sheet3'); % Test Inputs
9   TestTargets  = xlsread ('Example2','sheet4'); % Test Targets
10
11  %% Merge all samples
12  Inputs  = [TrainInputs;TestInputs];           % Inputs
13  Targets = [TrainTargets;TestTargets];         % Targets
```

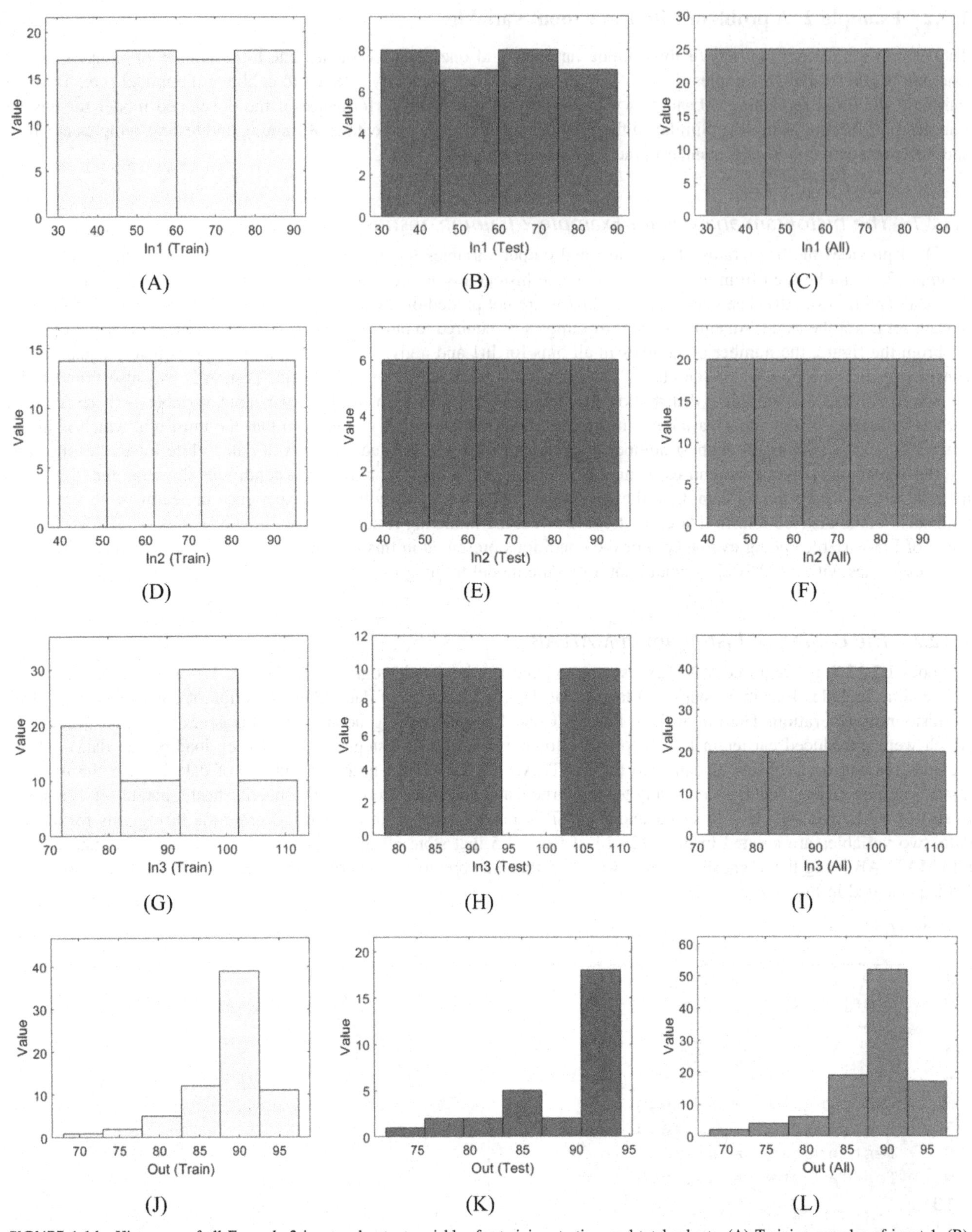

FIGURE 1.14 Histogram of all Example 2 input and output variables for training, testing, and total subsets. (A) Training samples of input 1, (B) testing samples of input 1, (C) all samples of input 1, (D) training samples of input 2, (E) testing samples of input 2, (F) all samples of input 2, (G) training samples of input 3, (H) testing samples of input 3, (I) all samples of input 3, (J) training samples of output, (K) testing samples of output, and (L) all samples of output.

Once the data has been read into MATLAB and the merged `Inputs` and `Targets` variables have been defined, the training, testing, and total subsets should be prepared for plotting. In this third step, datasets are defined using a standard format of "A_XN," where A is the name of the category (i.e., Train, Test, All), X is the type of variable (i.e., input or target), and N is the variable identifier (i.e., number). For Example, Train_In2 indicates the training samples of the second input (i.e., In2 provided in the Microsoft Excel file). Four variables are therefore created for each of the three data subsets, totaling twelve new variables generated (i.e., Train_In1, Train_In2, Train_In3, Train_Out, Test_In1, Test _In2, Test _In3, Test _Out, All_In1, All_In2, All_In3, All_Out). The schematic of Code 1.6 is provided as follows (Fig. 1.16):

Code 1.6

```
1   %% Prepare data for plotting
2   % Train
3   Train_In1 = TrainInputs(:,1); % Training samples for Input 1
4   Train_In2 = TrainInputs(:,2); % Training samples for Input 2
5   Train_In3 = TrainInputs(:,3); % Training samples for Input 3
6   Train_Out = TrainTargets;     % Training samples for Output
7
8   % Test
9   Test_In1 = TestInputs(:,1);   % Testing samples for Input 1
10  Test_In2 = TestInputs(:,2);   % Testing samples for Input 2
11  Test_In3 = TestInputs(:,3);   % Testing samples for Input 3
12  Test_Out = TestTargets;       % Testing samples for Output
13
14  % All
15  All_In1 = Inputs(:,1);        % All samples for Input 1
16  All_In2 = Inputs(:,2);        % All samples for Input 2
17  All_In3 = Inputs(:,3);        % All samples for Input 3
18  All_Out = Targets;            % All samples for Output
```

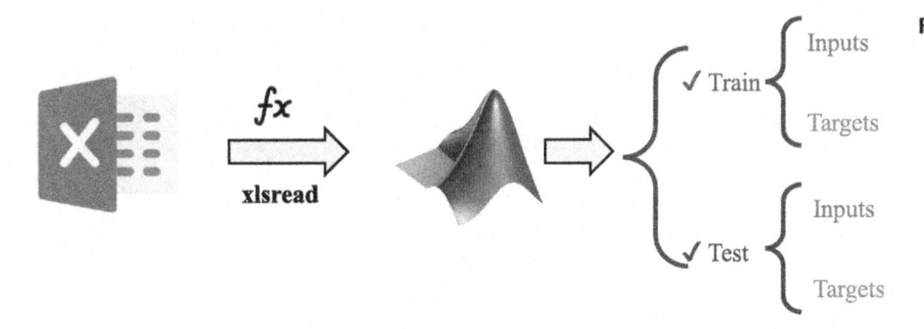

FIGURE 1.15 The schematic of Code 1.5.

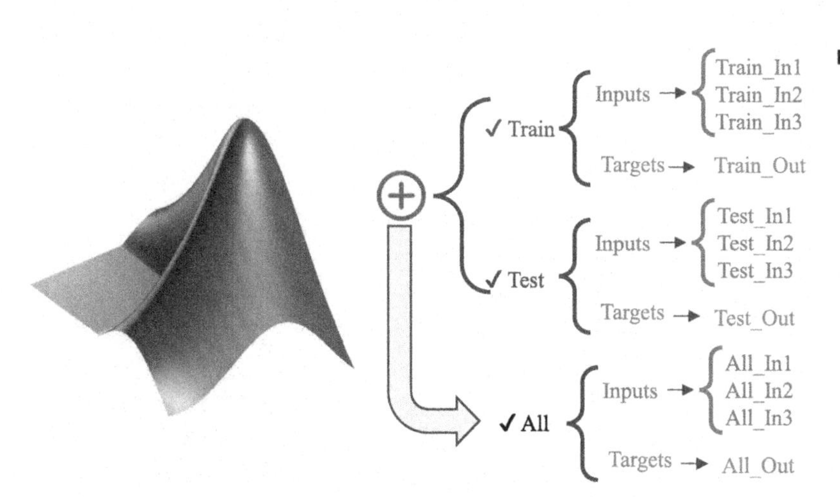

FIGURE 1.16 The schematic of Code 1.6.

The final step is to plot the histogram for all twelve newly generated variables. In Code 1.7 below (Code 1.7.A–Code 1.7.D for In1, In2, In3, and Out, respectively), the plot processing for the train, test, and all data is coded for one input (i.e., In1 for Code 1.7.A) and subsequently repeated for each of the remaining input and output variables. The plotting process includes three main steps: (i) employment of the `histogram` function, (ii) definition of histogram characteristics, and (iii) fine-tuning histogram axes. Using the `histogram(x)` command creates a histogram for the data contained within `x`. Initially, the function uses an algorithm that automatically selects the size and number of bins to use in the plot. These, and other features of the histogram, may be subsequently personalized by the modeler. In line 5 for Code 1.7.A, the number of bins is adjusted to 4 using the `h.NumBins` command. The colors of the bins are specified in lines 6–7, where the command `h.FaceColor` specifies the fill color of the bin (which in our case, is set to `y` for yellow) and `h.EdgeColour` specifies the color of the bin border (which is set to `k` for black). As we discussed previously in Code 1.4, the font size and font weight may be specified by the modeler using the `ax.FontSize` and `ax.FontWeight` commands (as seen in lines 9–10). Each point font corresponds to 1/72″—in our case, the font size is specified as 16 points. The font weight may be either normal or bold—here, we have selected normal. The commands `ylabel` and `xlabel` (lines 24–25) are used to specify the *y*- and *x*-axes titles, respectively, of the histogram, while `ylim` is used to define the range of the *y*-axis. The schematic description of Code 1.7 is shown in Fig. 1.17.

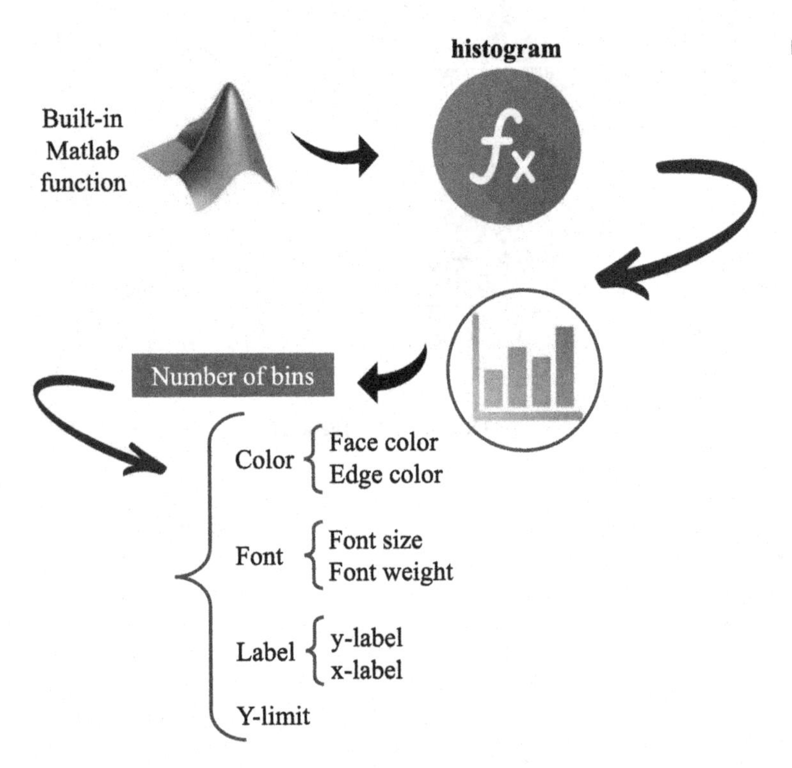

FIGURE 1.17 The schematic of Code 1.7.

Code 1.7.A

```
1   %% Histogram
2   %%%%%%%%%%%%%%%%%%%%%%%%%%%%%%%% In1 %%%%%%%%%%%%%%%%%%%%%%%%%%%%%%%%%%
3   % In1  - Train
4   h = histogram(Train_In1);
5   h.NumBins =  4;
6   h.FaceColor =  'y';
7   h.EdgeColor = 'k';
8   ax = gca;
9   ax.FontSize = 16;
10  ax.FontWeight='normal';
11  ylabel('Value')
12  xlabel('In1 (Train)')
13  ylim([0 1.2*max(h.Values)])
14  
15  % In1 - Test
16  figure;
17  h = histogram(Test_In1);
18  h.NumBins =  4;
19  h.FaceColor =  'b';
20  h.EdgeColor = 'k';
21  ax = gca;
22  ax.FontSize = 16;
23  ax.FontWeight='normal';
24  ylabel('Value')
25  xlabel('In1 (Test)')
26  ylim([0 1.2*max(h.Values)])
27  
28  % In1 - All
29  figure;
30  h = histogram(All_In1);
31  h.NumBins =  4;
32  h.FaceColor =  'r';
33  h.EdgeColor = 'k';
34  ax = gca;
35  ax.FontSize = 16;
36  ax.FontWeight='normal';
37  ylabel('Value')
38  xlabel('In1 (All)')
39  ylim([0 1.2*max(h.Values)])
```

Code 1.7.B

```
1   %%%%%%%%%%%%%%%%%%%%%%%%%%%%%%%% In2 %%%%%%%%%%%%%%%%%%%%%%%%%%%%%%%%%%
2   % In2  - Train
3   figure;
4   h = histogram(Train_In2);
5   h.NumBins =  5;
6   h.FaceColor =  'y';
7   h.EdgeColor = 'k';
8   ax = gca;
9   ax.FontSize = 16;
10  ax.FontWeight='normal';
11  ylabel('Value')
12  xlabel('In2 (Train)')
13  ylim([0 1.2*max(h.Values)])
14
15  % In2 - Test
16  figure;
17  h = histogram(Test_In2);
18  h.NumBins =  5;
19  h.FaceColor =  'b';
20  h.EdgeColor = 'k';
21  ax = gca;
22  ax.FontSize = 16;
23  ax.FontWeight='normal';
24  ylabel('Value')
25  xlabel('In2 (Test)')
26  ylim([0 1.2*max(h.Values)])
27
28  % In2 - All
29  figure;
30  h = histogram(All_In2);
31  h.NumBins =  5;
32  h.FaceColor =  'r';
33  h.EdgeColor = 'k';
34  ax = gca;
35  ax.FontSize = 16;
36  ax.FontWeight='normal';
37  ylabel('Value')
38  xlabel('In2 (All)')
39  ylim([0 1.2*max(h.Values)])
```

Code 1.7.C

```
1  %%%%%%%%%%%%%%%%%%%%%%%%%%%%%%%%% In3 %%%%%%%%%%%%%%%%%%%%%%%%%%%%%%%%%%
2  % In3  - Train
3  figure;
4  h = histogram(Train_In3);
5  h.NumBins =  6;
6  h.FaceColor =  'y';
7  h.EdgeColor = 'k';
8  ax = gca;
9  ax.FontSize = 16;
10 ax.FontWeight='normal';
11 ylabel('Value')
12 xlabel('In3 (Train)')
13 ylim([0 1.2*max(h.Values)])
14 
15 % In3 - Test
16 figure;
17 h = histogram(Test_In3);
18 h.NumBins =  6;
19 h.FaceColor =  'b';
20 h.EdgeColor = 'k';
21 ax = gca;
22 ax.FontSize = 16;
23 ax.FontWeight='normal';
24 ylabel('Value')
25 xlabel('In3 (Test)')
26 ylim([0 1.2*max(h.Values)])
27 
28 % In3 - All
29 figure;
30 h = histogram(All_In3);
31 h.NumBins =  6;
32 h.FaceColor =  'r';
33 h.EdgeColor = 'k';
34 ax = gca;
35 ax.FontSize = 16;
36 ax.FontWeight='normal';
37 ylabel('Value')
38 xlabel('In3 (All)')
39 ylim([0 1.2*max(h.Values)])
```

Code 1.7.D

```
1   %%%%%%%%%%%%%%%%%%%%%%%%%%%%%%%% Out %%%%%%%%%%%%%%%%%%%%%%%%%%%%%%%%%%%
2   % Out - Train
3   figure;
4   h = histogram(Train_Out);
5   h.NumBins =  6;
6   h.FaceColor =  'y';
7   h.EdgeColor = 'k';
8   ax = gca;
9   ax.FontSize = 16;
10  ax.FontWeight='normal';
11  ylabel('Value')
12  xlabel('Out (Train)')
13  ylim([0 1.2*max(h.Values)])
14  
15  % Out - Test
16  figure;
17  h = histogram(Test_Out);
18  h.NumBins =  6;
19  h.FaceColor =  'b';
20  h.EdgeColor = 'k';
21  ax = gca;
22  ax.FontSize = 16;
23  ax.FontWeight='normal';
24  ylabel('Value')
25  xlabel('Out (Test)')
26  ylim([0 1.2*max(h.Values)])
27  
28  % Out - All
29  figure;
30  h = histogram(All_Out);
31  h.NumBins =  6;
32  h.FaceColor =  'r';
33  h.EdgeColor = 'k';
34  ax = gca;
35  ax.FontSize = 16;
36  ax.FontWeight='normal';
37  ylabel('Value')
38  xlabel('Out (All)')
39  ylim([0 1.2*max(h.Values)])
```

1.3.3 Example 3: a problem with four input variables

The data set considered in Example 3 contains four input variables with 240 total sample points. 30% of all samples (i.e., 72 samples) are randomly selected as validation data for the testing stage, while the remaining 168 samples (70%) are employed to calibrate the ML-based model. To present the distribution of the inputs and output of Example 3, the box plot method is applied. Prior to detailing the MATLAB coding of boxplots for this example, a brief description of boxplot fundamentals is presented in the following subsection.

1.3.3.1 Boxplot fundamentals

Fig. 1.18 provides a schematic definition of a boxplot, which is defined based on five main values: the minimum, first quartile (or 25th percentile; Q1), median, third quartile (or 75th percentile; Q3), and outliers. Outliers are those data points which lie outside the interval defined by the minimum and maximum values. The difference between the Q1 and Q3 is defined as the interquartile range (IQR). For a data set with normal distribution, approximately 50% of all samples are located within the IQR. The minimum considered during the construction of a boxplot is not the minimum value of all samples, but rather is defined using the IQR and Q1 as Q1 minus 1.5*IQR. Similarly, the maximum of a boxplot is defined as Q3 plus 1.5 × IQR. Fig. 1.19 presents the idealized data distribution for a completely normal series plotted simultaneously with a boxplot. In this distribution, only 0.7% of data are considered outliers (0.35% below the minimum and 0.35% above the maximum), while 49.3% of data is evenly distributed between Q1 and Q3.

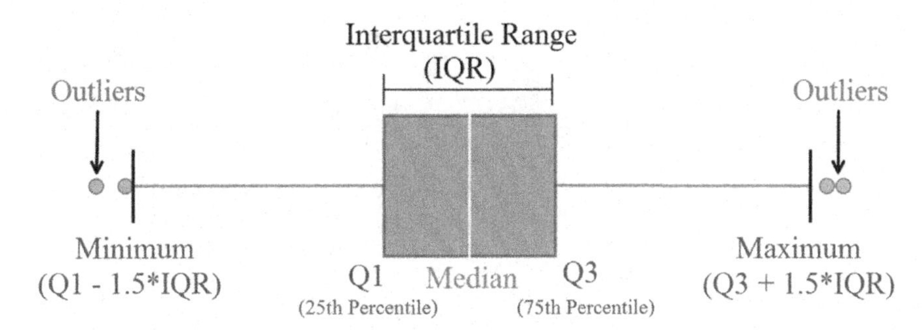

FIGURE 1.18 The definition of boxplot.

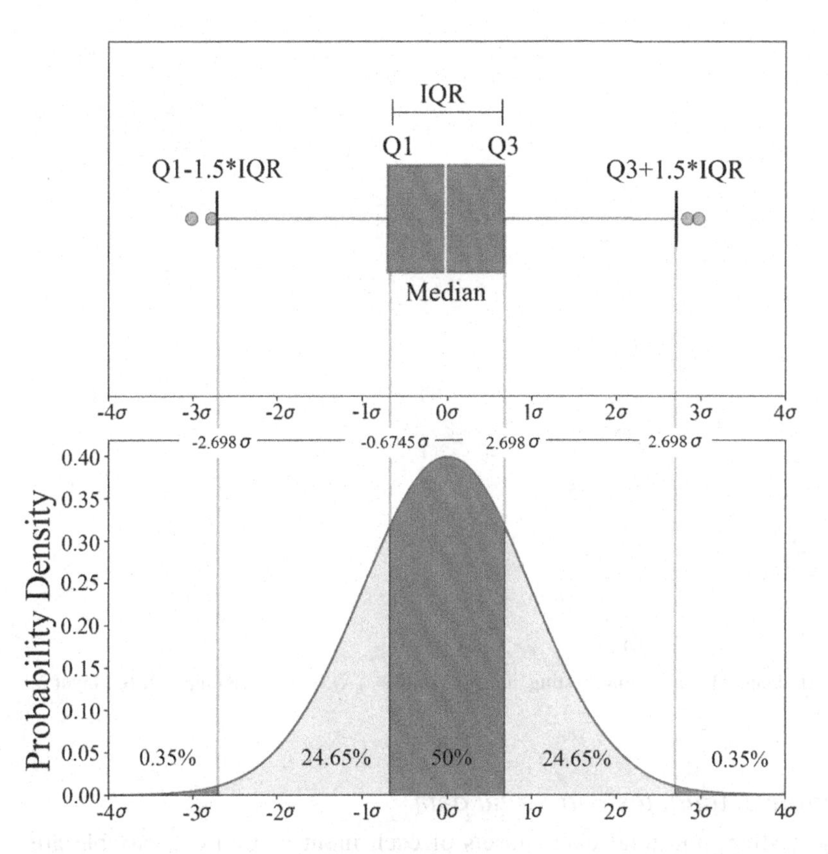

FIGURE 1.19 Boxplot along with corresponding probability density of the normal distribution.

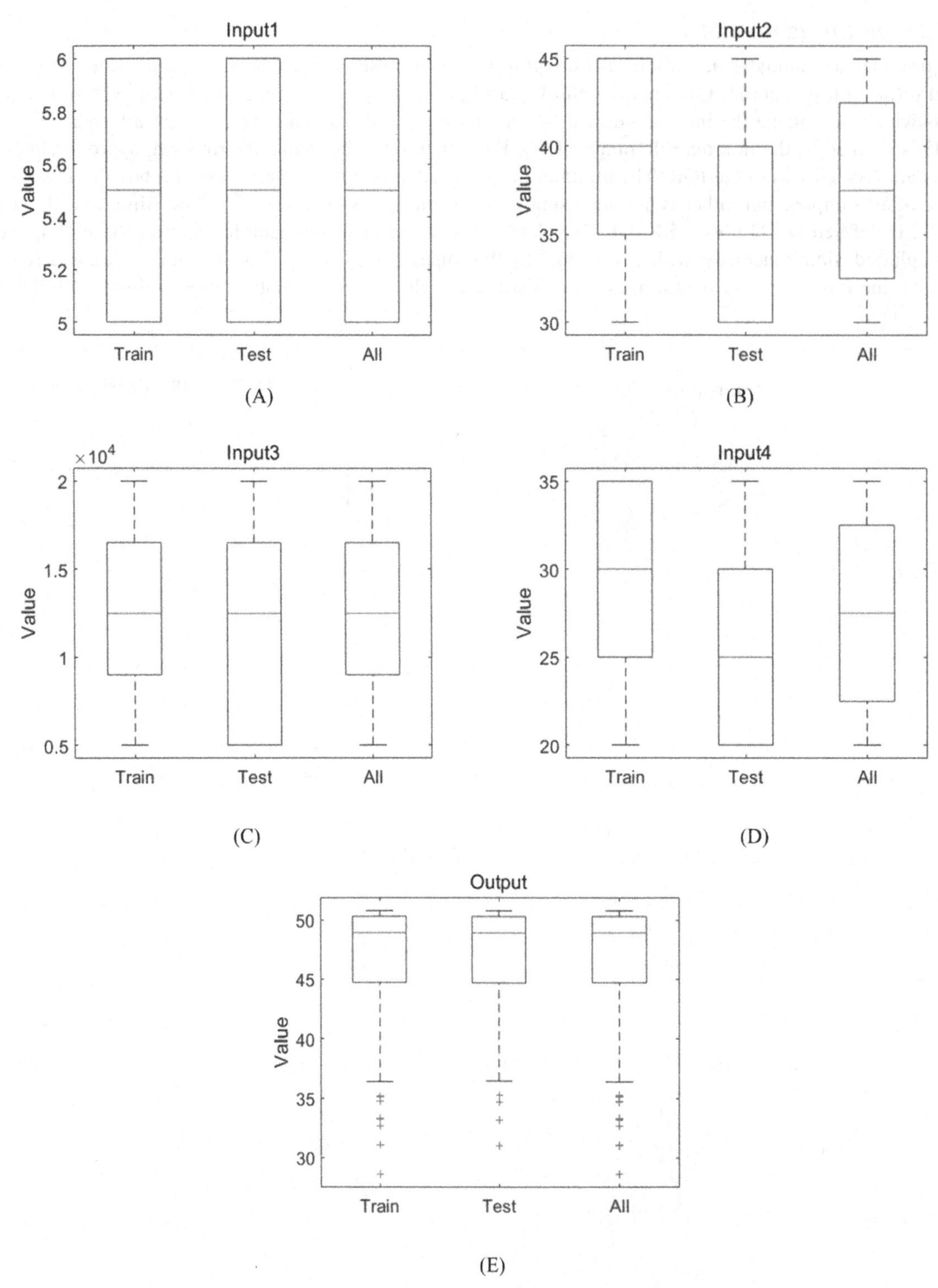

FIGURE 1.20 Input and output variable boxplots of Example 3 for training, testing, and all samples. (A) Input 1, (B) input 2, (C) input 3, (D) input 4, and (E) output.

1.3.3.2 The boxplot analysis of Example 3, train, test, and total data

In Fig. 1.20, the boxplots for the training, testing, and total data subsets of each input and output variable are presented for Example 3. The distribution of Input1 for all subsets of data (i.e., train, test, all) is similar so that the minimum, maximum, and median of the train, test, and all subsets are exactly the same. No outlying data points are observed in the three sets of data—in MATLAB, these are delimited by a red cross on the boxplot diagram. Similar to Input1, there are no outliers in all data subsets for Input2. For this variable, the maximum of the training samples is approximately equal to the maximum of other stages (i.e., Test and All). The maximum sample value of the training subset is contained in Q3, while for the testing and all subsets, the maximum sample value is out of Q3. For the testing stage, the minimum value is Q1, while the minimum value of the training and all samples is lower than Q1. For Input3, the minimum and maximum of this variable for all data subsets are equal. The minimum value of the test stage is at Q1, while it is lower than Q1 for the train and all samples. The distribution of Input4 at all stages is similar to Input2, however, the distribution of the output data differs from all inputs discussed. The main difference between the output and the inputs is that the output data contains outliers. There are no outliers observed higher than the maximum values for each data set, while a number of different outlying points can be observed below the minimum value computed for each data subset. The difference between Q1 and the sample minimum value is seen to be almost more than the difference between Q1 and maximum. The majority of samples have high values, while those outlying points in each subset have significantly lower values. The different distribution of inputs and output indicates that a strong ML method is needed to map input variables to output.

1.3.3.3 The MATLAB coding of a boxplot

To generate the boxplots seen in Fig. 1.20, Codes 1.8–1.12 detail the necessary MATLAB programming. The first step, as we saw when programming the bargraph plots, is to load the data into the MATLAB environment. In Code 1.8, the general `clc`, `clear`, and `close all` commands are used in lines 1–3 to wipe clean the Command Window. Following this, the built-in `xlsread` function is employed to read the data, which is contained within an Excel workbook. The Excel data is read from four different spreadsheets—sheet1, sheet2, sheet3, and sheet 4. The training inputs and targets are contained within the first two sheets, while the testing inputs and targets are categorized in the third and fourth sheets (i.e., sheet3 and sheet4). Given that, in addition to training and testing data, the total data is also plotted, the input and output data must be merged. The inputs and targets of all samples are stored in the `Inputs` and `Targets` variables, respectively, on lines 12–13. The schematic diagram of Code 1.8 is presented in Fig. 1.21.

Code 1.8

```
1   clc
2   clear
3   close all
4
5   %% load data
6   TrainInputs  = xlsread ('Example3','sheet1'); % Train Inputs
7   TrainTargets = xlsread ('Example3','sheet2'); % Train Targets
8   TestInputs   = xlsread ('Example3','sheet3'); % Test Inputs
9   TestTargets  = xlsread ('Example3','sheet4'); % Test Targets
10
11  %% Merge all samples
12  Inputs  = [TrainInputs;TestInputs];           % Inputs
13  Targets = [TrainTargets;TestTargets];         % Targets
```

In the second step, the five main values used to generate the boxplot must be computed, namely, the first and third quartiles (i.e., Q1 and Q3, respectively), the IQR (i.e., IQR = Q3-Q1), and the minimum and maximum values of all inputs and output for the train, test, and all data sets. These calculations are performed in Code 1.9 in six parts—the train, test, and total data sets for both the input and output data. The general computational process is identical for each of the size types of data, and so a detailed description will be provided for the training input. For other sections, it is sufficient to provide the code. The built-in MATLAB function `quantile(x,P)` is applied to calculate the first and third quantiles, where `x` is the data matrix under consideration, and `P` specifies the desired

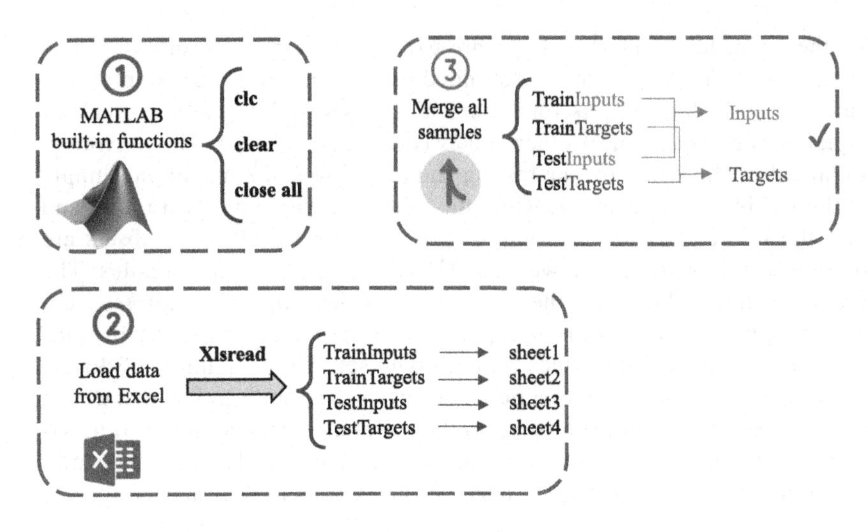

FIGURE 1.21 Schematic diagram of Code 1.8 function.

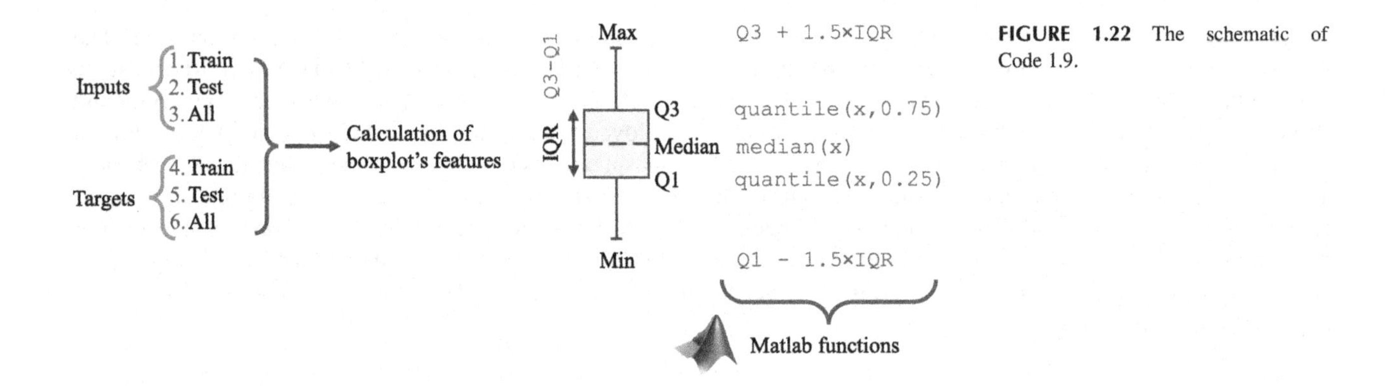

FIGURE 1.22 The schematic of Code 1.9.

quantile. For the first and third quintiles, `P` is equal to 0.25 and 0.75, respectively. In line 3 below, the lower quartile for the training inputs is computed by using `TrainInputs` as the `x` input argument and 0.25 as the `P` input argument. The built-in `median` function is applied in line 4 to determine the median of the data set, which was read to the `TrainInputs` variable from excel in Code 1.8. The median could alternatively be computed by using the `quantile(x,P)` function with `P` set to 0.5. The minimum and maximum whiskers used in the boxplot are not considered to be the minimum, and maximum values of the data samples, but rather is defined based on the IQR, which is the difference between the third and first quartiles (i.e., IQR = Q1-Q3). After computation of the IQR, the minimum and maximum whisker for the boxplot may be defined as $Q1-1.5\times IQR$ and $Q3+1.5\times IQR$, respectively. The associated coding for the IQR, minimum, and maximum is presented in lines 6–8 in Code 1.9 below. After calculating the minimum, first quartile, median, third quartile, and maximum, they are stored in a matrix by introducing the variable `TrIn1` for training inputs. This process is repeated for training outputs, testing inputs and outputs, and inputs and outputs of all samples. The schematic of Code 1.9 is provided in Fig. 1.22.

Code 1.9

```
1   %% Statistical indices of samples
2   % Training Inputs
3   TrIn_LQ = quantile(TrainInputs,0.25);
4   TrIn_Median=median(TrainInputs);
5   TrIn_UQ=quantile(TrainInputs,0.75);
6   TrIn_IQR = TrIn_UQ - TrIn_LQ;
7   TrIn_Min = TrIn_LQ - 1.5*TrIn_IQR;
8   TrIn_Max = TrIn_UQ + 1.5*TrIn_IQR;
9   TrIn1=[TrIn_Min;TrIn_LQ;TrIn_Median;TrIn_UQ;TrIn_Max;];
10
11  % Testing Inputs
12  TsIn_LQ = quantile(TestInputs,0.25);
13  TsIn_Median=median(TestInputs);
14  TsIn_UQ=quantile(TestInputs,0.75);
15  TsIn_IQR = TsIn_UQ - TsIn_LQ;
16  TsIn_Min = TsIn_LQ - 1.5*TsIn_IQR;
17  TsIn_Max = TsIn_UQ + 1.5*TsIn_IQR;
18  TsIn1=[TsIn_Min;TsIn_LQ;TsIn_Median;TsIn_UQ;TsIn_Max];
19
20  % All Inputs
21  AllIn_LQ = quantile(Inputs,0.25);
22  AllIn_Median=median(Inputs);
23  AllIn_UQ=quantile(Inputs,0.75);
24  AllIn_IQR = AllIn_UQ - AllIn_LQ;
25  AllIn_Min = AllIn_LQ - 1.5*AllIn_IQR;
26  AllIn_Max = AllIn_UQ + 1.5*AllIn_IQR;
27  AllIn1=[AllIn_Min;AllIn_LQ;AllIn_Median;AllIn_UQ;AllIn_Max];
28
29  % Training Targets
30  TrTa_LQ = quantile(TrainTargets,0.25);
31  TrTa_Median=median(TrainTargets);
32  TrTa_UQ=quantile(TrainTargets,0.75);
33  TrTa_IQR = TrTa_UQ - TrTa_LQ;
34  TrTa_Min = TrTa_LQ - 1.5*TrTa_IQR;
35  TrTa_Max = TrTa_UQ + 1.5*TrTa_IQR;
36  TrTa1=[TrTa_Min;TrTa_LQ;TrTa_Median;TrTa_UQ;TrTa_Max];
37
38  % Testing Targets
39  TsTa_LQ = quantile(TestTargets,0.25);
40  TsTa_Median=median(TestTargets);
41  TsTa_UQ=quantile(TestTargets,0.75);
42  TsTa_IQR = TsTa_UQ - TsTa_LQ;
43  TsTa_Min = TsTa_LQ - 1.5*TsTa_IQR;
44  TsTa_Max = TsTa_UQ + 1.5*TsTa_IQR;
45  TsTa1=[TsTa_Min;TsTa_LQ;TsTa_Median;TsTa_UQ;TsTa_Max];
46
47  % All Targets
48  AllTa_LQ = quantile(Targets,0.25);
49  AllTa_Median=median(Targets);
50  AllTa_UQ=quantile(Targets,0.75);
51  AllTa_IQR = AllTa_UQ - AllTa_LQ;
52  AllTa_Min = AllTa_LQ - 1.5*AllTa_IQR;
53  AllTa_Max = AllTa_UQ + 1.5*AllTa_IQR;
54  AllTa1=[AllTa_Min;AllTa_LQ;AllTa_Median;AllTa_UQ;AllTa_Max];
```

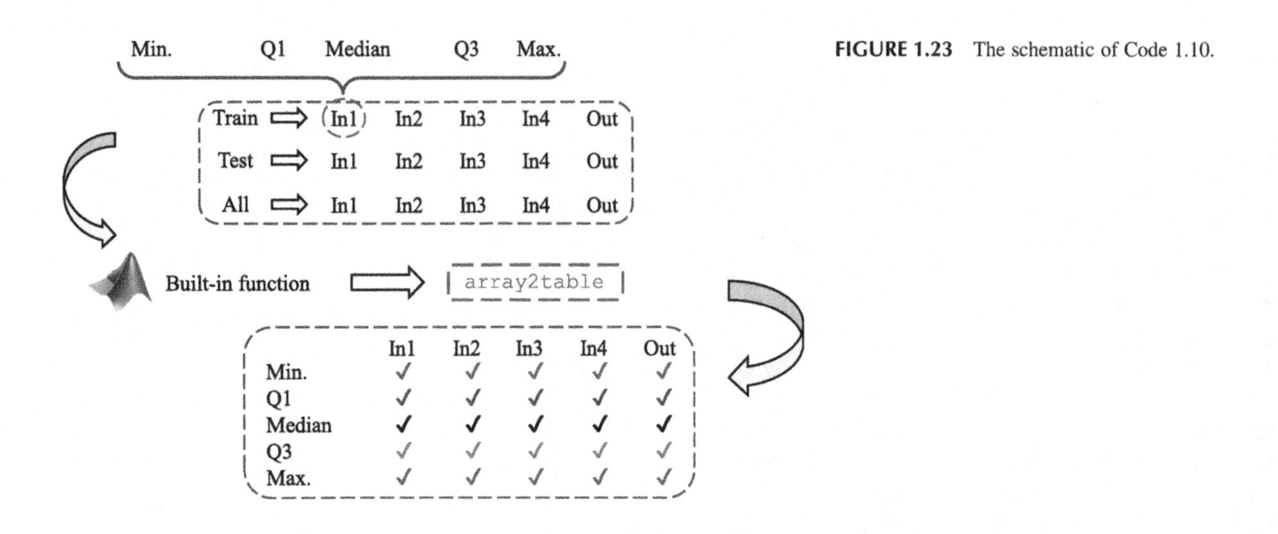

FIGURE 1.23 The schematic of Code 1.10.

Once these characteristic values have been computed for each training, testing, and all data sets, the results are output to the MATLAB Command Window using Code 1.10 for the modeler to view. In this code, the `array2table` function is applied to convert the homogeneous m × n array to an m × n table. The input arguments `VariableNames` and `RowNames` are specified in lines 3–4 to define the column and row titles, respectively. As can be seen in Fig. 1.11, the column titles are associated with the input variable, while the row titles provide the data type. Fig. 1.24 displays the results of this program, as seen in the MATLAB Command Window. The schematic of Code 1.10 is provided in Fig. 1.23.

Code 1.10

```
1   %% Show the results of indices in the command window
2   Train = array2table([TrIn1 TrTa1]);
3   Train.Properties.VariableNames = {'In1','In2','In3','In4','Out'};
4   Train.Properties.RowNames = {'Minimum','Lower Quartile','Median','Upper
    Quartile','Maximum'}
5
6   Test  = array2table([TsIn1 TsTa1]);
7   Test.Properties.VariableNames = {'In1','In2','In3','In4','Out'};
8   Test.Properties.RowNames = {'Minimum','Lower Quartile','Median','Upper
    Quartile','Maximum'}
9
10  All   = array2table([AllIn1 AllTa1]);
11  All.Properties.VariableNames = {'In1','In2','In3','In4','Out'};
12  All.Properties.RowNames = {'Minimum','Lower Quartile','Median','Upper
    Quartile','Maximum'}
13
```

The next step is data preparation for plotting. All inputs and targets for the train, test, and all data sets are defined into single variables at this step. First, training samples for In1, In2, In3, In4, and the output are stored into the new variables `Train_In1`, `Train_In2`, `Train_In3`, `Train_In4`, and `Train_Out`, respectively. This step is repeated for the testing samples as well as for all samples. Finally, the train (i.e., `Train_In1`), test (i.e., `Test_In1`), and All (i.e., `All_In1`) values of the first input are stored in the array `In1`. This is repeated for In2, In3, In4, and Out (Fig. 1.24). The schematic of Code 1.11 is presented as shown in Fig. 1.25.

Train =

(A)

5×5 table

	In1	In2	In3	In4	Out
Minimum	3.5	20	-2250	10	36.376
Lower Quartile	5	35	9000	25	44.725
Median	5.5	40	12500	30	48.921
Upper Quartile	6	45	16500	35	50.291
Maximum	7.5	60	27750	50	58.641

Test =

(B)

5×5 table

	In1	In2	In3	In4	Out
Minimum	3.5	15	-12250	5	36.301
Lower Quartile	5	30	5000	20	44.693
Median	5.5	35	12500	25	48.91
Upper Quartile	6	40	16500	30	50.288
Maximum	7.5	55	33750	45	58.68

All =

(C)

5×5 table

	In1	In2	In3	In4	Out
Minimum	3.5	17.5	-2250	7.5	36.376
Lower Quartile	5	32.5	9000	22.5	44.725
Median	5.5	37.5	12500	27.5	48.921
Upper Quartile	6	42.5	16500	32.5	50.291
Maximum	7.5	57.5	27750	47.5	58.641

FIGURE 1.24 Boxplot characteristics for different sample subsets of Example 3.

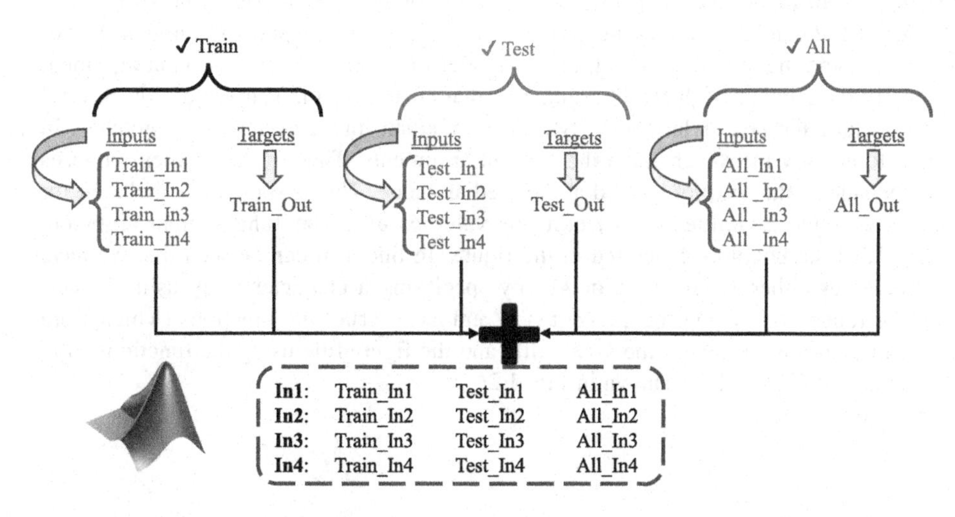

FIGURE 1.25 The schematic of Code 1.11.

Code 1.11

```
1  % Prepare data for plotting
2  % Train
3  Train_In1 = TrainInputs(:,1); % Training samples for Input 1
4  Train_In2 = TrainInputs(:,2); % Training samples for Input 2
5  Train_In3 = TrainInputs(:,3); % Training samples for Input 3
6  Train_In4 = TrainInputs(:,4); % Training samples for Input 4
7  Train_Out = TrainTargets;     % Training samples for Output
8
9  % Test
10 Test_In1 = TestInputs(:,1);  % Testing samples for Input 1
11 Test_In2 = TestInputs(:,2);  % Testing samples for Input 2
12 Test_In3 = TestInputs(:,3);  % Testing samples for Input 3
13 Test_In4 = TestInputs(:,4);  % Testing samples for Input 4
14 Test_Out = TestTargets;      % Testing samples for Output
15
16 % All
17 All_In1 = Inputs(:,1);       % All samples for Input 1
18 All_In2 = Inputs(:,2);       % All samples for Input 2
19 All_In3 = Inputs(:,3);       % All samples for Input 3
20 All_In4 = Inputs(:,4);       % All samples for Input 4
21 All_Out = Targets;           % All samples for Output
22
23 % All stages in one variable
24 In1 = [Train_In1; Test_In1; All_In1];
25 In2 = [Train_In2; Test_In2; All_In2];
26 In3 = [Train_In3; Test_In3; All_In3];
27 In4 = [Train_In4; Test_In4; All_In4];
28 Out = [Train_Out; Test_Out; All_Out];
```

Using newly created variables `In1`, `In2`, `In3`, `In4`, and `Out` from lines 24 to 28 of Code 1.11, the plotting may be performed using Code 1.12. The built-in `boxplot` command is employed to produce a boxplot for the inputs and outputs, and follows the generic syntax format of `boxplot(x,g,Name,Value)`. Because the number of samples for each subset of data (i.e., train, test, all) is not the same, the `g` input argument is introduced to group the variables. This input argument is a column vector, which has a number of rows equal to the total number of entries in `x`, the variable storing the data. The number of rows is the sum of the number of samples at the train, test, and all stages for `In1`. For In1 and Example 3, the number of training and testing samples are 168 and 72, respectively, resulting in 240 total data points. Consequently, the number of rows in `g` for the In1 of Example 3 is 480 (i.e., 168 + 72 + 240). Due to the number of samples at the train, test, and all stages for all inputs (i.e., In1, In2, In3, In4) and output are the same, the size of `g` is constant in all steps. In line 3 of Code 1.12, it can be seen that the `g` vector groups the training set with an identifier of 0, the testing subset with an identifier of 1, and the total data set with an identifier of 2. The `Name,Value` input argument of the boxplot function is used to specify the number of different arguments that customize the generated plot. Several examples of customizable parameters include the box style, color, style of the median line, and the use of notches. In line 4 of Code 1.12, we have selected to employ the `Notch` and `labels` input arguments. The `Notch` argument chooses whether or not to include a notch on the side of the boxplot located at the median value. This is sometimes included for comparison purposes, however, in this example is omitted by selecting its value as `off`. The `labels` input argument allows us to input a character array to label the boxplots generated in the figure. In line 4, it can be seen that we have specified the three boxplots to be labeled as either `Train`, `Test` or `All` by specifying a character array using braces. Lines 6–7 specify the font size and the font-weight using the `ax.FontSize` and `ax.FontWeight` functions (which were previously introduced), respectively, while lines 8–9 specify the y-axis title and the figure title using the functions `ylabel` and `title`, respectively. The schematic of Code 1.12 is shown in Fig. 1.26.

Code 1.12

```
1   %%%%%%%%%%%%%%%%%%%%%%%%%%%%% Input1 %%%%%%%%%%%%%%%%%%%%%%%%%%%%%%%
2   figure;
3   g = [zeros(length(Train_In1), 1); ones(length(Test_In1), 1);
    2*ones(length(All_In1), 1)];
4   boxplot(In1, g,'Notch','off','labels',{'Train','Test','All'})
5   ax = gca;
6   ax.FontSize = 16;
7   ax.FontWeight='normal';
8   ylabel('Value')
9   title('Input1','FontWeight','normal')
10
11  %%%%%%%%%%%%%%%%%%%%%%%%%%%%% Input2 %%%%%%%%%%%%%%%%%%%%%%%%%%%%%%%
12  figure;
13  g = [zeros(length(Train_In2), 1); ones(length(Test_In2), 1);
    2*ones(length(All_In2), 1)];
14  boxplot(In2, g,'Notch','off','labels',{'Train','Test','All'})
15  ax = gca;
16  ax.FontSize = 16;
17  ax.FontWeight='normal';
18  ylabel('Value')
19  title('Input2','FontWeight','normal')
20
21  %%%%%%%%%%%%%%%%%%%%%%%%%%%%% Input3 %%%%%%%%%%%%%%%%%%%%%%%%%%%%%%%
22  figure;
23  g = [zeros(length(Train_In3), 1); ones(length(Test_In3), 1);
    2*ones(length(All_In3), 1)];
24  boxplot(In3, g,'Notch','off','labels',{'Train','Test','All'})
25  ax = gca;
26  ax.FontSize = 16;
27  ax.FontWeight='normal';
28  ylabel('Value')
29  title('Input3','FontWeight','normal')
30
31  %%%%%%%%%%%%%%%%%%%%%%%%%%%%% Input4 %%%%%%%%%%%%%%%%%%%%%%%%%%%%%%%
32  figure;
33  g = [zeros(length(Train_In4), 1); ones(length(Test_In4), 1);
    2*ones(length(All_In4), 1)];
34  boxplot(In4, g,'Notch','off','labels',{'Train','Test','All'})
35  ax = gca;
36  ax.FontSize = 16;
37  ax.FontWeight='normal';
38  ylabel('Value')
39  title('Input4','FontWeight','normal')
40
41  %%%%%%%%%%%%%%%%%%%%%%%%%%%%%% Output %%%%%%%%%%%%%%%%%%%%%%%%%%%%%%%
42  figure;
43  g = [zeros(length(Train_Out), 1); ones(length(Test_Out), 1);
    2*ones(length(All_Out), 1)];
44  boxplot(Out, g,'Notch','off','labels',{'Train','Test','All'})
45  ax = gca;
46  ax.FontSize = 16;
47  ax.FontWeight='normal';
48  ylabel('Value')
49  title('Output','FontWeight','normal')
```

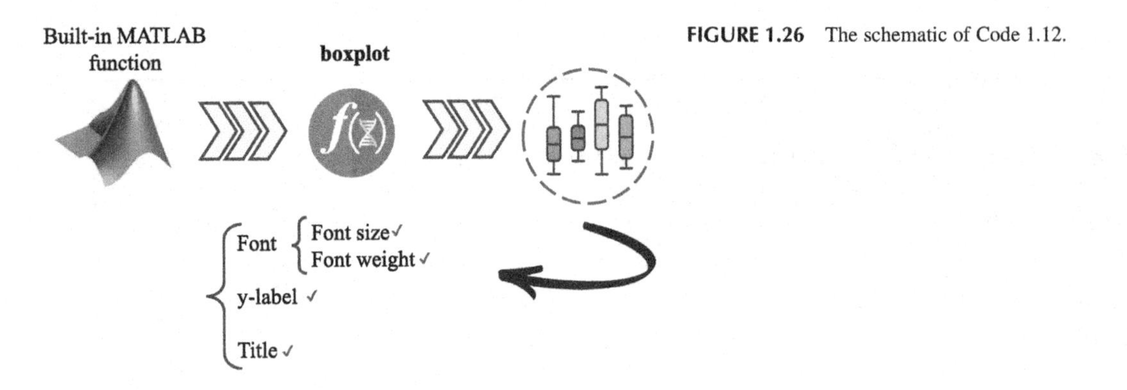

FIGURE 1.26 The schematic of Code 1.12.

1.3.4 Example 4: a problem with five input variables

The data contained within Example 4 consists of 250 sample points, of which 75 samples (i.e., 30% of all samples) are randomly selected as validation data for the testing stage. The remaining samples (i.e., 175 samples) are employed to calibrate the ML-based model. From Fig. 1.5, we know that this data series contains 5 input variables and a single output variable.

1.3.4.1 3D plot generation of training, testing, and all samples for Example 4

Fig. 1.27 shows the surface plots of the inputs and target of Example 4 for the train, test, and all data subsets. It is seen that each plot is generated by considering three variables. The output variable was constant among all plots on the *z*-axis, while the other two vectors were selected from the five input variables for use on the *x*- and *y*-axes. The number of inputs is odd; three figures were plotted for each data subset by considering In1 and In3, In2 and In5, and In2 and In4 as the two vectors coupled with the target variable (i.e., Out). Similar to previous examples, it can be clearly seen from the surface plots that the distribution of different input variables is almost identical. It is clear that the distribution of different input variables at each stage (i.e., Train, test, all) are almost similar.

1.3.4.2 The MATLAB coding of 3D plot

Codes 1.13–1.15 detail the MATLAB programming required to generate Fig. 1.27. This code contains four main sections, including data loading, merging of samples, preparation for plotting, and generation of the surface plot. The first two sections (i.e., data loading and merging of samples) have been introduced and applied in the resolution of previous examples using the bargraph, histogram, and boxplot. However, the required coding is provided again in Code 1.13. Similar to most MATLAB-based codes, three well-known commands are employed at the beginning to clear the Command Window (i.e., `clc`), clear all preexisting variables (i.e., `clear`), and close all opened files, including existing figures (i.e., `close all`). To load the data into the MATLAB environment, the `xlsread` command is used to read the data contained in a Microsoft Excel file (i.e., Example 4). Lastly, the training and testing samples for both inputs and targets are merged so that all inputs and targets are stored in the variables `Inputs`, and `Targets`, respectively. The schematic of Code 1.13 is shown in Fig. 1.28.

Code 1.13

```
1   clc
2   clear
3   close all
4   
5   %% load data
6   TrainInputs  = xlsread ('Example4','sheet1'); % Train Inputs
7   TrainTargets = xlsread ('Example4','sheet2'); % Train Targets
8   TestInputs   = xlsread ('Example4','sheet3'); % Test Inputs
9   TestTargets  = xlsread ('Example4','sheet4'); % Test Targets
10  
11  %% Merge all samples
12  Inputs  = [TrainInputs;TestInputs];           % Inputs
13  Targets = [TrainTargets;TestTargets];         % Targets
```

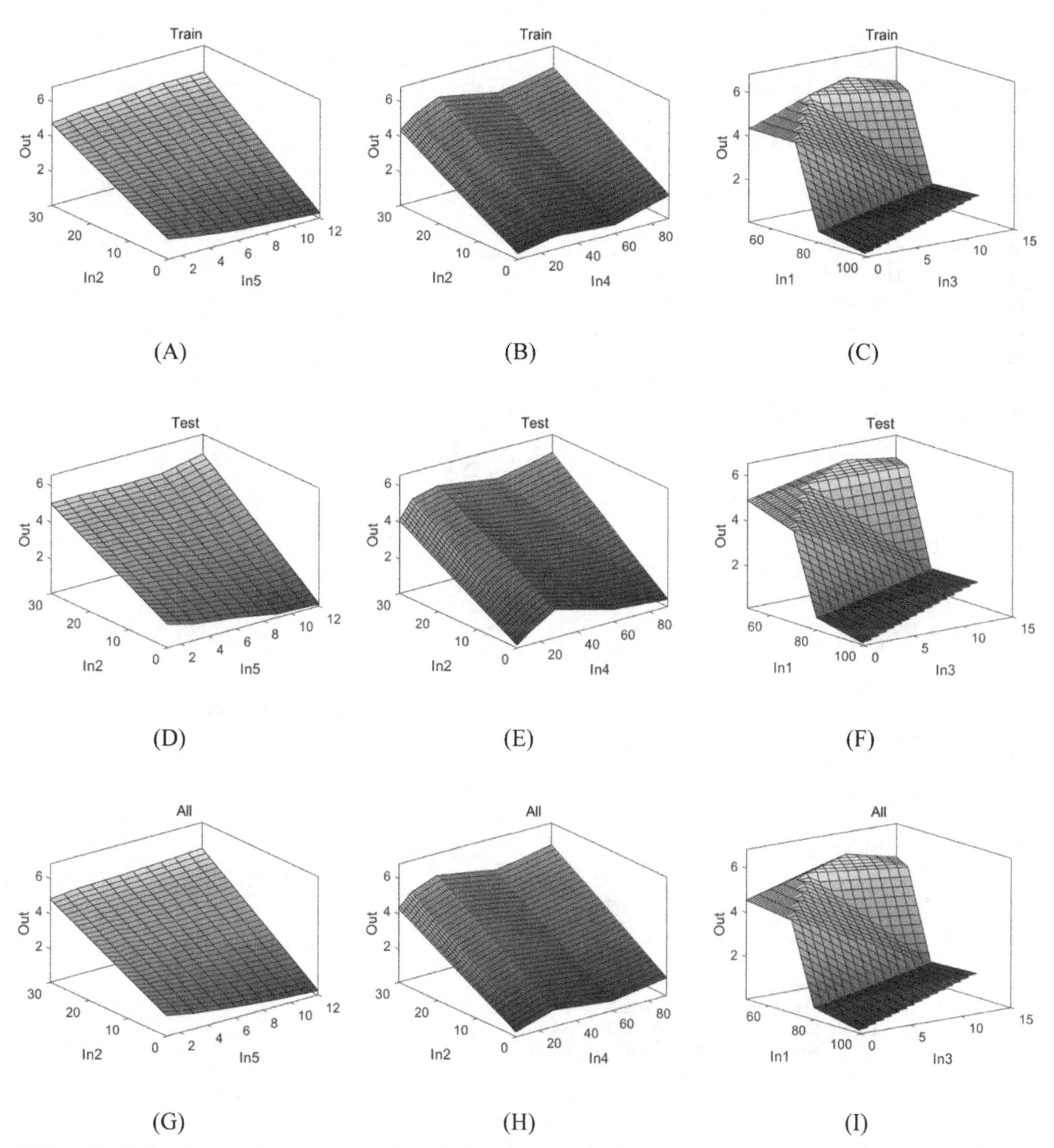

FIGURE 1.27 Surface plot of the inputs and target of Example 4 at train, test, and all stages. (A) distribution of inputs 2 and 5 along with output at training stage, (B) distribution of inputs 2 and 4 along with output at training stage, (C) distribution of inputs 1 and 3 along with output at training stage, (D) distribution of inputs 2 and 5 along with output at testing stage, (E) distribution of inputs 2 and 4 along with output at testing stage, (F) distribution of inputs 1 and 3 along with output at testing stage, (G) distribution of inputs 2 and 5 along with output for all samples, (H) distribution of inputs 2 and 4 along with output for all samples, and (I) distribution of inputs 1 and 3 along with output for all samples.

Once the dataset has been read into the MATLAB environment and the inputs and targets have been merged, the next step is to prepare the data for plotting. In this step, for each of the five inputs and the single target value, a new variable is defined for each of the training, testing, and total data subsets, resulting in eighteen new variables. For example, in line 3 of Code 1.14, `Train_In1` is the new variable created to store the data for the first input (i.e., In1) in the training data subset. The schematic of Code 1.14 is shown in Fig. 1.29.

Code 1.14

```
1   %% Prepare data for plotting
2   % Train
3   Train_In1 = TrainInputs(:,1); % Training samples for Input 1
4   Train_In2 = TrainInputs(:,2); % Training samples for Input 2
5   Train_In3 = TrainInputs(:,3); % Training samples for Input 3
6   Train_In4 = TrainInputs(:,4); % Training samples for Input 4
7   Train_In5 = TrainInputs(:,5); % Training samples for Input 5
8   Train_Out = TrainTargets;     % Training samples for Output
9
10  % Test
11  Test_In1 = TestInputs(:,1);  % Testing samples for Input 1
12  Test_In2 = TestInputs(:,2);  % Testing samples for Input 2
13  Test_In3 = TestInputs(:,3);  % Testing samples for Input 3
14  Test_In4 = TestInputs(:,4);  % Testing samples for Input 4
15  Test_In5 = TestInputs(:,5);  % Testing samples for Input 5
16  Test_Out = TestTargets;      % Testing samples for Output
17
18  % All
19  All_In1 = Inputs(:,1);       % All samples for Input 1
20  All_In2 = Inputs(:,2);       % All samples for Input 2
21  All_In3 = Inputs(:,3);       % All samples for Input 3
22  All_In4 = Inputs(:,4);       % All samples for Input 4
23  All_In5 = Inputs(:,5);       % All samples for Input 5
24  All_Out = Targets;           % All samples for Output
```

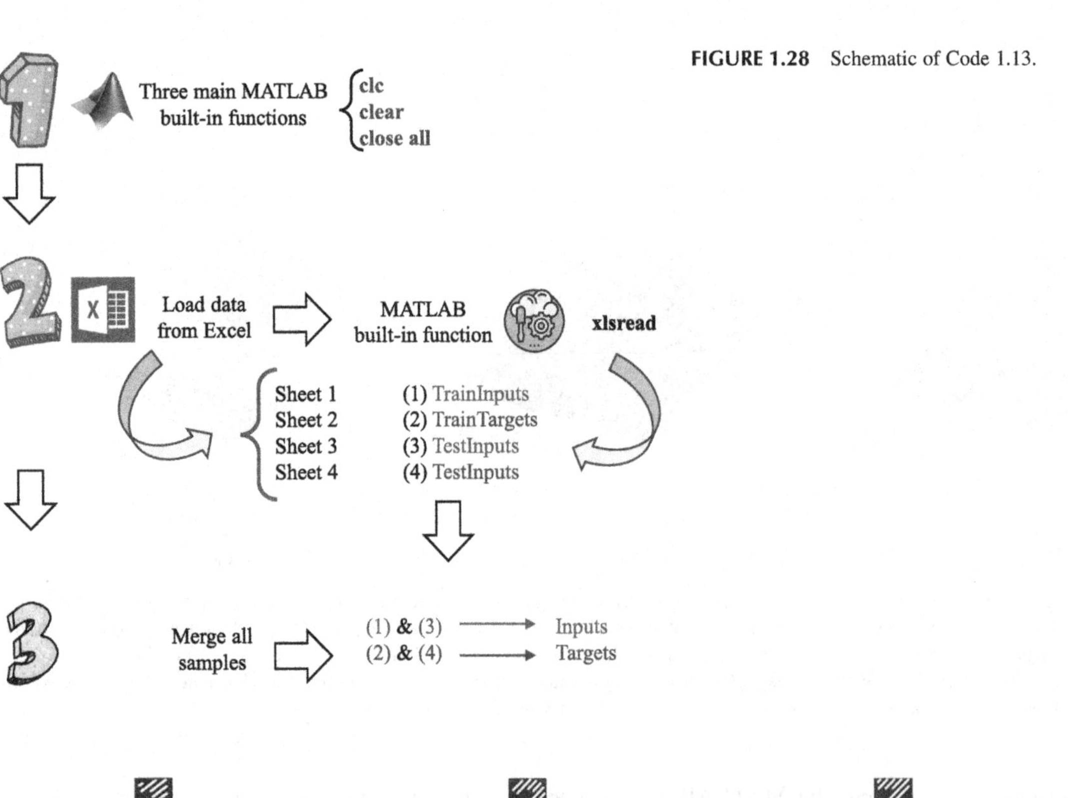

FIGURE 1.28 Schematic of Code 1.13.

TrainInputs
Train_In1
Train_In2
Train_In3
Train_In4
Train_In5
TrainTargets
Train_Out

TestInputs
Test_In1
Test_In2
Test_In3
Test_In4
Test_In5
TestTargets
Test_Out

Inputs
All_In1
All_In2
All_In3
All_In4
All_In5
Targets
All_Out

FIGURE 1.29 Schematic of Code 1.14.

After data preparation, the surface plots are generated and their characteristics personalized according to Codes 1.15.A–1.15.C. As previously mentioned, each surface plot is generated by considering three axes. The z-axis is associated with the target variable, while two of the five input variables are considered for the x- and y-axes. In this example, In1 and In3, In2 and In5, and In2 and In4 were selected as the combinations of two input variables to generate the surface plot in combination with the target variable. Therefore a total of nine figures are generated—three input combinations are considered for each of the training, testing, and total data subsets. The first step is the definition of the axes. In lines 5–8 of Code 1.15.A, the variables `a`, `b`, and `c` are used to define the x, y, and z axes of each surface plot. For example, the values of `a`, `b`, and `c` for the first figure can be defined as `Train_In1`, `Train_In3`, and `Train_Out`, respectively. In lines 8–9, the range of the x and y axes are specified according to the minimum and maximum values contained within `Train_In1` and `Train_In3`. In line 10, the`meshgrid` command is used to generate a two-dimensional grid defined based on the ranges contained within `x` and`y` from lines 8 to 9. The result of the `meshgrid` command is stored at `[X,Y]`. Using the `a`, `b`, and `c` values, the result of the `meshgrid` command (i.e., `[X,Y]`) and the `griddata` command, `Z` is defined. The `griddata` command fits a surface of the form of $k = f(a, b)$ to the scattered data in the vectors (`a`, `b`, `c`). The `griddata` command interpolates the surface at the query points specified by `[X,Y]` and returns the interpolated values, `Z`. The `surf` command is applied to generate a three-dimensional surface plot of three variables (i.e., `a`, `b`, `c`). In `surf(a,b,c)`, the `a`, `b`, and `c` values are the x, y, and z axes, respectively. The color data is defined based on the value of the "z" axis. In lines 13–22, several characteristics of the surface plot are defined, including the range of each axis, the font size, the font-weight, the chart grid, axes labels, and the chart label. To define the axis limits, the built-in `axis` function is used in line 13. Its input arguments consist of the minimum and maximum values of `a`, `b`, and `c`. Font size and font weight are specified in lines 15–16, and have previously been discussed in detail for the other examples considered in this chapter. `Box` in line 17 is used to display the outline of the current axes by setting its value to `on`, while `grid` in line 18 allows the modeler to display the grids of the axes. When its value is set to `off` then the grids are prevented from being shown. To define the axis label, the `xlabel`, `ylabel`, and `zlabel` commands are applied in lines 19–21. Finally, a title is added to the entire plot by using the built-in `title` command. This process is repeated for the other eight-figures. The schematic of Code 1.15 is shown in Fig. 1.30.

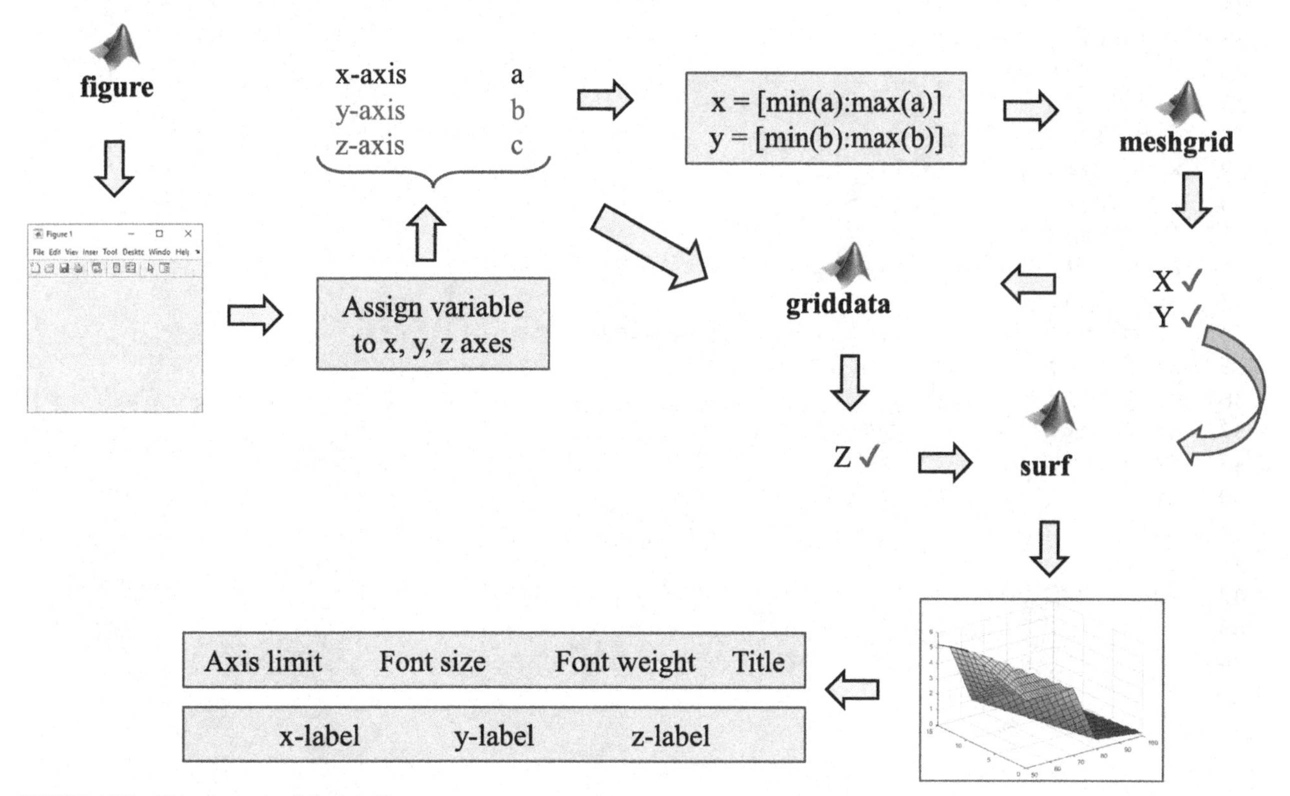

FIGURE 1.30 The schematic of Code 1.15.

Code 1.15.A

```
1   %% Surface plot
2   %%%%%%%%%%%%%%%%%%%%%%%%%%%%%%%% Train %%%%%%%%%%%%%%%%%%%%%%%%%%%%%%%%
3   % In1 - In3
4   figure;
5   a = Train_In1;
6   b = Train_In3;
7   c = Train_Out;
8   x = [min(a):max(a)];
9   y = [min(b):max(b)];
10  [X,Y]=meshgrid(x,y);
11  Z=griddata(a,b,c,X,Y);
12  surf(X,Y,Z)
13  axis([min(a),max(a),min(b),max(b),min(c),max(c)])
14  ax = gca;
15  ax.FontSize = 16;
16  ax.FontWeight='normal';
17  ax.Box='on';
18  grid off
19  xlabel('In1')
20  ylabel('In3')
21  zlabel('Out')
22  title('Train','FontWeight','normal')
23
24  % In2 - In4
25  figure;
26  a = Train_In4;
27  b = Train_In2;
28  c = Train_Out;
29  x = [min(a):max(a)];
30  y = [min(b):max(b)];
31  [X,Y]=meshgrid(x,y);
32  Z=griddata(a,b,c,X,Y);
33  surf(X,Y,Z)
34  axis([min(a),max(a),min(b),max(b),min(c),max(c)])
35  ax = gca;
36  ax.FontSize = 16;
37  ax.FontWeight='normal';
38  ax.Box='on';
39  grid off
40  xlabel('In4')
41  ylabel('In2')
42  zlabel('Out')
43  title('Train','FontWeight','normal')
44
```

```
45  % In2 - In5
46  figure;
47  a = Train_In5;
48  b = Train_In2;
49  c = Train_Out;
50  x = [min(a):max(a)];
51  y = [min(b):max(b)];
52  [X,Y]=meshgrid(x,y);
53  Z=griddata(a,b,c,X,Y);
54  surf(X,Y,Z)
55  axis([min(a),max(a),min(b),max(b),min(c),max(c)])
56  ax = gca;
57  ax.FontSize = 16;
58  ax.FontWeight='normal';
59  ax.Box='on';
60  grid off
61  xlabel('In5')
62  ylabel('In2')
63  zlabel('Out')
64  title('Train','FontWeight','normal')
```

Code 1.15.B

```
1   %%%%%%%%%%%%%%%%%%%%%%%%%%%%%%%%% Test %%%%%%%%%%%%%%%%%%%%%%%%%%%%%%%%%
2   % In1 - In3
3   figure;
4   a = Test_In1;
5   b = Test_In3;
6   c = Test_Out;
7   x = [min(a):max(a)];
8   y = [min(b):max(b)];
9   [X,Y]=meshgrid(x,y);
10  Z=griddata(a,b,c,X,Y);
11  surf(X,Y,Z)
12  axis([min(a),max(a),min(b),max(b),min(c),max(c)])
13  ax = gca;
14  ax.FontSize = 16;
15  ax.FontWeight='normal';
16  ax.Box='on';
17  grid off
18  xlabel('In1')
19  ylabel('In3')
20  zlabel('Out')
21  title('Test','FontWeight','normal')
22
23  % In2 - In4
24  figure;
25  a = Test_In4;
26  b = Test_In2;
27  c = Test_Out;
28  x = [min(a):max(a)];
29  y = [min(b):max(b)];
30  [X,Y]=meshgrid(x,y);
31  Z=griddata(a,b,c,X,Y);
32  surf(X,Y,Z)
33  axis([min(a),max(a),min(b),max(b),min(c),max(c)])
34  ax = gca;
35  ax.FontSize = 16;
36  ax.FontWeight='normal';
37  ax.Box='on';
38  grid off
39  xlabel('In4')
40  ylabel('In2')
41  zlabel('Out')
42  title('Test','FontWeight','normal')
43
```

```
44  % In2 - In5
45  figure;
46  a = Test_In5;
47  b = Test_In2;
48  c = Test_Out;
49  x = [min(a):max(a)];
50  y = [min(b):max(b)];
51  [X,Y]=meshgrid(x,y);
52  Z=griddata(a,b,c,X,Y);
53  surf(X,Y,Z)
54  axis([min(a),max(a),min(b),max(b),min(c),max(c)])
55  ax = gca;
56  ax.FontSize = 16;
57  ax.FontWeight='normal';
58  ax.Box='on';
59  grid off
60  xlabel('In5')
61  ylabel('In2')
62  zlabel('Out')
63  title('Test','FontWeight','normal')
```

Code 1.15.C

```
1   %%%%%%%%%%%%%%%%%%%%%%%%%%%%%%%%%% All %%%%%%%%%%%%%%%%%%%%%%%%%%%%%%%%%%
2   % In1 - In3
3   figure;
4   a = All_In1;
5   b = All_In3;
6   c = All_Out;
7   x = [min(a):max(a)];
8   y = [min(b):max(b)];
9   [X,Y]=meshgrid(x,y);
10  Z=griddata(a,b,c,X,Y);
11  surf(X,Y,Z)
12  axis([min(a),max(a),min(b),max(b),min(c),max(c)])
13  ax = gca;
14  ax.FontSize = 16;
15  ax.FontWeight='normal';
16  ax.Box='on';
17  grid off
18  xlabel('In1')
19  ylabel('In3')
20  zlabel('Out')
21  title('All','FontWeight','normal')
22  
23  % In2 - In4
24  figure;
25  a = All_In4;
26  b = All_In2;
27  c = All_Out;
28  x = [min(a):max(a)];
29  y = [min(b):max(b)];
30  [X,Y]=meshgrid(x,y);
31  Z=griddata(a,b,c,X,Y);
32  surf(X,Y,Z)
33  axis([min(a),max(a),min(b),max(b),min(c),max(c)])
34  ax = gca;
35  ax.FontSize = 16;
36  ax.FontWeight='normal';
37  ax.Box='on';
38  grid off
39  xlabel('In4')
40  ylabel('In2')
41  zlabel('Out')
42  title('All','FontWeight','normal')
43  
```

```
44	% In2 - In5
45	figure;
46	a = All_In5;
47	b = All_In2;
48	c = All_Out;
49	x = [min(a):max(a)];
50	y = [min(b):max(b)];
51	[X,Y]=meshgrid(x,y);
52	Z=griddata(a,b,c,X,Y);
53	surf(X,Y,Z)
54	axis([min(a),max(a),min(b),max(b),min(c),max(c)])
55	ax = gca;
56	ax.FontSize = 16;
57	ax.FontWeight='normal';
58	ax.Box='on';
59	grid off
60	xlabel('In5')
61	ylabel('In2')
62	zlabel('Out')
63	title('All','FontWeight','normal')
```

1.3.5 Example 5: a problem with two input variables

This example has the lowest number of input variables but the highest number of samples amongst the examples considered so far in this text. The number of inputs in this example is two, while the total number of observations is 1095. This example is a time series-based problem. In time series problems, the random selection of the training and testing subsets cannot be performed as we must use past observations to forecast the future. To split samples into training and testing subsets, the first 70% (i.e., 767 samples) of samples employed for training and the remaining 328 samples are considered for testing and verification of generalizability.

1.3.5.1 The time series of the training, testing, and all samples of Example 5

Fig. 1.31 demonstrates the time series plot of the input variables and target variable of Example 5, as well as the distinction between training and testing data. In all subfigures, the first 767 samples are related to the training stage, while the 768th to 1095th samples are related to the testing stage. From the figure, it can be seen that both inputs and output follow a periodic pattern. The maximum value of In1 is lower than both In2 and the output. Additionally, In1 consists entirely of positive values, while In2 and the output contain some negative values. The range and period of the In2 and output are very close together, but they are not the same. Therefore, as a preliminary conclusion, it can be said that the effect of In2 on the output could be much greater than In1. However, the precise impact of each one should be verified through the modeling phase.

1.3.5.2 The MATLAB coding of time series

The plotting process of Fig. 1.31 and its detailed MATLAB code are provided in Code 1.16 below. This plotting process can be summarized according to three steps: (1) data loading, (2) preparation of data for plotting, and (3) plotting. The first two steps are similar to what has been described in detail in the previous examples. First, the data is read from a Microsoft Excel file entitled "Example5" using the built-in `xlsread` function. This file consists of four sheets (sheet1–sheet4), which contain the training inputs, training outputs, testing inputs, and testing outputs, respectively. In the second step, six new variables are defined and named `Train_In1`, `Train_In2`, `Train_Out`, `Test_In1`, `Test_In2`, `Test_Out`. These represent the data at the training stage, the testing stage, and the total data for In1, In2, and the output. The schematic of Code 1.16 is shown in Fig. 1.32.

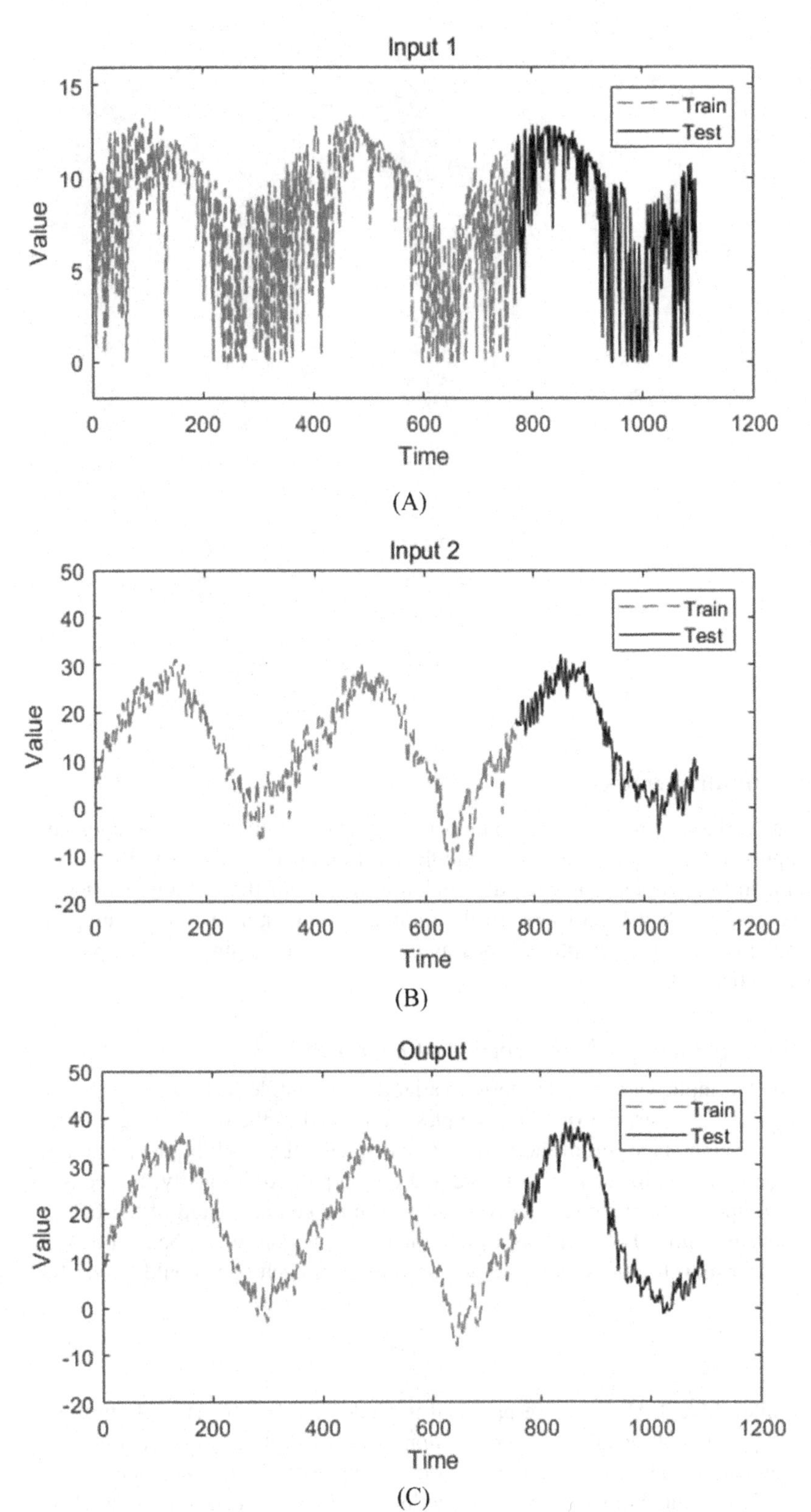

FIGURE 1.31 Time series plot of the inputs and target of Example 5 during the training and testing stages. (A) Input 1, (B) input 2, and (C) output.

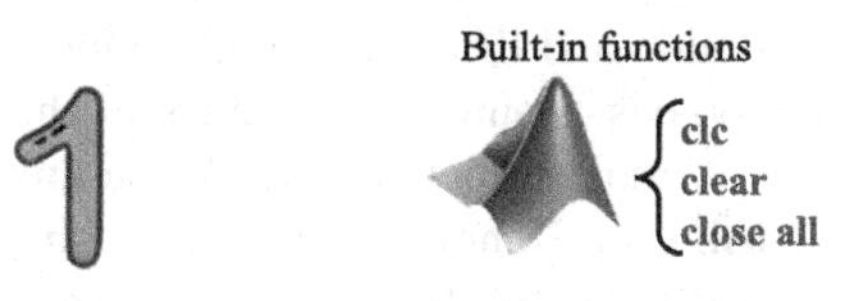

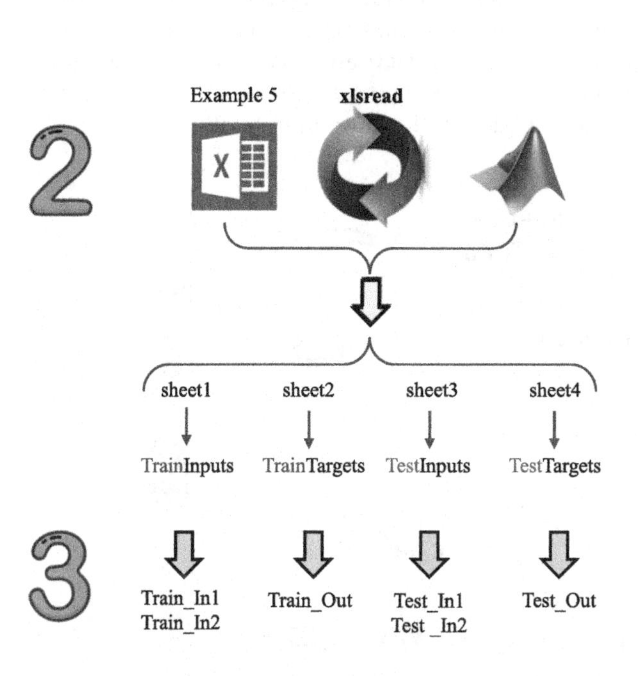

FIGURE 1.32 The schematic of Code 1.16.

Code 1.16

```
1    clc
2    clear
3    close all
4
5    %% Load data
6    TrainInputs  = xlsread ('Example5','sheet1'); % Train Inputs
7    TrainTargets = xlsread ('Example5','sheet2'); % Train Targets
8    TestInputs   = xlsread ('Example5','sheet3'); % Test Inputs
9    TestTargets  = xlsread ('Example5','sheet4'); % Test Targets
10
11   %% Prepare data for plotting
12   % Train
13   Train_In1 = TrainInputs(:,1); % Training samples for Input 1
14   Train_In2 = TrainInputs(:,2); % Training samples for Input 2
15   Train_Out = TrainTargets;      % Training samples for Output
16
17   % Test
18   Test_In1 = TestInputs(:,1);    % Testing samples for Input 1
19   Test_In2 = TestInputs(:,2);    % Testing samples for Input 2
20   Test_Out = TestTargets;        % Training samples for Output
```

In the next step, the detailed coding required to plot the time series plots of In1, In2, and the output are provided. To apply the `plot` command in MATLAB, the *x* and *y*-axes must be defined. The *y*-axis is simply the values of the desired variables (i.e., In1, In2, or Output at the training or testing stage), while the "*x*" must be defined. To define this axis for In1, the number of training samples is calculated in line 4 of Code 1.17 using the `length(Train_In1)` command. The *x*-axis domain is then defined from 1 to `length(Train_In1)`. To plot the test data on the same set of axes, the `hold` command is set to `on`, which retains the current plot while a new one is added. Because this is time series data, the testing subset spans from the end of the training data to the end of the total data series. To code this, the *x*-axis domain of the testing samples is specified as `length(Train_In1) + 1:1:length(Train_In1) + length(Test_In1)` in line 7. This process is repeated for In2 and the output. The schematic of Code 1.17 is shown in Fig. 1.33.

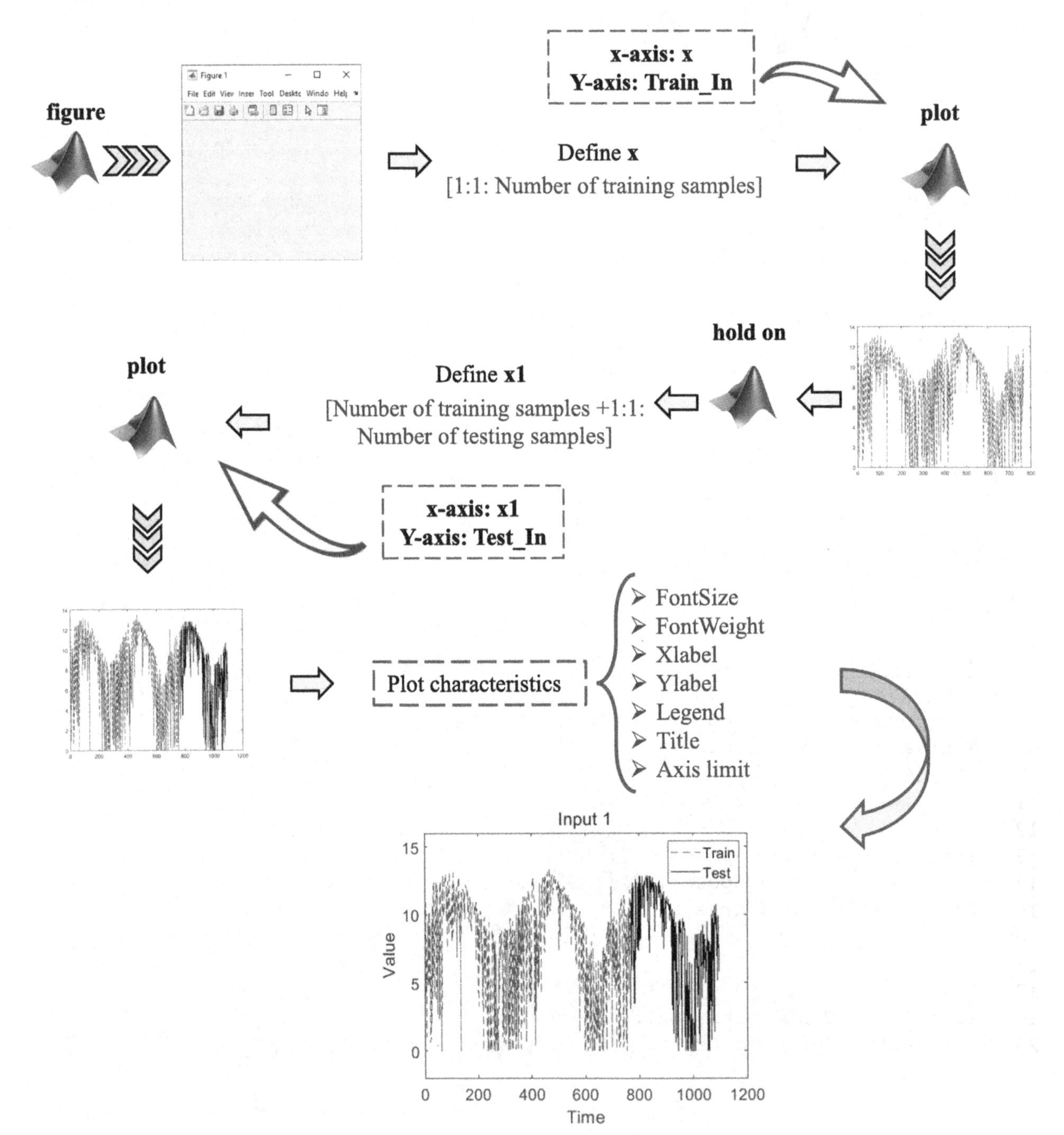

FIGURE 1.33 The schematic of Code 1.17.

Code 1.17

```
1   %% Plot
2   %%%%%%%%%%%%%%%%%%%%%%%%%%%%%% In1 %%%%%%%%%%%%%%%%%%%%%%%%%%%%%%
3   figure;
4   x = 1:1:length(Train_In1);
5   plot(x',Train_In1,'-r')
6   hold on
7   x1 = length(Train_In1)+1:1:length(Train_In1)+length(Test_In1);
8   plot(x1',Test_In1,'-k')
9
10  ax = gca;
11  ax.FontSize = 16;
12  ax.FontWeight='normal';
13  xlabel('Time')
14  ylabel('Value')
15  legend ('Train','Test')
16  title('Input 1','FontWeight','normal')
17  axis ([0,1200,-2,16])
18  %%%%%%%%%%%%%%%%%%%%%%%%%%%%%% In2 %%%%%%%%%%%%%%%%%%%%%%%%%%%%%%
19  figure;
20  x = 1:1:length(Train_In2);
21  plot(x',Train_In2,'-r')
22  hold on
23  x1 = length(Train_In2)+1:1:length(Train_In2)+length(Test_In2);
24  plot(x1',Test_In2,'-k')
25
26  ax = gca;
27  ax.FontSize = 16;
28  ax.FontWeight='normal';
29  xlabel('Time')
30  ylabel('Value')
31  legend ('Train','Test')
32  title('Input 2','FontWeight','normal')
33  axis ([0,1200,-20,50])
34  %%%%%%%%%%%%%%%%%%%%%%%%%%%%%% Out %%%%%%%%%%%%%%%%%%%%%%%%%%%%%%
35  figure;
36  x = 1:1:length(Train_Out);
37  plot(x',Train_Out,'-r')
38  hold on
39  x1 = length(Train_Out)+1:1:length(Train_Out)+length(Test_Out);
40  plot(x1',Test_Out,'-k')
41
42  ax = gca;
43  ax.FontSize = 16;
44  ax.FontWeight='normal';
45  xlabel('Time')
46  ylabel('Value')
47  legend ('Train','Test')
48  title('Output','FontWeight','normal')
49  axis ([0,1200,-20,50])
```

1.4 Summary

This chapter provided a full conceptual and mathematical description of the modeling process in problems that are solved using ML without considering a specific technique in the book. Five different problems in the real world were investigated. In choosing these problems, we tried to consider problems with a wide variety in the number of samples (i.e., 100–1000) and input variables (i.e., 2–6). It is now possible for the reader to understand employed problems in the following chapters after completing this chapter.

Appendix 1A Supporting information

Examples 1–5 can be found in the online version at doi:10.1016/B978-0-443-15284-9.00002-1.

References

Azimi, H., Bonakdari, H., Ebtehaj, I., Talesh, S. H. A., Michelson, D. G., & Jamali, A. (2016). Evolutionary Pareto optimization of an ANFIS network for modeling scour at pile groups in clear water condition. *Fuzzy Sets and Systems*, *319*, 50–69. Available from https://doi.org/10.1016/j.fss.2016.10.010.

Bagheri, S., Kabiri-Samani, A. R., & Heidarpour, M. (2014). Discharge coefficient of rectangular sharp-crested side weirs Part II: Domínguez's method. *Flow Measurement and Instrumentation*, *35*, 116–121. Available from https://doi.org/10.1016/j.flowmeasinst.2013.10.006.

Bonakdari, H., & Zeynoddin, M. (2022). Goodness-of-fit & precision criteria. *Journal: Stochastic Modeling*, 187–264. Available from https://doi.org/10.1016/B978-0-323-91748-3.00003-3.

Bonakdari, H., Moeeni, H., Ebtehaj, I., Zeynoddin, M., Mahoammadian, A., & Gharabaghi, B. (2019). New insights into soil temperature time series modeling: Linear or nonlinear? *Theoretical and Applied Climatology*, *135*(3), 1157–1177. Available from https://doi.org/10.1007/s00704-018-2436-2.

Cheong, H. (1991). Discharge coefficient of lateral diversion from trapezoidal channel. *Journal of Irrigation and Drainage Engineering*, *117*(4), 461–475. Available from https://doi.org/10.1061/(ASCE)0733-9437(1991)117:4(461).

Ebtehaj, I., & Bonakdari, H. (2014). Comparison of genetic algorithm and imperialist competitive algorithms in predicting bed load transport in clean pipe. *Water Science and Technology*, *70*(10), 1695–1701. Available from https://doi.org/10.2166/wst.2014.434.

Ebtehaj, I., & Bonakdari, H. (2016a). Bed load sediment transport estimation in a clean pipe using multilayer perceptron with different training algorithms. *KSCE Journal of Civil Engineering*, *20*(2), 581–589. Available from https://doi.org/10.1007/s12205-015-0630-7.

Ebtehaj, I., & Bonakdari, H. (2016b). Assessment of evolutionary algorithms in predicting non-deposition sediment transport. *Urban Water Journal*, *13*(5), 499–510. Available from https://doi.org/10.1080/1573062X.2014.994003.

Ebtehaj, I., Bonakdari, H., & Gharabaghi, B. (2019). A reliable linear method for modeling lake level fluctuations. *Journal of Hydrology*, *570*, 236–250. Available from https://doi.org/10.1016/j.jhydrol.2019.01.010.

Ebtehaj, I., Bonakdari, H., & Shamshirband, S. (2016). Extreme learning machine assessment for estimating sediment transport in open channels. *Engineering with Computers*, *32*(4), 691–704. Available from https://doi.org/10.1007/s00366-016-0446-1.

Ebtehaj, I., Bonakdari, H., Safari, M. J. S., Gharabaghi, B., Zaji, A. H., Madavar, H. R., & Mehr, A. D. (2020). Combination of sensitivity and uncertainty analyses for sediment transport modeling in sewer pipes. *International Journal of Sediment Research*, *35*(2), 157–170. Available from https://doi.org/10.1016/j.ijsrc.2019.08.005.

Ebtehaj, I., Bonakdari, H., Zaji, A. H., Azimi, H., & Khoshbin, F. (2015). GMDH-type neural network approach for modeling the discharge coefficient of rectangular sharp-crested side weirs. *Engineering Science and Technology, an International Journal*, *18*(4), 746–757. Available from https://doi.org/10.1016/j.jestch.2015.04.012.

Ebtehaj, I., Sattar, A. M., Bonakdari, H., & Zaji, A. H. (2017). Prediction of scour depth around bridge piers using self-adaptive extreme learning machine. *Journal of Hydroinformatics*, *19*(2), 207–224. Available from https://doi.org/10.2166/hydro.2016.025.

Ferdinandy, B., Gerencsér, L., Corrieri, L., Perez, P., Újváry, D., Csizmadia, G., & Miklósi, Á. (2020). Challenges of machine learning model validation using correlated behaviour data: Evaluation of cross-validation strategies and accuracy measures. *PLoS One*, *15*(7), e0236092. Available from https://doi.org/10.1371/journal.pone.0236092.

Gholami, A., Bonakdari, H., Ebtehaj, I., Shaghaghi, S., & Khoshbin, F. (2017). Developing an expert group method of data handling system for predicting the geometry of a stable channel with a gravel bed. *Earth Surface Processes and Landforms*, *42*(10), 1460–1471. Available from https://doi.org/10.1002/esp.4104.

Gómez-Escalonilla, V., Martínez-Santos, P., & Martín-Loeches, M. (2022). Preprocessing approaches in machine-learning-based groundwater potential mapping: An application to the Koulikoro and Bamako regions, Mali. *Hydrology and Earth System Sciences*, *26*(2), 221–243. Available from https://doi.org/10.5194/hess-26-221-2022.

Hariri-Ardebili, M. A., & Pourkamali-Anaraki, F. (2018a). Support vector machine based reliability analysis of concrete dams. *Soil Dynamics and Earthquake Engineering*, *104*, 276–295. Available from https://doi.org/10.1016/j.soildyn.2017.09.016.

Hariri-Ardebili, M. A., & Pourkamali-Anaraki, F. (2018b). Simplified reliability analysis of multi hazard risk in gravity dams via machine learning techniques. *Archives of Civil and Mechanical Engineering*, *18*(2), 592–610. Available from https://doi.org/10.1016/j.acme.2017.09.003.

Herrera, P. A., Marazuela, M. A., & Hofmann, T. (2022). Parameter estimation and uncertainty analysis in hydrological modeling. *Wiley Interdisciplinary Reviews: Water*, *9*(1), e1569. Available from https://doi.org/10.1002/wat2.1569.

Hu, Z., Karami, H., Rezaei, A., DadrasAjirlou, Y., Piran, M. J., Band, S. S., Chau, K. W., & Mosavi, A. (2021). Using soft computing and machine learning algorithms to predict the discharge coefficient of curved labyrinth overflows. *Engineering Applications of Computational Fluid Mechanics*, *15*(1), 1002–1015. Available from https://doi.org/10.1080/19942060.2021.1934546.

Ivanyuk, V., & Soloviev, V. (2019). Efficiency of neural networks in forecasting problems. In: *2019 twelfth international conference "Management of large-scale system development"(MLSD)* (pp. 1–4). IEEE. https://doi.org/10.1109/MLSD.2019.8911046.

Jato-Espino, D., Sillanpää, N., & Pathak, S. (2019). Flood modelling in sewer networks using dependence measures and learning classifier systems. *Journal of Hydrology*, *578*, 124013. Available from https://doi.org/10.1016/j.jhydrol.2019.124013.

Kim, J., Han, H., Johnson, L. E., Lim, S., & Cifelli, R. (2019). Hybrid machine learning framework for hydrological assessment. *Journal of Hydrology*, *577*, 123913. Available from https://doi.org/10.1016/j.jhydrol.2019.123913.

Moeeni, H., Bonakdari, H., & Ebtehaj, I. (2017). Monthly reservoir inflow forecasting using a new hybrid SARIMA genetic programming approach. *Journal of Earth System Science*, *126*(2), 18. Available from https://doi.org/10.1007/s12040-017-0798-y.

Niu, W. J., Feng, Z. K., Yang, W. F., & Zhang, J. (2020). Short-term streamflow time series prediction model by machine learning tool based on data preprocessing technique and swarm intelligence algorithm. *Hydrological Sciences Journal*, *65*(15), 2590–2603. Available from https://doi.org/10.1080/02626667.2020.1828889.

Obaid, H.S., Dheyab, S.A., & Sabry, S.S. (2019). The impact of data pre-processing techniques and dimensionality reduction on the accuracy of machine learning. In: *2019 9th annual information technology, electromechanical engineering and microelectronics conference (IEMECON)* (pp. 279–283). IEEE. https://doi.org/10.1109/IEMECONX.2019.8877011.

Qasem, S. N., Ebtehaj, I., & Bonakdari, H. (2017). Potential of radial basis function network with particle swarm optimization for prediction of sediment transport at the limit of deposition in a clean pipe. *Sustainable Water Resources Management*, *3*(4), 391–401. Available from https://doi.org/10.1007/s40899-017-0104-9.

Sharafati, A., Haji Seyed Asadollah, S. B., Motta, D., & Yaseen, Z. M. (2020). Application of newly developed ensemble machine learning models for daily suspended sediment load prediction and related uncertainty analysis. *Hydrological Sciences Journal*, *65*(12), 2022–2042. Available from https://doi.org/10.1080/02626667.2020.1786571.

Zeynoddin, M., Bonakdari, H., Ebtehaj, I., Azari, A., & Gharabaghi, B. (2020). A generalized linear stochastic model for lake level prediction. *Science of The Total Environment*, *723*, 138015. Available from https://doi.org/10.1016/j.scitotenv.2020.138015.

Zeynoddin, M., Ebtehaj, I., & Bonakdari, H. (2020). Development of a linear based stochastic model for daily soil temperature prediction: One step forward to sustainable agriculture. *Computers and Electronics in Agriculture*, *176*, 105636. Available from https://doi.org/10.1016/j.compag.2020.105636.

Zeynoddin, M., Bonakdari, H., Ebtehaj, I., Esmaeilbeiki, F., Gharabaghi, B., & Haghi, D. Z. (2019). A reliable linear stochastic daily soil temperature forecast model. *Soil and Tillage Research*, *189*, 73–87. Available from https://doi.org/10.1016/j.still.2018.12.023.

Zhang, J., Zhu, Y., Zhang, X., Ye, M., & Yang, J. (2018). Developing a long short-term memory (LSTM) based model for predicting water table depth in agricultural areas. *Journal of hydrology*, *561*, 918–929. Available from https://doi.org/10.1016/j.jhydrol.2018.04.065.

Chapter 2

Preprocessing approaches

2.1 Normalization

This section provides the MATLAB® code and theoretical details relating to how normalization can be applied to transform data into a number of different ranges. Before commencing any MATLAB coding, it is important to first review the mathematical formulation of normalization and why/when it should be applied.

2.1.1 The min–max normalization

Normalization is one of the methods used in data science to transform features in a dataset to the same scale. The normalization process provided in this section is the min–max normalization, sometimes referred to as feature scaling, which is well-known in different fields of science (Ebtehaj & Bonakdari, 2016; Ebtehaj et al., 2020a; Qasem et al., 2017a; Riahi-Madvar & Gharabaghi, 2022; Safari, 2019; Salih et al., 2020). In this type of normalization, each feature is linearly transformed to a scale that ranges between b and $b + a$ [Eq. (2.1)]. Normalization makes the datasets homogenized, resulting in a more accurate and faster modeling process (Ebtehaj & Bonakdari, 2013; Wang et al., 2017).

The general form of the min–max normalization is as follows:

$$xn_i = a\left(\frac{x_i - x_{\min}}{x_{\max} - x_{\min}}\right) + b \tag{2.1}$$

where x_i is the ith sample of data that will be normalized, $x_{\min}$ and $x_{\max}$ are the minimum and maximum of the data, xn_i is the normalized value of the ith sample of data (x_i), and a and b are constants. Different values of the constants a and b can result in different ranges of normalized data [b, $a + b$]. Accordingly, there is no universally accepted or standard normalization range (Dawson & Wilby, 1998), and different types of normalization have been historically employed by scholars, including [0, 1], [0.2, 8], [− 1, 1], etc. (Arefinia et al., 2022; Cigizoglu, 2003; Ebtehaj & Bonakdari, 2014).

While normalized values of a data series are used throughout the training and model development phases, the resulting output of the developed model must be de-normalized to have any significance to the modeler/decision maker. The outputs must be returned to their true values. To achieve this, Eq. (2.2) is used to de-normalize the data.

$$x_i = \left(\frac{xn_i - b}{a}\right) \times (x_{\max} - x_{\min}) + x_{\min} \tag{2.2}$$

2.1.2 The MATLAB coding of min–max normalization

In Code 2.1, the detailed MATLAB coding for normalization and de-normalization of a dataset is provided. In lines 1–3, the general cleaning of the workspace is performed by use of `clc`, `clear`, and `close all`. Following this, the data set should be read into the MATLAB environment. In line 6, the `xlsread` command is used to load the data series, which is contained within sheet1 of a Microsoft Excel file named "Data." It is noteworthy that data normalization can be done both before and after the data separation step (i.e., categorizing all samples into training and testing samples). The only difference is that the minimum and maximum values used in both cases may not be equal, although this likely will not affect the modeling process. The variable `N` is introduced in line 7 to store the number of variables contained within the loaded data set.

Machine Learning in Earth, Environmental and Planetary Sciences. DOI: https://doi.org/10.1016/B978-0-443-15284-9.00003-3

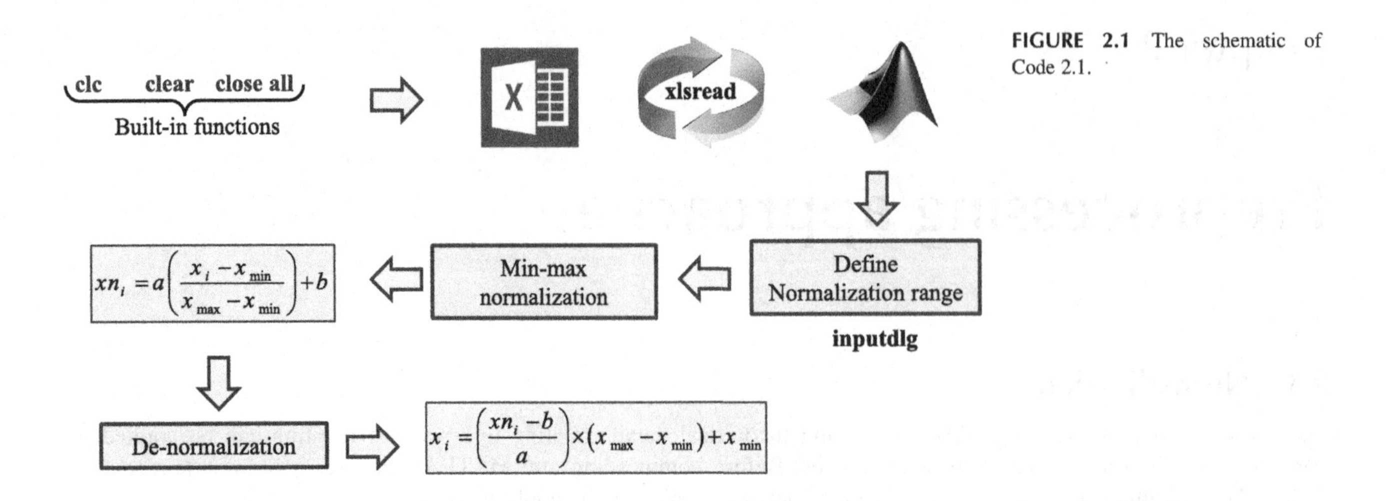

FIGURE 2.1 The schematic of Code 2.1.

Once the data has been fully loaded, a dialog box prompts the user to enter the minimum and maximum values of the normalized data range. To generate the dialog box, the `inputdlg(prompt,dlgtitle,dims,definput)` command is employed, as seen in line 13 of Code 2.1. `Prompt`, as seen in line 10, is used to specify a character string to label the fields that are to be edited by the user. `dlgtitle` is specified in line 11 using the variable `title` and is the title assigned to the dialog box. `dims` is specified in line 13 as [1,50], which means that the height of the edit fields is one line of text high and 50 characters in length. The `definput` argument is related to the default values, which for this example are specified in line 12 as [0.2,0.8]. The array of character values returned from the dialog box and stored in `A` and `B` is transformed in line 17 using the `str2num` function. These are then saved under the variable `NormRange`. The normalization procedure is performed in lines 21–25. A loop over 1 to `N` (recall `N` is the number of variables) is considered, which identifies the minimum (i.e., `MinX`) and maximum (i.e., `MaxX`) values of the data contained within the variable (lines 22–23). Subsequently, in line 24, a nested for loop is used to transform all the data contained within the variable to the defined normalization range (i.e., `NormRange`) using the approach outlined in Eq. (2.1). Following the normalization, all variables generated in the program are removed with the exception ofx, `xN`, `NormRange`, `MinX`, `MaxX`, and `N` using the function `clearvars -except`.

Once the modeling process has been completed, the performance of the model must be assessed by comparing the estimated variables with the validation data in their actual range (i.e., the range of the actual samples and not the transformed range). It is therefore required to apply de-normalization to the data output from the model. To achieve this, the inverse of operations from Eq. (2.1) is applied [i.e., Eq. (2.2)] in line 34. The schematic of Code 2.1 is shown in Fig. 2.1.

Code 2.1

```
1   clc;
2   clear;
3   close all;
4   
5   %% load data
6   x = xlsread ('Data','sheet1');
7   N = size(x,2);
8   
9   %% Min-Max normalization
10  Prompt={'Minimum = ...','Maximum = ...'};
11      Title='Normalization Range';
12      DefaultValues={'0.2','0.8'};
13      PARAMS=inputdlg(Prompt,Title,[1, 50],DefaultValues);
14      A=PARAMS{1};
15      B=PARAMS{2};
16  
17  NormRange =[str2num(A)  str2num(B)];
18  
19  
20  %% Normalization
21  for j = 1:N
22      MinX(1,j) = min(x(:,j));
23      MaxX(1,j) = max(x(:,j));
24      for I = 1:length(x)
25          xN(I,j) = (NormRange(1,2)-NormRange(1,1))*((x(I,j) - MinX(1,j)) /
    (MaxX(1,j) - MinX(1,j)))+NormRange(1,1);
26      end
27  end
28  
29  clearvars -except x xN NormRange MinX  MaxX N
30  
31  %% De-Normalization
32  for j = 1:N
33      for I = 1:length(xN)
34          xR(I,j) = ((xN(I,j)-NormRange(1,1))/(NormRange(1,2)-
    NormRange(1,1)))*(MaxX(1,j) - MinX(1,j))+MinX(1,j);
35      end
36  end
37  
38  clearvars -except x xN xR NormRange MinX  MaxX
```

2.1.2.1 The effect of normalization on data distribution

Fig. 2.2 shows the distribution of all inputs and outputs of five different datasets. For Example 1, it can be seen that there is a significant difference between the range of the different inputs, so that this range is [0, 2] for all inputs except In3, while the maximum of the samples presented in In3 is about 10. In addition, the output range of Example 1 is also significantly different from In3. Given that the different inputs have such varied ranges, it is expected that the values of the optimized weights (or other parameters) during the modeling process for In3 may differ significantly from other inputs, which could affect the model's performance. Observing the normalized values, it can be seen that although the range of all inputs and output is [0.2, 0.8], the distribution of samples related to inputs and output is similar to the actual data.

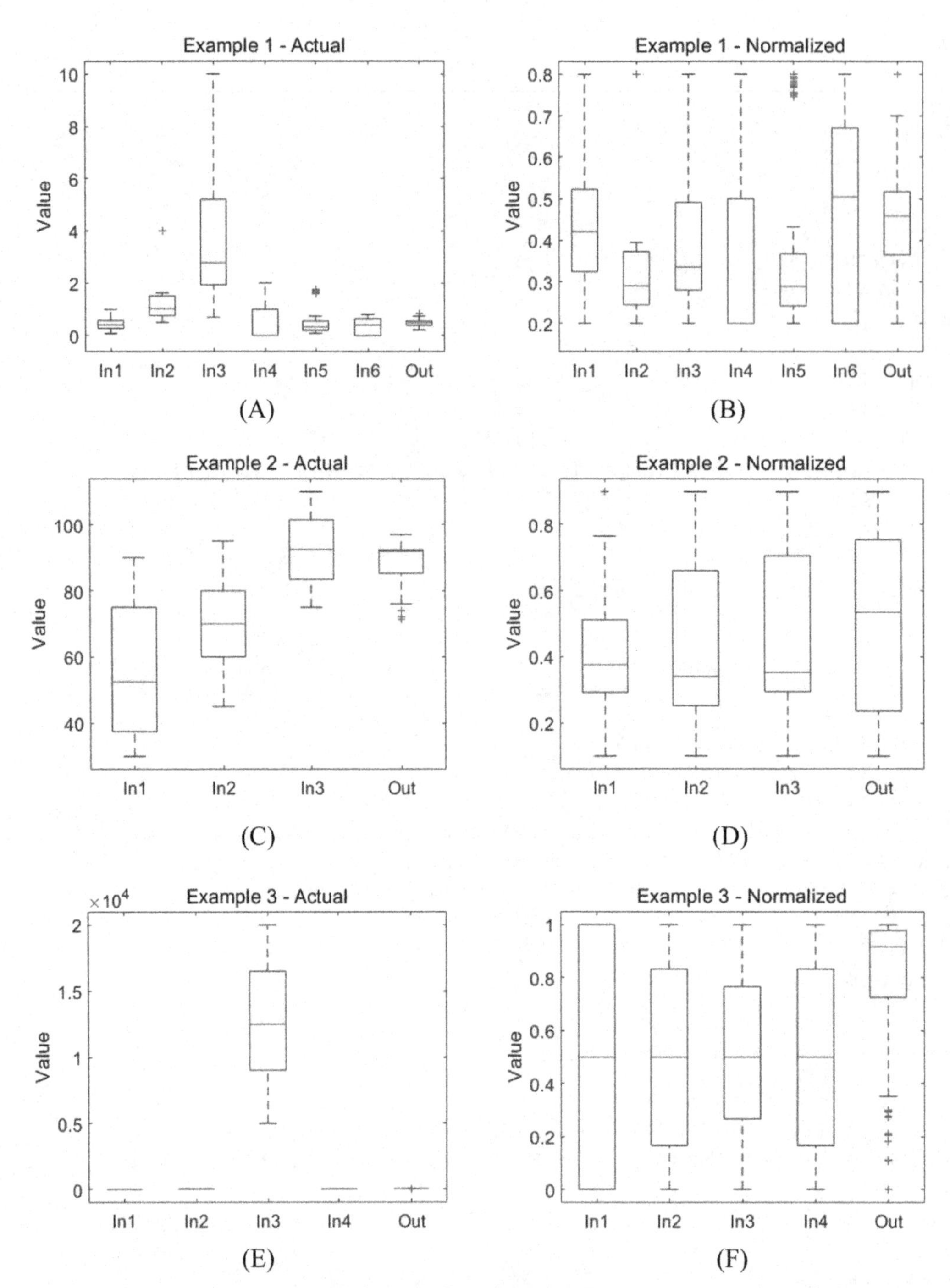

FIGURE 2.2 The actual and normalized values of all inputs in Examples 1–5. ((A) Example 1 – Actual, (B) Example 1 – Normalized, (C) Example 2 – Actual, (D) Example 2 – Normalized, (E) Example 3 – Actual, (F) Example 3 – Normalized, (G) Example 4 – Actual, (H) Example 4 – Normalized, (I) Example 5 – Actual, (J) Example 5 – Normalized).

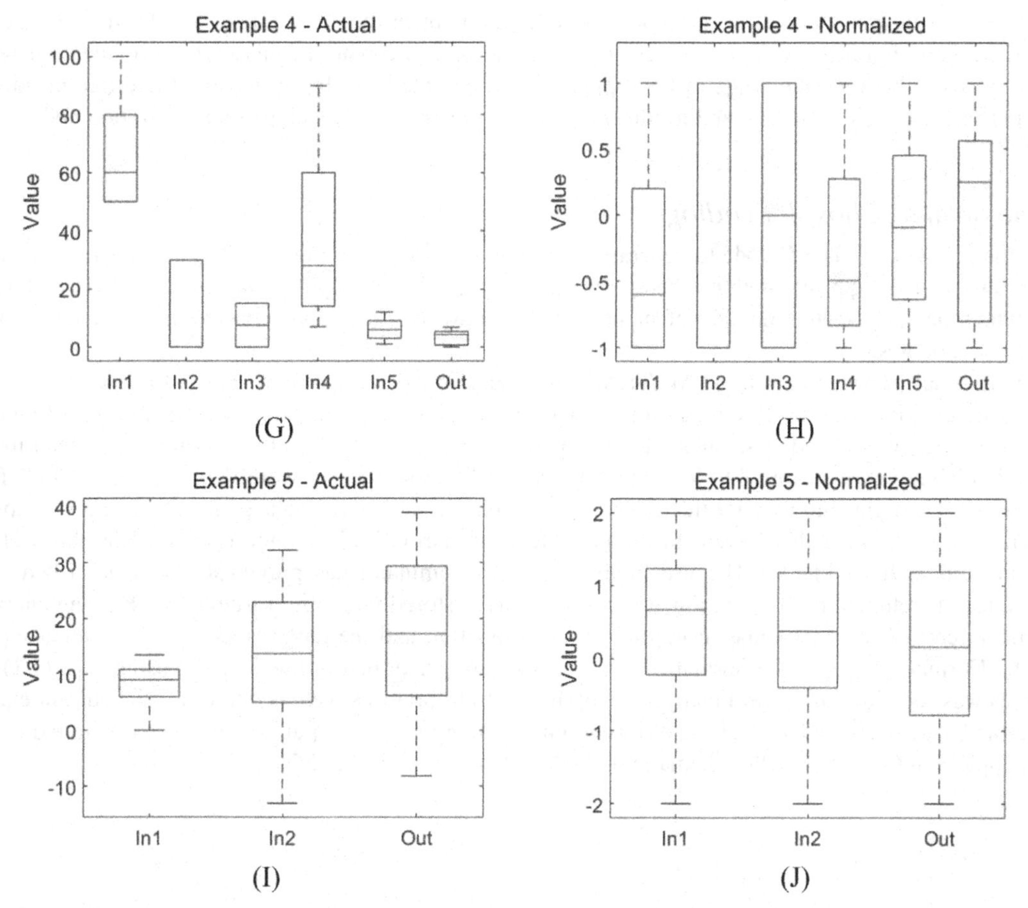

FIGURE 2.2 Continued

For Example 2, the largest range of samples is for In1, where the difference between the smallest and largest samples is about 60 units. When this difference is considered for the remaining inputs, it is found to be about 50 for In2 and less than 40 for In3. In addition, the range of the output variable in this example is given by [71.3, 97.01]. Indeed, the difference between the minimum and maximum range defined for output is less than 30, which is smaller than all inputs. Moreover, the maximum values for In1, In2, and In3 are 90, 95, and 110, respectively. Therefore, to overcome the differences observed in the range of the inputs, the data is normalized. The results of data normalization show that the data are normalized in the range [0.1, 0.9] so that the distribution of each input, as well as the output of the problem, are different from each other.

For Example 3, the range of values presented in In3 is very different from other inputs as well as the output of the problem. From the figure, it can be seen that In3 is significantly greater than all other input variables and output variables, to the point where their boxplots are not even discriminable on the plot. Using data normalization in the range [0, 1], the significant differences in the range of the inputs are well resolved. For Example 4, the range of the inputs varies in the range [0, 100], while In1 is in the range [50, 100] and In3 is in the range [0, 15]. Also, the maximum output value is 12, which is smaller than the maximum value of all inputs. Normalization for the data of this example is performed in the range [− 1, 1] to transform all of the inputs as well as output in the same range. Similar to other examples, in Example 5, there is a significant difference between the existing two inputs. In this case, normalization was able to transform all values to between the range of [− 2,2]. It should be noted for the reader that the different normalization ranges used for each example were selected to illustrate the flexibility of the developed normalization code.

2.1.2.2 *The details of boxplot coding*

In Code 2.2, the detailed MATLAB code to generate the boxplots is shown in Fig. 2.2. The process for each example is identical, the only factor requiring attention being the changing number of input variables. This code is developed for Example 1, which has the greatest number of input variables, however, it is applicable to any data set by addition or deletion of input variables.

As was done in all previous codes, the MATLAB workspace is cleared using the `clc`, `close`, and `clear all` commands. Following this, the data must be read into the workspace. Since normalization is generally done before the data separation to the training and testing samples, the data used in this step includes all the samples. To read the data, the `xlsread` function is employed in line 7 to load the data from a Microsoft Excel workbook entitled "Data." The `N` variable shown in line 8 is defined to identify the number of variables in the loaded dataset. In the current example with six inputs and one output, `N` is equal to seven. In the next step, the normalization range is defined by the user using the `inputdlg` command in lines 11–16. The use of the `inputdlg` command has previously been described. The range entered by the user is returned to the program as a character array stored under the variable `PARAMS`. The character array values are converted to numerical values using the `str2num` function and the range is stored in the `NormRange` variable in line 18. The Normalization and De-normalization processes are performed in lines 10–38 using Eqs. (2.1) and (2.2), respectively, as was seen in Code 1.18. Finally, the box plots of all inputs as well as output for the current examples are plotted using the `boxplot` command. The use of this command along with its input arguments was detailed in the explanation of Example 3 in Code 1.12. The schematic of Code 2.2 is shown in Fig. 2.3.

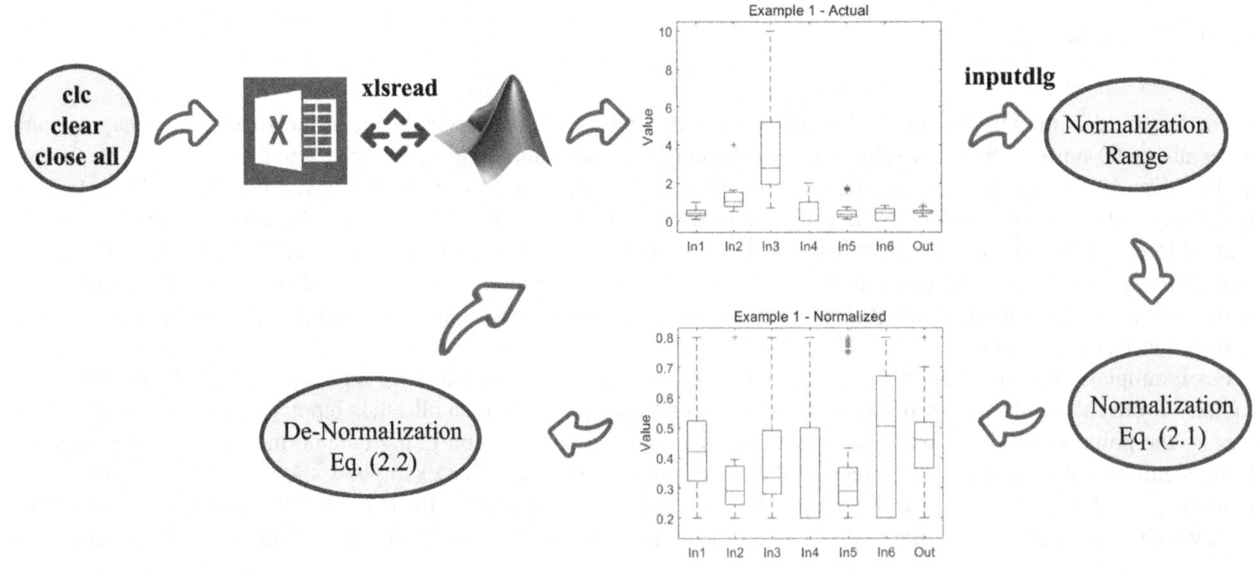

FIGURE 2.3 The schematic of Code 2.2.

Code 2.2

```
1   clc;
2   clear;
3   close all;
4
5   %% Example 1
6   % load data
7   x = xlsread ('Data','sheet1'); % Inputs and outputs of training and
    testing stages
8   N = size(x,2);
9
10  % Min-Max normalization
11  Prompt={'Minimum = ...','Maximum = ...'};
12      Title='Normalization Range';
13      DefaultValues={'0.2','0.8'};
14      PARAMS=inputdlg(Prompt,Title,[1, 50],DefaultValues);
15      A=PARAMS{1};
16      B=PARAMS{2};
17
18  NormRange =[str2num(A)  str2num(B)];
19
20  % Normalization
21  for j = 1:N
22      MinX(1,j) = min(x(:,j));
23      MaxX(1,j) = max(x(:,j));
24      for i = 1:length(x)
25          xN(i,j) = (NormRange(1,2)-NormRange(1,1))*((x(i,j) - MinX(1,j)) /
    (MaxX(1,j) - MinX(1,j)))+NormRange(1,1);
26      end
27  end
28
29  clearvars -except x xN NormRange MinX  MaxX N
30
31  % De-Normalization
32  for j = 1:N
33      for i = 1:length(xN)
34          xR(i,j) = ((xN(i,j)-NormRange(1,1))/(NormRange(1,2)-
    NormRange(1,1)))*(MaxX(1,j) - MinX(1,j))+MinX(1,j);
35      end
36  end
37
38  clearvars -except x xN xR NormRange MinX  MaxX
39
```

```
40  In1 = x(:,1); % Actual Input 1 - Example 1
41  In2 = x(:,2); % Actual Input 2 - Example 1
42  In3 = x(:,3); % Actual Input 3 - Example 1
43  In4 = x(:,4); % Actual Input 4 - Example 1
44  In5 = x(:,5); % Actual Input 5 - Example 1
45  In6 = x(:,6); % Actual Input 6 - Example 1
46  Out = x(:,7); % Actual Output  - Example 1
47  Example1   = [In1 In2 In3 In4 In5 In6 Out];
48  
49  figure;
50  boxplot(Example1,'Notch','off','labels',{'In1','In2','In3','In4','In5','I
    n6','Out'})
51  ax = gca;
52  ax.FontSize = 16;
53  ax.FontWeight='normal';
54  ylabel('Value')
55  title('Example 1 - Actual','FontWeight','normal')
56  
57  In1 = xN(:,1); % Normalized Input 1 - ExNample 1
58  In2 = xN(:,2); % Normalized Input 2 - ExNample 1
59  In3 = xN(:,3); % Normalized Input 3 - ExNample 1
60  In4 = xN(:,4); % Normalized Input 4 - ExNample 1
61  In5 = xN(:,5); % Normalized Input 5 - ExNample 1
62  In6 = xN(:,6); % Normalized Input 6 - ExNample 1
63  Out = xN(:,7); % Normalized Output  - Example 1
64  ExNample1 = [In1 In2 In3 In4 In5 In6 Out];
65  
66  figure;
67  boxplot(ExNample1,'Notch','off','labels',{'In1','In2','In3','In4','In5','
    In6','Out'})
68  ax = gca;
69  ax.FontSize = 16;
70  ax.FontWeight='normal';
71  ylabel('Value')
72  title('Example 1 - Normalized','FontWeight','normal')
```

Data on normalization can be found at Appendix 2A.

2.2 Standardization

2.2.1 The standardization concept

Another preprocessing method that can be applied to the raw data is standardization (Katipoğlu et al., 2020; Mohamad & Usman, 2013; Yaraghi et al., 2020). In this method, the raw data are converted to a new space using the average and standard deviation of all samples. Indeed, the raw data is transformed into a new space that has a mean of zero and a standard deviation of one. The transformation is performed according to Eq. (2.3):

$$xStd_i = \left(\frac{x_i - x_{avg}}{SD}\right) \tag{2.3}$$

where x_i is the ith sample of data that will be standardized, x_{avg} is the average of all samples, SD is the standard deviation of all samples, and $xStd_i$ is the ith standardized value of the samples of data (x_i). As was seen with normalization, it is required to transform the standardized data back to the original space following the modeling exercise so that its significance may be interpreted by the modeler. To do so, Eq. (2.4) is applied.

$$x_i = (xStd_i - SD) + x_{avg} \tag{2.4}$$

2.2.2 The MATLAB coding of standardization

The detailed MATLAB coding of the standardization algorithms is provided for Example 1 in Code 2.3. After closing all open figures using the `close all` command and removing the contents of the Command Window and Workspace using the `clc` and `clear` commands, the data may be loaded. The data for all five examples are contained within a singular Microsoft Excel workbook (i.e., Data.xls), each saved within their own sheet. The sample data of Examples 1–5 are situated in sheet1 to sheet5, respectively. For the case of Example 1, the `xlsread` function is used to read the saved data from "sheet1" of Data.xls, where it is stored in the variable `X1`. This `X1` variable is a matrix with a number of rows equal to the number of observations and a number `N1` of columns equal to the number of variables. The number of all variables is computed in line 8 using the `size(X1,2)` command, which stores the value in the variable `N1`. To transform all samples of each variable to a standardized form, a `for` loop is utilized. For each variable (there are `N1` different variables), the mean and standard deviation are calculated using the built-in MATLAB `mean` and `std` functions—this can be seen in lines 12 and 13. Following the computation of the mean and standard deviation, Eq. (2.3) is applied to transform the actual data into the transformed data in a nested `for` loop, and saved under the variable `XSTD1`. Similar to standardization, the de-standardization process is performed for all `N1` variables using calculated mean and standard deviation and Eq. (2.4). This can be seen in line 23 of Code 2.3. This process can be easily applied to the other examples by changing the `xlsread` input argument from sheet1 to sheet2, sheet3, sheet4, and sheet5. Codes 2.4–2.7 detail the standardization coding for Examples 2–5, respectively. The schematic of Codes 2.4–2.7 is provided in Fig. 2.4.

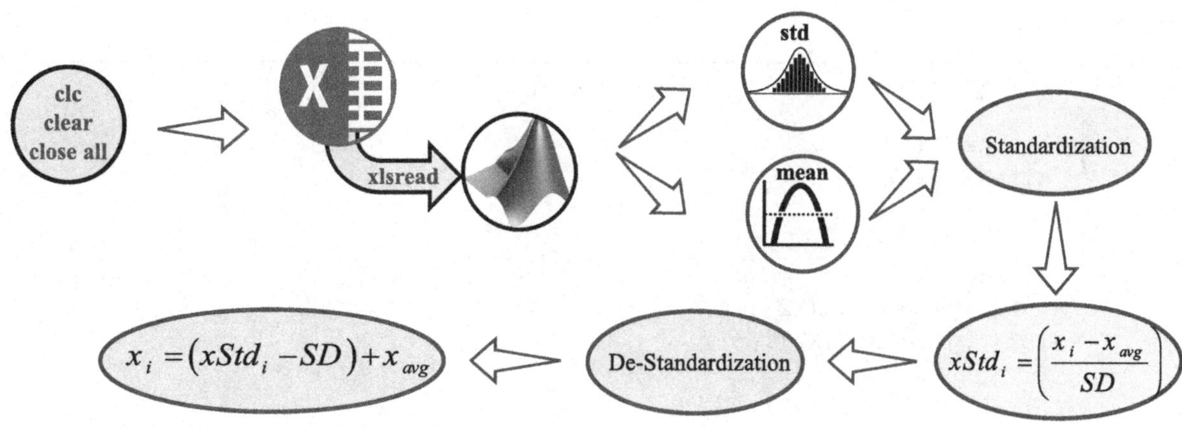

FIGURE 2.4 The schematic of Codes 2.4–2.7.

Code 2.3

```
1   clc;
2   clear;
3   close all;
4
5   %% Example 1
6   % load data
7   X1 = xlsread ('Data','sheet1');
8   N1 = size(X1,2);
9
10  % Standardization
11  for j = 1:N1
12      Avg1(1,j) = mean(X1(:,j));
13      STD1(1,j) = std(X1(:,j));
14
15      for i = 1:length(X1)
16          XSTD1(i,j) = (X1(i,j) - Avg1(1,j))./STD1(1,j);
17      end
18  end
19
20  % De-Standardization
21  for j = 1:N1
22      for i = 1:length(X1)
23          XA1(i,j) = (XSTD1(i,j).*STD1(1,j))+Avg1(1,j);
24      end
25  end
```

Code 2.4

```
1   %% Example 2
2   % load data
3   X2 = xlsread ('Data','sheet2');
4   N2 = size(X2,2);
5
6   % Standardization
7   for j = 1:N2
8       Avg2(1,j) = mean(X2(:,j));
9       STD2(1,j) = std(X2(:,j));
10
11      for i = 1:length(X2)
12          XSTD2(i,j) = (X2(i,j) - Avg2(1,j))./STD2(1,j);
13      end
14  end
15
16  % De-Standardization
17  for j = 1:N2
18      for i = 1:length(X2)
19          XA2(i,j) = (XSTD2(i,j).*STD2(1,j))+Avg2(1,j);
20      end
21  end
```

Code 2.5

```
1   %% Example 3
2   % load data
3   X3 = xlsread ('Data','sheet3');
4   N3 = size(X3,2);
5
6   % Standardization
7   for j = 1:N3
8       Avg3(1,j) = mean(X3(:,j));
9       STD3(1,j) = std(X3(:,j));
10
11      for i = 1:length(X3)
12          XSTD3(i,j) = (X3(i,j) - Avg3(1,j))./STD3(1,j);
13      end
14  end
15
16  % De-Standardization
17  for j = 1:N3
18      for i = 1:length(X3)
19          XA3(i,j) = (XSTD3(i,j).*STD3(1,j))+Avg3(1,j);
20      end
21  end
```

Code 2.6

```
1   %% Example 4
2   % load data
3   X4 = xlsread ('Data','sheet4');
4   N4 = size(X4,2);
5
6   % Standardization
7   for j = 1:N4
8       Avg4(1,j) = mean(X4(:,j));
9       STD4(1,j) = std(X4(:,j));
10
11      for i = 1:length(X4)
12          XSTD4(i,j) = (X4(i,j) - Avg4(1,j))./STD4(1,j);
13      end
14  end
15
16  % De-Standardization
17  for j = 1:N4
18      for i = 1:length(X4)
19          XA4(i,j) = (XSTD4(i,j).*STD4(1,j))+Avg4(1,j);
20      end
21  end
```

Code 2.7

```
1   %% Example 5
2   % load data
3   X5 = xlsread ('Data','sheet5');
4   N5 = size(X5,2);
5
6   % Standardization
7   for j = 1:N5
8       Avg5(1,j) = mean(X5(:,j));
9       STD5(1,j) = std(X5(:,j));
10
11      for i = 1:length(X5)
12          XSTD5(i,j) = (X5(i,j) - Avg5(1,j))./STD5(1,j);
13      end
14  end
15
16  % De-Standardization
17  for j = 1:N5
18      for i = 1:length(X5)
19          XA5(i,j) = (XSTD5(i,j).*STD5(1,j))+Avg5(1,j);
20      end
21  end
```

2.2.2.1 *The effect of standardization on data distribution*

To examine the changes applied to different variables in each example, the mean and standard deviation values for all variables before and after standardization should be examined. Indeed, it should be verified whether the standardization has been performed correctly or not (i.e., the mean and standard deviation values for each variable after standardization should be equal to zero and one, respectively). Codes 2.3–2.7 presented the standardization and de-standardization algorithms for each example dataset. The next step is to examine the changes applied to the data prior to and following standardization, which is detailed in Code 2.8.

The mean and standard deviation values of the actual data (before standardization) are already calculated through the standardization process and, therefore, must only be calculated for the transformed data (data after standardization). The built-in MATLAB `mean` and `std` commands are again employed to compute the mean and standard deviation of all the transformed data. These calculations are coded in lines 3–7 and 10–14 of Code 2.8, respectively. The indices for the actual data and the transformed data may be plotted in a column chart for ease of comparison. In Code 2.8, this is coded in lines 16–48. Detailed explanations of the implementation of bar charts in MATLAB have previously been provided in Section 1.3.1.2. The schematic of Code 2.8 is shown in Fig. 2.5.

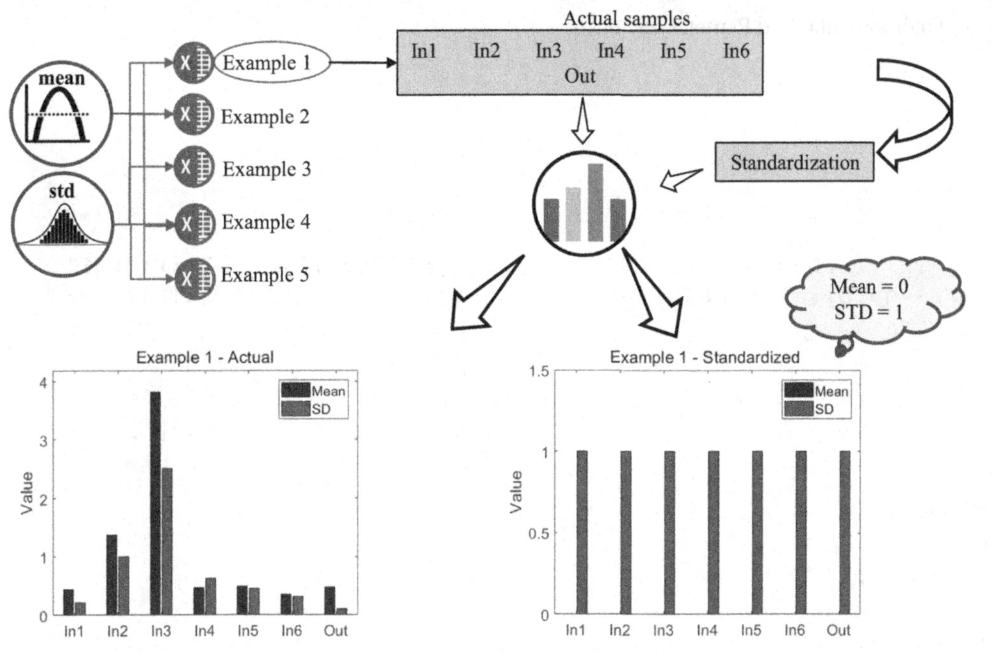

FIGURE 2.5 Schematic of Code 2.8.

Code 2.8

```
1   %% Calculation of statistical indices for Standardized Samples
2   % Average - Standardized samples
3   AvgSTD1 = mean(XSTD1);          % Average of Example 1 - Standardized
4   AvgSTD2 = mean(XSTD2);          % Average of Example 2 - Standardized
5   AvgSTD3 = mean(XSTD3);          % Average of Example 3 - Standardized
6   AvgSTD4 = mean(XSTD4);          % Average of Example 4 - Standardized
7   AvgSTD5 = mean(XSTD5);          % Average of Example 5 - Standardized
8
9   % Standard deviation - Standardized samples
10  STDSTD1 = std(XSTD1);  % Standard deviation of Example 1 - Standardized
11  STDSTD2 = std(XSTD2);  % Standard deviation of Example 2 - Standardized
12  STDSTD3 = std(XSTD3);  % Standard deviation of Example 3 - Standardized
13  STDSTD4 = std(XSTD4);  % Standard deviation of Example 4 - Standardized
14  STDSTD5 = std(XSTD5);  % Standard deviation of Example 5 - Standardized
15
16  %% Plot results
17
18  % Example #1 - Actual
    In1 =
19  [Avg1(1),STD1(1);Avg1(2),STD1(2);Avg1(3),STD1(3);Avg1(4),STD1(4);Avg1(5),
    STD1(5);Avg1(6),STD1(6);Avg1(7),STD1(7)];
    xx =
20  categorical({'In1','In1';'In2','In2';'In3','In3';'In4','In4';'In5','In5';
    'In6','In6';'Out','Out'});
21
22  figure
23  B = bar(xx,In1);
24  B(1).FaceColor = 'b';
25  B(2).FaceColor = 'r';
26  ax = gca;
27  ax.FontSize = 16;
28  ax.FontWeight='normal';
29  ylabel('Value')
30  legend('Mean','SD')
31  ylim([0 1.1*max(max(In1))])
32  title('Example 1 - Actual','FontWeight','normal')
```

```
33
34  % Example #1 - Standardized
    InSTD1 =
35  [AvgSTD1(1),STDSTD1(1);AvgSTD1(2),STDSTD1(2);AvgSTD1(3),STDSTD1(3);AvgSTD
    1(4),STDSTD1(4);AvgSTD1(5),STDSTD1(5);AvgSTD1(6),STDSTD1(6);AvgSTD1(7),ST
    DSTD1(7)];
    xx =
36  categorical({'In1','In1';'In2','In2';'In3','In3';'In4','In4';'In5','In5';
    'In6','In6';'Out','Out'});
37
38  figure
39  B = bar(xx,InSTD1);
40  B(1).FaceColor = 'b';
41  B(2).FaceColor = 'r';
42  ax = gca;
43  ax.FontSize = 16;
44  ax.FontWeight='normal';
45  ylabel('Value')
46  legend('Mean','SD')
47  ylim([0 1.5*max(max(InSTD1))])
48  title('Example 1 - Standardized ','FontWeight','normal')
```

In Codes 2.9–2.12, the detailed coding to generate the bar plots for Examples 2–5 using the mean and standard deviation values from Code 2.8 is provided.

Code 2.9

```
1   % Example #2 - Actual
2   In2 = [Avg2(1),STD2(1);Avg2(2),STD2(2);Avg2(3),STD2(3);Avg2(4),STD2(4)];
3   xx = categorical({'In1','In1';'In2','In2';'In3','In3';'Out','Out'});
4   
5   figure
6   B = bar(xx,In2);
7   B(1).FaceColor = 'b';
8   B(2).FaceColor = 'r';
9   ax = gca;
10  ax.FontSize = 16;
11  ax.FontWeight='normal';
12  ylabel('Value')
13  legend('Mean','SD')
14  ylim([0 1.1*max(max(In2))])
15  title('Example 2 - Actual','FontWeight','normal')
16  
17  % Example #2 - Standardized
    InSTD2 =
18  [AvgSTD2(1),STDSTD2(1);AvgSTD2(2),STDSTD2(2);AvgSTD2(3),STDSTD2(3);AvgSTD
    2(4),STDSTD2(4)];
19  xx = categorical({'In1','In1';'In2','In2';'In3','In3';'Out','Out'});
20  
21  figure
22  B = bar(xx,InSTD2);
23  B(1).FaceColor = 'b';
24  B(2).FaceColor = 'r';
25  ax = gca;
26  ax.FontSize = 16;
27  ax.FontWeight='normal';
28  ylabel('Value')
29  legend('Mean','SD')
30  ylim([0 1.5*max(max(InSTD2))])
31  title('Example 2 - Standardized','FontWeight','normal')
```

Code 2.10

```
1   % Example #3 - Actual
2   In3 = [Avg3(1),STD3(1);Avg3(2),STD3(2);Avg3(3),STD3(3);Avg3(4),STD3(4);Avg3(5),STD3(5)];
3   xx = categorical({'In1','In1';'In2','In2';'In3','In3';'In4','In4';'Out','Out'});
4   
5   figure
6   B = bar(xx,In3);
7   B(1).FaceColor = 'b';
8   B(2).FaceColor = 'r';
9   ax = gca;
10  ax.FontSize = 16;
11  ax.FontWeight='normal';
12  ylabel('Value')
13  legend('Mean','SD')
14  ylim([0 1.1*max(max(In3))])
15  title('Example 3 - Actual','FontWeight','normal')
16  
17  % Example #3 - Standardized
18  InSTD3 = [AvgSTD3(1),STDSTD3(1);AvgSTD3(2),STDSTD3(2);AvgSTD3(3),STDSTD3(3);AvgSTD3(4),STDSTD3(4);AvgSTD3(5),STDSTD3(5)];
19  
20  figure
21  B = bar(xx,InSTD3);
22  B(1).FaceColor = 'b';
23  B(2).FaceColor = 'r';
24  ax = gca;
25  ax.FontSize = 16;
26  ax.FontWeight='normal';
27  ylabel('Value')
28  legend('Mean','SD')
29  ylim([0 1.5*max(max(InSTD3))])
30  title('Example 3 - Standardized','FontWeight','normal')
```

Code 2.11

```
1   % Example #4 - Actual
    In4 =
2   [Avg4(1),STD4(1);Avg4(2),STD4(2);Avg4(3),STD4(3);Avg4(4),STD4(4);Avg4(5),
    STD4(5);Avg4(6),STD4(6)];
    xx =
3   categorical({'In1','In1';'In2','In2';'In3','In3';'In4','In4';'In5','In5';
    'Out','Out'});
4
5   figure
6   B = bar(xx,In4);
7   B(1).FaceColor = 'b';
8   B(2).FaceColor = 'r';
9   ax = gca;
10  ax.FontSize = 16;
11  ax.FontWeight='normal';
12  ylabel('Value')
13  legend('Mean','SD')
14  ylim([0 1.1*max(max(In4))])
15  title('Example 4 - Actual','FontWeight','normal')
16
17  % Example #4 - Standardized
    InSTD4 =
18  [AvgSTD4(1),STDSTD4(1);AvgSTD4(2),STDSTD4(2);AvgSTD4(3),STDSTD4(3);AvgSTD
    4(4),STDSTD4(4);AvgSTD4(5),STDSTD4(5);AvgSTD4(6),STDSTD4(6)];
    xx =
19  categorical({'In1','In1';'In2','In2';'In3','In3';'In4','In4';'In5','In5';
    'Out','Out'});
20
21  figure
22  B = bar(xx,InSTD4);
23  B(1).FaceColor = 'b';
24  B(2).FaceColor = 'r';
25  ax = gca;
26  ax.FontSize = 16;
27  ax.FontWeight='normal';
28  ylabel('Value')
29  legend('Mean','SD')
30  ylim([0 1.5*max(max(InSTD4))])
31  title('Example 4 - Standardized','FontWeight','normal')
```

Code 2.12

```
1   % Example #5 - Actual
2   In5 = [Avg5(1),STD5(1);Avg5(2),STD5(2);Avg5(3),STD5(3)];
3   xx = categorical({'In1','In1';'In2','In2';'Out','Out'});
4   
5   figure
6   B = bar(xx,In5);
7   B(1).FaceColor = 'b';
8   B(2).FaceColor = 'r';
9   ax = gca;
10  ax.FontSize = 16;
11  ax.FontWeight='normal';
12  ylabel('Value')
13  legend('Mean','SD')
14  ylim([0 1.1*max(max(In5))])
15  title('Example 5 - Actual','FontWeight','normal')
16  
17  % Example #5 - Standardized
18  InSTD5 =
    [AvgSTD5(1),STDSTD5(1);AvgSTD5(2),STDSTD5(2);AvgSTD5(3),STDSTD5(3)];
19  xx = categorical({'In1','In1';'In2','In2';'Out','Out'});
20  
21  figure
22  B = bar(xx,InSTD5);
23  B(1).FaceColor = 'b';
24  B(2).FaceColor = 'r';
25  ax = gca;
26  ax.FontSize = 16;
27  ax.FontWeight='normal';
28  ylabel('Value')
29  legend('Mean','SD')
30  ylim([0 1.5*max(max(InSTD5))])
31  title('Example 5 - Standardized','FontWeight','normal')
```

2.2.2.2 The details of barplot coding

Fig. 2.6 presents the results of the bar graph analysis from Codes 2.8–2.12 where the mean and standard deviation values are provided for all input and outputs of Examples 1–5. From the figure, it can be seen that each variable of all the example datasets has different mean values. The different range of the input variables of a problem may result in a different range of optimized model parameters for each input variable. Therefore the effect of some input variables may be considered negligible compared to other variables during the modeling process. Indeed, the ability of the model to predict the target variable by different input variables is impacted by optimized parameters for each variable. Large differences in magnitude may present the ML method with difficulty in achieving the optimal result.

In Example 1, the mean and standard deviation of In3 are significantly different from the other variables, while In1, In4, In5, and In6 are almost identical. For Example 2, the effect of the range of different variables is not very significant, while for Example 3, In3 has an average of about 12,000. The mean and standard deviation of the other variables compared to those of In3 are negligible and cannot be discriminated on the plot. In Example 4, the mean and standard deviation of the In1 and In4 are significantly different from other variables, while for Example 5, the difference in magnitude is relatively small. Significant differences between some or all variables may not be observed in all cases, however, standardization can still be used to test the model's ability to estimate objective variables.

The standardization results for all examples show that the variables are transformed so that the standard deviation and their mean are equal to one and zero (respectively). Using this transformation, ML-based model(s) can be expected to perform better than the case where the input variables have significantly different mean and standard deviation values.

Data on Standardization can be found at Appendix 2A.

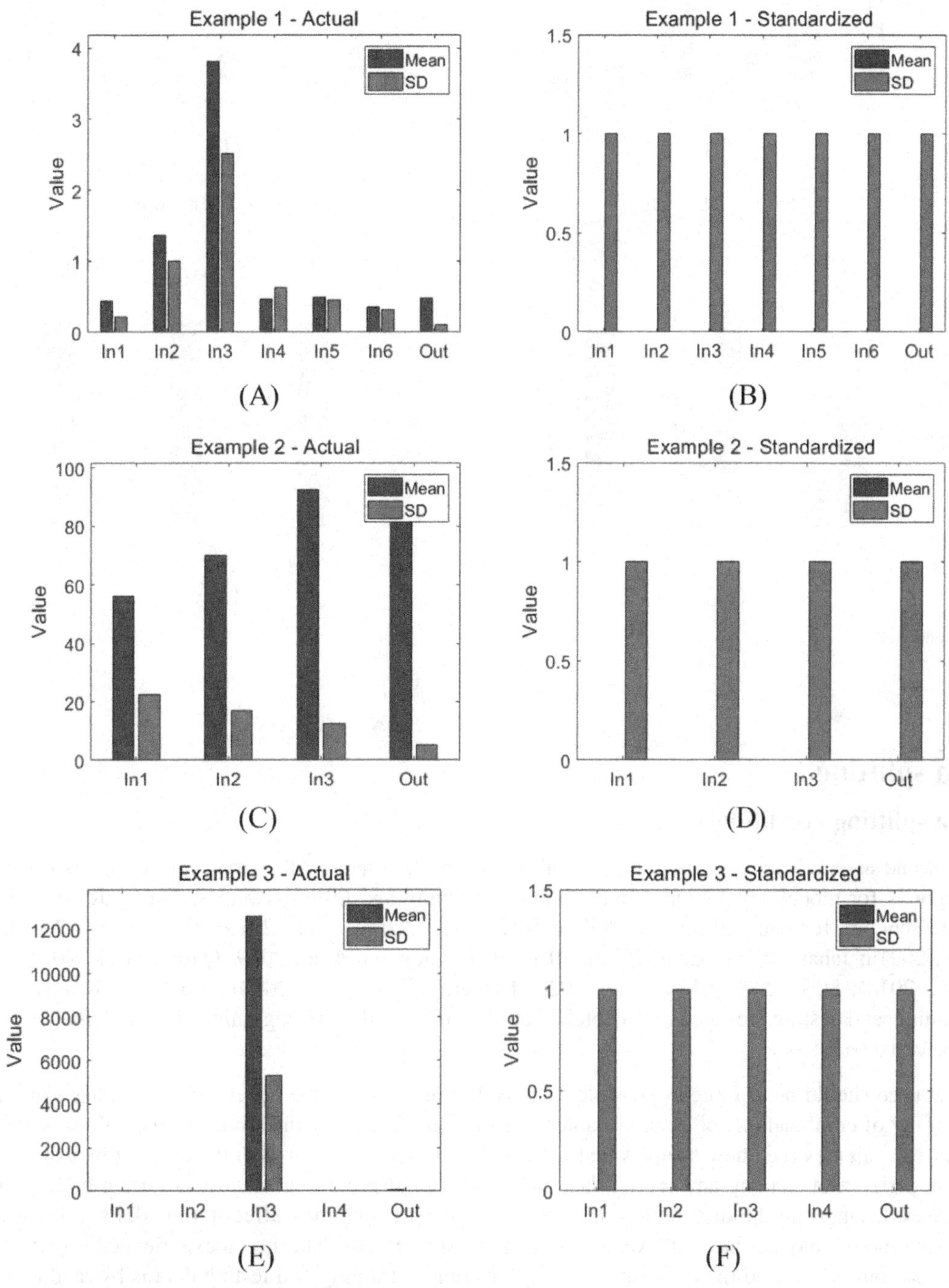

FIGURE 2.6 The actual and standardized values of all inputs and output in Examples 1–5. ((A) Example 1 – Actual, (B) Example 1 – Standardized, (C) Example 2 – Actual, (D) Example 2 – Standardized, (E) Example 3 – Actual, (F) Example 3 – Standardized, (G) Example 4 – Actual, (H) Example 4 – Standardized, (I) Example 5 – Actual, (J) Example 5 – Standardized).

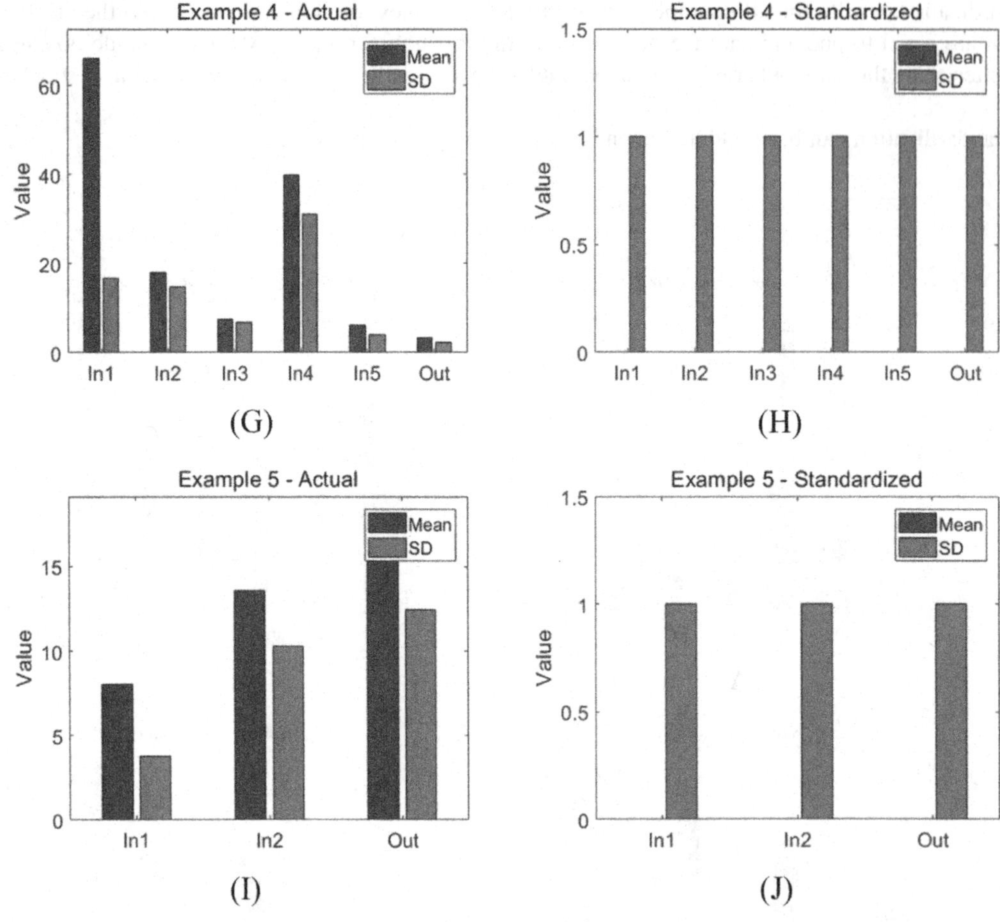

FIGURE 2.6 Continued

2.3 Data splitting

2.3.1 Data splitting conditions

A fundamental and essential issue in modeling a real-world problem using ML-based approaches is categorizing data into two categories for model development and validation: training and testing samples. There are several percentages applied by the scholars for data splitting, including 50%–50% (Azimi et al., 2019; Zhang et al., 2018), 60%–40% (Ebehaj et al., 2021a; Jahani & Rayegani, 2020), 70%–30% (Gholami et al., 2019; Qasem et al., 2017b), 75%–25% (Ebtehaj et al., 2018); 80%–20% (Chen et al., 2020; Halpern-Wight et al., 2020), and 90%–10% (Erbir & Ünver, 2021) for training and testing, respectively (Ebtehaj et al., 2020b). Before beginning the modeling process, several things must be considered:

1. The dataset used should be as large as possible. This is desirable so that the developed model has experienced a significant variety of combinations of input variables to gain good flexibility in estimating the values of the target variable for unseen samples (i.e., new samples that models didn't experience through the training phase).
2. Correct categorizing of training and testing data. A model should provide good results during both the training and testing phases. Changes in the distribution of training and testing data may affect the model's training process, and the developed model may not have an excellent ability to estimate test data (i.e., inexperienced samples). One of the best and most commonly used methods for fair categorization of training and testing data is by random selection.

2.3.2 The MATLAB coding of data splitting

In the discussion next, the MATLAB coding required for random categorization of samples into training and testing data is provided. Code 2.13 details the MATLAB code for Example 1 data. Initially, as seen in all previous codes, the MATLAB workspace is cleaned using the `clc`, `clear`, and `close all` commands. Next, the data should be loaded into the MATLAB environment. The data set is read into the variable `A1` in line 7 using the `xlsread` command to import the data from the Microsoft Excel file named "Data." This excel workbook contains the data for all examples, such that sheet1 to sheet5 represent the data for Examples 1–5. In each sheet, the final column is related to the output variables, and the other columns detail the input variables. In line 9, the `Inputs` variable is defined considering that the inputs extend to the second last column of `A1`. The `Targets` variable is defined considering that the output data is contained within the last column of the sheet. The code prompts the user with a dialog box to enter the percentage of data considered during the training phase. The coding for this is specified in lines 18–22 using the `inputdlg` command, which has previously been defined. The training percentage specified by the user is converted from character string to numerical value and stored in the variable `Pr` in line 22. The number of training samples (i.e., `nTrain`) is computed in line 25 by rounding the product of `Pr` and the number of samples. The schematic of Code 2.13 is provided in Fig. 2.7.

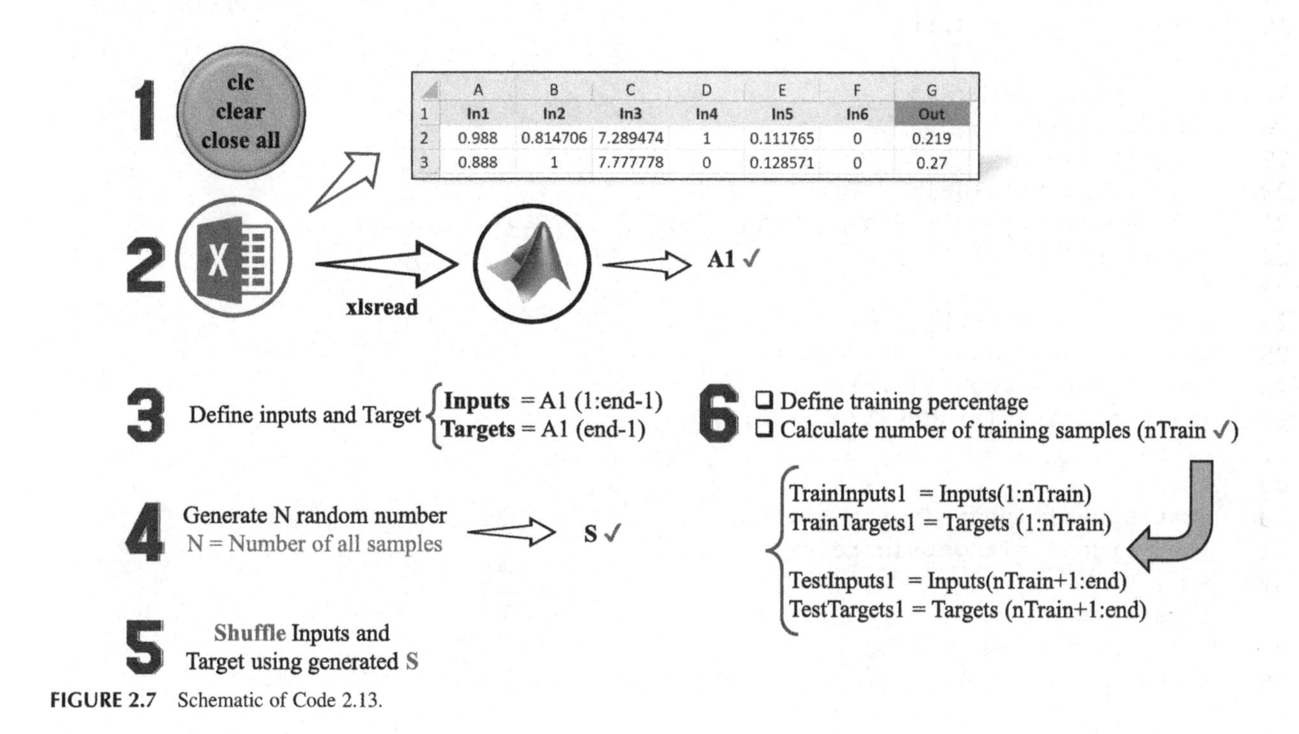

FIGURE 2.7 Schematic of Code 2.13.

Code 2.13

```
1     clc
2     clear
3     close all
4
5     %% Example 1
6     % load data
7     A1 = xlsread ('Data','sheet1');
8
9     Inputs = A1(:,1:end-1);
10    Targets = A1(:,end);
11
12    nSample=size(Inputs,1);
13    S=randperm(nSample);
14    Inputs=Inputs(S,:);
15    Targets=Targets(S,:);
16
17    % Train Data
18    Prompt={'Training percentage is ....'};
19        Title='Percent';
20        DefaultValues={'0.7'};
21        PARAMS=inputdlg(Prompt,Title,[1, 30],DefaultValues);
22        Pr=PARAMS{1};
23
24    pTrain=str2num(Pr);
25    nTrain=round(pTrain*nSample);
26    TrainInputs1=Inputs(1:nTrain,:);
27    TrainTargets1=Targets(1:nTrain,:);
28
29    % Test Data
30    TestInputs1=Inputs(nTrain+1:end,:);
31    TestTargets1=Targets(nTrain+1:end,:);
32
33    clearvars -except TrainInputs1 TrainTargets1 TestInputs1 TestTargets1
34
35    save("Example 1")
```

In Codes 2.14–2.17, the detailed coding required to split the data for Examples 2–5 is presented.

Code 2.14

```
1   %% Example 2
2   % load data
3   A2 = xlsread ('Data','sheet2');
4
5   Inputs = A2(:,1:end-1);
6   Targets = A2(:,end);
7
8   nSample=size(Inputs,1);
9   S=randperm(nSample);
10  Inputs=Inputs(S,:);
11  Targets=Targets(S,:);
12
13  % Train Data
14  Prompt={'Training percentage is ....'};
15      Title='Percent';
16      DefaultValues={'0.7'};
17      PARAMS=inputdlg(Prompt,Title,[1, 30],DefaultValues);
18      Pr=PARAMS{1};
19
20  pTrain=str2num(Pr);
21  nTrain=round(pTrain*nSample);
22  TrainInputs2=Inputs(1:nTrain,:);
23  TrainTargets2=Targets(1:nTrain,:);
24
25  % Test Data
26  TestInputs2=Inputs(nTrain+1:end,:);
27  TestTargets2=Targets(nTrain+1:end,:);
28
29  clearvars -except TrainInputs1 TrainTargets1 TestInputs1 TestTargets1 ...
30      TrainInputs2 TrainTargets2 TestInputs2 TestTargets2
31
32  save("Example 2")
```

Code 2.15

```
1   %% Example 3
2   % load data
3   A3 = xlsread ('Data','sheet3');
4   
5   Inputs = A3(:,1:end-1);
6   Targets = A3(:,end);
7   
8   nSample=size(Inputs,1);
9   S=randperm(nSample);
10  Inputs=Inputs(S,:);
11  Targets=Targets(S,:);
12  
13  % Train Data
14  Prompt={'Training percentage is ....'};
15      Title='Percent';
16      DefaultValues={'0.7'};
17      PARAMS=inputdlg(Prompt,Title,[1, 30],DefaultValues);
18      Pr=PARAMS{1};
19  
20  pTrain=str2num(Pr);
21  nTrain=round(pTrain*nSample);
22  TrainInputs3=Inputs(1:nTrain,:);
23  TrainTargets3=Targets(1:nTrain,:);
24  
25  % Test Data
26  TestInputs3=Inputs(nTrain+1:end,:);
27  TestTargets3=Targets(nTrain+1:end,:);
28  
29  clearvars -except TrainInputs1 TrainTargets1 TestInputs1 TestTargets1 ...
30      TrainInputs2 TrainTargets2 TestInputs2 TestTargets2 ...
31      TrainInputs3 TrainTargets3 TestInputs3 TestTargets3
32  
33  save("Example 3")
```

Code 2.16

```
1   %% Example 4
2   % load data
3   A4 = xlsread ('Data','sheet4');
4   
5   Inputs = A4(:,1:end-1);
6   Targets = A4(:,end);
7   
8   nSample=size(Inputs,1);
9   S=randperm(nSample);
10  Inputs=Inputs(S,:);
11  Targets=Targets(S,:);
12  
13  % Train Data
14  Prompt={'Training percentage is ....'};
15      Title='Percent';
16      DefaultValues={'0.7'};
17      PARAMS=inputdlg(Prompt,Title,[1, 30],DefaultValues);
18      Pr=PARAMS{1};
19  
20  pTrain=str2num(Pr);
21  nTrain=round(pTrain*nSample);
22  TrainInputs4=Inputs(1:nTrain,:);
23  TrainTargets4=Targets(1:nTrain,:);
24  
25  % Test Data
26  TestInputs4=Inputs(nTrain+1:end,:);
27  TestTargets4=Targets(nTrain+1:end,:);
28  
29  clearvars -except TrainInputs1 TrainTargets1 TestInputs1 TestTargets1 ...
30      TrainInputs2 TrainTargets2 TestInputs2 TestTargets2 ...
31      TrainInputs3 TrainTargets3 TestInputs3 TestTargets3 ...
32      TrainInputs4 TrainTargets4 TestInputs4 TestTargets4
33  
34  save("Example 4")
```

Code 2.17

```
1    %% Example 5
2    % load data
3    A5 = xlsread ('Data','sheet5');
4
5    Inputs = A5(:,1:end-1);
6    Targets = A5(:,end);
7
8    nSample=size(Inputs,1);
9    S=randperm(nSample);
10   Inputs=Inputs(S,:);
11   Targets=Targets(S,:);
12
13   % Train Data
14   Prompt={'Training percentage is ....'};
15       Title='Percent';
16       DefaultValues={'0.7'};
17       PARAMS=inputdlg(Prompt,Title,[1, 30],DefaultValues);
18       Pr=PARAMS{1};
19
20   pTrain=str2num(Pr);
21   nTrain=round(pTrain*nSample);
22   TrainInputs5=Inputs(1:nTrain,:);
23   TrainTargets5=Targets(1:nTrain,:);
24
25   % Test Data
26   TestInputs5=Inputs(nTrain+1:end,:);
27   TestTargets5=Targets(nTrain+1:end,:);
28
29   clearvars -except TrainInputs1 TrainTargets1 TestInputs1 TestTargets1 ...
30       TrainInputs2 TrainTargets2 TestInputs2 TestTargets2 ...
31       TrainInputs3 TrainTargets3 TestInputs3 TestTargets3 ...
32       TrainInputs4 TrainTargets4 TestInputs4 TestTargets4 ...
33       TrainInputs5 TrainTargets5 TestInputs5 TestTargets5
34
35   save("Example 5")
```

Data on Data splitting can be found at Appendix 2A.

2.4 Cross-validation

Cross-validation (CV) is a statistical method that uses a resampling procedure to evaluate the performance of an ML approach developed with limited data or to check its ability to model with different ranges of inputs and output variables. Indeed, the CV is applied in modeling real-world problems using ML techniques to check the generalizability of the desired method under different conditions. Application of this method in modeling with ML techniques is common (Azimi et al., 2021; Bonakdari & Ebtehaj, 2021) because it is easy to implement and understand. Furthermore, the results of models developed using CV generally have a lower bias compared to other approaches.

2.4.1 K-fold cross-validation

One of the most well-known CV techniques is the K-fold CV (Akhbari et al., 2019; Ebtehaj et al., 2021b). In this method, datasets are split into k subsamples so that one of the subsamples is considered testing data, and the rest of the

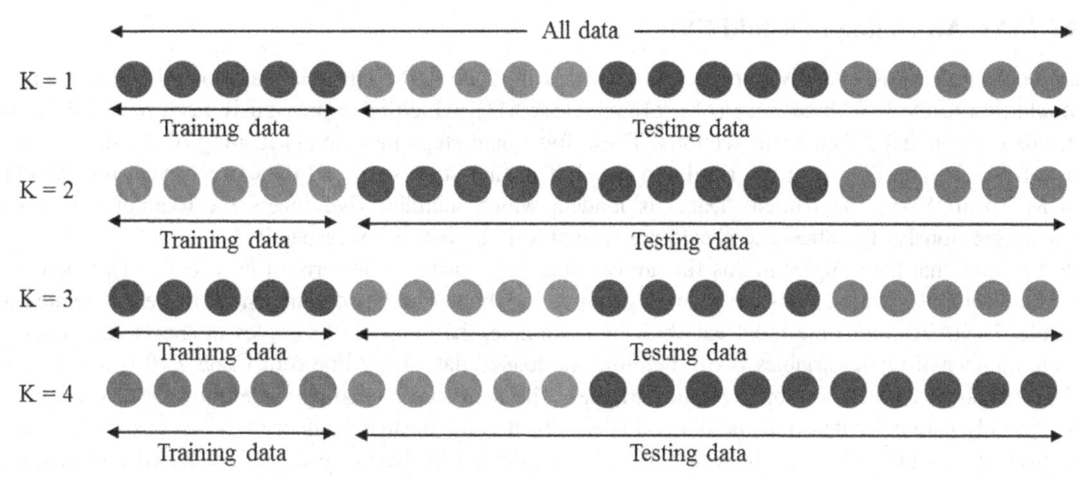

FIGURE 2.8 An example of K-fold cross-validation for $K = 4$.

subsamples (i.e., $k-1$ subsamples) are considered training samples. The modeling process is performed k times so that each subsample is used at least once as testing sample. A conceptual diagram of the data splitting using the K-fold approach ($K = 4$) is provided in the following figure. Given that the value of K is assumed to be equal to 4, all samples are randomly split into four groups. At each iteration, one of the newly generated subsamples is considered the testing data, and the rest are considered training samples (Fig. 2.8).

The modeling process of ML techniques using the K-fold CV approach can be described using pseudo-code 1. In this code, N and K denote the number of samples and all subgroups, respectively. To define different subgroups, the samples of each subgroup must be selected randomly. Therefore all samples are shuffled randomly at the first step. Using the shuffled samples, K different subgroups are defined. In the third step, an iterative process is repeated for all subgroups (i.e., K different subgroups). This process includes four main steps: (1) select one un-selected sample as the test data, (2) take the remaining subsamples as training data, (3) implement the ML model, and (4) evaluate the performance of the developed model using training and testing samples. After implementing all four steps, the final step of K-fold CV summarizes the performance of the desired ML model by averaging its performance for each K.

The value of K can be selected within the range of [2, N]. Suppose the value of this parameter was considered equal to one, then all data will be selected as training data, and there is no data to check the model's performance for the unseen samples. From this, the minimum value of this parameter was considered equal to 2. For $K = 2$, 50% of the samples are considered for the training stage and 50% as test data. Moreover, when $K = N$, K-fold CV is termed the leave-one-out CV method. Indeed, the modeling is implemented for N times so that the number of training and testing samples are $N-1$ and one, respectively. Based on the number of samples, the number of K folds in recent studies was considered 3 (Akhbari et al., 2019), 5 (Ebtehaj et al., 2016; Walton et al., 2019) and 10 (Akhbari et al., 2017; Ebtehaj et al., 2018, 2021b). Based on the author's experience, for a dataset with more than 200 or 300 samples, $K = 10$ is suggested, and $K = 5$ is recommended for lower samples.

Pseudo-code 1

1	N is the number of all samples, K is the number of all sub-groups
2	Shuffle the dataset randomly.
3	***for*** i = 1:K
4	Selection of N/K of all samples to generate K^{th} sub-sample.
5	***end***
6	***for*** i = 1:K
7	Take one that has not been selected yet as sub-samples as the test data.
8	Take the rest of the sub-samples as training data.
9	Implement the machine learning model.
10	Evaluate the performance of the developed model using training and testing samples.
11	***end***
12	Summarize the performance of the desired machine learning model by averaging its performance for each K.

2.4.2 The MATLAB coding of K-fold CV

The detailed coding of the K-fold CV approach is provided in Code 2.18. The schematic of this code is presented in Fig. 2.9. In addition to the first three lines (`clc`, `clear`, `close all`), which have previously been presented in all codes in this text, this code includes four main sections. These four main steps include (1) loading of the data, (2) CV parameters, (3) K-fold CV, and (4) exporting results to excel. The first step is to load data from Excel into MATLAB. In this step, a Microsoft Excel file named "Data" is loaded, which includes five sheets for Examples 1–5 that were defined in detail previously. The sheet1 to sheet5 are related to Examples 1–5, respectively.

It should be noted that the samples in this file are considered a matrix in the form of $N \times (\mathrm{InV} + 1)$, where N denotes the number of all samples, while InV is the number of input variables. The last column in each sheet is related to output variables, while the InV columns are input variables. For example, the number of samples in sheet1 (i.e., Example 1) is 161, while the number of input variables is 6. Therefore the loaded data using line 6 of Code 2.19 is a matrix with 161 rows and 7 columns (i.e., 6 InV and one Output variable). The next step of coding is defining the CV parameters. For K-Fold CV, the only parameter that must be defined is K, which is the desired number of subgroups. As a recommendation for the user, it was proposed that the value of K be considered in the range of [2, 10] based on the sample size. Besides, another way for K-fold CV without or with the random selection of samples is defined in this book (i.e., Code 2.21). Therefore the user could choose to shuffle data before categorizing them into K subgroups or to categorize the data without any shuffling. To select whether shuffling is performed prior to data categorization, the modeler can insert 1 to shuffle the data or insert 2 for categorizing data without shuffling. It is highly recommended to shuffle data before splitting them. To prompt the user to insert the value of these two parameters (i.e., K and shuffling condition), the `inputdlg` command is employed. The structure of this command is `inputdlg(Prompt,Title,size,DefaultValues)`. The `Prompt`, `Title`, and `DefaultValues` are defined in lines 9–12, respectively. Besides, the `size` was considered [1, 55] directly in line 13 of Code 2.19. These input arguments have all previously been discussed in detail. Using the `str2double`

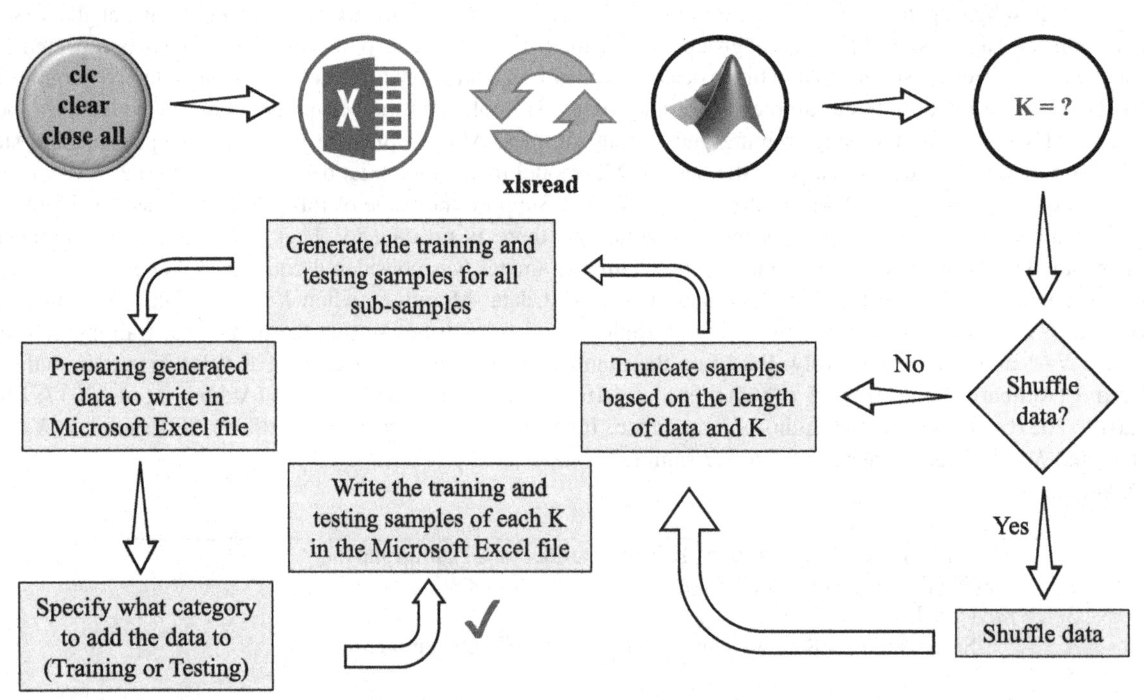

FIGURE 2.9 Schematic of code 2.18.

command, the user-defined values for K and the shuffling samples are converted to numerical values in lines 17–18 of Code 2.19. If the user inserts 1 to shuffle the data prior to splitting, the randomization of the data is done in lines 21–23 of Code 2.19. The randomization process is defined as easily as possible. The number of samples is read using the `size (x1,1)` command in line 21 and stored in the variable `nSample`. Using the `randperm` command with an input argument of `nSample`, `nSample` random numbers are generated and stored in the variable **S**. The data which was loaded into MATLAB from Excel using the `xlsread` command in line 6 (i.e., x1) is resorted based on the newly generated random data stored in the **S**. If the length of the data (i.e., *N*) is not perfectly divisible by the desired number of the fold (i.e., K), the data will be truncated. For instance, in Example 1, the number of all samples is 161. For the $K=5$, the number of samples in each group is 32, and one sample will be removed. It should be noted that for the *K* value in the range [2, 10], the maximum number of samples that may be removed is nine, which is small in comparison to the total number of samples (for example, 161 in Example 1). In addition, the number of deleted samples can be added to different categories. For example, if 3 samples are truncated for $K=5$, one of the samples is added to the categories $K=1$, $K=2$, and $K=3$. These additions could be made to the training or testing samples.

To find the number of samples that will be truncated, the `rem` command is applied (i.e., line 27 of Code 2.19). The inputs of this command are `L1` and `K`. The variable `L1` is calculated using the `size` command in line 26 of Code 2.19 from the variable `x1` calculated in line 6, while **K** is defined by the user. `L1` and `K` are the number of training samples and the number of subgroups, respectively. If the result of the `rem (L1,K)` command is greater than zero, a message is printed in the command window to indicate the number of samples removed from the total samples. This message for Example 1 with 161 samples and $K=5$ is shown in Fig. 2.10. In line 29 of Code 2.19, the data is truncated using the results of `the rem (L1,K)`. Following this, the values of the `L` and `N` are recalculated at line 31 of Code 2.19.

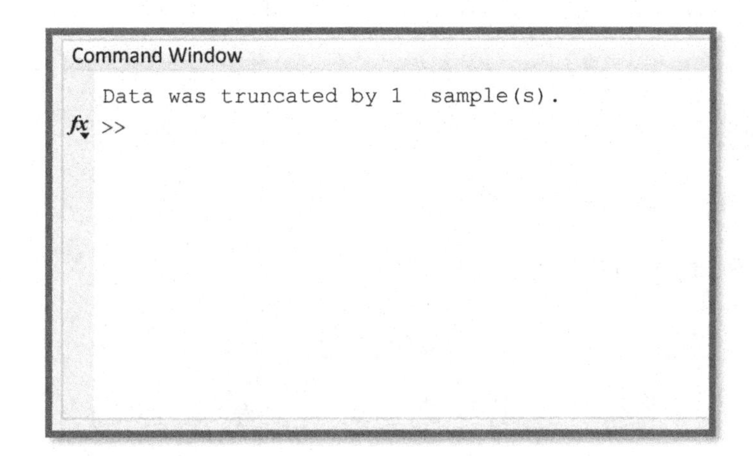

FIGURE 2.10 Command Window output message for truncated data.

Code 2.19

```
1   clc
2   clear
3   close all
4
5   %% load data
6   x1 = xlsread ('Data','sheet1');
7
8   %% Cross-Validation parameters
9   Prompt={'Please insert a number in the range of [2, 10] for K',...
10      'Please insert 1 if you want to shuffle data randomly, Otherwise,
    insert 2'};
11      Title='k-fold';
12      DefaultValues={'5','1'};
13      PARAMS=inputdlg(Prompt,Title,[1, 55],DefaultValues);
14      A=PARAMS{1};
15      B=PARAMS{2};
16
17  K=str2double(A);  % Convert to number
18  B=str2double(B);  % Convert to number
19
20  if B==1
21      nSample=size(x1,1);
22      S=randperm(nSample);
23      x1=x1(S,:);
24  end
25
26  [L1,N] = size(x1);
27  if rem(L1,K) > 0
28      fprintf('Data was truncated by %d',rem(L1,K)),fprintf('
    sample(s).\n')
29      x=x1;
30      x(L1-rem(L1,K)+1:end,:) = [];
31      [L,N] = size(x);
32  end
```

The third step of Code 2.18 is K-fold CV (i.e., Code 2.20). To categorize all samples into K subsamples, the `test-data` variable is defined in the first step. Following this, the matrix size for the training and testing data must be defined. To do this, the `cell(K,N)`command is applied in lines 3 and 4 of Code 2.20.

Suppose the number of subgroups is 5 ($K=5$), and the number of input variables is 7 (i.e., $N=7$). Then the defined structure for the training and testing samples are as shown in Fig. 2.11. It is evident from this figure that the structure of both samples (i.e., training and testing) would be entirely the same. The main difference between training and testing samples in K-fold CV is related to the number of samples in each cell which, at this step, are empty. Through two loops, the samples are assigned to each cell. In line 8 of Code 2.20, using the `false(L,1)` command, the `logicSet` variable is defined, which is a matrix with one column and `L` rows. All values of this matrix are zero. For $N \times K$ times (where N is the number of input variables and K is the number of subgroups), the `logicSet` variable is modified using the command written on line 11 of Code 2.20. After modifying this variable at each iteration, the `TestingData` and `TrainingData`, which were predefined as a matrix with the dimension of `K` $\times$ `N`, are assigned as presented in lines 12–13 of Code 2.20.

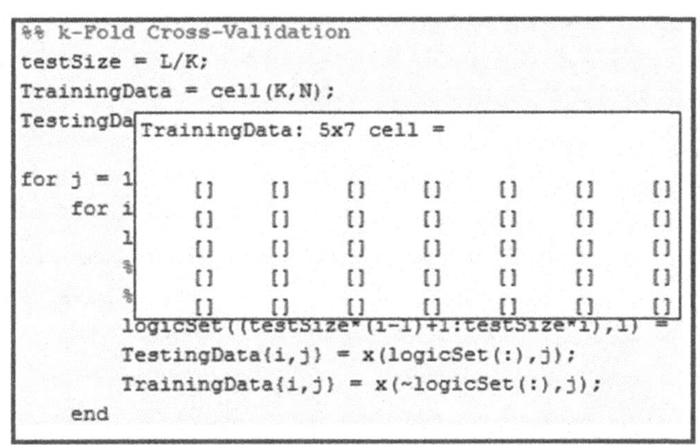

FIGURE 2.11 Structure of training and testing data with five subsamples and seven input variables.

Code 2.20

```
1   %% k-Fold Cross-Validation
2   testSize = L/K;
3   TrainingData = cell(K,N);
4   TestingData = cell(K,N);
5
6   for j = 1:N
7       for i = 1:K
8           logicSet = false(L,1);
9           % logical matrix determines which data to be used in training and
10          % testing each iteration, ensuring all points get used.
11          logicSet((testSize*(i-1)+1:testSize*i),1) = true;
12          TestingData{i,j} = x(logicSet(:),j);
13          TrainingData{i,j} = x(~logicSet(:),j);
14      end
15  end
```

After the implementation of Codes 2.19 and 2.20, the training and testing samples for all subsamples have been defined and saved in the `TestingData` and `TrainingData` variables, which can be seen in the workspace window (Fig. 2.12). According to this figure, the number of samples for all input variables in the training stage is 128, while for the testing stage, it is 32. The newly generated `TestingData` and `TrainingData` could be employed as the new dataset for modeling. It should be noted that there are no truncated samples in this data.

Another way to apply the split samples in training and testing samples for all subsamples is by importing them into the Microsoft Excel file. In this step, not only are the training and testing samples for each subsample saved in a Microsoft Excel file, but also the truncated samples are added to the training or testing subsamples based on the user selection. To do that, two variables, including `Test` and `Train`, are predefined with three dimensions. The first dimension of the `Test` variable is the number of testing samples (i.e., `testsize`), while for the `Train` variable, the first dimension is defined as `testSize*(K-1)`. The second and third dimensions of both variables (i.e., `Train` and `Test`) are the same (i.e., number of input variables, *N*, and the number of subsamples, *K*). Through two loops for all values of *K* and *N*, the `Test` and `Train` are defined using the `TestingData` and `TrainingData` variables, respectively. Using the `questdlg` command, the user may select to assign any truncated data into the training or testing samples. The general form of this command is `questdlg(qstring,title,str1,str2,str3,default)`. The input argument `qstring` is the question string, the `title` is the title of the dialog box that will be open to present the question, `str1,str2,` and `str3` are three options that the use may select, and the `default` is the default answer selected from one of the three options. The details of this

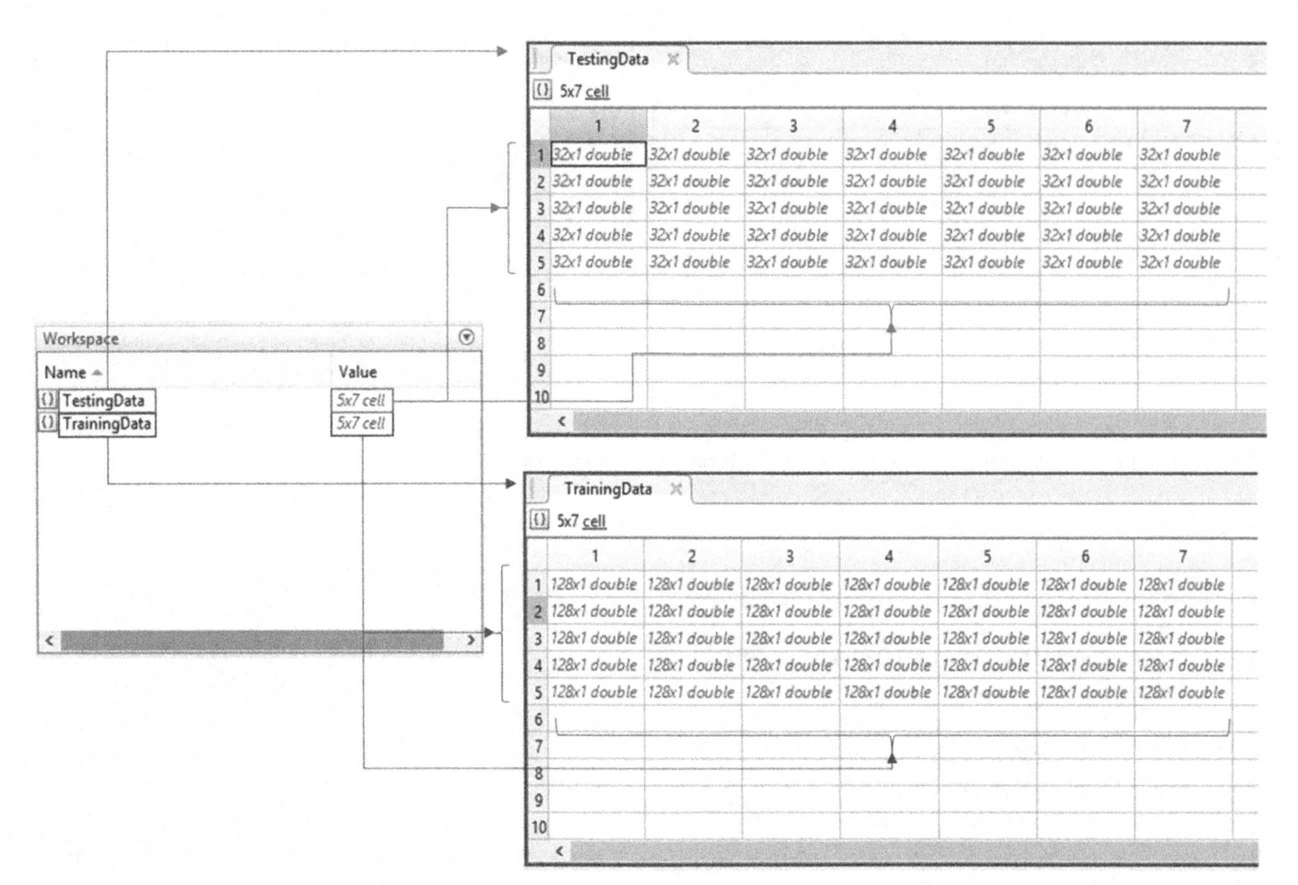

FIGURE 2.12 The results of the K-fold CV for a data set with $K = 5$ and 7 input variables.

are provided in lines 12–21 of Code 2.21. In line 22, the truncated samples are saved in the variable `x_truncated`. In lines 23–33, based on the modelers selection of whether to add the truncated samples to the training or testing samples, they are added to the corresponding subsamples. The `for` loop in lines 35–48 is used to write the newly defined variables (i.e., `Train` and `Test`) in two Microsoft Excel files (i.e., `TrainingData` and `TestingData`) using the `xlswrite` command. In lines 38 and 43, the rows with all zero values are removed before saving them in the Microsoft Excel file. In both generated Microsoft Excel files, `K` different sheets are defined so that the values of sheet 1 in the `TrainingData` include input variables and target variables of the training samples, while the values stored in the `TestingData` are input variables and output variable of the testing samples.

Code 2.21

```
1   %% Write results in Excel files
2   Test  = zeros(testSize,N,K);
3   Train = zeros(testSize*(K-1),N,K);
4   
5   for i = 1:K
6       for j=1:N
7           Test(:,j,i) = TestingData{i,j};
8           Train(:,j,i)= TrainingData{i,j};
9       end
10  end
11  % Select to add truncated samples to train or test samples
12  Option{1}='Train';
13  Option{2}='Test';
14  ANSWER=questdlg('Adding truncated samples to training or testing samples
    ?','Adding truncated samples',Option{1},Option{2},Option{1});
15  pause(0.1);
16  switch ANSWER
17      case Option{1}
18          TS = 1;      % Add truncated samples to the training samples
19      case Option{2}
20          TS = 2;      % Add truncated samples to the testing samples
21  end
22  x_truncated = x(end-rem(L1,K)+1:end,:); % Truncated samples
23  if TS == 1
24      Ltr = length(Train);
25      for i=1:rem(L1,K)
26          Train(Ltr+i,:,i) = x_truncated(i,:);
27      end
28  else
29      Lts = length(Test);
30      for i=1:rem(L1,K) %
31          Test(Lts+i,:,i) = x_truncated(i,:);
32      end
33  end
34  
35  for i = 1:K
36      if TS == 1
37          A = Train(:,:,i);
38          A( all(~A,2),:) = [];
39          xlswrite('TrainingData.xls',A,i)
40          xlswrite('TestingData.xls',Test(:,:,i),i)
41      else
42          A = Test(:,:,i);
43          A( all(~A,2),:) = [];
44          xlswrite('TrainingData.xls',Train(:,:,i),i)
45          xlswrite('TestingData.xls',A,i)
46      end
47      clear A
48  end
49  
50  clearvars -except TrainingData TestingData
```

Data on Cross-validation can be found at Appendix 2A.

2.5 Summary

ML models require preprocessing prior to application, as this chapter explained in a conceptual and mathematical manner. In addition, MATLAB software was made available for coding the preprocessing techniques line by line. The preprocessing techniques discussed included normalization, standardization, data splitting, and CVs. Standardization and normalization were applied to transform all input(s) and output with different dimensions into a single dimension. Additionally, data splitting and CV techniques were presented to generate the training and testing samples. Following the completion of this chapter, the reader now possesses the fundamental skills necessary to analyze the input and modeled data. The reader is also versed in different ways to separate training and testing samples, as well as how to transfer them to different domains to improve modeling results.

Appendix 2A Supporting information

Data on preprocessing techniques can be found in the online version at https://doi.org/10.1016/B978-0-443-15284-9.00003-3.

References

Akhbari, A., Bonakdari, H., & Ebtehaj, I. (2017). Evolutionary prediction of electrocoagulation efficiency and energy consumption probing. *Desalination and Water Treatment*, *64*, 54–63. Available from https://doi.org/10.5004/dwt.2017.20235.

Akhbari, A., Ibrahim, S., Zinatizadeh, A. A., Bonakdari, H., Ebtehaj, I., S. Khozani, Z., & Gharabaghi, B. (2019). Evolutionary prediction of biohydrogen production by dark fermentation. *CLEAN–Soil, Air, Water*, *47*(1), 1700494. Available from https://doi.org/10.1002/clen.201700494.

Arefinia, A., Bozorg-Haddad, O., Akhavan, M., Baghbani, R., Heidary, A., Zolghadr-Asli, B., & Chang, H. (2022). *Using support vector machine (SVM) in Modeling Water Resources Systems. Computational intelligence for water and environmental sciences* (pp. 179–199). Singapore: Springer. Available from https://doi.org/10.1007/978-981-19-2519-1_9.

Azimi, H., Bonakdari, H., & Ebtehaj, I. (2019). Design of radial basis function-based support vector regression in predicting the discharge coefficient of a side weir in a trapezoidal channel. *Applied Water Science*, *9*(4), 78. Available from https://doi.org/10.1007/s13201-019-0961-5.

Azimi, H., Bonakdari, H., & Ebtehaj, I. (2021). Gene expression programming-based approach for predicting the roller length of a hydraulic jump on a rough bed. *ISH Journal of Hydraulic Engineering*, *27*(sup1), 77–87. Available from https://doi.org/10.1080/09715010.2019.1579058.

Bonakdari, H., & Ebtehaj, I. (2021). Discussion of "time-series prediction of streamflows of Malaysian rivers using data-driven techniques" by Siraj Muhammed Pandhiani, Parveen Sihag, Ani Bin Shabri, Balraj Singh, and Quoc Bao Pham. *Journal of Irrigation and Drainage Engineering*, *147*(9), 07021014. Available from https://doi.org/10.1061/(ASCE)IR.1943-4774.0001602.

Chen, H., He, L., Qian, W., & Wang, S. (2020). Multiple aerodynamic coefficient prediction of airfoils using a convolutional neural network. *Symmetry*, *12*(4), 544. Available from https://doi.org/10.3390/sym12040544.

Cigizoglu, H. K. (2003). Estimation, forecasting and extrapolation of river flows by artificial neural networks. *Hydrological Sciences Journal*, *48*(3), 349–361. Available from https://doi.org/10.1623/hysj.48.3.349.45288.

Dawson, C. W., & Wilby, R. (1998). An artificial neural network approach to rainfall-runoff modelling. *Hydrological Sciences Journal*, *43*(1), 47–66. Available from https://doi.org/10.1080/02626669809492102.

Ebtehaj, I., & Bonakdari, H. (2013). Evaluation of sediment transport in sewer using artificial neural network. *Engineering Applications of Computational Fluid Mechanics*, *7*(3), 382–392. Available from https://doi.org/10.1080/19942060.2013.11015479.

Ebtehaj, I., & Bonakdari, H. (2014). Performance evaluation of adaptive neural fuzzy inference system for sediment transport in sewers. *Water Resources Management*, *28*(13), 4765–4779. Available from https://doi.org/10.1007/s11269-014-0774-0.

Ebtehaj, I., & Bonakdari, H. (2016). Bed load sediment transport estimation in a clean pipe using multilayer perceptron with different training algorithms. *KSCE Journal of Civil Engineering*, *20*(2), 581–589. Available from https://doi.org/10.1007/s12205-015-0630-7.

Ebtehaj, I., Bonakdari, H., Moradi, F., Gharabaghi, B., & Khozani, Z. S. (2018). An integrated framework of extreme learning machines for predicting scour at pile groups in clear water condition. *Coastal Engineering*, *135*, 1–15. Available from https://doi.org/10.1016/j.coastaleng.2017.12.012.

Ebtehaj, I., Bonakdari, H., Safari, M. J. S., Gharabaghi, B., Zaji, A. H., Madavar, H. R., & Mehr, A. D. (2020b). Combination of sensitivity and uncertainty analyses for sediment transport modeling in sewer pipes. *International Journal of Sediment Research*, *35*(2), 157–170. Available from https://doi.org/10.1016/j.ijsrc.2019.08.005.

Ebtehaj, I., Bonakdari, H., & Shamshirband, S. (2016). Extreme learning machine assessment for estimating sediment transport in open channels. *Engineering with. Computers*, *32*, 691–704. Available from https://doi.org/10.1007/s00366-016-0446-1.

Ebtehaj, I., Bonakdari, H., Zaji, A. H., & Gharabaghi, B. (2021b). Evolutionary optimization of neural network to predict sediment transport without sedimentation. *Complex & Intelligent Systems*, *7*(1), 401–416. Available from https://doi.org/10.1007/s40747-020-00213-9.

Ebtehaj, I., Sammen, S. S., Sidek, L. M., Malik, A., Sihag, P., Al-Janabi, A. M. S., & Bonakdari, H. (2021a). Prediction of daily water level using new hybridized GS-GMDH and ANFIS-FCM models. *Engineering Applications of Computational Fluid Mechanics*, *15*(1), 1343–1361. Available from https://doi.org/10.1080/19942060.2021.1966837.

Ebtehaj, I., Zeynoddin, M., & Bonakdari, H. (2020a). Discussion of "comparative assessment of time series and artificial intelligence models to estimate monthly streamflow: A local and external data analysis approach" by Saeid Mehdizadeh, Farshad Fathian, Mir Jafar Sadegh Safari and Jan F. Adamowski. *Journal of Hydrology*, *583*, 124614. Available from https://doi.org/10.1016/j.jhydrol.2020.124614.

Erbir, M. A., & Ünver, H. M. (2021). The do's and don'ts for increasing the accuracy of face recognition on VGGFace2 dataset. *Arabian Journal for Science and Engineering*, *46*(9), 8901–8911. Available from https://doi.org/10.1007/s13369-021-05693-6.

Gholami, A., Bonakdari, H., Akhtari, A. A., & Ebtehaj, I. (2019). A combination of computational fluid dynamics, artificial neural network, and support vectors machines models to predict flow variables in curved channel. *Scientia Iranica*, *26*(2), 726–741. Available from https://doi.org/10.24200/sci.2017.4520.

Halpern-Wight, N., Konstantinou, M., Charalambides, A.G., & Reinders, A. (2020). Training and testing of a single-layer LSTM network for near-future solar forecasting. *Applied Sciences*, *10*(17), 5873. Available from https://doi.org/10.3390/app10175873.

Jahani, A., & Rayegani, B. (2020). Forest landscape visual quality evaluation using artificial intelligence techniques as a decision support system. *Stochastic Environmental Research and Risk Assessment*, *34*(10), 1473–1486. Available from https://doi.org/10.1007/s00477-020-01832-x.

Katipoğlu, O. M., Acar, R., & Şengül, S. (2020). Comparison of meteorological indices for drought monitoring and evaluating: a case study from Euphrates basin, Turkey. *Journal of Water and Climate Change*, *11*(S1), 29–43. Available from https://doi.org/10.2166/wcc.2020.171.

Mohamad, I. B., & Usman, D. (2013). Standardization and its effects on K-means clustering algorithm. *Research Journal of Applied Sciences, Engineering and Technology*, *6*(17), 3299–3303. Available from https://doi.org/10.19026/rjaset.6.3638.

Qasem, S. N., Ebtehaj, I., & Bonakdari, H. (2017a). Potential of radial basis function network with particle swarm optimization for prediction of sediment transport at the limit of deposition in a clean pipe. *Sustainable Water Resources Management*, *3*(4), 391–401. Available from https://doi.org/10.1007/s40899-017-0104-9.

Qasem, S. N., Ebtehaj, I., & Riahi Madavar, H. (2017b). Optimizing ANFIS for sediment transport in open channels using different evolutionary algorithms. *Journal of Applied Research in Water and Wastewater*, *4*(1), 290–298. Available from https://doi.org/10.22126/arww.2017.773.

Riahi-Madvar, H., & Gharabaghi, B. (2022). *Pre-processing and input vector selection techniques in computational soft computing models of water engineering. Computational intelligence for water and environmental sciences* (pp. 429–447). Singapore: Springer. Available from https://doi.org/10.1007/978-981-19-2519-1_20.

Safari, M. J. S. (2019). Decision tree (DT), generalized regression neural network (GR) and multivariate adaptive regression splines (MARS) models for sediment transport in sewer pipes. *Water Science and Technology*, *79*(6), 1113–1122. Available from https://doi.org/10.2166/wst.2019.106.

Salih, S. Q., Sharafati, A., Khosravi, K., Faris, H., Kisi, O., Tao, H., & Yaseen, Z. M. (2020). River suspended sediment load prediction based on river discharge information: application of newly developed data mining models. *Hydrological Sciences Journal*, *65*(4), 624–637. Available from https://doi.org/10.1080/02626667.2019.1703186.

Walton, R., Binns, A., Bonakdari, H., Ebtehaj, I., & Gharabaghi, B. (2019). Estimating 2-year flood flows using the generalized structure of the group method of data handling. *Journal of Hydrology*, *575*, 671–689. Available from https://doi.org/10.1016/j.jhydrol.2019.05.068.

Wang, Y. F., Huai, W. X., & Wang, W. J. (2017). Physically sound formula for longitudinal dispersion coefficients of natural rivers. *Journal of Hydrology*, *544*, 511–523. Available from https://doi.org/10.1016/j.jhydrol.2016.11.058.

Yaraghi, N., Ronkanen, A. K., Haghighi, A. T., Aminikhah, M., Kujala, K., & Kløve, B. (2020). Impacts of gold mine effluent on water quality in a pristine sub-Arctic river. *Journal of Hydrology*, *589*, 125170. Available from https://doi.org/10.1016/j.jhydrol.2020.125170.

Zhang, B., Li, W., Li, X. L., & Ng, S. K. (2018). Intelligent fault diagnosis under varying working conditions based on domain adaptive convolutional neural networks. *Ieee Access*, *6*, 66367–66384. Available from https://doi.org/10.1109/ACCESS.2018.2878491.

Chapter 3

Postprocessing approaches

3.1 Introduction

Machine learning (ML) is a subset of "artificial intelligence" (AI). The terms ML and AI are sometimes used interchangeably to obtain a better understanding of how software programs may grow more effective at predicting events without being expressly designed to do so. Scholars from a wide variety of fields, such as earth and planetary sciences, and environmental science, have been attracted to ML application to address different practical issues. It is the process of creating an intelligent model capable of resolving complex problems that people would not be able to solve on their own. The aim of this chapter is to introduce qualitative and quantitative tools needed to present and discriminate the results generated by ML models (Fig. 3.1). The quantitative approach uses tools such as statistical indices analysis (i.e., correlation-based, dimensional, nondimensional, and hybrid indices), uncertainty analysis, and reliability analysis, whereas the qualitative tools presented include the Taylor diagram, scatter plot, barplot, histogram, and box plot. For the postprocessing tools, not only the theoretical concepts are presented but also the details of the MATLAB® coding are provided and applied to the five real-world sample problems presented in the first chapter.

3.2 Quantitative tools

The quantitative tools presented in this section are categorized into three main groups: statistical indices, UA, and reliability analysis (Fig. 3.2). The details of each tool, along with their MATLAB codings, are provided in the following subsections.

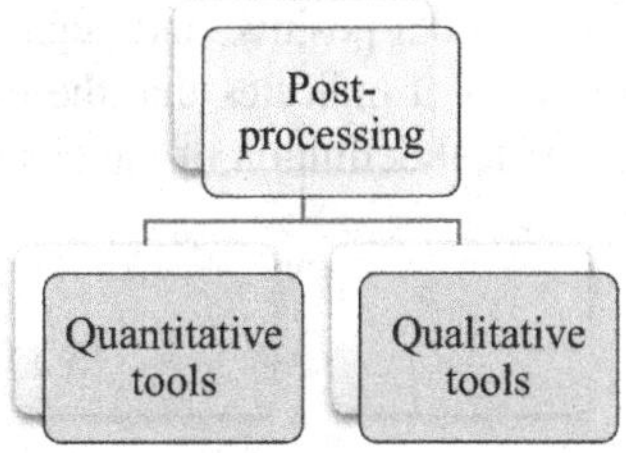

Post-processing

Quantitative tools

Qualitative tools

FIGURE 3.1 The classification of the postprocessing tools.

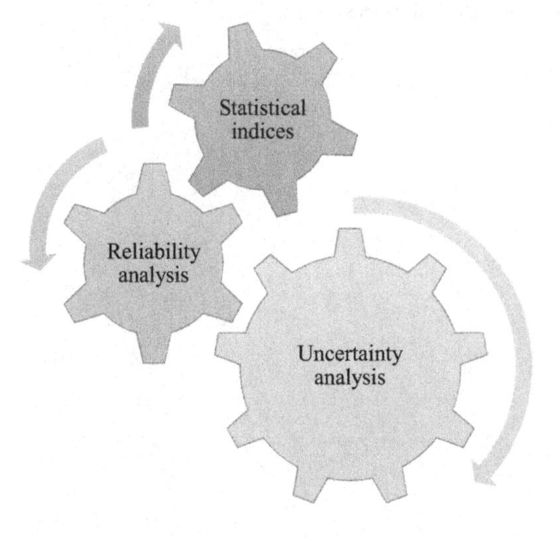

FIGURE 3.2 The presented quantitative tools.

Machine Learning in Earth, Environmental and Planetary Sciences. DOI: https://doi.org/10.1016/B978-0-443-15284-9.00006-9

3.2.1 Statistical indices

One of the most important steps in evaluating the performance of a model is calculating different statistical indices and comparing their values with those from other models. Certainly, the performance of a given ML model will depend on the values provided for various numerical indicators. In this subsection, eleven statistical indices, all of which are applied in the analysis of the examples throughout this test, are provided. These eleven indices include the correlation coefficient (R), the variance accounted for (VAF), the Root Mean Square Error (RMSE), the Normalized Root Mean Square Error (NRMSE), the Mean Absolute Error (MAE), the Mean Absolute Percentage Error (MARE), the Root Mean Squared Relative Error (RMSRE), the Mean Relative Error (MRE), the BIAS, the Akaike Information Criteria (AIC), and the Nash Sutcliffe Efficiency (NSE). These indices are categorized into four main groups: correlation-based indices, dimensional indices, nondimensional indices, and hybrid indices. Correlation-based indices include the R, VAF, and NSE. Dimensional indices include the RMSE, MAE, and BIAS, while nondimensional indices include the NRMSE, MRE, MARE, and RMSRE. Finally, the AIC index is considered a hybrid index (Fig. 3.3). The categorization of the AIC as a hybrid index is because this index is composed of two different terms. In the first term, the amount of error is calculated as an absolute index, while in the second term, the effect of the number of optimized parameters during the training process is calculated. Indeed, due to the effect of the second component of the AIC index, more complicated models will be resulted in a higher AIC value.

3.2.1.1 Correlation-based indices

The correlation coefficient (R) is the most well-known statistical index and has been applied in many different fields of science, such as blasting (Mojtahedi et al., 2019), curved open channel flow (Bonakdari et al., 2021; Gholami et al., 2017), sewer systems (Khozani et al., 2017), side weirs (Azimi et al., 2017), Stage discharge (Shukla et al., 2022), streamflow (Dehghani et al., 2020), compound narrow channels (Bonakdari et al., 2018), river ice thickness (Barzegar et al., 2019), and side orifices (Bonakdari et al., 2020) to name a few. The R index provides an indication of the direction and strength of the linear relationship between the actual and estimated values through the interpretation of its sign (i.e., positive or negative) and magnitude (which has a range of $[-1, 1]$). It is an intuitive and easy way to judge the correlation between the actual and predicted values. When $R = \pm 1$, a perfect relationship exists between actual and estimated samples. The difference between the sign of the R index (i.e., $R = +1$ and $R = -1$) is the orientation, or slope, of the linear relationship. For $R = 1$, all samples must fall on an upward positive line, while for $R = -1$, all samples fall on a downward negative line. It should be mentioned that although $R = -1$ may indicate a perfect fit of the regressed line with the data, it does not confirm the acceptable performance of the model due to its negative sign. Examples of different types of correlations are provided in Fig. 3.4. Based on this figure, we can see that weak, moderate, and robust correlations can occur in both positive and negative values of the R. If the correlation coefficient between actual values and estimated values is zero, it indicates that these two variables (i.e., actual values and estimated values) are independent of each other. The mathematical definition of the correlation coefficient (R) is as follows:

$$R = \frac{\sum_{i=1}^{N} (A_i - \overline{A}) \sum_{i=1}^{N} (E_i - \overline{E})}{\sqrt{\sum_{i=1}^{N} (A_i - \overline{A})^2 \sum_{i=1}^{N} (E_i - \overline{E})^2}} \tag{3.1}$$

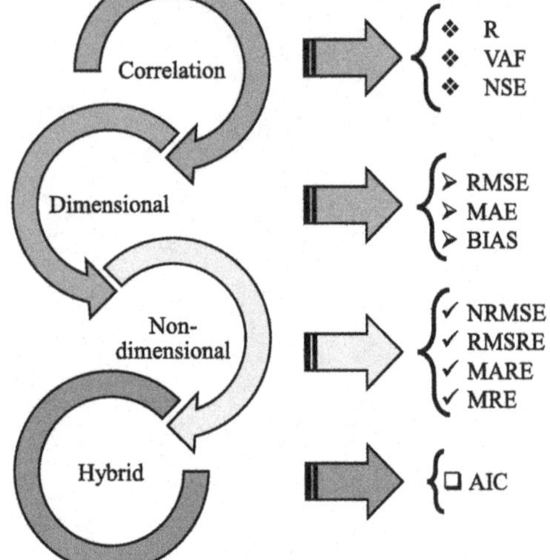

FIGURE 3.3 Different types of the statistical indices.

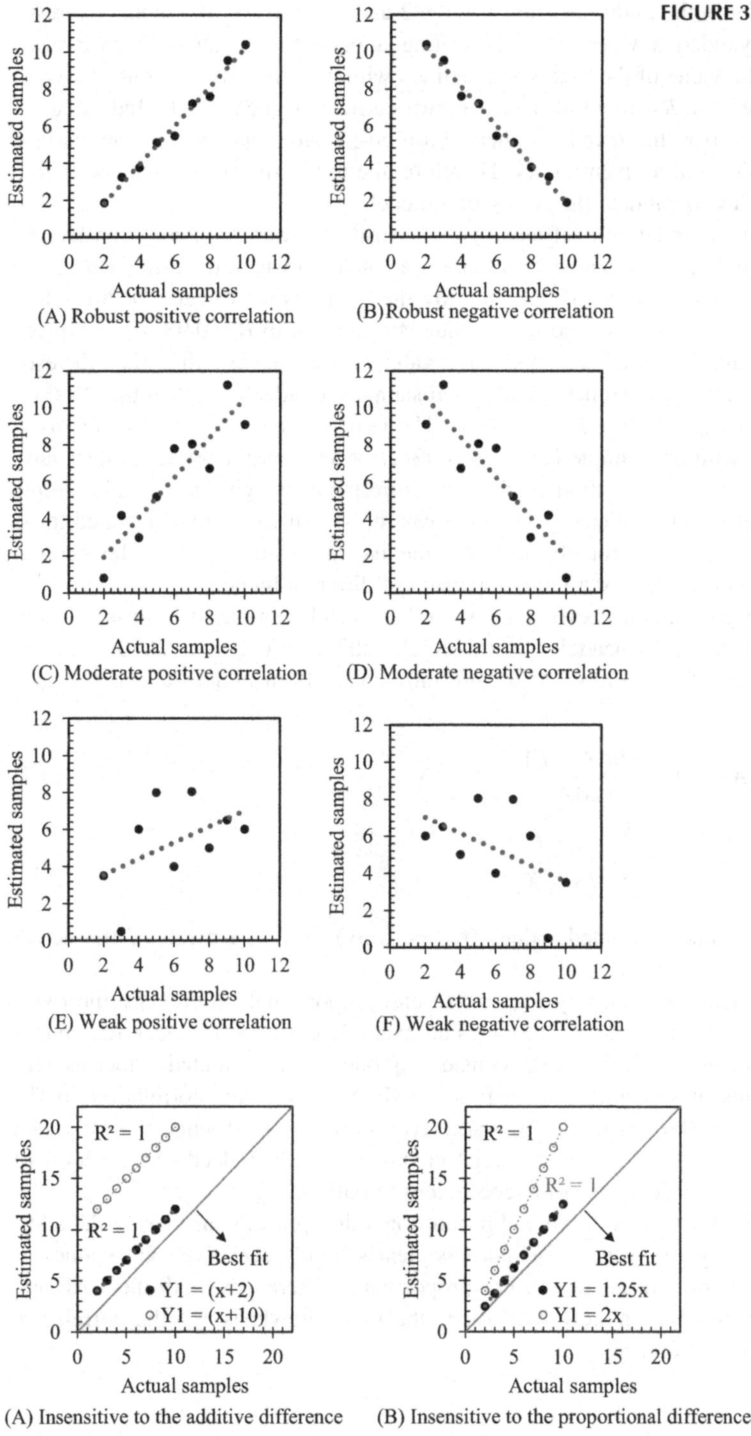

FIGURE 3.4 Different types of correlations.

FIGURE 3.5 An Example of the main limitations of the correlation coefficient regarding its insensitivity to additive and proportional differences.

where A_i and E_i are the ith samples of the actual and estimated values (respectively), $\overline{A}$ and $\overline{E}$ are the mean of the actual and estimated samples (respectively), and N is the number of samples. If we want to estimate R for the training stage, we need to consider N equal to the number of training samples, but if we want to estimate the R of the training stage, we should consider N as the number of training samples.

Although most researchers have considered the use of this index, it also has key limitations which may lead to incorrect conclusions regarding the performance of a model. The main limitations of this index are: (1) It is insensitive to additive and proportional difference, and (2) it is oversensitive to outliers. These key limitations are demonstrated in Fig. 3.5. Regarding its insensitivity to the additive difference, values of two and ten were added to each of the values on the x-axis for the provided data series. It could be seen that for both conditions (i.e., $Y = x + 2$ and $Y = x + 10$), the model overestimated the actual values, although the R index yielded a value of 1, indicating a perfect correlation in each case. Regarding insensitivity to the additive difference, the value of the y-axis was defined with 25% and 100% error relative to the corresponding values of the x-axis. In both cases, the R-value indicated a robust correlation (i.e., $R = 1$). Indeed, even though the model produced 25% and 100% relative error, the R index indicated a robust performance which can confuse the user when making a final assessment of the validity of a given model. Therefore, the use of the correlation coefficient must be carefully considered when applying this index to validate the ability of a model.

Fig. 3.6 provides an example of a case where the correlation coefficient is oversensitive toward outliers—considering an example with nine samples ranging in value from 2 to 10. It can be seen that for a condition where the estimated values are identical to actual ones (i.e., Fig. 3.6A), $R = 1$. When the estimated value by the model is set to zero for the actual data point of 2, the average relative error of the series is 11.11%, and the R-value is decreased to $R = 0.98$. If we replace the estimated model value of the 3, 4, 5, 6, 7, 8, 9, and 10 actual samples with a value of zero one at a time (i.e., the estimated value for the actual sample of 3 is zero while all other estimated and actual samples are identical) then the R values are 0.96, 0.92, 0.87, 0.81, 0.72, 0.62, 0.48, 0.31, respectively (see Fig. 3.6C–J). The remarkable point is that in all cases, the average relative error value is equal to 11.11%, with one sample having a 100% error and the remaining samples having a relative error of 0%. It is seen that lower magnitude correlation coefficients correspond to higher magnitude sample data so that the lowest value of the correlation coefficient is related to the case where the value of 10 is estimated to be zero. In all cases, however, a single data point has a relative error of 100%, and the average relative error of the samples is identical at 11.11%. This drawback also must be considered for a fair comparison of different models.

Due to the abovementioned limitations of the correlation coefficient, two other correlation-based indices, namely the variance account for (VAF) (Cui et al., 2022; Mohsenzadeh Karimi et al., 2020; Ray et al., 2020) and the Nash–Sutcliffe efficiency (NSE) (Gharib & Davies, 2021; Siddik, 2022), are introduced. Both indices are mathematically defined next and have ranges of $(-\infty, 1]$.

$$VAF = 1 - \frac{\mathrm{var}(A - E)}{\mathrm{var}(A)} \tag{3.2}$$

$$NSE = 1 - \frac{\sum_{i=1}^{N} (A_i - E_i)^2}{\sum_{i=1}^{N} (A_i - \overline{A})^2} \tag{3.3}$$

where A_i and E_i are the ith samples of the actual and estimated values (respectively), $\overline{A}$ is the mean of the actual samples, var() is the variance, and N is the number of samples.

The limitations seen for the correlation coefficient (insensitivity to additive and proportional differences and oversensitivity to outliers) are investigated for VAF and NSE. Table 3.1 provides an example designed to check the behavior of VAF and NSE in the presence of outliers. According to this table, considering one of the estimated values as zero and the remaining equal to the actual values results in a decrease in VAF and NSE. Similar to the correlation coefficient, the lowest values of the NSE and VAF (i.e., -0.67 and -0.48, respectively) are observed when the estimated zero value is located at the highest magnitude actual sample value (i.e., 10 in this example). Indeed, the reaction of these indices to outliers is very similar to the reaction of the correlation coefficient to outliers.

Fig. 3.7 shows the sensitivity of the NSE and VAF to the additive and proportional difference. Considering the additive difference, the VAF demonstrates an insensitivity while the NSE index is clearly highly sensitive, with values of 0.4 and -14 when the data is shifted by 2 and 10 units, respectively. For proportional differences, both the VAF and NSE indices are extremely sensitive. Indeed, the NSE completely overcomes the main drawbacks of the correlation coefficient associated with additive and proportional differences.

3.2.1.2 Dimensional indices

The dimensional indices provided in this subsection include the RMSE, MAE, MRE, and the BIAS, which are mathematically defined as follows:

$$\mathrm{RMSE} = \sqrt{\frac{1}{N}\sum_{i=1}^{N} (A_i - E_i)^2} \tag{3.4}$$

$$\mathrm{MAE} = \frac{1}{N}\sum_{i=1}^{N} |A_i - E_i| \tag{3.5}$$

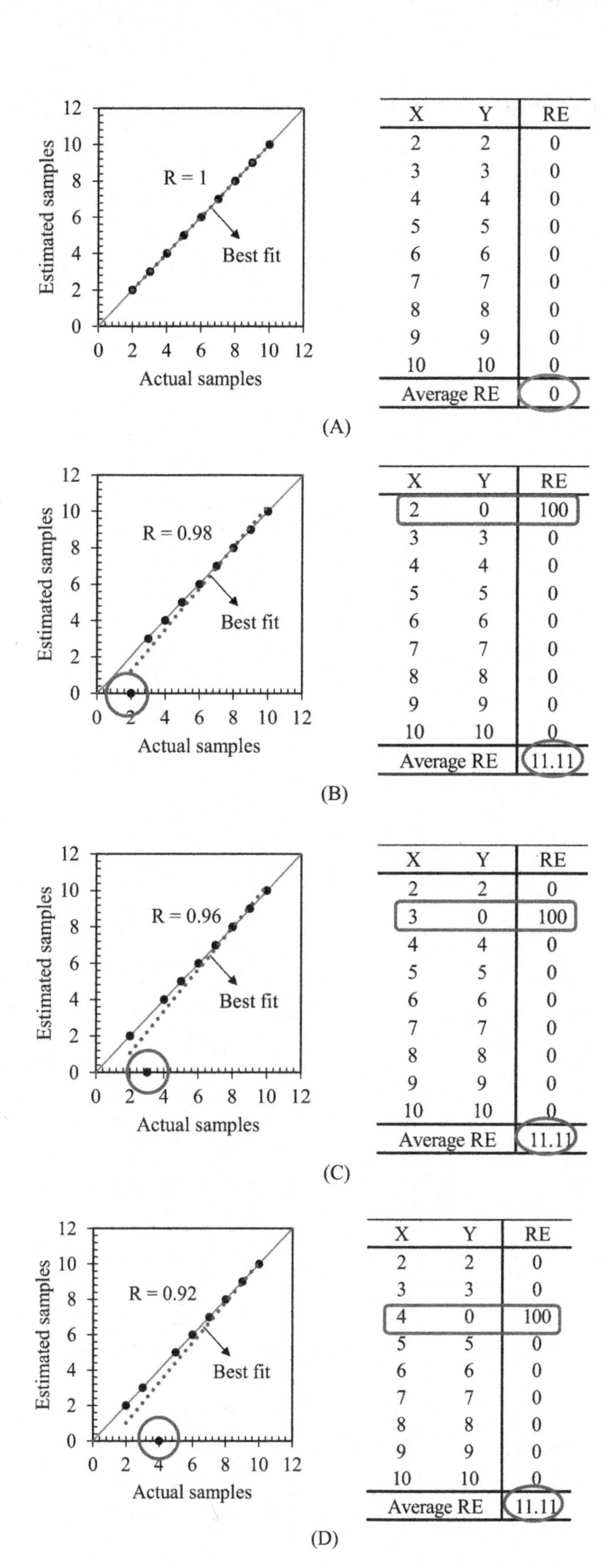

FIGURE 3.6 An example of the correlation coefficient to indicate it is oversensitive to outliers. (A) R = 1, (B) R = 0.98, (C) R = 0.96, (D) R = 0.92, (E) R = 0.87, (F) R = 0.81, (G) R = 0.72, (H) R = 0.62, (I) R = 0.48, and (J) R = 0.31.

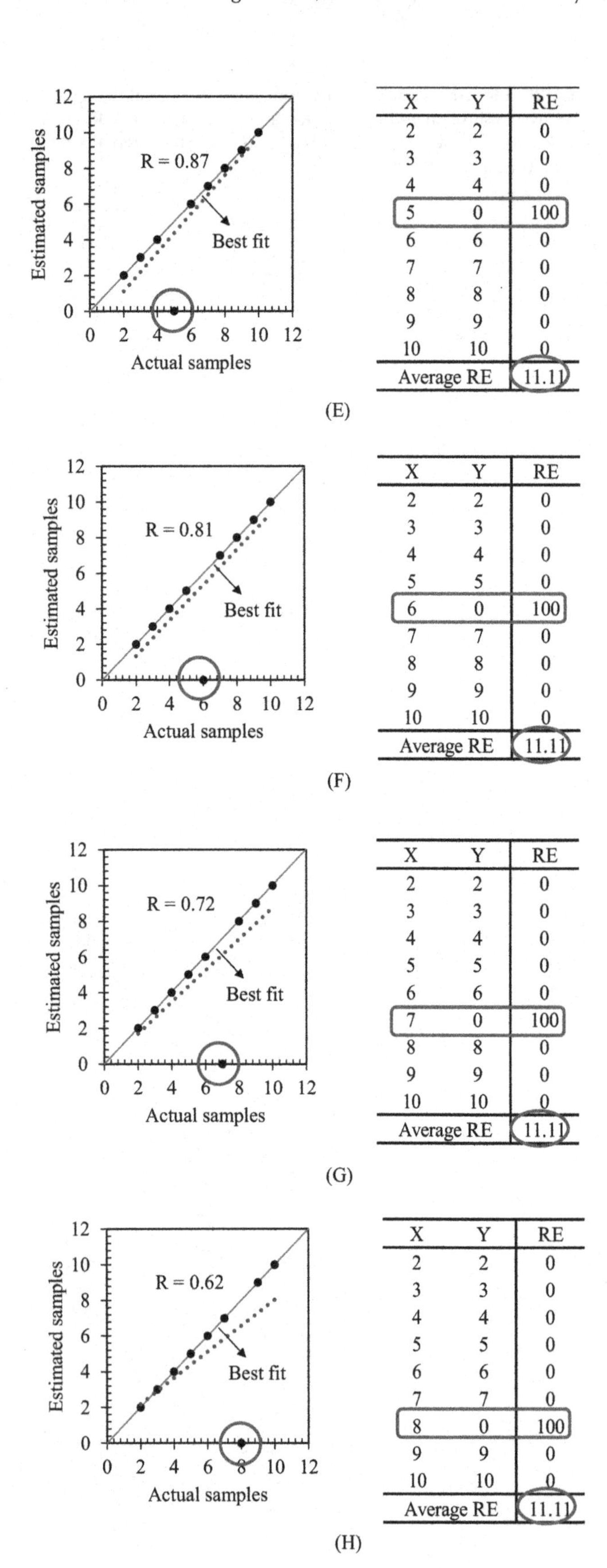

(E)

X	Y	RE
2	2	0
3	3	0
4	4	0
5	0	100
6	6	0
7	7	0
8	8	0
9	9	0
10	10	0
Average RE		11.11

(F)

X	Y	RE
2	2	0
3	3	0
4	4	0
5	5	0
6	0	100
7	7	0
8	8	0
9	9	0
10	10	0
Average RE		11.11

(G)

X	Y	RE
2	2	0
3	3	0
4	4	0
5	5	0
6	6	0
7	0	100
8	8	0
9	9	0
10	10	0
Average RE		11.11

(H)

X	Y	RE
2	2	0
3	3	0
4	4	0
5	5	0
6	6	0
7	7	0
8	0	100
9	9	0
10	10	0
Average RE		11.11

FIGURE 3.6 (Continued).

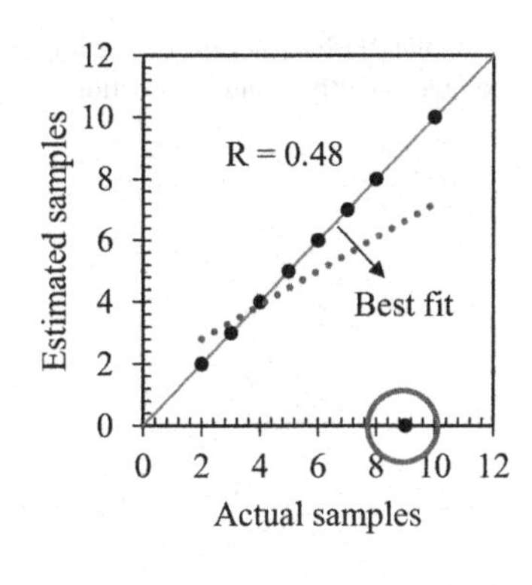

X	Y	RE
2	2	0
3	3	0
4	4	0
5	5	0
6	6	0
7	7	0
8	8	0
9	0	100
10	10	0
Average RE		11.11

(I)

R = 0.31
Best fit
Estimated samples
Actual samples

X	Y	RE
2	2	0
3	3	0
4	4	0
5	5	0
6	6	0
7	7	0
8	8	0
9	9	0
10	0	100
Average RE		11.11

(J)

FIGURE 3.6 (Continued).

TABLE 3.1 An example to check the behavior of variance account for and Nash–Sutcliffe efficiency in the presence of outliers.

Oversensitive to outliers										
X	Y1	Y2	Y3	Y4	Y5	Y6	Y7	Y8	Y9	Y10
2	2	**0**	2	2	2	2	2	2	2	2
3	3	3	**0**	3	3	3	3	3	3	3
4	4	4	4	**0**	4	4	4	4	4	4
5	5	5	5	5	**0**	5	5	5	5	5
6	6	6	6	6	6	**0**	6	6	6	6
7	7	7	7	7	7	7	**0**	7	7	7
8	8	8	8	8	8	8	8	**0**	8	8
9	9	9	9	9	9	9	9	9	**0**	9
10	10	10	10	10	10	10	10	10	10	**0**
VAF	1.00	0.94	0.87	0.76	0.63	0.47	0.27	0.05	− 0.20	− 0.48
NSE	1.00	0.93	0.85	0.73	0.58	0.40	0.18	− 0.07	− 0.35	− 0.67
RE (%)	0.00	11.11	11.11	11.11	11.11	11.11	11.11	11.11	11.11	11.11

(In)sensitive to the additive difference

X	Y=X+2	Y = X + 10
2	4	12
3	5	13
4	6	14
5	7	15
6	8	16
7	9	17
8	10	18
9	11	19
10	12	20
VAF	1	1
NSE	0.4	-14

Sensitive to the proportional difference

X	Y=1.25X	Y = 2X
2	2.50	4
3	3.75	6
4	5.00	8
5	6.25	10
6	7.50	12
7	8.75	14
8	10.00	16
9	11.25	18
10	12.50	20
VAF	0.94	0
NSE	0.60	-5.4

FIGURE 3.7 The sensitivity of the Nash–Sutcliffe efficiency and variance account for to the additive and proportional difference.

TABLE 3.2 Different examples of the errors and their corresponding MAE, RMSE, and variance.

	Case 1	Case 2	Case 3	Case 4	Case 5	Case 6
ID	Error	Error	Error	Error	Error	Error
1	2	1	0	0	10	6
2	2	1	0	0	0	8
3	2	1	0	0	10	6
4	2	1	0	0	0	8
5	2	1	10	0	10	6
6	2	3	0	0	0	8
7	2	3	0	0	10	6
8	2	3	0	0	0	8
9	2	3	0	0	10	6
10	2	3	10	20	0	8
Variance	0.00	1.11	17.78	40.00	27.78	1.11
MAE	2.00	2.00	2.00	2.00	5.00	7.00
RMSE	4.00	5.00	20.00	40.00	7.07	7.07

$$\mathrm{BIAS} = \frac{1}{N}\sum_{i=1}^{N}(A_i - E_i) \tag{3.6}$$

where A_i and E_i are the ith samples of the actual and estimated values (respectively), and N is the number of samples.

In the RMSE, the errors are squared. If the magnitude of the error values is less than one, the RMSE yields a relatively low value, while if the error is greater than the magnitude of one, then the RMSE value will be high. As a result of the squared errors used in this index, a relatively high weight is assigned to the significant errors. Therefore, this index is particularly useful for problems where large errors are remarkably undesirable. The range of the RMSE is $[0, \infty)$ with the best value of this index being zero.

The MAE calculates the average difference between the actual and estimated samples. The range of the MAE is also $[0, \infty)$, with the best value of this index again being zero. The main advantage of the MAE is that it provides a linear weighting between small and large errors. This is unlike the RMSE, which considers squared error.

Table 3.2 provides a number of different examples for a comparison of the RMSE and MAE. According to the error values presented in the table, the variance and RMSE increase from case 1 to case 4, while the MAE of all cases is equal. Therefore, it could be concluded that for these cases, a higher degree of variance results in a higher RMSE. In contrast, however, the RMSE values of cases 5 and 6 were found to be the same, although the MAE of case six was found to be greater than that of case 5. Consequently, it is not possible to draw a definite conclusion about the relationship between variance and these two indices (i.e., RMSE and MAE).

The final dimensional index presented in this test is BIAS which is a measure used to check the performance of the model as a whole. According to the sign provided for this index, it can be decided whether the model has a tendency to overestimate (positive sign) or underestimate (negative sign) the target variable.

3.2.1.3 Nondimensional indices

The nondimensional statistical indices considered in this text include the MRE, MARE, NRMSE, and NRMSE, which are mathematically defined as follows:

$$\mathrm{MRE} = \frac{1}{N}\sum_{i=1}^{N}\frac{A_i - E_i}{A_i} \tag{3.7}$$

$$\mathrm{MARE} = \frac{1}{N}\sum_{i=1}^{N}\left|\frac{A_i - E_i}{A_i}\right| \tag{3.8}$$

$$\mathrm{NRMSE} = \frac{\sqrt{\frac{1}{N}\sum_{i=1}^{N}(A_i - E_i)^2}}{\frac{1}{N}\sum_{i=1}^{N}A_i} \tag{3.9}$$

$$\mathrm{RMSRE} = \sqrt{\frac{1}{N}\sum_{i=1}^{N}\left(\frac{A_i - E_i}{A_i}\right)^2} \tag{3.10}$$

where A_i and E_i are the ith samples of the actual and estimated values (respectively), and N is the number of samples.

The main weakness of the MRE is its ignorance of the relative error sign of each sample. For example, suppose there exist two samples with very large errors. One of the samples is larger than the actual (true) value, while the other sample is smaller in magnitude compared to the actual value. In this case, the sum of the relative errors of these two numerical variables will be close to zero. The presence of a large number of these samples, which have different signs, may show a very small mediated result which can lead to a decision error when evaluating the performance of a model. To overcome this limitation of the MRE, the MARE index is introduced. The structure of the MARE index is entirely the same as the MRE, except that the MARE formula considers the "absolute value" of the relative errors. The range of the MARE and MRE are $[0, \infty)$ and $(-\infty, \infty)$, respectively.

Table 3.3 presents an example demonstrating the effect of absolute errors on the assessment of model performance. From this table, the relative errors are in the range of $[-10, 10]$. Applying Eqs. (3.7) and (3.8) the $\mathrm{MRE} = 0$ and $\mathrm{MARE}\ (\%) = 5.6\%$. If the absolute relative error of the MARE had not been considered, then the results of a model with a relative error of more than 5% are assumed to be equal to zero.

Given that the value of actual samples is in the denominator of the fraction, if the denominator of the fraction is zero, using this index is not possible. It should be noted that if the value of at least one of the actual samples is equal to zero, one of the following two solutions should be considered: (1) delete all samples whose actual value is zero, and (2) consider the amount of estimated samples at the denominator of the fraction instead of actual ones. If the second solution is chosen, we have Z samples with 100% error, where Z is the number of samples whose actual values are zero. Therefore, according to the desired problem, the right decision should be made about how to use this index.

An important scenario that may be encountered with this index is the case where the sample data has very different magnitudes. For example, this situation is often encountered in rainfall and runoff modeling. Torrential rainfall and flood events yield much higher rainfall and discharge values, respectively, than the average condition. In such a case, the use of the MARE can be misleading, and the use of relative indices similar to MARE may not lead to descriptive conclusions. It is recommended to use this index only for modeling extreme values.

The third nondimensional index is the NRMSE, which is the dimensionless version of the RMSE. The NRMSE is not a relative-based index, but the use of this index may provide a better understanding in some cases of model

TABLE 3.3 An example for comparison of the MRE and MARE.

ID	Actual	Estimated	Relative Error	Absolute relative error
1	5	5.5	−0.10	0.10
2	10	9	0.10	0.10
3	15	15.75	−0.05	0.05
4	20	19	0.05	0.05
5	25	26	−0.04	0.04
6	30	28.8	0.04	0.04
7	35	35.7	−0.02	0.02
8	40	39.2	0.02	0.02
9	45	48.15	−0.07	0.07
10	50	46.5	0.07	0.07
MRE			0.00	—
MARE			—	0.056

performance than a dimensional index like RMSE. The fourth dimensionless index is the relative version of the RMSRE. It could be applied as an alternative index for the RMSE in a problem where the relative error is more important than the absolute one.

3.2.1.4 Hybrid index

The fourth group of statistical indices is the so-called hybrid index, the best known of which is the Akaike Information Criteria (AIC). The reason that this index is termed "hybrid" is that these indices have more than one term so that one calculates the model error while the second calculates the model complexity. Indeed, the use of hybrid indices has the advantage that model error, as well as complexity, may be evaluated simultaneously when several models with different structures are compared with each other.

The classical AIC is a combination of the mean square error (MSE) with a penalty term that considers the complexity of the model. It is defined as follows:

$$\mathrm{AIC} = N \times \log\left(\sqrt{\frac{1}{N}\sum\nolimits_{i=1}^{N} (A_i - E_i)^2}\right) + 2 \times k \tag{3.11}$$

where A_i and E_i are the ith samples of the actual and estimated values (respectively), N is the number of samples, and K is the number of tuned parameters through the training phase. However, while the AIC has recently been applied in a number of different studies (Ebtehaj & Bonakdari, 2016; Ebtehaj, Bonakdari, et al., 2020; Ebtehaj, Zeynoddin, et al., 2020; Safari et al., 2019), the corrected version of this index is also becoming more popular (Azari et al., 2021; Moeeni et al., 2017a, 2017b; Zeynoddin, Bonakdari, et al., 2020; Zeynoddin, Ebtehaj, et al., 2020). The mathematical definition of the AICc is as follows:

$$\mathrm{AICc} = N\ln\left(\sigma_{\varepsilon}^{2}\right) + \frac{2KN}{N-K-1} \tag{3.12}$$

where σ_{ε}^{2} is the residuals' standard deviations.

It can be seen that the differences between these two models extend to both the error and complexity terms. In the classical method, the error value is calculated using MSE, while in AICc, residual standard deviations are used. In the second term, the value of 2k is considered as the complexity of the classical AIC, while in AICc, the effect of K is reduced compared to the classical method. Indeed, the modified method not only uses the standard deviation of errors instead of MSE but also reduces the intensity of the K effect.

3.2.1.5 Coding of the statistical indices

In this subsection, the detailed MATLAB coding for all statistical indices are provided by considering the five example datasets previously introduced. As always, the first step is applying the three main commands, including `clc`, `clear`, and `close all`, which clear the command window, clear all variables in the workspace, and close all open figures, respectively. Following this, the training and testing samples must be loaded (read) into the MATLAB environment from an Excel spreadsheet. The samples of each stage (i.e., train and test) include two columns, target and estimated values. In line 11 of code 3.1.A, the `Train_x1` and `Train_x2` variables contain the actual and estimated samples of Example 1, respectively. In line 13, the number of samples for the training stage is calculated using the `size` command. The first input of this command is the desired file (i.e., `Train_x1` in line 13), while the second input argument is the number of the column of interest. Since the prepared data in the Microsoft Excel file is stored in a column-based arrangement, the column number is defined as one to calculate the number of training samples (which is equivalent to the number of rows). Lines 15–17 are identical to lines 11–13, with the difference that they are related to the test stage. In line 19, the number of tuned variables during the training phase must be defined as `k`. This value depends on the developed model. For example, in the following simple equation, the number of tuned parameters, `k`, is 3.

$$y = ax^2 + bx + c \tag{3.13}$$

where a, b, and c are calculated through the training phase. Then, `k` is 3

In lines 22–23, the `Statistical_Indices` function is applied to calculate the value of different indices for the training and testing stages. To use this function for the training stage, `Train_x1`, `Train_x2`, `Train_N`, and `k` are considered as inputs and correspond to the actual training samples, the estimated training samples, the number of training samples, and the number of tuned variables through the training phase, respectively. Finally, in lines 26–27, all of the generated variables except the statistical indices of the training and testing samples for Examples 1 (i.e., `SI_Ex1_Tr` and `SI_Ex1_Ts`) are removed using the `clearvars -except` command. The coding of lines 7–27 for Example 1 is repeated for Examples 2–5 as Code 3.1.B to 1.36E. The schematic of Code 3.1 is provided in Fig. 3.8.

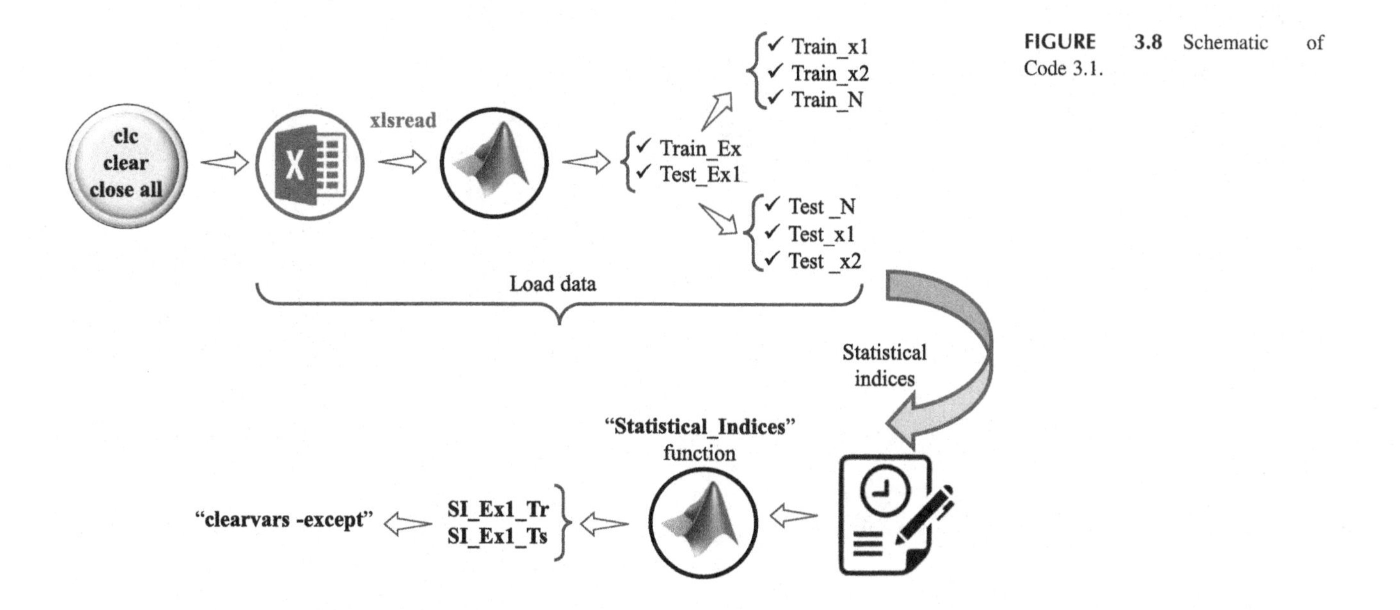

FIGURE 3.8 Schematic of Code 3.1.

Code 3.1.A

```
1   clc
2   clear
3   close all
4
5   %% Example 1
6   % load data
7   Train_Ex1 = xlsread('Data','sheet1');
8   Test_Ex1  = xlsread('Data','sheet2');
9
10  % definition of the input variables for the main function
11  Train_x1 = Train_Ex1(:,1);
12  Train_x2 = Train_Ex1(:,2);
13  Train_N  = size(Train_x1,1);
14
15  Test_x1 = Test_Ex1(:,1);
16  Test_x2 = Test_Ex1(:,2);
17  Test_N  = size(Test_x1,1);
18
19  k = 10; % It is not constant.It could be defined based on the desired
    problem.
20
21  % Statistical indices calculation
22  [SI_Ex1_Tr] = Statistical_Indices(Train_x1,Train_x2,Train_N,k);
23  [SI_Ex1_Ts] = Statistical_Indices(Test_x1,Test_x2,Test_N,k);
24
25  % clear
26  clearvars -except SI_Ex1_Tr ...
27                    SI_Ex1_Ts
```

Code 3.1.B

```
1   %% Example 2
2   % load data
3   Train_Ex2 = xlsread('Data','sheet3');
4   Test_Ex2  = xlsread('Data','sheet4');
5
6   % definition of the input variables for the main function
7   Train_x1 = Train_Ex2(:,1);
8   Train_x2 = Train_Ex2(:,2);
9   Train_N  = size(Train_x1,1);
10
11  Test_x1 = Test_Ex2(:,1);
12  Test_x2 = Test_Ex2(:,2);
13  Test_N  = size(Test_x1,1);
14
15  k = 10; % It is not constant.It could be defined based on the desired
    problem.
16
17  % Statistical indices calculation
18  [SI_Ex2_Tr] = Statistical_Indices(Train_x1,Train_x2,Train_N,k);
19  [SI_Ex2_Ts] = Statistical_Indices(Test_x1,Test_x2,Test_N,k);
20
21  % clear
22  clearvars -except SI_Ex1_Tr SI_Ex2_Tr ...
23                    SI_Ex1_Ts SI_Ex2_Ts
```

Code 3.1.C

```
1   %% Example 3
2   % load data
3   Train_Ex3 = xlsread('Data','sheet5');
4   Test_Ex3  = xlsread('Data','sheet6');
5
6   % definition of the input variables for the main function
7   Train_x1 = Train_Ex3(:,1);
8   Train_x2 = Train_Ex3(:,2);
9   Train_N  = size(Train_x1,1);
10
11  Test_x1 = Test_Ex3(:,1);
12  Test_x2 = Test_Ex3(:,2);
13  Test_N  = size(Test_x1,1);
14
15  k = 10; % It is not constant.It could be defined based on the desired
    problem.
16
17  % Statistical indices calculation
18  [SI_Ex3_Tr] = Statistical_Indices(Train_x1,Train_x2,Train_N,k);
19  [SI_Ex3_Ts] = Statistical_Indices(Test_x1,Test_x2,Test_N,k);
20
21  % clear
22  clearvars -except SI_Ex1_Tr SI_Ex2_Tr SI_Ex3_Tr ...
23                    SI_Ex1_Ts SI_Ex2_Ts SI_Ex3_Ts
```

Code 3.1.D

```
1   %% Example 4
2   % load data
3   Train_Ex4 = xlsread('Data','sheet7');
4   Test_Ex4  = xlsread('Data','sheet8');
5
6   % definition of the input variables for the main function
7   Train_x1 = Train_Ex4(:,1);
8   Train_x2 = Train_Ex4(:,2);
9   Train_N  = size(Train_x1,1);
10
11  Test_x1 = Test_Ex4(:,1);
12  Test_x2 = Test_Ex4(:,2);
13  Test_N  = size(Test_x1,1);
14
15  k = 10; % It is not constant.It could be defined based on the desired
    problem.
16
17  % Statistical indices calculation
18  [SI_Ex4_Tr] = Statistical_Indices(Train_x1,Train_x2,Train_N,k);
19  [SI_Ex4_Ts] = Statistical_Indices(Test_x1,Test_x2,Test_N,k);
20
21  % clear
22  clearvars -except SI_Ex1_Tr SI_Ex2_Tr SI_Ex3_Tr SI_Ex4_Tr ...
23                    SI_Ex1_Ts SI_Ex2_Ts SI_Ex3_Ts SI_Ex4_Ts
```

Code 3.1.E

```
1   %% Example 5
2   % load data
3   Train_Ex5 = xlsread('Data','sheet9');
4   Test_Ex5  = xlsread('Data','sheet10');
5
6   % definition of the input variables for the main function
7   Train_x1 = Train_Ex5(:,1);
8   Train_x2 = Train_Ex5(:,2);
9   Train_N  = size(Train_x1,1);
10
11  Test_x1 = Test_Ex5(:,1);
12  Test_x2 = Test_Ex5(:,2);
13  Test_N  = size(Test_x1,1);
14
15  k = 10; % It is not constant.It could be defined based on the desired
    problem.
16
17  % Statistical indices calculation
18  [SI_Ex5_Tr] = Statistical_Indices(Train_x1,Train_x2,Train_N,k);
19  [SI_Ex5_Ts] = Statistical_Indices(Test_x1,Test_x2,Test_N,k);
20
21  % clear
22  clearvars -except SI_Ex1_Tr SI_Ex2_Tr SI_Ex3_Tr SI_Ex4_Tr SI_Ex5_Tr ...
23                    SI_Ex1_Ts SI_Ex2_Ts SI_Ex3_Ts SI_Ex4_Ts SI_Ex5_Ts
```

After calculation of all indices for each Example, `Index_Train` and `Index_Test` are predefined as zeros matrices in lines 3 and 12 of Code 3.1.F. The index of the training samples is calculated in lines 5–9 of Code 3.1.F for Examples 1–5, respectively. This process is repeated for the testing samples of Examples 1–5 in lines 14–18.

Code 3.1.F

```
1   %% All in one
2   % Train
3   Index_Train = zeros(5,12);
4
5   Index_Train(1,:) = SI_Ex1_Tr;
6   Index_Train(2,:) = SI_Ex2_Tr;
7   Index_Train(3,:) = SI_Ex3_Tr;
8   Index_Train(4,:) = SI_Ex4_Tr;
9   Index_Train(5,:) = SI_Ex5_Tr;
10
11  % Test
12  Index_Test  = zeros(5,12);
13
14  Index_Test(1,:) = SI_Ex1_Ts;
15  Index_Test(2,:) = SI_Ex2_Ts;
16  Index_Test(3,:) = SI_Ex3_Ts;
17  Index_Test(4,:) = SI_Ex4_Ts;
18  Index_Test(5,:) = SI_Ex5_Ts;
```

As mentioned in Code 3.1 (i.e., Codes 3.1.A–3.1.E for Examples 1–5), the statistical indices are calculated using the `Statistical_Indices` function. The details of this function are provided in Code 3.2. As mentioned before, this code has four inputs, including the actual samples (i.e., `x1`), the estimated samples (i.e., `y1`), the number of samples (i.e., `N`), and the number of tuned variables through the training phase (i.e., `k`). It can be seen that Eqs. (3.1–3.12) are easily defined in this code. The values of all indices are stored in the variable `A` which is reported as the output of this function. The schematic of Code 3.2 is provided in Fig. 3.9.

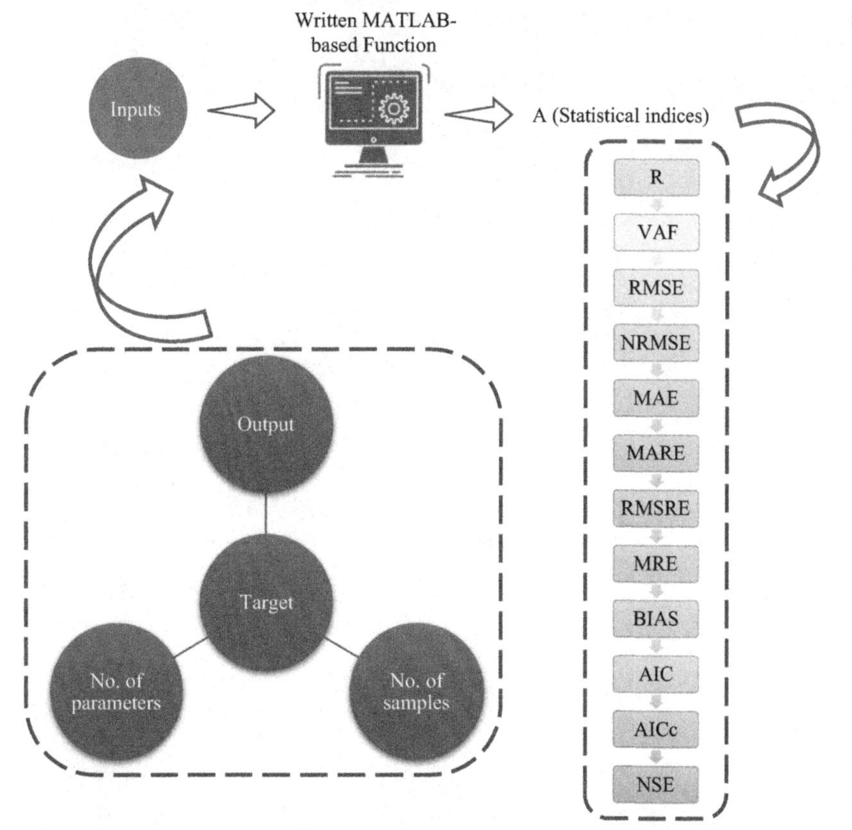

FIGURE 3.9 Schematic of Code 3.2.

Code 3.2

```
1   function [A] = Statistical_Indices(x1,y1,N,k)
2   % x1 = Targets
3   % y1 = Outputs
4   % N  = Number of samples
5   % k  = Number of tunned variables
6
7   Diff=x1-y1;
8   % Correlation coefficient
9   R=((corr2(x1,y1)));
10  % variance accounted for (VAF)
11  VarDiff1=var(Diff);
12  VarX1=var(x1);
13  VAF=(1-((VarDiff1)/(VarX1)));
14  % Root Mean Square Error (RMSE)
15  RMSE=sqrt(mse(x1,y1));
16  % Normalized Root Mean Square Error (NRMSE)
17  NRMSE=(sqrt(mse(x1,y1)))/mean(x1);
18  % Mean Absolute Error (MAE)
19  MAE=mean(abs(x1-y1));
20  % Mean Absolute Percentage Error (MARE)
21  xx=[];
22  for i=1:N
23      if x1(i)==0
24          xx(i)=y1(i);
25      else
26           xx(i)=x1(i);
27      end
28  end
29  xx=xx';
30  MARE=mean(abs(Diff./(abs(xx))));
```

```
31   % Root Mean Squared Relative Error (RMSRE)
32   RMSRE=sqrt(mean((Diff./(xx)).^2));
33   % Mean Relative Error (MRE)
34   MRE=mean(Diff./(xx));
35   % BIAS
36   BIAS=mean(x1-y1);
37   % Akaike Information Criteria (AIC)
38   AIC=N*log10(mse(x1,y1))+2*k;
39   % Corrected Akaike Information Criteria (AICc)
40   AICc=(2*k*N)/(N-k-1)+N*log(var(Diff));
41   % Nash Sutcliffe Efficiency (NSE)
42   AA=mean(Diff.^2);
43   BB=mean((x1-mean(x1)).^2);
44   NSE=1-((AA)/(BB));
45
46   A=zeros;
47   A(1,1)=R;
48   A(1,2)=VAF;
49   A(1,3)=RMSE;
50   A(1,4)=NRMSE;
51   A(1,5)=MAE;
52   A(1,6)=MARE;
53   A(1,7)=RMSRE;
54   A(1,8)=MRE;
55   A(1,9)=BIAS;
56   A(1,10)=AIC ;
57   A(1,11)=AICc;
58   A(1,12)=NSE ;
```

Data on statistical indices can be found at Appendix 3A.

3.2.2 Uncertainty analysis

In addition to statistical indices, UA is also a quantitative analysis method that evaluates the uncertainties of the estimation generated using ML models. The primary objective of UA is to limit the expected range in which the actual values of the result lies. UA is typically presented as an estimated range, which is known as an uncertainty interval. This interval is determined using the calculated errors between the observed values and those estimated by the developed model. One of the most well-known approaches to calculating the uncertainty interval is the U95 method (Ai et al., 2022; Azimi et al., 2018; Bonakdari et al., 2020; Zhang et al., 2022). The concept of U95 can be expressed in such a way that if we repeat an experiment for a large number of times, the true value of the result obtained from that experiment is present within the uncertainty interval is approximately 95 times out of 100 trials. The U95 is defined as follows:

$$U95 = \left(\frac{1.96}{N}\right)\sqrt{\sum_{i=1}^{N}(A_i-\overline{A})^2 - \sum_{i=1}^{N}(A_i-E_i)^2} \tag{3.14}$$

where A_i and E_i are the ith samples of the actual and estimated values (respectively), $\overline{A}$ is the mean of the actual samples, and N is the number of samples.

The details of the UA are provided in Code 3.3. To clear the command window and all variables in the workspace as well as to close all open figures, the `clc`, `clear`, and `close all` commands are applied in lines 1–3. In lines 5–30, UA beginning with loading the dataset and finishing with the calculation of U95, is provided for both the train and test stages of Example 1. Using the `xlsread` command, the training and testing data of Example 1 are loaded (read) from a Microsoft Excel file with a filename of "Data." The data contained within this file consists of ten different sheets, so that sheet 1 is related to the actual and estimated values of the training samples of Example 1, and sheet 2 includes of actual and estimated values of the testing samples of Example 1. In the same way, sheets 3 and 4 are related to Example 2, sheets 5 and 6 are related to Example 3, sheets 7 and 8 are related to Example 4, and sheets 9 and 10 are related to Example 5. The training and testing data of Example 1 are stored in the variables `Train_Ex1` and `Test_Ex1`, respectively. The newly generated variables contain two columns, the first of which are the actual values, while the second is the estimated ones. Using these explanations, the `Train_x1`, `Train_x2`, `Test_x1`, and `Test_x2` variables are defined in lines 11, 12, 15, 16, respectively. The number of training and testing samples are calculated using the `size` command as provided in lines 13 and 17, respectively. The `size` command includes two inputs: (1) the name of the desired variables (i.e., `Train_x1` as training samples of Example 1, or `Test_x1` as testing samples of Example 1), and (2) the number of the row or column to be considered (i.e., 1 or 2 for Example 1). Because the Microsoft Excel file (i.e., "Data") is a matrix with two columns and a number of rows equal to the number of samples, the second input of the `size` command was considered as 1.

After loading data and separating the actual and estimated values at both training and test stages, the U95 is calculated. Eq. (3.14) is coded as lines 20–26 to calculate U95 for the training and testing stages. Finally, all of the generated variables except `UA_Train_Ex1` and `UA_Test_Ex1` are removed using the `clearvars -except` command. This process is repeated for Examples 2–5 in Code 3.3.B to 1.38.E.

After calculating the U95 for all examples in both training and testing stages, the results are saved in one variable (i.e., `UA_All`), which contains two rows and five columns. The first row is for the U95 at the training stage, while the second one is for the U95 at the testing stage. The first to fifth columns are related to Examples 1–5, respectively. The schematic of Code 3.3 is provided in Fig. 3.10.

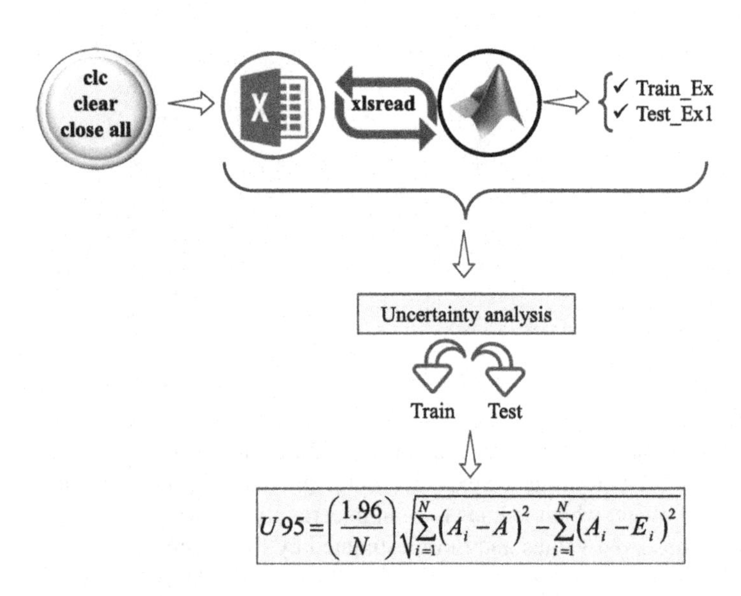

FIGURE 3.10 Schematic of Code 3.3.

Code 3.3.A

```
1  clc
2  clear
3  close all
4  
5  %% Example 1
6  % load data
7  Train_Ex1 = xlsread('Data','sheet1'); % Train data
8  Test_Ex1  = xlsread('Data','sheet2'); % Test data
9  
10 % definition of the input variables  for the main function
11 Train_x1 = Train_Ex1(:,1);
12 Train_x2 = Train_Ex1(:,2);
13 Train_N  = size(Train_x1,1);
14 
15 Test_x1 = Test_Ex1(:,1);
16 Test_x2 = Test_Ex1(:,2);
17 Test_N  = size(Test_x1,1);
18 
19 % Uncertainty analysis
20 A = sum((Train_x1-mean(Train_x1)).^2);
21 B = sum((Train_x1-Train_x2).^2);
22 UA_Train_Ex1 = (1.96/Train_N)*sqrt((A)+(B));
23 
24 A = sum((Test_x1-mean(Test_x1)).^2);
25 B = sum((Test_x1-Test_x2).^2);
26 UA_Test_Ex1 = (1.96/Test_N)*sqrt((A)+(B));
27 
28 % clear
29 clearvars -except UA_Train_Ex1 ...
30                   UA_Test_Ex1
```

Code 3.3.B

```
1   %% Example 2
2   % load data
3   Train_Ex2 = xlsread('Data','sheet3'); % Train data
4   Test_Ex2  = xlsread('Data','sheet4'); % Test data
5
6   % definition of the input variables for the main function
7   Train_x1 = Train_Ex2(:,1);
8   Train_x2 = Train_Ex2(:,2);
9   Train_N  = size(Train_x1,1);
10
11  Test_x1 = Test_Ex2(:,1);
12  Test_x2 = Test_Ex2(:,2);
13  Test_N  = size(Test_x1,1);
14
15  % Uncertainty analysis
16  A = sum((Train_x1-mean(Train_x1)).^2);
17  B = sum((Train_x1-Train_x2).^2);
18  UA_Train_Ex2 = (1.96/Train_N)*sqrt((A)+(B));
19
20  A = sum((Test_x1-mean(Test_x1)).^2);
21  B = sum((Test_x1-Test_x2).^2);
22  UA_Test_Ex2 = (1.96/Test_N)*sqrt((A)+(B));
23
24  % clear
25  clearvars -except UA_Train_Ex1 UA_Train_Ex2 ...
26                    UA_Test_Ex1  UA_Test_Ex2
```

Code 3.3.C

```
1   %% Example 3
2   % load data
3   Train_Ex3 = xlsread('Data','sheet5'); % Train data
4   Test_Ex3  = xlsread('Data','sheet6'); % Test data
5
6   % definition of the input variables for the main function
7   Train_x1 = Train_Ex3(:,1);
8   Train_x2 = Train_Ex3(:,2);
9   Train_N  = size(Train_x1,1);
10
11  Test_x1 = Test_Ex3(:,1);
12  Test_x2 = Test_Ex3(:,2);
13  Test_N  = size(Test_x1,1);
14
15  % Uncertainty analysis
16  A = sum((Train_x1-mean(Train_x1)).^2);
17  B = sum((Train_x1-Train_x2).^2);
18  UA_Train_Ex3 = (1.96/Train_N)*sqrt((A)+(B));
19
20  A = sum((Test_x1-mean(Test_x1)).^2);
21  B = sum((Test_x1-Test_x2).^2);
22  UA_Test_Ex3 = (1.96/Test_N)*sqrt((A)+(B));
23
24  clearvars -except UA_Train_Ex1 UA_Train_Ex2 UA_Train_Ex3 ...
25                    UA_Test_Ex1  UA_Test_Ex2  UA_Test_Ex3
```

Code 3.3.D

```
1   %% Example 4
2   % load data
3   Train_Ex4 = xlsread('Data','sheet7'); % Train data
4   Test_Ex4  = xlsread('Data','sheet8'); % Test data
5
6   % definition of the input variables for the main function
7   Train_x1 = Train_Ex4(:,1);
8   Train_x2 = Train_Ex4(:,2);
9   Train_N  = size(Train_x1,1);
10
11  Test_x1 = Test_Ex4(:,1);
12  Test_x2 = Test_Ex4(:,2);
13  Test_N  = size(Test_x1,1);
14
15  % Uncertainty analysis
16  A = sum((Train_x1-mean(Train_x1)).^2);
17  B = sum((Train_x1-Train_x2).^2);
18  UA_Train_Ex4 = (1.96/Train_N)*sqrt((A)+(B));
19
20  A = sum((Test_x1-mean(Test_x1)).^2);
21  B = sum((Test_x1-Test_x2).^2);
22  UA_Test_Ex4 = (1.96/Test_N)*sqrt((A)+(B));
23
24  clearvars -except UA_Train_Ex1 UA_Train_Ex2 UA_Train_Ex3 UA_Train_Ex4 ...
25                    UA_Test_Ex1  UA_Test_Ex2  UA_Test_Ex3  UA_Test_Ex4
```

Code 3.3.E

```
1   %% Example 5
2   % load data
3   Train_Ex5 = xlsread('Data','sheet9');  % Train data
4   Test_Ex5  = xlsread('Data','sheet10'); % Test data
5
6   % definition of the input variables for the main function
7   Train_x1 = Train_Ex5(:,1);
8   Train_x2 = Train_Ex5(:,2);
9   Train_N  = size(Train_x1,1);
10
11  Test_x1 = Test_Ex5(:,1);
12  Test_x2 = Test_Ex5(:,2);
13  Test_N  = size(Test_x1,1);
14
15  % Uncertainty analysis
16  A = sum((Train_x1-mean(Train_x1)).^2);
17  B = sum((Train_x1-Train_x2).^2);
18  UA_Train_Ex5 = (1.96/Train_N)*sqrt((A)+(B));
19
20  A = sum((Test_x1-mean(Test_x1)).^2);
21  B = sum((Test_x1-Test_x2).^2);
22  UA_Test_Ex5 = (1.96/Test_N)*sqrt((A)+(B));
23
24  clearvars -except UA_Train_Ex1 UA_Train_Ex2 UA_Train_Ex3 UA_Train_Ex4
    UA_Train_Ex5 ...
25                    UA_Test_Ex1  UA_Test_Ex2  UA_Test_Ex3  UA_Test_Ex4
    UA_Test_Ex5
```

Code 3.3.F

```
1  %% All in one
2  % Train
3  UA_All = zeros(2,5);
4
5  UA_All(1,1) = UA_Train_Ex1; % Train - Ex. 1
6  UA_All(1,2) = UA_Train_Ex2; % Train - Ex. 2
7  UA_All(1,3) = UA_Train_Ex3; % Train - Ex. 3
8  UA_All(1,4) = UA_Train_Ex4; % Train - Ex. 4
9  UA_All(1,5) = UA_Train_Ex5; % Train - Ex. 5
10
11 UA_All(2,1) = UA_Test_Ex1; % Test - Ex. 1
12 UA_All(2,2) = UA_Test_Ex2; % Test - Ex. 2
13 UA_All(2,3) = UA_Test_Ex3; % Test - Ex. 3
14 UA_All(2,4) = UA_Test_Ex4; % Test - Ex. 4
15 UA_All(2,5) = UA_Test_Ex5; % Test - Ex. 5
```

Data on uncertainty analysis can be found at Appendix 3A.

3.2.3 Reliability analysis

Another quantitative statistical method that could be applied to measure the consistency of a model is reliability analysis. Indeed, the result of this analysis shows whether the model has reached an acceptable level of performance or not.

$$\mathrm{RA} = \left(\frac{\sum_{i=1}^{N} k_i}{N}\right) \tag{3.15}$$

where RA is the reliability analysis, N is the number of samples, and k is calculated as follows:

$$\begin{cases} k_i = 1 \ \mathrm{RAE}_i \leq \Delta \\ k_i = 0 \ \text{Otherwise} \end{cases} \tag{3.16}$$

where RAE is the relative average error, Δ is the acceptable threshold value of the desired parameter. The value of Δ is one that could be changed depending on the nature of the different problems encountered. For example, the value of Δ is 0.2 based on the Chinese standards in water quality parameters (Saberi-Movahed et al., 2020). The RAE is calculated as follows:

$$\mathrm{RAE}_i = \left|\frac{A_i - E_i}{A_i}\right| \tag{3.17}$$

where A_i and E_i are the ith samples of the actual and estimated values, respectively.

The coding details of the reliability analysis are provided in Code 3.4. Similar to all of the provided MATLAB-based code in this book, the `clc`, `clear`, and `close all` commands are applied to clear the command window, clear all variables in the workspace, and clear all open figures, respectively (i.e., lines 1–3 Code 3.4.A). The first step after data loading is the definition of the Δ. Since the value of this variable is not constant in all problems, the user should assign its value before proceeding with the reliability analysis. To define the value of Δ, the `inputdlq` command is employed (i.e., line 9 of Code 3.4.A) to generate a dialog box where the user may enter the desired values. The first input argument,`Prompt`, is used to specify the character array, which describes the edit field for the user. In line 6, it can be seen that the character array is defined as "Please define the threshold value of the acceptable desired parameter." In lines 7 and 8, the `Title` and `defaultValues` are defined. The default input is identified in the `DefaultValues` variable in line 8. The [1, 72] seen in line 9 indicates that each entry field is one line in height with a length of 72 characters. The window that the user is prompted by Code 3.4.A is provided in Fig. 3.11. The value input by the user is stored in the variable `Delta` using the `str2double` command (i.e., line 12 of Code 3.4.A). The schematic of Code 3.4.A is presented in Fig. 3.12.

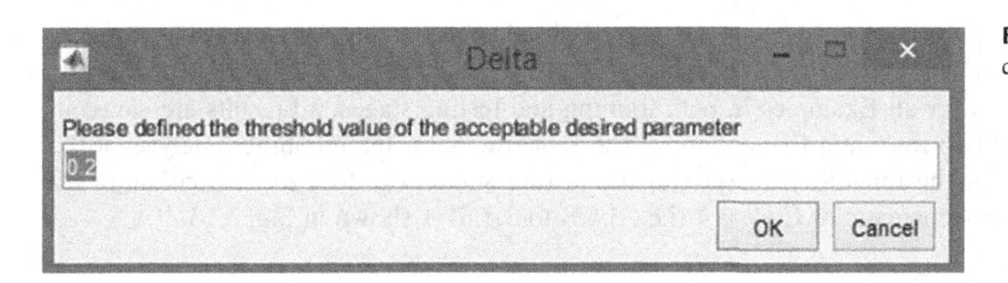

FIGURE 3.11 The results of the `inputdlq` command for reliability analysis.

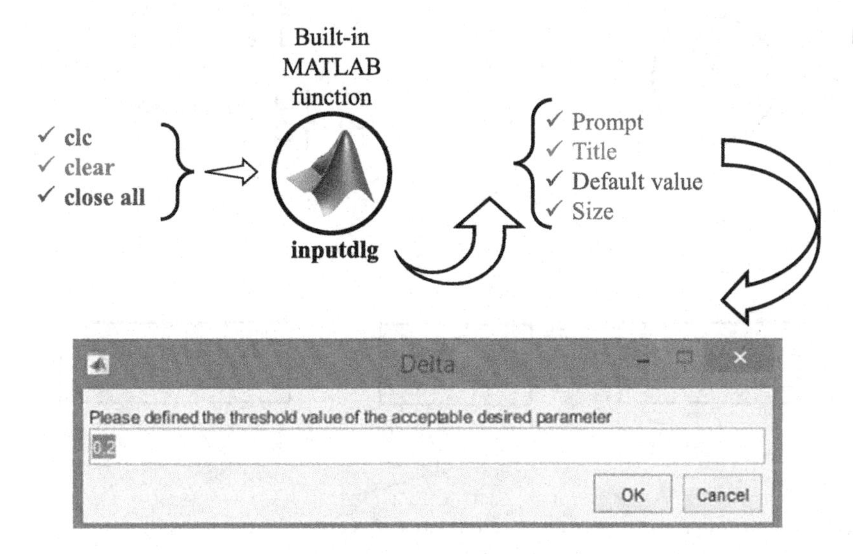

FIGURE 3.12 Schematic of Code 3.4.A.

Code 3.4.A

```
1   clc
2   clear
3   close all
4
5   %% Delta
6   Prompt={'Please define the threshold value of the acceptable desired
    parameter'};
7       Title='Delta';
8       DefaultValues={'0.2'};
9       PARAMS=inputdlg(Prompt,Title,[1, 72],DefaultValues);
10      A=PARAMS{1};
11
12  Delta = str2double(A);
```

The next step is data loading. Based on Code 3.4.B, the data are read from a Microsoft Excel file using the `xlsread` function. In lines 3 and 4, the training and testing samples are loaded from sheet1 and sheet2 (respectively) of the Microsoft Excel file (i.e., "Data") and stored in `Train_Ex1` and `Test_Ex1`, respectively. As can be seen in lines 7–10, the first column of `Train_Ex1` and `Test_Ex1` is related to the target samples, while the second column in both variables is related to the outputs (i.e., the results of the modeling).

For the reliability analysis step, the RAE is first calculated using Eq. (3.17). Using the RAE, the value of k in Eq. (3.15) is calculated using the condition outlined by Eq. (3.16) through a `for` loop in lines 29–35. Finally, `R_Train_Ex1` is calculated as the reliability of the training samples of Example 1. From lines 39–50, the reliability of the testing samples of Example 1 is also calculated in the same manner as the training samples by replacing the input training sample data with the testing samples. After calculation of the reliability for the training and testing stages, all of the generated variables except `R_Train_Ex1` and `R_Test_Ex1` are removed using the `clearvars -except` command. This process is repeated for Examples 2–5.

After calculating the reliability analysis for all Examples in both training and testing stages, all results are saved in one variable (i.e., `R_All`), which contains two rows and five columns. The first row is for the reliability analysis at the training stage, while the second one is for the reliability analysis at the testing stage. The first to fifth columns are related to Examples 1–5, respectively. The schematic of Code 3.4 (i.e., 3.4.B to 3.4.F) is shown in Fig. 3.13.

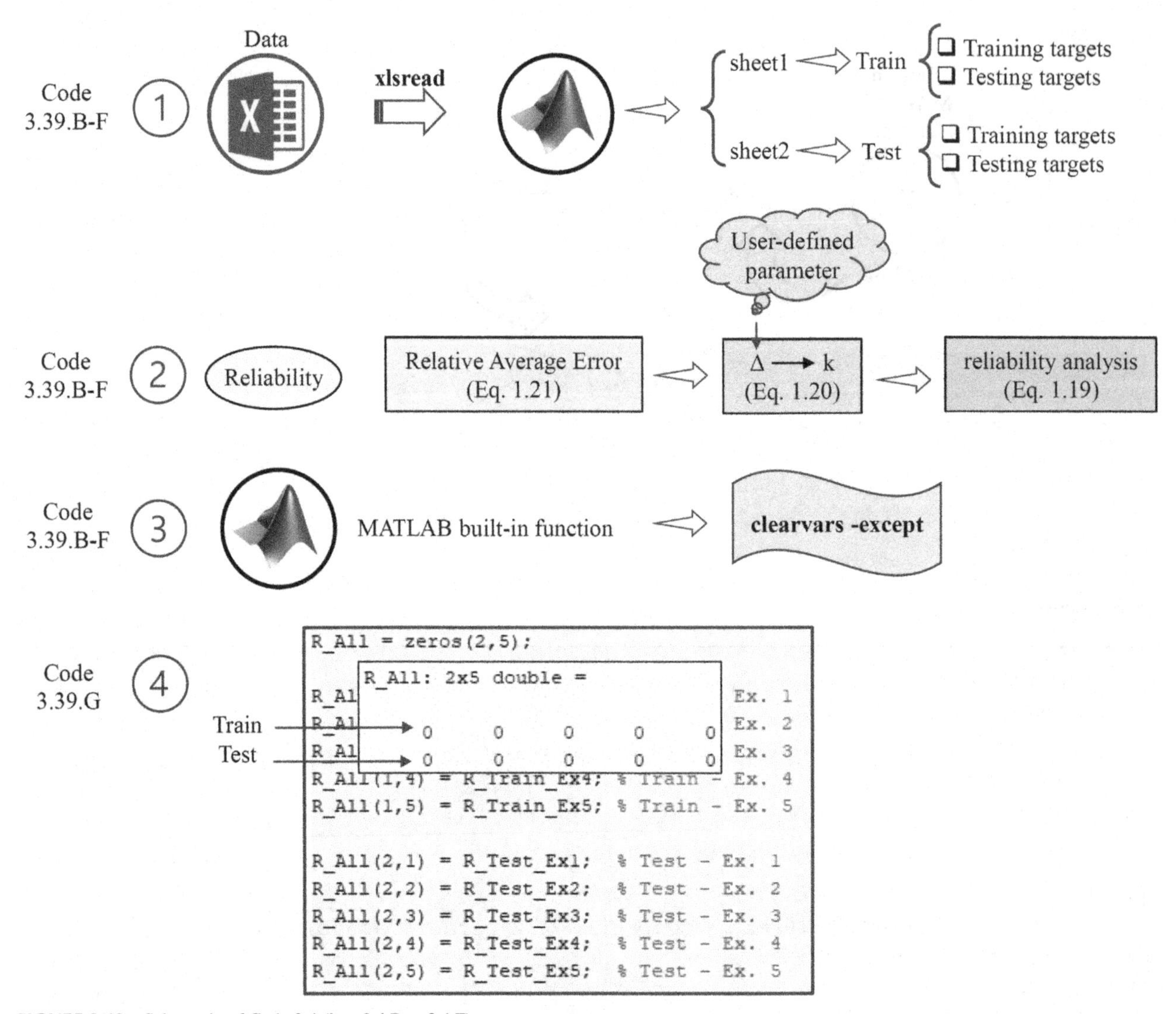

FIGURE 3.13 Schematic of Code 3.4 (i.e., 3.4.B to 3.4.F).

Code 3.4.B

```
1   %% Example 1
2   % load data
3   Train_Ex1 = xlsread('Data','sheet1'); % Train data
4   Test_Ex1  = xlsread('Data','sheet2'); % Test data
5
6   % definition of the input variables  for the main function
7   Train_x1 = Train_Ex1(:,1); % Training Targets
8   Train_x2 = Train_Ex1(:,2); % Training Outputs
9   Test_x1 = Test_Ex1(:,1);   % Testing Targets
10  Test_x2 = Test_Ex1(:,2);   % Testing Outputs
11
12  % Reliability
13  RAE = abs((Train_x1-Train_x2)./Train_x1);
14
15  k=zeros(length(RAE),1);
16  for i=1:length(RAE)
17      if RAE(i) <= Delta
18          k(i) = 1;
19      else
20          k(i) = 0;
21      end
22  end
23
24  R_Train_Ex1 = mean(k)*100;
25
26  RAE = abs((Test_x1-Test_x2)./Test_x1);
27
28  k=zeros(length(RAE),1);
29  for i=1:length(RAE)
30      if RAE(i) <= Delta
31          k(i) = 1;
32      else
33          k(i) = 0;
34      end
35  end
36
37  R_Test_Ex1 = mean(k)*100;
38
39  % clear
40  clearvars -except R_Train_Ex1 ...
41                    R_Test_Ex1 ...
42                    Delta
```

Code 3.4.C

```
1   %% Example 2
2   % load data
3   Train_Ex2 = xlsread('Data','sheet3'); % Train data
4   Test_Ex2  = xlsread('Data','sheet4'); % Test data
5
6   % definition of the input variables  for the main function
7   Train_x1 = Train_Ex2(:,1); % Training Targets
8   Train_x2 = Train_Ex2(:,2); % Training Outputs
9   Test_x1 = Test_Ex2(:,1);    % Testing Targets
10  Test_x2 = Test_Ex2(:,2);    % Testing Outputs
11
12  % Reliability
13  RAE = abs((Train_x1-Train_x2)./Train_x1);
14
15  k=zeros(length(RAE),1);
16  for i=1:length(RAE)
17      if RAE(i) <= Delta
18          k(i) = 1;
19      else
20          k(i) = 0;
21      end
22  end
23
24  R_Train_Ex2 = mean(k)*100;
25
26  RAE = abs((Test_x1-Test_x2)./Test_x1);
27
28  k=zeros(length(RAE),1);
29  for i=1:length(RAE)
30      if RAE(i) <= Delta
31          k(i) = 1;
32      else
33          k(i) = 0;
34      end
35  end
36
37  R_Test_Ex2 = mean(k)*100;
38
39  % clear
40  clearvars -except R_Train_Ex1 R_Train_Ex2 ...
41                    R_Test_Ex1  R_Test_Ex2 ...
42                    Delta
```

Code 3.4.D

```
1   %% Example 3
2   % load data
3   Train_Ex3 = xlsread('Data','sheet5'); % Train data
4   Test_Ex3  = xlsread('Data','sheet6'); % Test data
5
6   % definition of the input variables  for the main function
7   Train_x1 = Train_Ex3(:,1); % Training Targets
8   Train_x2 = Train_Ex3(:,2); % Training Outputs
9   Test_x1 = Test_Ex3(:,1);    % Testing Targets
10  Test_x2 = Test_Ex3(:,2);    % Testing Outputs
11
12  % Reliability
13  RAE = abs((Train_x1-Train_x2)./Train_x1);
14
15  k=zeros(length(RAE),1);
16  for i=1:length(RAE)
17      if RAE(i) <= Delta
18          k(i) = 1;
19      else
20          k(i) = 0;
21      end
22  end
23
24  R_Train_Ex3 = mean(k)*100;
25
26  RAE = abs((Test_x1-Test_x2)./Test_x1);
27
28  k=zeros(length(RAE),1);
29  for i=1:length(RAE)
30      if RAE(i) <= Delta
31          k(i) = 1;
32      else
33          k(i) = 0;
34      end
35  end
36
37  R_Test_Ex3 = mean(k)*100;
38
39  % clear
40  clearvars -except R_Train_Ex1 R_Train_Ex2 R_Train_Ex3 ...
41                    R_Test_Ex1  R_Test_Ex2  R_Test_Ex3 ...
42                    Delta
```

Code 3.4.E

```
1   %% Example 4
2   % load data
3   Train_Ex4 = xlsread('Data','sheet7'); % Train data
4   Test_Ex4  = xlsread('Data','sheet8'); % Test data
5   
6   % definition of the input variables  for the main function
7   Train_x1 = Train_Ex4(:,1); % Training Targets
8   Train_x2 = Train_Ex4(:,2); % Training Outputs
9   Test_x1 = Test_Ex4(:,1);    % Testing Targets
10  Test_x2 = Test_Ex4(:,2);    % Testing Outputs
11  
12  % Reliability
13  RAE = abs((Train_x1-Train_x2)./Train_x1);
14  
15  k=zeros(length(RAE),1);
16  for i=1:length(RAE)
17      if RAE(i) <= Delta
18          k(i) = 1;
19      else
20          k(i) = 0;
21      end
22  end
23  
24  R_Train_Ex4 = mean(k)*100;
25  
26  RAE = abs((Test_x1-Test_x2)./Test_x1);
27  
28  k=zeros(length(RAE),1);
29  for i=1:length(RAE)
30      if RAE(i) <= Delta
31          k(i) = 1;
32      else
33          k(i) = 0;
34      end
35  end
36  
37  R_Test_Ex4 = mean(k)*100;
38  
39  % clear
40  clearvars -except R_Train_Ex1 R_Train_Ex2 R_Train_Ex3 R_Train_Ex4 ...
41                    R_Test_Ex1  R_Test_Ex2  R_Test_Ex3  R_Test_Ex4 ...
42                    Delta
```

Code 3.4.F

```
1   %% Example 5
2   % load data
3   Train_Ex5 = xlsread('Data','sheet9');  % Train data
4   Test_Ex5  = xlsread('Data','sheet10'); % Test data
5
6   % definition of the input variables  for the main function
7   Train_x1 = Train_Ex5(:,1); % Training Targets
8   Train_x2 = Train_Ex5(:,2); % Training Outputs
9   Test_x1 = Test_Ex5(:,1);    % Testing Targets
10  Test_x2 = Test_Ex5(:,2);    % Testing Outputs
11
12  % Reliability
13  RAE = abs((Train_x1-Train_x2)./Train_x1);
14
15  k=zeros(length(RAE),1);
16  for i=1:length(RAE)
17      if RAE(i) <= Delta
18          k(i) = 1;
19      else
20          k(i) = 0;
21      end
22  end
23
24  R_Train_Ex5 = mean(k)*100;
25
26  RAE = abs((Test_x1-Test_x2)./Test_x1);
27
28  k=zeros(length(RAE),1);
29  for i=1:length(RAE)
30      if RAE(i) <= Delta
31          k(i) = 1;
32      else
33          k(i) = 0;
34      end
35  end
36
37  R_Test_Ex5 = mean(k)*100;
38
39  % clear
40  clearvars -except R_Train_Ex1 R_Train_Ex2 R_Train_Ex3 R_Train_Ex4
    R_Train_Ex5 ...
41                    R_Test_Ex1  R_Test_Ex2  R_Test_Ex3  R_Test_Ex4
    R_Test_Ex5 ...
42                    Delta
```

Code 3.4.G

```
1    %% All in one
2    % Train
3    R_All = zeros(2,5);
4
5    R_All(1,1) = R_Train_Ex1; % Train - Ex. 1
6    R_All(1,2) = R_Train_Ex2; % Train - Ex. 2
7    R_All(1,3) = R_Train_Ex3; % Train - Ex. 3
8    R_All(1,4) = R_Train_Ex4; % Train - Ex. 4
9    R_All(1,5) = R_Train_Ex5; % Train - Ex. 5
10   % Test
11   R_All(2,1) = R_Test_Ex1;  % Test - Ex. 1
12   R_All(2,2) = R_Test_Ex2;  % Test - Ex. 2
13   R_All(2,3) = R_Test_Ex3;  % Test - Ex. 3
14   R_All(2,4) = R_Test_Ex4;  % Test - Ex. 4
15   R_All(2,5) = R_Test_Ex5;  % Test - Ex. 5
```

Data on reliability analysis can be found at Appendix 3A.

3.3 Qualitative tools

The qualitative approaches presented in this section are categorized into five groups: Taylor diagram, scatter plot, barplot, histogram, and boxplot (Fig. 3.14). The details of each approach, along with their MATLAB codings, are provided in the following subsections.

3.3.1 Taylor diagram

The Taylor diagram introduced by Taylor (2001) is a graphical method to compare different models in terms of standard deviation, RMSE, and correlation coefficient (*R*) simultaneously (Anwar et al., 2019; Zeynolabedin et al., 2021). Often, the comparison of several models using a number of different statistical indices simultaneously may not provide a clear answer as to the superiority of a particular model—the Taylor diagram was proposed to help solve this problem. For example, consider the situation where three indices *x*, *y*, and *z* are used to compare models A, B, and C. Comparing

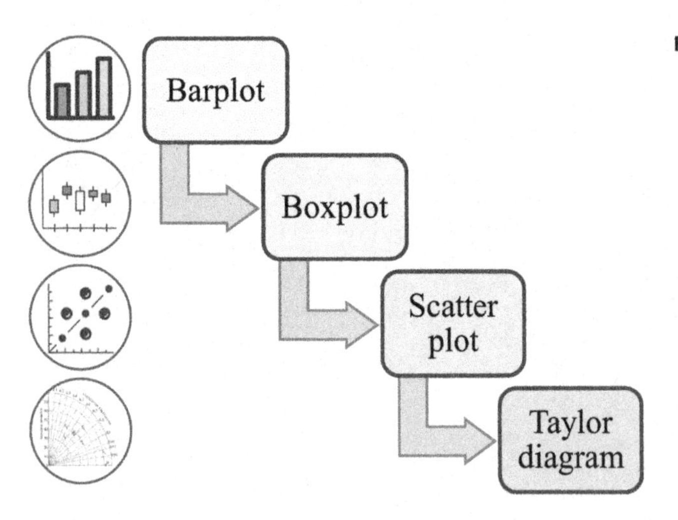

FIGURE 3.14 The presented qualitative approaches.

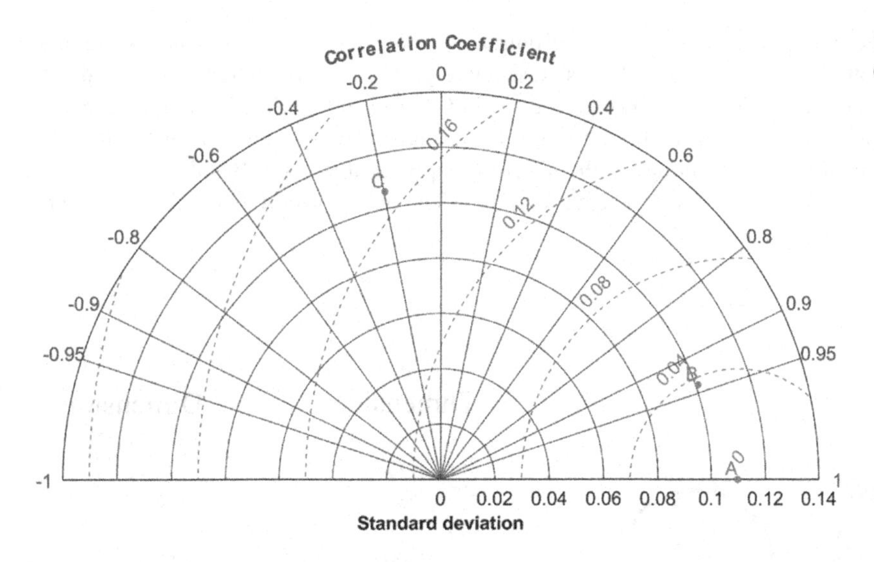

FIGURE 3.15 An example of the Taylor diagram.

these three models using the expressed indices may not give the same result. For example, model A has the best performance using the y index, and also models B and C has the best performance using the *z* and *x* indices. Indeed, each model offers the best performance in only one index. To help determine the overall best model, the Taylor diagram considers the three different indices simultaneously.

The Taylor diagram is structured as a semicircle, with each quarter circle showing negative and positive correlations. The standard deviation values are plotted as concentric circles from the center of the diagram, while the RMSE is plotted as concentric circles centered on the reference point on the x-axis. The reference point indicates the position of the observation based on its standard deviation. Because the value of RMSE and the correlation coefficient of observations in comparison with itself will be zero and one, respectively, the position of the reference point on the horizontal axis will be determined based on the value of the standard deviation. The positions of the studied data are plotted on the diagram based on RMSE, standard deviation, and correlation coefficient, with observations closer to the reference point, indicating a more accurate estimated target parameter and, therefore, a more appropriate model. An example of a Taylor diagram is plotted in Fig. 3.15. In this is the reference point, while the numbers 0, 0.04, 0.08, 0.12, etc., in green, are related to the RMSE value. Points B and C are the corresponding value of two models with positive and negative (respectively) correlation coefficients.

The coding details required to generate Taylor diagrams are provided in Code 3.5. After clearing the command window and workspace as well as closing all opened figures using the `clc`, `clear` and `close all` commands, the first step is data loading, which is performed using the `xlsread` command (line 7). The Excel filename and sheetname are referred to as "Data" and "Sheet1," respectively. Contained within each sheet are the actual (observed) values as well as the results of the four models. Sheets 1–5 are related to Examples 1–5, respectively.

In line 8, the loaded dataset from the Excel file "Data" is transposed and saved into the variable `BUOY`. Indeed, in the "Data" file, the data was stored as a matrix with NV columns and NS rows, where NV and NS are the numbers of variables and the number of samples, respectively. In contrast, `BUOY` is a matrix with NV rows and NS columns. In lines 11–16, the average, RMSE, correlation coefficient, and standard deviation of all models are calculated using the `allstats` functions. In lines 18–31, the `inputdlg` command is used to read the maximum values and intervals of the standard deviation and RMSE, which is assigned by the modeler. The details of the`inputdlg` command were provided in the previous section (i.e., Section 1.6.1.3). The final step, plotting the Taylor diagram, is provided in lines 33–40. Since the number of models that may be compared within the Taylor diagram could be different for any given problem, the `alphab` is defined from A to Z, allowing the diagram to support up to 25 different models. The first entry of the `alphab` is related to observed values, while the rest (i.e., B to Z) are applied as the name of each model under consideration. The Taylor diagram is finally plotted using the `taylordiag` function. The main format of this function is `[hp ht axl]=taylordiag(STDs,RMSs,CORs,['option',value])`. The `hp`, `ht`, and `axl` are the output of the function, while `STDs`, `RMSs`, `CORs`, and `['option',value]` are its inputs. `STDs`, `RMSs`, and `CORs`, are the standard deviation, RMSE, and correlation coefficient, respectively. All of these indices, which are defined as the inputs of the `taylordiag` function, have the same length and are one-dimensional. The first value of each index is related to the reference point (i.e., actual samples). As we see in line 12, the `for` loop begins in position 2. The outputs of this function, `ht`, `hp`, and `axl` return handles of the text legend of points, plotted points, and axis labels, respectively. These steps, which are contained within lines 5–40, are repeated for the other Examples (i.e., Examples 2–5) in Codes 3.5.B–3.5.E. It should be noted that the main challenge in using the Taylor diagram code is the definition of the maximum and interval of the RMSE and standard deviations by the user that could find simplicity by them before applying this code. Besides, the results of them are also calculated by the code in line 13 by the `allstats` function. The schematic of Code 3.5 is shown in Fig. 3.16.

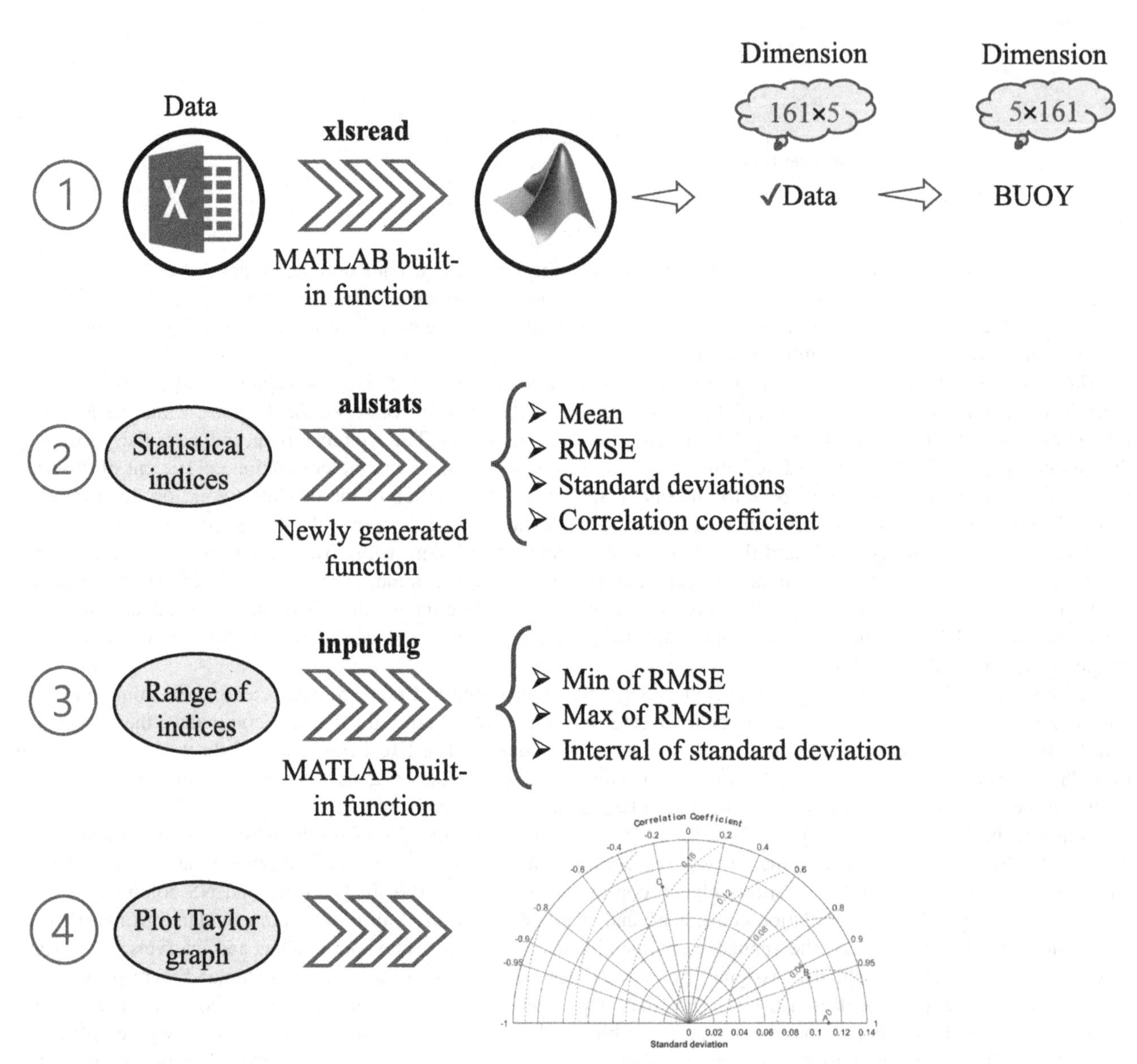

FIGURE 3.16 Schematic of Code 3.5.

Code 3.5.A

```
1   clc
2   clear
3   close all
4   
5   %% Example 1
6   % Load data
7   Data = xlsread('Data','sheet1'); % Example 1
8   BUOY=Data';
9   
10  % Get statistics from the dataset
11  statm = zeros(size(Data,2),4);
12  for ii = 2:size(BUOY,1)
13      C = allstats(BUOY(1,:),BUOY(ii,:));
14      statm(ii,:) = C(:,2);
15  end
16  statm(1,:) = C(:,1);
17  
18  % Assign range of RMSE, Correlation coefficient and standard deviation
19  Prompt={'Max. RMSE','Interval of the RMSE','Max. of the standard
    deviation','Interval of the standard deviation'};
20      Title='Max. of statistical indices';
21      DefaultValues={'0.2','0.01','0.14','0.02'};  % 0.2','0.01
22      PARAMS=inputdlg(Prompt,Title,[1, 50],DefaultValues);
23      A=PARAMS{1};
24      B=PARAMS{2};
25      C=PARAMS{3};
26      D=PARAMS{4};
27  
28  RMSE_Max = str2double(A);
29  RMSE_Interval = str2double(B);
30  STD_Max = str2double(C);
31  STD_Interval = str2double(D);
32  
33  % Plot:
34  figure
35  alphab = 'ABCDEFGHIJKLMNOPQRSTUVWXYZ';
36  
37  [pp, tt, axl] =
    taylordiag(squeeze(statm(:,2)),squeeze(statm(:,3)),squeeze(statm(:,4)),..
    .
38  'tickRMS',0:RMSE_Interval:RMSE_Max,'titleRMS',RMSE_Max,'tickRMSangle',135
    ,'titleRMS',0,'showlabelsRMS',0,'widthRMS',1,...
39              'tickSTD',0:STD_Interval:STD_Max,'limSTD',STD_Max,...
40              'tickCOR',[.1:.1:.9 .95 0.97 0.98
    .99],'showlabelsCOR',1,'titleCOR',1);
```

Code 3.5.B

```
1   %% Example 2
2   % Load data
3   Data = xlsread('Data','sheet2');  % Example 2
4   BUOY=Data';
5
6   % Get statistics from the dataset
7   statm = zeros(size(Data,2),4);
8   for ii = 2:size(BUOY,1)
9       C = allstats(BUOY(1,:),BUOY(ii,:));
10      statm(ii,:) = C(:,2);
11  end
12  statm(1,:) = C(:,1);
13
14  % Assign range of RMSE, Correlation coefficient and standard deviation
15  Prompt={'Max. RMSE','Interval of the RMSE','Max. of the standard
    deviation','Interval of the standard deviation'};
16      Title='Max. of statistical indices;
17      DefaultValues={'14','1','14','2'};
18      PARAMS=inputdlg(Prompt,Title,[1, 50],DefaultValues);
19      A=PARAMS{1};
20      B=PARAMS{2};
21      C=PARAMS{3};
22      D=PARAMS{4};
23
24  RMSE_Max = str2double(A);
25  RMSE_Interval = str2double(B);
26  STD_Max = str2double(C);
27  STD_Interval = str2double(D);
28
29  % Plot:
30  figure
31  alphab = 'ABCDEFGHIJKLMNOPQRSTUVWXYZ';
32
    [pp, tt, axl] =
33  taylordiag(squeeze(statm(:,2)),squeeze(statm(:,3)),squeeze(statm(:,4)),..
    .

34  'tickRMS',0:RMSE_Interval:RMSE_Max,'titleRMS',RMSE_Max,'tickRMSangle',135
    ,'titleRMS',0,'showlabelsRMS',0,'widthRMS',1,...
35              'tickSTD',0:STD_Interval:STD_Max,'limSTD',STD_Max,...
36              'tickCOR',[.1:.1:.9 .95 0.97 0.98
    .99],'showlabelsCOR',1,'titleCOR',1);
```

Code 3.5.C

```
1   %% Example 3
2   % Load data
3   Data = xlsread('Data','sheet3');  % Example 3
4   BUOY=Data';
5   
6   % Get statistics from the dataset
7   statm = zeros(size(Data,2),4);
8   for ii = 2:size(BUOY,1)
9       C = allstats(BUOY(1,:),BUOY(ii,:));
10      statm(ii,:) = C(:,2);
11  end
12  statm(1,:) = C(:,1);
13  
14  % Assign range of RMSE, Correlation coefficient and standard deviation
15  Prompt={'Max. RMSE','Interval of the RMSE','Max. of the standard
    deviation','Interval of the standard deviation'};
16      Title='Max. of statistical indices;
17      DefaultValues={'14','1','10','1'};
18      PARAMS=inputdlg(Prompt,Title,[1, 50],DefaultValues);
19      A=PARAMS{1};
20      B=PARAMS{2};
21      C=PARAMS{3};
22      D=PARAMS{4};
23  
24  RMSE_Max = str2double(A);
25  RMSE_Interval = str2double(B);
26  STD_Max = str2double(C);
27  STD_Interval = str2double(D);
28  
29  % Plot:
30  figure
31  alphab = 'ABCDEFGHIJKLMNOPQRSTUVWXYZ';
32  
    [pp, tt, axl] =
33  taylordiag(squeeze(statm(:,2)),squeeze(statm(:,3)),squeeze(statm(:,4)),..
    .
34  'tickRMS',0:RMSE_Interval:RMSE_Max,'titleRMS',RMSE_Max,'tickRMSangle',135
    ,'titleRMS',0,'showlabelsRMS',0,'widthRMS',1,...
35              'tickSTD',0:STD_Interval:STD_Max,'limSTD',STD_Max,...
36              'tickCOR',[.1:.1:.9 .95 0.97 0.98
    .99],'showlabelsCOR',1,'titleCOR',1);
```

Code 3.5.D

```
1   %% Example 4
2   % Load data
3   Data = xlsread('Data','sheet4');  % Example 4
4   BUOY=Data';
5   
6   % Get statistics from the dataset
7   statm = zeros(size(Data,2),4);
8   for ii = 2:size(BUOY,1)
9       C = allstats(BUOY(1,:),BUOY(ii,:));
10      statm(ii,:) = C(:,2);
11  end
12  statm(1,:) = C(:,1);
13  
14  % Assign range of RMSE, Correlation coefficient and standard deviation
15  Prompt={'Max. RMSE','Interval of the RMSE','Max. of the standard
    deviation','Interval of the standard deviation'};
16      Title='Max. of statistical indices;
17      DefaultValues={'4','0.2','2.5','0.25'};
18      PARAMS=inputdlg(Prompt,Title,[1, 50],DefaultValues);
19      A=PARAMS{1};
20      B=PARAMS{2};
21      C=PARAMS{3};
22      D=PARAMS{4};
23  
24  RMSE_Max = str2double(A);
25  RMSE_Interval = str2double(B);
26  STD_Max = str2double(C);
27  STD_Interval = str2double(D);
28  
29  % Plot:
30  figure
31  alphab = 'ABCDEFGHIJKLMNOPQRSTUVWXYZ';
32  
    [pp, tt, axl] =
33  taylordiag(squeeze(statm(:,2)),squeeze(statm(:,3)),squeeze(statm(:,4)),..
    .

34  'tickRMS',0:RMSE_Interval:RMSE_Max,'titleRMS',RMSE_Max,'tickRMSangle',135
    ,'titleRMS',0,'showlabelsRMS',0,'widthRMS',1,...
35              'tickSTD',0:STD_Interval:STD_Max,'limSTD',STD_Max,...
36              'tickCOR',[.1:.1:.9 .95 0.97 0.98
    .99],'showlabelsCOR',1,'titleCOR',1);
```

Code 3.5.E

```
1   %% Example 5
2   % Load data
3   Data = xlsread('Data','sheet5');  % Example 5
4   BUOY=Data';
5   
6   % Get statistics from the dataset
7   statm = zeros(size(Data,2),4);
8   for ii = 2:size(BUOY,1)
9       C = allstats(BUOY(1,:),BUOY(ii,:));
10      statm(ii,:) = C(:,2);
11  end
12  statm(1,:) = C(:,1);
13  
14  % Assign range of RMSE, Correlation coefficient and standard deviation
15  Prompt={'Max. RMSE','Interval of the RMSE','Max. of the standard
    deviation','Interval of the standard deviation'};
16      Title='Max. of statistical indices;
17      DefaultValues={'14','1','14','2'};
18      PARAMS=inputdlg(Prompt,Title,[1, 50],DefaultValues);
19      A=PARAMS{1};
20      B=PARAMS{2};
21      C=PARAMS{3};
22      D=PARAMS{4};
23  
24  RMSE_Max = str2double(A);
25  RMSE_Interval = str2double(B);
26  STD_Max = str2double(C);
27  STD_Interval = str2double(D);
28  
29  % Plot:
30  figure
31  alphab = 'ABCDEFGHIJKLMNOPQRSTUVWXYZ';
32  
    [pp, tt, axl] =
33  taylordiag(squeeze(statm(:,2)),squeeze(statm(:,3)),squeeze(statm(:,4)),..
    .
34  'tickRMS',0:RMSE_Interval:RMSE_Max,'titleRMS',RMSE_Max,'tickRMSangle',135
    ,'titleRMS',0,'showlabelsRMS',0,'widthRMS',1,...
35              'tickSTD',0:STD_Interval:STD_Max,'limSTD',STD_Max,...
36              'tickCOR',[.1:.1:.9 .95 0.97 0.98
    .99],'showlabelsCOR',1,'titleCOR',1);
```

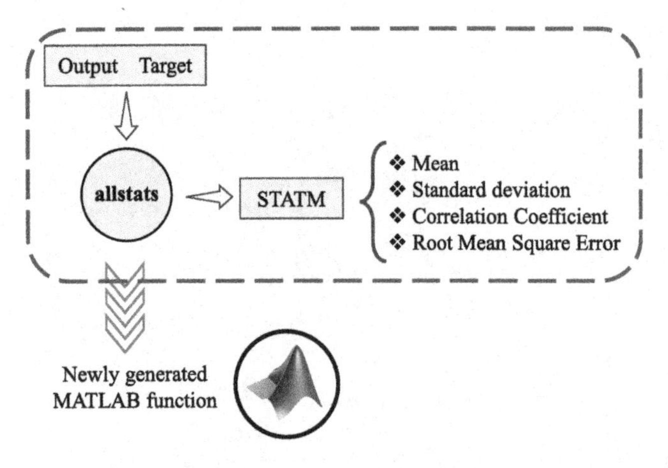

FIGURE 3.17 Schematic of Code 3.6.

In executing Code 3.5, two different functions are called, namely, `allstats` and `taylordiag`. The `allstats` function is defined in Code 3.6. From line 13 of Code 3.5, we can see that the `allstats` function has two inputs: the first one is the observations, and the second one is the model results. In lines 3 and 4 of Code 3.6, the variables `Cr` and `Cf` are defined as the first and second inputs (respectively) of the `allstats` function. In line 7, it is verified whether the size of `Cr` and `Cf` is equal—if not, then the message in line 8 is displayed to the user. In line 12, the number of samples with zero value (i.e., `NaN`) is estimated for both samples using the `isnan` and `find` commands with the result being stored in the variable `iok`. If there are any `NaN`s, they must be found and removed from all samples. In lines 21–22, 26–27, 30–31, and 34–35, the standard deviation, average, RMSE, and correlation coefficients of both `Cr` and `Cf` are calculated. Finally, these indices are stored in the first to fourth rows of the `STATM` variable and reported as the output of this function. The schematic of Code 3.6 is shown in Fig. 3.17.

Code 3.6

```
1   function STATM = allstats(varargin)
2
3   Cr = varargin{1}; Cr = Cr(:);
4   Cf = varargin{2}; Cf = Cf(:);
5
6   %%% Check size:
7   if length(Cr) ~= length(Cf)
8       error('Cr and Cf must be of same length');
9   end
10
11  %%% Check NaNs:
12  iok = find(isnan(Cr)==0 & isnan(Cf)==0);
13  if length(iok) ~= length(Cr)
14      warning('Found NaNs in inputs, removed them to compute statistics');
15  end
16  Cr   = Cr(iok);
17  Cf   = Cf(iok);
18  N    = length(Cr);
19
20  %%% STD:
21  st(1) = sqrt(sum(   (Cr-mean(Cr) ).^2) / N );
22  st(2) = sqrt(sum(   (Cf-mean(Cf) ).^2) / N );
23
24
25  %%% MEAN:
26  me(1) = mean(Cr);
27  me(2) = mean(Cf);
28
29  %%% RMSE:
30  rms(1) = sqrt(sum(   ( ( Cr-mean(Cr) )-( Cr-mean(Cr) )).^2)   /N);
31  rms(2) = sqrt(sum(   ( ( Cf-mean(Cf) )-( Cr-mean(Cr) )).^2)   /N);
32
33  %%% CORRELATIONS:
34  co(1) = sum( ( ( Cr-mean(Cr) ).*( Cr-mean(Cr) )))/N/st(1)/st(1);
35  co(2) = sum( ( ( Cf-mean(Cf) ).*( Cr-mean(Cr) )))/N/st(2)/st(1);
36
37
38  %%% OUTPUT
39  STATM(1,:) = me;
40  STATM(2,:) = st;
41  STATM(3,:) = rms;
42  STATM(4,:) = co;
43
44  end
```

With the detailed MATLAB coding required to plot the Taylor diagram defined, the code was executed to plot the performance of four models in estimating our five example datasets. The Taylor diagrams for all five examples are shown in Fig. 3.18. Due to this figure, for Example 1, the best performance is related to Model B (closest to reference point), while Models C, D, and E are ranked second to fourth, respectively. The same conclusions can be drawn from Fig. 3.18B and C for Examples 2 and 3, although the magnitudes of the standard deviations of their observations are significantly higher than the standard deviation of observations in Example 1. The results of the Taylor diagram for Examples 4 and 5 indicated that the differences in performance between the various models are small so that we cannot discriminate them without a higher resolution.

In fact, in such cases, the use of the Taylor diagram confirms the remarkable closeness of different models. In such a case, we must resort to methods that consider factors other than only the accuracy of the model, such as the Akaike Information Criterion, which was previously described in full.

Data on Taylor diagram can be found at Appendix 3A.

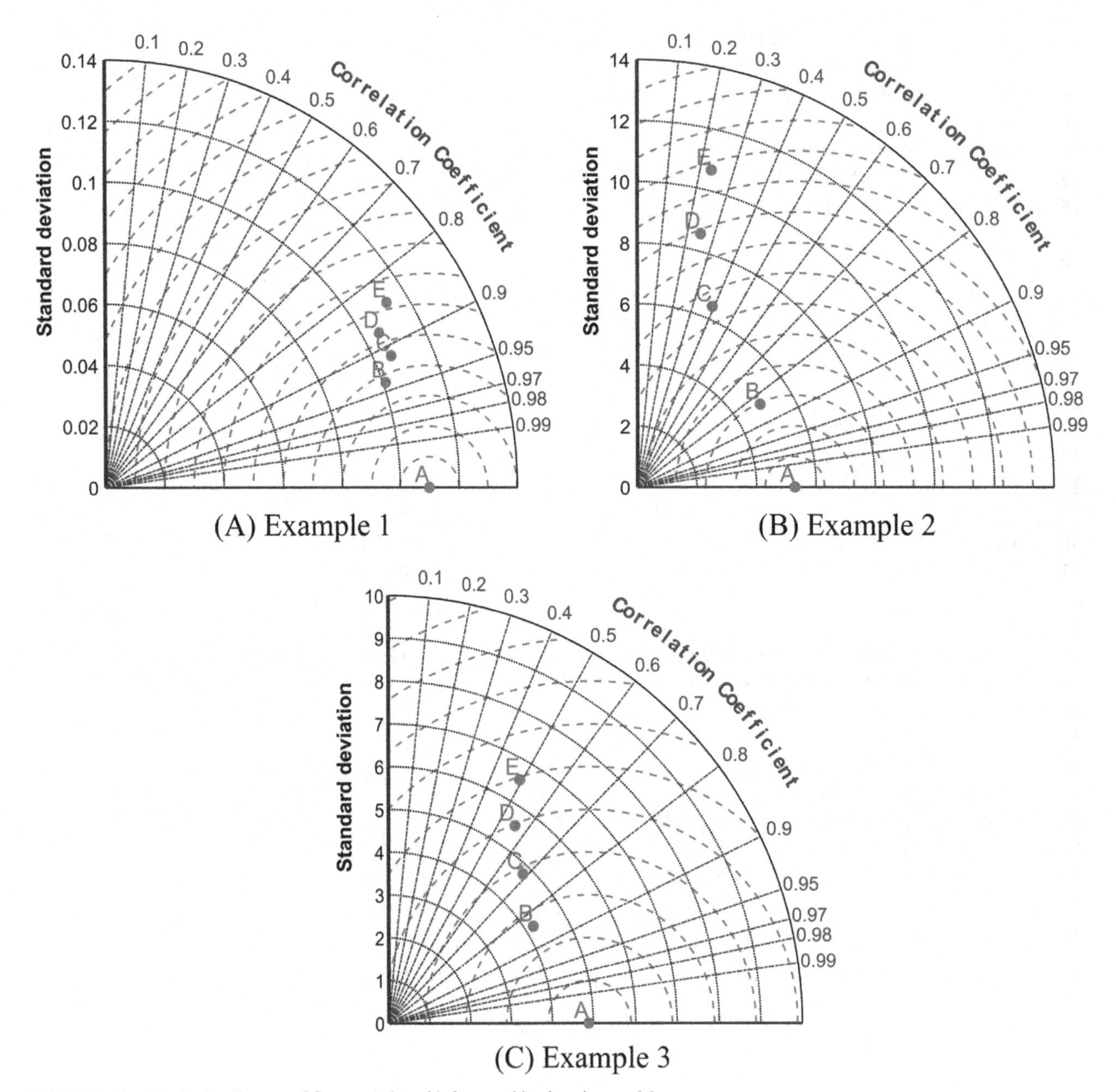

FIGURE 3.18 The Taylor diagram of five examples with four machine learning models.

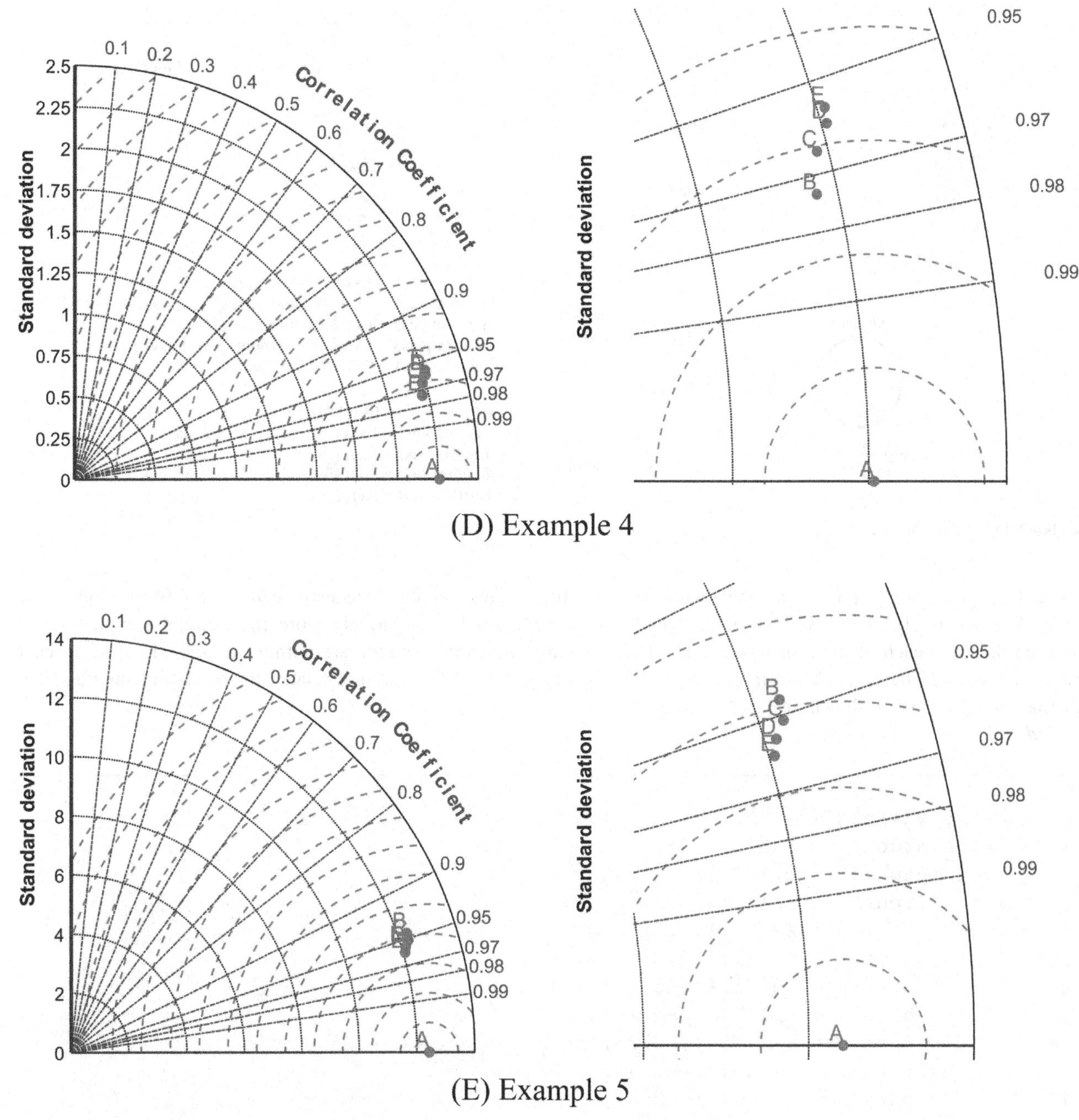

FIGURE 3.18 (Continued)

3.3.2 Scatter plot

The scatter plot is a well-known statistical data visualization tool used to demonstrate the correlation of two variables. Because of this, it can also be applied to check the correlation of the observed and estimated values in an ML model development process. The higher the correlation of the observed and estimated samples, the higher the accuracy of the model. The scatter plot can provide information on the direction of the correlation between two variables (i.e., positive or negative correlations) as well as the strength (i.e., weak or robust). Different types of correlations were previously illustrated in Fig. 3.4.

The details of the MATLAB coding required for scatter plot generation are in the following codes. In Code 3.7, all of the data related to Example 1 is stored in the variable `Ex1` using the `xlsread` command with "Data" as the filename and "sheet1" as the sheet location. In this case, sheet1 (data of Example 1) includes five columns with the first being

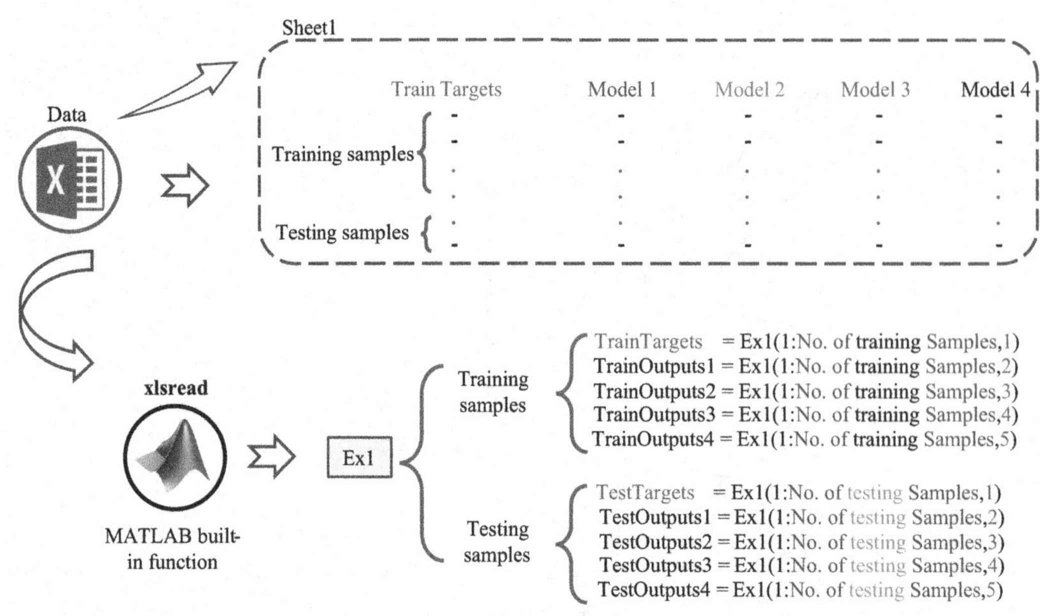

FIGURE 3.19 Schematic of Code 3.7.

related to actual observations while the second to the fifth columns contain the estimated values from Model 1 to Model 4, respectively. The variables defined in liens 3–7 are used to separately store the training samples of the observed data and each of the four models. The training samples in this example are within the first 113 rows of each column in sheet1. The remaining samples from row 114 to the end of the data are related to the testing samples. The schematic of Code 3.7 is shown in Fig. 3.19.

Code 3.7

```
1   %% Load data
2   Ex1 = xlsread ('Data','sheet1');   % Example 1
3   TrainTargets   = Ex1(1:113,1);     % Train Targets
4   TrainOutputs1  = Ex1(1:113,2);     % Train Outputs - Model 1
5   TrainOutputs2  = Ex1(1:113,3);     % Train Outputs - Model 2
6   TrainOutputs3  = Ex1(1:113,4);     % Train Outputs - Model 3
7   TrainOutputs4  = Ex1(1:113,5);     % Train Outputs - Model 4
8   TestTargets    = Ex1(114:end,1);   % Test Targets
9   TestOutputs1   = Ex1(114:end,2);   % Test Outputs  - Model 1
10  TestOutputs2   = Ex1(114:end,3);   % Test Outputs  - Model 2
11  TestOutputs3   = Ex1(114:end,4);   % Test Outputs  - Model 3
12  TestOutputs4   = Ex1(114:end,5);   % Test Outputs  - Model 4
```

After loading data using Code 3.7, the main step of the scatter plot generation and its parameters are defined in Code 3.8. Using the `figure` command in line 2, a figure is opened, and the results of the`plot` command are added to this figure. The first and second input arguments of the `plot` command are the actual and estimated samples (respectively). The third input argument is related to the line style, marker, and color. The different types of line styles, markers, and colors are presented in Tables 3.4–3.6, respectively. The fourth, fifth, and sixth input arguments are `MarkerSize`, `MarkerEdgeColor`, and `MarkerFaceColor`, respectively. In line 5, variable `A` is defined based on the minimum and maximum values of all actual and estimated values (i.e., `Ex1`). This variable is used to generate the plot of best fit across the range of actual and estimated values using the command shown in line 6 of Code 3.8. The label of the y and x axes are defined in lines 8 and 9, respectively, while the title and legend are also defined in lines 10 and 11, respectively.

The minimum and maximum of each axis are defined using the `ylim` and `xlim` commands. It is clear that they are selected using the main data (`Ex1`), and there is no need to change it for other problems. It is automatically changed

TABLE 3.4 **Different types of line styles for the `plot` command.**

Line Style	Description
-	Solid line[a]
-.	Dash-dot line
:	Dotted line
–	Dashed line

[a] *It is the default line style.*

TABLE 3.5 **Different types of markers for the `plot` command.**

Marker	Description
o	Circle
*	Asterisk
x	Cross
d	Diamond
h	Hexagram
p	Pentagram
+	Plus sign
s	Square
.	Point
v	Downward-pointing triangle
^	Upward-pointing triangle
<	Left-pointing triangle
>	Right-pointing triangle

TABLE 3.6 **Different types of colors for the `plot` command.**

Color	Description
b	blue
c	cyan
g	green
k	black
m	magenta
r	red
w	white
y	yellow

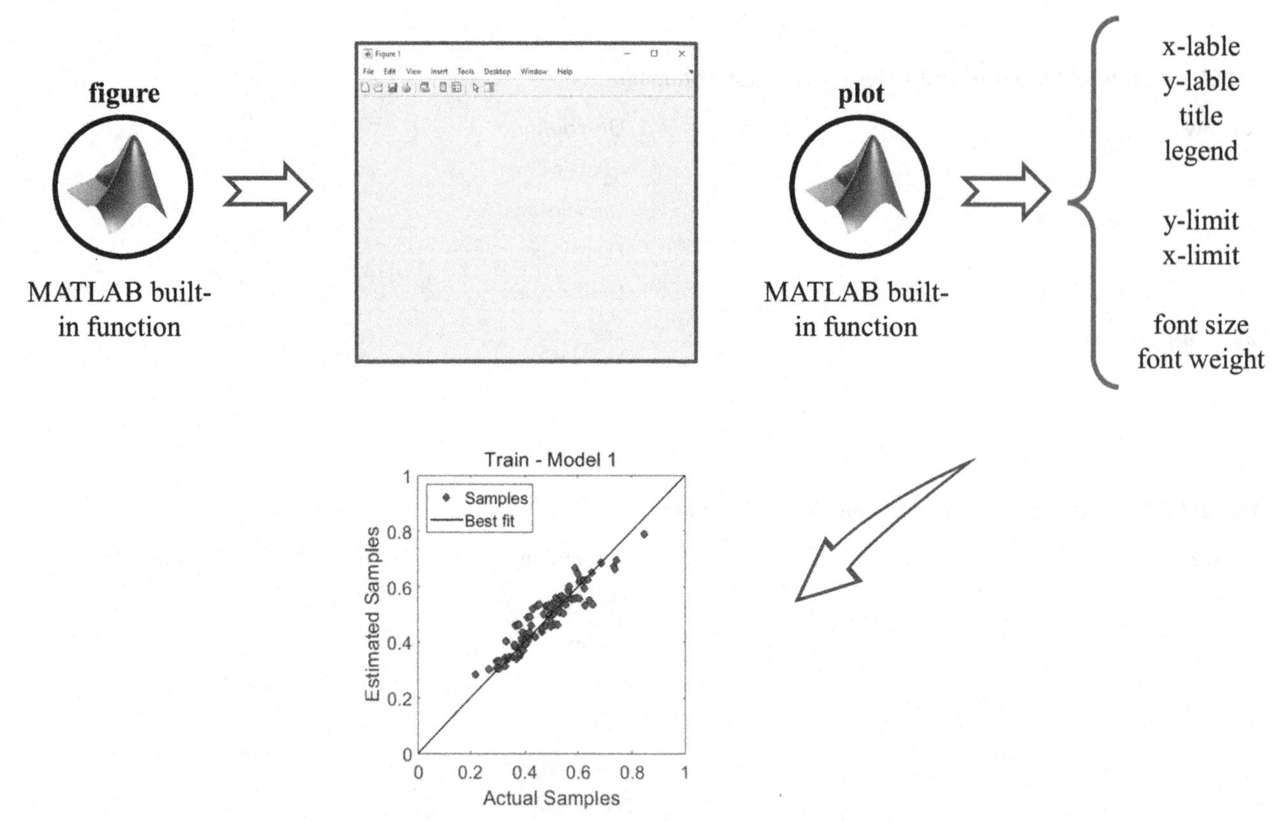

FIGURE 3.20 The schematic of Code 3.8.

based on the applied dataset (actual and estimated ones). Similar to these commands, the axis divisions, which are defined using the `xticks` and `yticks` commands, are also defined based on the minimum and maximum values of the main data (i.e., `Ex1`). Indeed, it is also automatically changed after you change the Microsoft Excel file. It should be noted that in lines 15 and 16, there are only six divisions generated because the value applied in the denominator of the fraction is 5. Based on the user's needs for different problems, the denominator may be changed to implement more or less divisions. Finally, the size and type of the fonts are defined in lines 20 and 21, respectively. Code 3.8 is repeated for Models 2–4 as Codes 3.9–3.11, respectively. The main differences between the syntax of these codes are the defined types and color of the markers. The general forms of Codes 3.8–3.11 are the same and can be represented by the schematic shown in Fig. 3.20.

Code 3.8

```
1   %% Plot Results - TrainOutputs 1
2   figure;
3   plot(TrainTargets,TrainOutputs1,'o','MarkerSize',6,'MarkerEdgeColor','k',
    'MarkerFaceColor','r');
4   hold on
5   A=round(min(min(Ex1))):(round(max(max(Ex1)))-
    round(min(min(Ex1))))/5:round(max(max(Ex1)));
6   plot(A,A,'-k','LineWidth',1.5)
7   
8   ylabel('Estimated Samples')
9   xlabel('Actual Samples')
10  title('Train - Model 1','FontWeight','normal')
11  legend('Samples','Best fit','Location','NorthWest')
12  
13  ylim([round(min(min(Ex1))) round(max(max(Ex1)))])
14  xlim([round(min(min(Ex1))) round(max(max(Ex1)))])
15  xticks(round(min(min(Ex1))):(round(max(max(Ex1)))-
    round(min(min(Ex1))))/5:round(max(max(Ex1))));
16  yticks(round(min(min(Ex1))):(round(max(max(Ex1)))-
    round(min(min(Ex1))))/5:round(max(max(Ex1))));
17  axis square
18  
19  ax = gca;
20  ax.FontSize = 16;
21  ax.FontWeight='normal';
```

Code 3.9

```
1   %% Plot Results - TrainOutputs 2
2   figure;
3   plot(TrainTargets,TrainOutputs2,'.','MarkerSize',20,'MarkerEdgeColor','g'
    ,'MarkerFaceColor','r');
4   hold on
5   A=round(min(min(Ex1))):(round(max(max(Ex1)))-
    round(min(min(Ex1))))/5:round(max(max(Ex1)));
6   plot(A,A,'-k','LineWidth',1.5)
7   
8   ylabel('Estimated Samples')
9   xlabel('Actual Samples')
10  title('Train - Model 2','FontWeight','normal')
11  legend('Samples','Best fit','Location','NorthWest')
12  
13  ylim([round(min(min(Ex1))) round(max(max(Ex1)))])
14  xlim([round(min(min(Ex1))) round(max(max(Ex1)))])
15  xticks(round(min(min(Ex1))):(round(max(max(Ex1)))-
    round(min(min(Ex1))))/5:round(max(max(Ex1))));
16  yticks(round(min(min(Ex1))):(round(max(max(Ex1)))-
    round(min(min(Ex1))))/5:round(max(max(Ex1))));
17  axis square
18  
19  ax = gca;
20  ax.FontSize = 16;
21  ax.FontWeight='normal';
```

Code 3.10

```
1   %% Plot Results - TrainOutputs 3
2   figure;
3   plot(TrainTargets,TrainOutputs3,'*','MarkerSize',6,'MarkerEdgeColor','b',
    'MarkerFaceColor','r');
4   hold on
5   A=round(min(min(Ex1))):(round(max(max(Ex1)))-
    round(min(min(Ex1))))/5:round(max(max(Ex1)));
6   plot(A,A,'-k','LineWidth',1.5)
7
8   ylabel('Estimated Samples')
9   xlabel('Actual Samples')
10  title('Train - Model 3','FontWeight','normal')
11  legend('Samples','Best fit','Location','NorthWest')
12
13  ylim([round(min(min(Ex1))) round(max(max(Ex1)))])
14  xlim([round(min(min(Ex1))) round(max(max(Ex1)))])
15  xticks(round(min(min(Ex1))):(round(max(max(Ex1)))-
    round(min(min(Ex1))))/5:round(max(max(Ex1))));
16  yticks(round(min(min(Ex1))):(round(max(max(Ex1)))-
    round(min(min(Ex1))))/5:round(max(max(Ex1))));
17  axis square
18
19  ax = gca;
20  ax.FontSize = 16;
21  ax.FontWeight='normal';
```

Code 3.11

```
1   %% Plot Results - TrainOutputs 4
2   figure;
3   plot(TrainTargets,TrainOutputs4,'d','MarkerSize',8,'MarkerEdgeColor','k',
    'MarkerFaceColor','y');
4   hold on
5   A=round(min(min(Ex1))):(round(max(max(Ex1)))-
    round(min(min(Ex1))))/5:round(max(max(Ex1)));
6   plot(A,A,'-k','LineWidth',1.5)
7
8   ylabel('Estimated Samples')
9   xlabel('Actual Samples')
10  title('Train - Model 4','FontWeight','normal')
11  legend('Samples','Best fit','Location','NorthWest')
12
13  ylim([round(min(min(Ex1))) round(max(max(Ex1)))])
14  xlim([round(min(min(Ex1))) round(max(max(Ex1)))])
15  xticks(round(min(min(Ex1))):(round(max(max(Ex1)))-
    round(min(min(Ex1))))/5:round(max(max(Ex1))));
16  yticks(round(min(min(Ex1))):(round(max(max(Ex1)))-
    round(min(min(Ex1))))/5:round(max(max(Ex1))));
17  axis square
18
19  ax = gca;
20  ax.FontSize = 16;
21  ax.FontWeight='normal';
```

Codes 3.9–3.11 produce scatter plots of the observed and estimated samples during the training stage by Models 1 to 4, respectively, using Example 1 data. These codes can also be easily applied to the testing samples by replacing the training samples with the testing ones. For example, to plot the scatter plot during the testing stage for Model 1 using Example 1 data, Code 3.8 could be applied by replacing `TrainTargets` and `TrainOutputs1` with `TestTargets` and `TestOutputs1`, respectively. The scatter plot of testing samples could further be applied for other examples by replacing `Ex1` with `Ex2`, `Ex3`, `Ex4`, or `Ex5`. The modeler must note that when applying Code 3.8 to the data for other examples, the syntax of Code 3.7 must also be revised because the loaded data in line 2 is related to Example 1. The results of the estimated samples and actual samples for both the training and testing samples are stored within the same Microsoft Excel file (i.e., Data), but in sheet2 to sheet5. For lines 3–12 that split the training and testing samples, the reader is referred back to Fig. 1.2, where the number of training and testing samples are defined. In all cases, 70% of the samples were considered as training samples, while the remaining 30% of samples were used as testing/validation data. The number of samples for Examples 1–5 are 161, 100, 240, 250, and 1095, respectively. Therefore, the number of training samples for Examples 1–5 are 113, 70, 168, 175, and 767, respectively, and the number of testing samples for Examples 1–5 are 48, 30, 72, 75, and 328, respectively. All of the prepared codes for training and testing stages related to all Examples are provided in the packages of the MATLAB-based codes.

Fig. 3.21 presents the scatter plots of the observed and estimated data by Models 1–4 for the five example datasets for both the training and testing stages. According to this figure, the performance of Models 1–4 in both the training and testing stages are very similar to one another. As mentioned previously, the plots are generated using Codes 3.8–3.11 for Models 1–4, respectively. It could be seen that the differences between these figures are not only the estimated values but also the types and colors of markers for each model is different from others.

Data on scatter plot can be found at Appendix 3A.

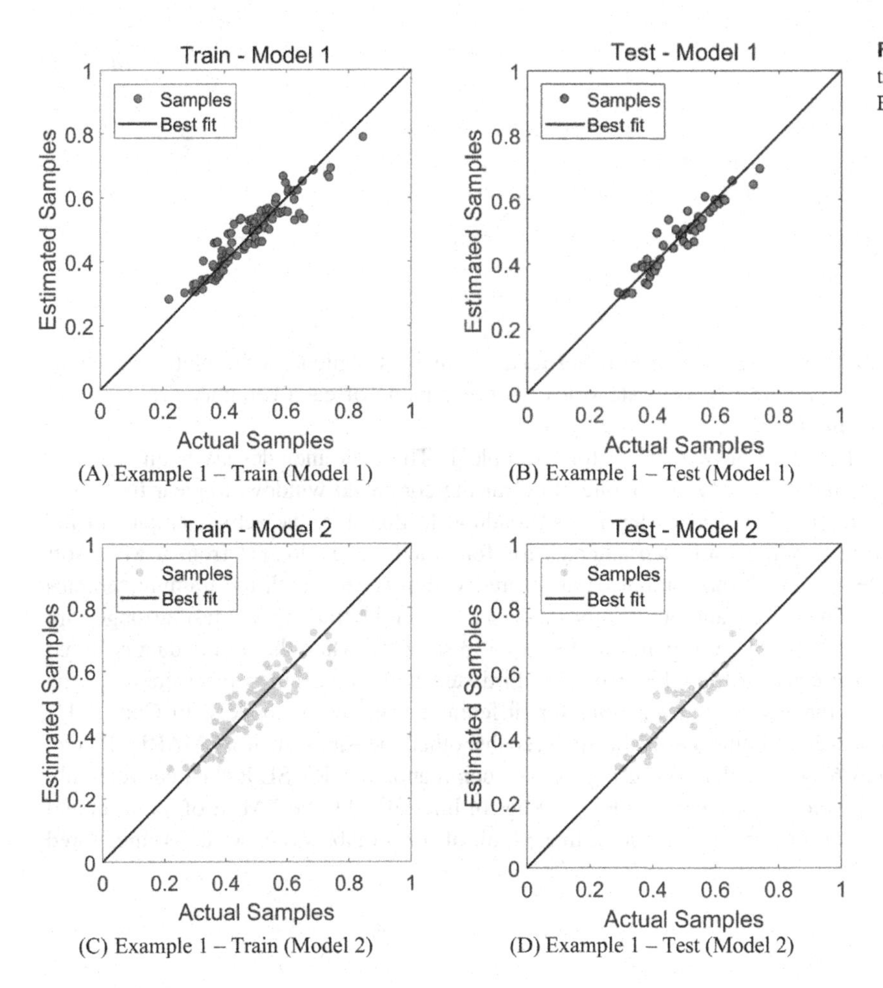

FIGURE 3.21 The scatter plots of the observations and estimated values by Models 1–4 of Example 1 at both the training and testing stages.

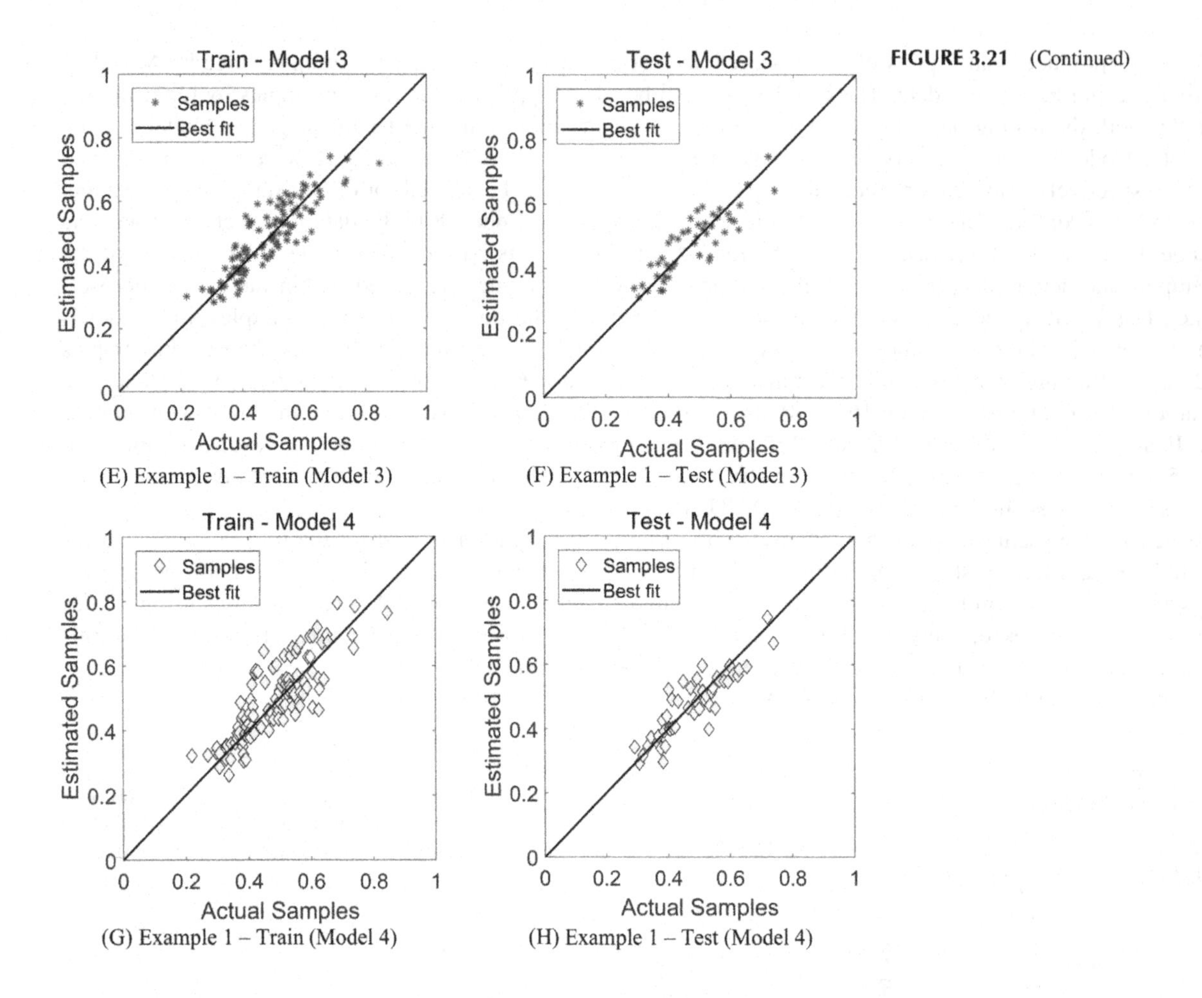

(E) Example 1 – Train (Model 3) (F) Example 1 – Test (Model 3)
(G) Example 1 – Train (Model 4) (H) Example 1 – Test (Model 4)

FIGURE 3.21 (Continued)

3.3.3 The barplot

A barplot is a graph that indicates the values of categorical data using rectangular bars, which can be plotted horizontally or vertically. This graph is applied to compare discrete categories, with the name of each category presented on one axis and the value of each category on another.

Code 3.12 details the required MATLAB coding of the barplot for Example 1. The code includes six main sections. First, the three main commands `clc`, `clear`, and `close all`, are applied to clear the command window, to clear the workspace, and to close all open figures, respectively. The second step is data loading. In this step, the actual samples of the training and testing stages, as well as corresponding estimated samples for four models, are loaded from a Microsoft Excel file. All samples related to Example 1 (Actual and estimated) are stored within sheet1, with the training samples being stored first above the testing samples. The targets and outputs for models 1 to 4 in the training and testing stages are defined in lines 7–16. In lines 7–11, the training data corresponds to the first 113 samples, while the testing data is stored in samples 114 to the end of the data (as indicated in lines 12–16). The third step is the calculation of errors (e.g., the RMSE). The barplot is applied to compare the values of a measure for different categories or models. In Code 3.12, the RMSE was applied as a measure, however, it could easily be replaced by other measures such as MARE, R, etc. The RMSE of the variables defined in lines 6–16 is calculated using the `mse` command. The RMSE is then conveniently calculated using `sqrt(mse)`. The fourth step is to prepare the data for plotting. In lines 30–33, the RMSE of the train and test stages for all models are stored in `M1` to `M4`, respectively, while in line 34, all of the variables (i.e., `M1` to `M4`) are stored in a new variable (i.e., `All`). The RMSEs of training and testing are first stored together in a single variable because they are to be plotted together. These newly generated variables (i.e., `M1` to `M4`) are finally stored in the variable `All` so as to create only a barplot with 4 main categories, each of which is composed of two bars showing the RMSE values for train and test. In line 37, the categorical command is used to define a set of discrete categories for the four main variables (M1, M2, M3, M4), each of which consists of two inputs (i.e., for example, Train1 and Test1 for Model 1). Using the variables `xx` (line 37) and `All` (line 34) as the categorial names and the values, respectively, the `barplot` command is applied to plot the RMSE of each model for both the training and testing stages. In lines 40–50, the options of the barplot, including the face color of each variable, the size and weights of font, the y-axis label, the legend, and the title of this figure, are defined. Finally, the `clearvars -except` command is applied to clear all variables except `M1`, `M2`, `M3`, `M4`, and `All`. This process is repeated for all other Examples. The code of the other Examples is provided in the MATLAB package code. The schematic of Code 3.12 is shown in Fig. 3.22.

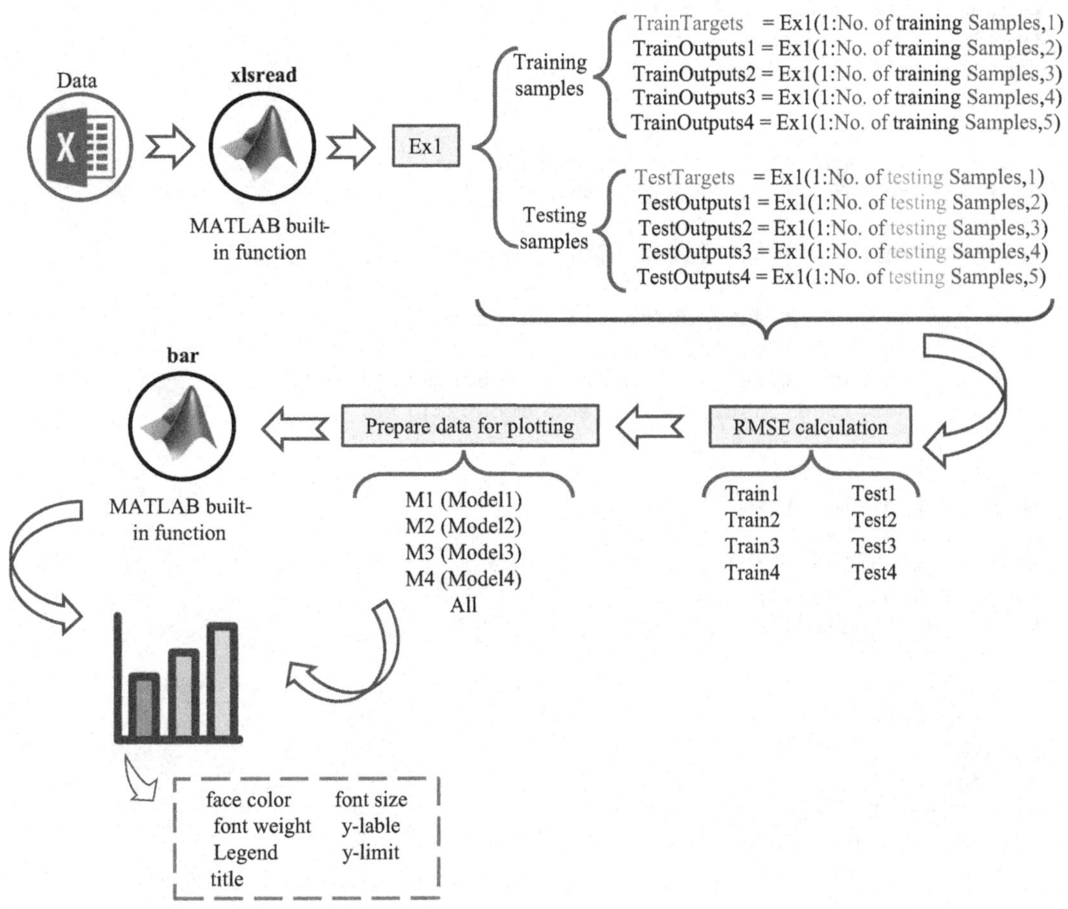

FIGURE 3.22 Schematic of Code 3.12.

Code 3.12

```
1   clc
2   clear
3   close all
4
5   %% Load data
6   Ex1 = xlsread ('Data','sheet1');  % Example 1
7   TrainTargets   = Ex1(1:113,1);    % Train Targets
8   TrainOutputs1 = Ex1(1:113,2);     % Train Outputs - Model 1
9   TrainOutputs2 = Ex1(1:113,3);     % Train Outputs - Model 2
10  TrainOutputs3 = Ex1(1:113,4);     % Train Outputs - Model 3
11  TrainOutputs4 = Ex1(1:113,5);     % Train Outputs - Model 4
12  TestTargets    = Ex1(114:end,1);  % Test Targets
13  TestOutputs1   = Ex1(114:end,2);  % Test Outputs  - Model 1
14  TestOutputs2   = Ex1(114:end,3);  % Test Outputs  - Model 2
15  TestOutputs3   = Ex1(114:end,4);  % Test Outputs  - Model 3
16  TestOutputs4   = Ex1(114:end,5);  % Test Outputs  - Model 4
17
18  %% Calculation of Errors (For Example: RMSE)
19  Train1 = sqrt(mse(TrainTargets,TrainOutputs1)); % Train - Model 1
20  Train2 = sqrt(mse(TrainTargets,TrainOutputs2)); % Train - Model 2
21  Train3 = sqrt(mse(TrainTargets,TrainOutputs3)); % Train - Model 3
22  Train4 = sqrt(mse(TrainTargets,TrainOutputs4)); % Train - Model 4
23
24  Test1  = sqrt(mse(TestTargets,TestOutputs1));    % Test - Model 1
25  Test2  = sqrt(mse(TestTargets,TestOutputs2));    % Test - Model 2
26  Test3  = sqrt(mse(TestTargets,TestOutputs3));    % Test - Model 3
27  Test4  = sqrt(mse(TestTargets,TestOutputs4));    % Test - Model 4
28
29  %% Prepare data for plotting
30  M1 = [Train1,Test1]; % Model 1
31  M2 = [Train2,Test2]; % Model 2
32  M3 = [Train3,Test3]; % Model 3
33  M4 = [Train4,Test4]; % Model 4
34  All = [M1;M2;M3;M4]; % All models in one variable
35
36  %% Barplot
37  xx = categorical({'M1','M1';'M2','M2';'M3','M3';'M4','M4'});
38  figure
39  B = bar(xx,All);
40  B(1).FaceColor = 'w';
41  B(2).FaceColor = 'g';
42
43  ax = gca;
44  ax.FontSize = 16;
45  ax.FontWeight='normal';
46
47  ylabel('RMSE')
48  legend('Train','Test','Location','NorthWest')
49  ylim([0 1.1*max(max(All))])
50  title('Example 1','FontWeight','normal')
51
52  %% Clear all except mandatory information
53  clearvars -except M1 M2 M3 M4 All
```

Fig. 1.28 presents the RMSE barplot for all models during the training and testing stages of Examples 1–5. For Example 1, the lowest RMSE is for M1, while the highest value is for M4. The remarkable thing about this model is that the value of RMSE in the training stage shows a larger value than the testing stage. Indeed, the results of this model show that the model performs better in the testing stage than in training, which is unexpected. Generally, a model's accuracy in the training stage should be more than the testing stage because the model was calibrated using training samples, while the test data had no role in calibrating the model. However, there may be cases similar to Example 1 in Fig. 3.23, the reasons for which can be explained as follows:

(1) The data used in the test stage had very similar samples to the training data. Indeed, the model has experienced almost the same conditions as the test data.
(2) The data splitting was not well performed, so a more favorable set may be used for testing. It is recommended that the data be selected randomly or that the data be split using a cross-validation approach.
(3) The modeling may not have been executed well, and the reason for the better performance of the model in the test mode is completely random. For example, in the Extreme Learning Machine (ELM) method, the two matrices of input weights and bias of hidden neurons, which contain more than 60% of the adjustable parameters, are determined completely randomly. Therefore, it is necessary to perform the modeling repeatedly to eliminate the effect of random determination of parameters.

For Example 2, the difference in RMSE values between the train and test stages is very small, although, in M1 and M2, the model performs better during testing than in training. Similar to Example 1, Example 4 also experiences better model performance (in terms of RMSE) in the testing stage compared to the training phase, while for Examples 3 and 5, this problem is almost completely resolved.

Data on barplot can be found at Appendix 3A.

3.3.4 The histogram

The histogram is a representation that approximates the distribution of the numerical values. In this representation tool, the range of values is divided into the number of bins defined by the user. Therefore, a series of intervals are defined based on the number of bins, and the number of values considered in each interval is counted and reported as the final histogram.

Code 3.13 presents the details of the histogram in MATLAB. In this code, the loading of the data (i.e., Code 3.13. A) is similar to Code 3.12, and therefore the reader is directed to the discussion on Code 3.12 for detailed information relating to coding. The histogram can be applied for the actual (or estimated) values of a variable, or it could be applied to study the distribution of a certain measure. In this code, the distribution of the relative error with the sign is examined for all estimated samples by Models 1– 4. The relative error for all models in both the training and testing stages is calculated in Code 3.13.B. For all eight estimated datasets (i.e., Training and testing samples estimated by Models 1–4), the plotting approach of the histogram is completely the same. In Code 3.13.C, the `figure` command is used in line 3 to open a figure. In line 4, the `histogram` command is employed with its single input argument being the variable that contains the desired values to be plotted (i.e., for Example 1, `RE_Train1` is the relative error of Model 1 in the training stage). Following the introduction of the histogram command, its options must be defined by the user. The first option is the number of bins, which is defined using the `h.NumBins` command. In this example, the number of bins is set to 10, although this value could be set lower or higher by the user. If greater detail (sensitivity) regarding the data distribution or relative error values is needed, the number of bins could be increased. The face and edge color are defined using the `h.FaceColor` and `h.EdgeColor` commands, respectively. The size and weight of the font, the *y*-axis label, the *x*-axis label, the title, and the maximum values of the *y*-axis are defined using the `ax.FontSize`, `ax.FontWeight`, `ylabel`, `xlabel`, `title`, and`ylim` commands, respectively. The codings outlined in lines 31–44 are repeated for `RE_Test1`, `RE_Train2`, `RE_Test2`, `RE_Train3`, `RE_Test3`, `RE_Train4`, and `RE_Test4`. The main difference between the plotted histograms is the applied color, which is used to distinguish each model. For Model 1, the face color is defined as yellow (i.e., `y`), while for Models 2, 3, and 4, the face color is defined as green (i.e., `g`), blue (i.e., `b`), and magenta (i.e., `m`). The schematic of Code 3.13 is shown in Fig. 3.24.

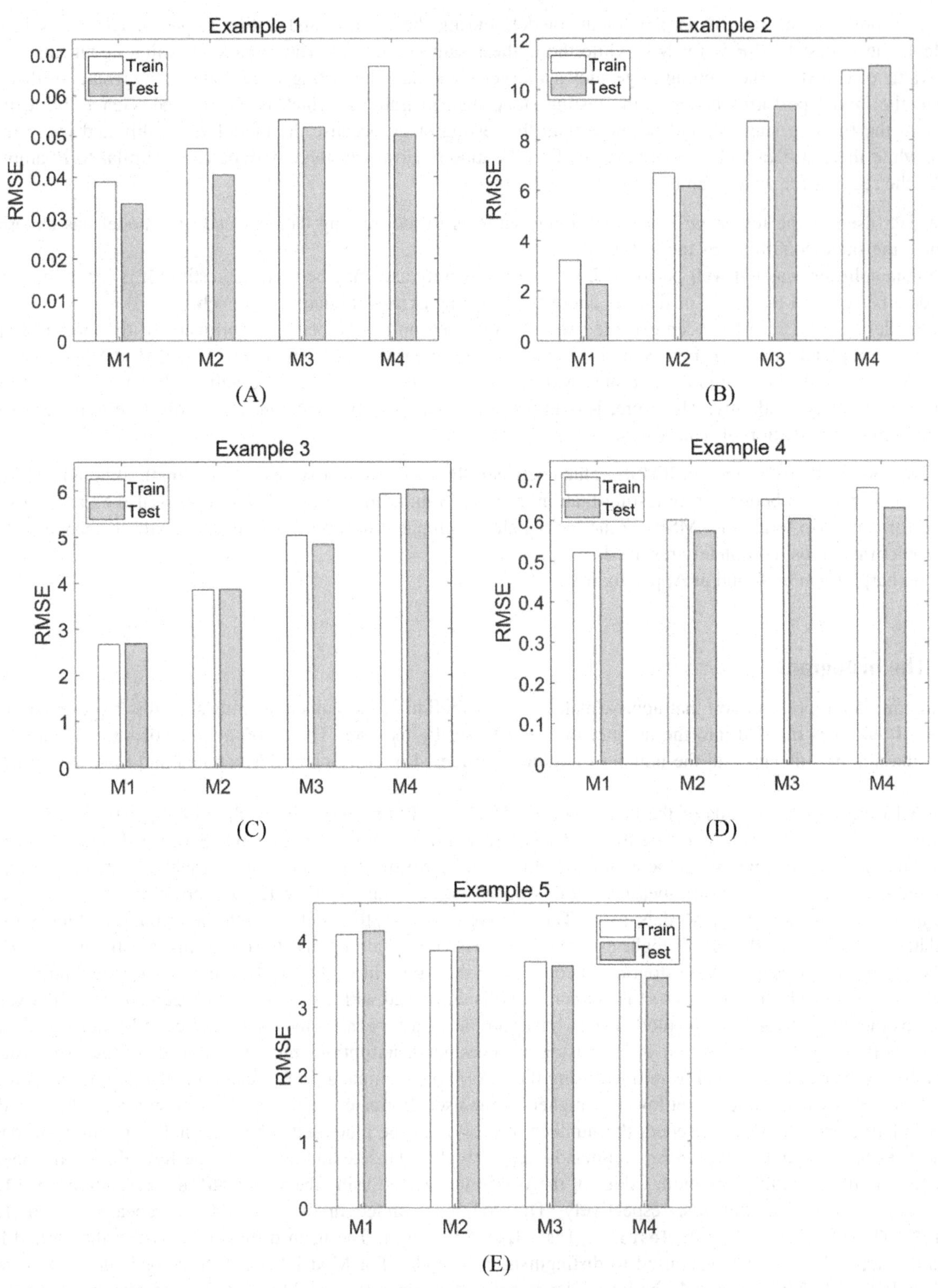

FIGURE 3.23 The RMSE for all models at both training and testing stages of Examples 1–5.

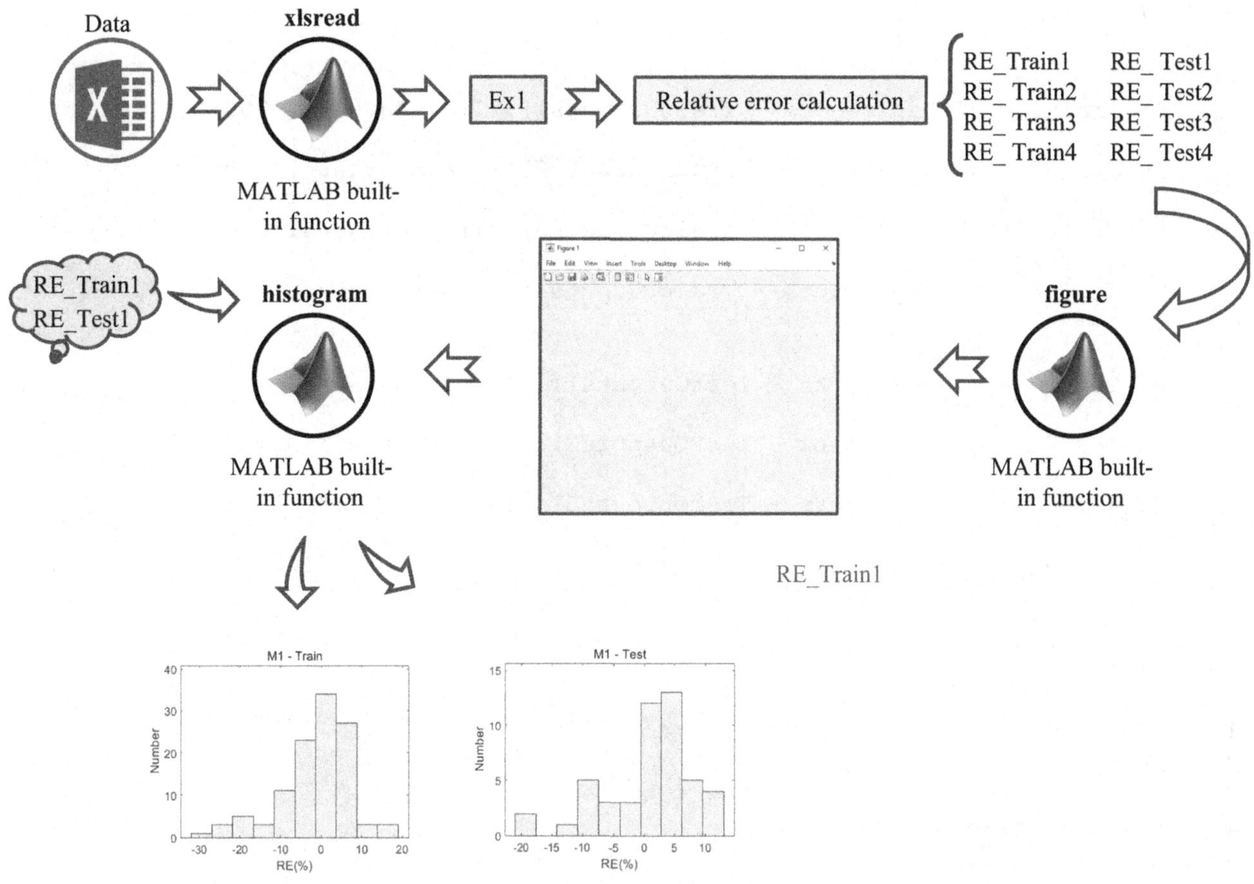

FIGURE 3.24 Schematic of Code 3.13.

Code 3.13.A

```
1   clc
2   clear
3   close all
4
5   %% load data
6   Ex1 = xlsread ('Data','sheet1');   % Example 1
7   TrainTargets   = Ex1(1:113,1);     % Train Targets
8   TrainOutputs1 = Ex1(1:113,2);      % Train Outputs - Model 1
9   TrainOutputs2 = Ex1(1:113,3);      % Train Outputs - Model 2
10  TrainOutputs3 = Ex1(1:113,4);      % Train Outputs - Model 3
11  TrainOutputs4 = Ex1(1:113,5);      % Train Outputs - Model 4
12  TestTargets    = Ex1(114:end,1);   % Test Targets
13  TestOutputs1   = Ex1(114:end,2);   % Test Outputs  - Model 1
14  TestOutputs2   = Ex1(114:end,3);   % Test Outputs  - Model 2
15  TestOutputs3   = Ex1(114:end,4);   % Test Outputs  - Model 3
16  TestOutputs4   = Ex1(114:end,5);   % Test Outputs  - Model 4
```

Code 3.13.B

```
1   %% Calculation of the Relative Error (%)
2   RE_Train1 = (TrainTargets - TrainOutputs1)./TrainTargets.*100; % Train -
    Model 1
3   RE_Train2 = (TrainTargets - TrainOutputs2)./TrainTargets.*100; % Train -
    Model 2
4   RE_Train3 = (TrainTargets - TrainOutputs3)./TrainTargets.*100; % Train -
    Model 3
5   RE_Train4 = (TrainTargets - TrainOutputs4)./TrainTargets.*100; % Train -
    Model 4
6
7   RE_Test1  = (TestTargets - TestOutputs1)./TestTargets.*100;     % Test -
    Model 1
8   RE_Test2  = (TestTargets - TestOutputs2)./TestTargets.*100;     % Test -
    Model 2
9   RE_Test3  = (TestTargets - TestOutputs3)./TestTargets.*100;     % Test -
    Model 3
10  RE_Test4  = (TestTargets - TestOutputs4)./TestTargets.*100;     % Test -
    Model 4
```

Code 3.13.C

```
1   %%%%%%%%%%%%%%%%%%%%%%%%%%%%%% Model 1 %%%%%%%%%%%%%%%%%%%%%%%%%%%%%%
2   % Model 1  - Train
3   figure;
4   h = histogram(RE_Train1);
5   h.NumBins =  10;
6   h.FaceColor =  'y';
7   h.EdgeColor = 'k';
8   ax = gca;
9   ax.FontSize = 16;
10  ax.FontWeight='normal';
11  ylabel('Number')
12  xlabel('RE(%)')
13
14  title('M1 - Train','FontWeight','normal')
15  ylim([0 1.2*max(h.Values)])
16
17  % Model 1 - Test
18  figure;
19  h = histogram(RE_Test1);
20  h.NumBins =  10;
21  h.FaceColor =  'y';
22  h.EdgeColor = 'k';
23  ax = gca;
24  ax.FontSize = 16;
25  ax.FontWeight='normal';
26  ylabel('Number')
27  xlabel('RE(%)')
28  title('M1 - Test','FontWeight','normal')
29  ylim([0 1.2*max(h.Values)])
```

Code 3.13.D

```
1   %%%%%%%%%%%%%%%%%%%%%%%%%%%%%% Model 2 %%%%%%%%%%%%%%%%%%%%%%%%%%%%%%
2   % Model 2  - Train
3   figure;
4   h = histogram(RE_Train2);
5   h.NumBins =  10;
6   h.FaceColor =  'g';
7   h.EdgeColor = 'k';
8   ax = gca;
9   ax.FontSize = 16;
10  ax.FontWeight='normal';
11  ylabel('Number')
12  xlabel('RE(%)')
13  title('M2 - Train','FontWeight','normal')
14  ylim([0 1.2*max(h.Values)])
15  
16  % Model 2  - Test
17  figure;
18  h = histogram(RE_Test2);
19  h.NumBins =  10;
20  h.FaceColor =  'g';
21  h.EdgeColor = 'k';
22  ax = gca;
23  ax.FontSize = 16;
24  ax.FontWeight='normal';
25  ylabel('Number')
26  xlabel('RE(%)')
27  title('M2 - Test','FontWeight','normal')
28  ylim([0 1.2*max(h.Values)])
```

Code 3.13.E

```
1   89   %%%%%%%%%%%%%%%%%%%%%%%%%%%%%% Model 3 %%%%%%%%%%%%%%%%%%%%%%%%%%%%%%
2   90   % Model 3  - Train
3   91   figure;
4   92   h = histogram(RE_Train3);
5   93   h.NumBins =  10;
6   94   h.FaceColor =  'b';
7   95   h.EdgeColor = 'k';
8   96   ax = gca;
9   97   ax.FontSize = 16;
10  98   ax.FontWeight='normal';
11  99   ylabel('Number')
12  100  xlabel('RE(%)')
13  101  title('M3 - Train','FontWeight','normal')
14  102  ylim([0 1.2*max(h.Values)])
15  103  
16  104  % Model 3  - Test
17  105  figure;
18  106  h = histogram(RE_Test3);
19  107  h.NumBins =  10;
20  108  h.FaceColor =  'b';
21  109  h.EdgeColor = 'k';
22  110  ax = gca;
23  111  ax.FontSize = 16;
24  112  ax.FontWeight='normal';
25  113  ylabel('Number')
26  114  xlabel('RE(%)')
27  115  title('M3 - Test','FontWeight','normal')
28  116  ylim([0 1.2*max(h.Values)])
```

Code 3.13.F

```
1  %%%%%%%%%%%%%%%%%%%%%%%%%%%%%%%% Model 4 %%%%%%%%%%%%%%%%%%%%%%%%%%%%%%%%
2  % Model 4  - Train
3  figure;
4  h = histogram(RE_Train4);
5  h.NumBins =  10;
6  h.FaceColor =  'm';
7  h.EdgeColor = 'k';
8  ax = gca;
9  ax.FontSize = 16;
10 ax.FontWeight='normal';
11 ylabel('Number')
12 xlabel('RE(%)')
13 title('M4 - Train','FontWeight','normal')
14 ylim([0 1.2*max(h.Values)])
15 
16 % Model 1  - Test
17 figure;
18 h = histogram(RE_Test1);
19 h.NumBins =  10;
20 h.FaceColor =  'm';
21 h.EdgeColor = 'k';
22 ax = gca;
23 ax.FontSize = 16;
24 ax.FontWeight='normal';
25 ylabel('Number')
26 xlabel('RE(%)')
27 title('M4 - Test','FontWeight','normal')
28 ylim([0 1.2*max(h.Values)])
29 
30 %% Clear all except mandatory information
31 clearvars -except RE_Train1 RE_Train2 RE_Train3 RE_Train4 ...
32                   RE_Test1  RE_Test2  RE_Test3  RE_Test4
```

Fig. 3.25 shows the histogram of the relative error for all models in both the training and testing stages of Example 1. For Model 1, during the training stage, most RE is distributed $\pm$ 10%, while for the testing stage, the distribution is much more spread out. Additionally, in the training stage, there are some RE values greater than 25%, which is greater than what was observed in the testing stage. For Model 2, the distribution of negative RE for both training and testing stages is almost identical. The maximum positive RE for training is 15%, while it is 20% for the testing stage. For Models 3 and 4, the maximum negative and positive RE in the training stage is higher than in the testing stage, however, it should be noted that the RE in the range of $\pm$ 10% for both models in the training stage is much higher than the testing stage. As can be seen, this index shows the distribution of values and can not be used to assess the superiority or not of one model over other models.

Data on histogram can be found at Appendix 3A.

3.3.5 The boxplot

The boxplot is a standardized representative tool which indicates the distribution of samples using five parameters, including the minimum, the maximum, the first quartile (Q1), the third quartile (Q3), and the median. Using the boxplot, it is possible to identify the outliers within a dataset. Furthermore, this plot provides an indication of how an applied dataset may be skewed or tightly grouped. Outliers are considered to be samples that are outside of the minimum and maximum values defined in the boxplot. The minimum is defined using the interquartile range (IQR) as $Q1-1.5 \times IQR$, while the maximum is defined as $Q3 + 1.5 \times IQR$. For data with normal distributions, about 50% of all samples are located between Q3 and Q1 (Fig. 1.19).

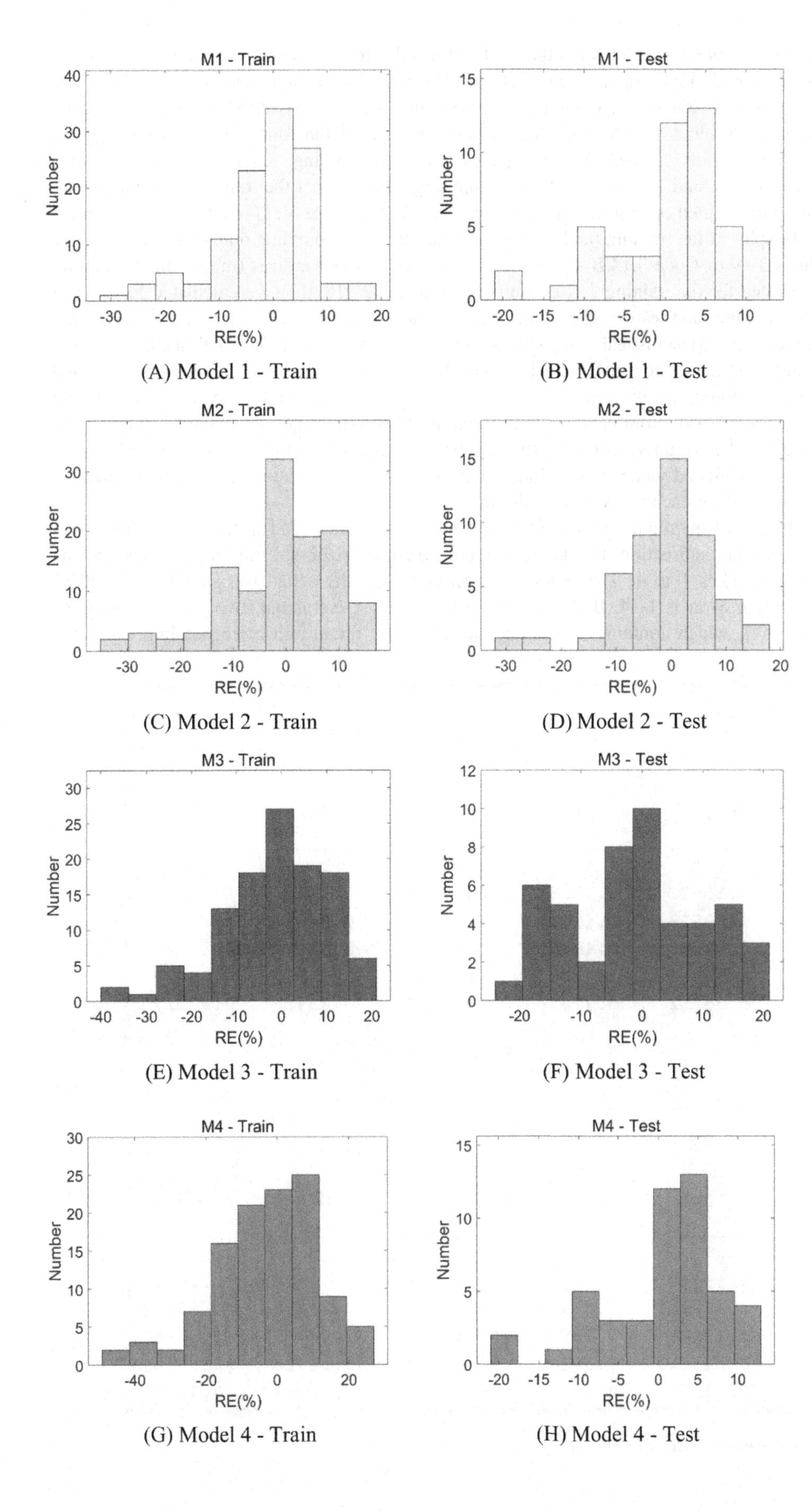

(A) Model 1 - Train

(B) Model 1 - Test

(C) Model 2 - Train

(D) Model 2 - Test

(E) Model 3 - Train

(F) Model 3 - Test

(G) Model 4 - Train

(H) Model 4 - Test

FIGURE 3.25 The histogram of the relative error for all models at both training and testing stages of Example 1.

The detailed MATLAB coding to generate a boxplot for actual and estimated samples (generated using four different models) is provided in Code 3.14. This code includes six main sections. The first two sections, clearing the MATLAB command window and workspace as well as loading of the data, have been detailed previously. After loading the data, variables are defined to store the training and testing targets and outputs for each of the four models; `TrainTargets`, `TrainOutputs1`, `TrainOutputs2`, `TrainOutputs3`, and `TrainOutputs4` for the training stage and `TestTargets`, `TestOutputs1`, `TestOutputs2`, `TestOutputs3`, and `TestOutputs4` for testing stage. Notice that the train and test targets are the same for each model and so are only defined once using the variables `TrainTargets` and`TestTargets`. The reason for defining these variables is that the boxplot of the training (and testing) samples and corresponding outputs by Models 1 to 4 are plotted simultaneously. In lines 3–9 of Code 3.14.B, the values of the five required indices (minimum, maximum, median, Q1, Q3, and IQR) are calculated for the training stage. Similarly, in lines 12–18, they are calculated for the test stage. The value of these indices for training and testing stages are stored in the variables `Tr` and `Ts`, respectively. Using the `array2table` command, the value of all five statistical indices for actual and estimated values for both the training and testing stages are calculated and presented in the command window. The `Train1.Properties.VariableNames` command is applied to assign the name of each column and the `Train1.Properties.RowNames` command is applied to assign the name of each row. The results of the `array2table` command for both training and testing stages are provided in Fig. 3.26. According to this figure, we can see that there are five columns, the first being related to the target values, while the second to fifth columns correspond to the estimated values from Models 1 to 4. Besides, the minimum, maximum, median, Q1, Q3, and maximum values are specified for each model (i.e., column).

The `boxplot` generation consists of two main subsections, train and test, which are coded in the same manner but consider different data inputs. For the train subsection, the data considered include the actual training samples and the corresponding estimated values from models 1 to 4. The test subsection considers the actual test samples as well as the corresponding estimated values from Models 1–4. Before applying the `boxplot` command, two variables must be defined. The first variable is `Train_All`, which contains all training samples (i.e., target and corresponding estimated

```
Train1 =

  5×5 table

                     Target      M1         M2         M3        M4
                     _______    _______    _______    ______    _______

    Minimum          0.15429    0.17567    0.17258    0.1228    0.14395
    Lower Quartile   0.39175    0.40093    0.40275    0.3898    0.39345
    Median           0.48953    0.49782    0.49256    0.4805    0.47589
    Upper Quartile   0.55006     0.5511     0.5562    0.5678    0.55978
    Maximum          0.78752    0.77636    0.78637    0.8348    0.80927


Test1 =

  5×5 table

                     Target      M1         M2         M3         M4
                     _______    _______    _______    _______    _______

    Minimum          0.15429    0.17567    0.17258     0.1228    0.14395
    Lower Quartile    0.3925    0.38907    0.39702    0.40817    0.39853
    Median           0.48828    0.48563    0.49004     0.4935    0.48447
    Upper Quartile   0.54842    0.54238    0.54667    0.54925     0.5469
    Maximum          0.78231    0.77234    0.77116    0.76087    0.76945
```

FIGURE 3.26 The results of the `array2table` command for Example 1.

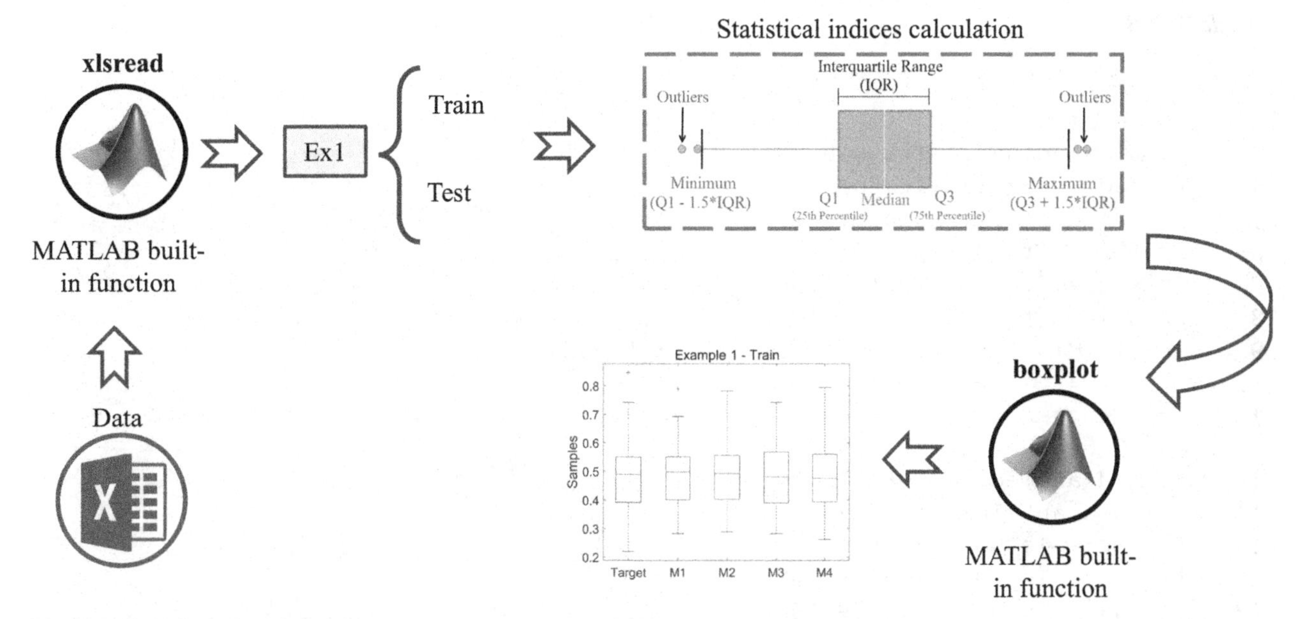

FIGURE 3.27 Schematic of Code 3.14.

values by Models 1–4) in a column. The second variable isg, which is a column-based matrix with zero values. The box plots are plotted using these two newly generated variables (i.e., `Train_All` and `g`) as input arguments. The other input arguments of the `boxplot` command include `Notch` and `labels`. The `Notch` argument can be set to "on" or "off" depending on whether the user wants a notch in the plot at the median value. The`labels` input argument is defined as a character with the categories of data located in the `Train_All` variable. Finally, the size and weight of the font, the y-axis label, and the plot title are defined. This process is repeated for the test subsection. Finally, at line 75, all variables except `Tr` and `Ts` are removed using `clearvars -except` command. The codes of the other examples are provided in the MATLAB-based package. The schematic of Code 3.14 is shown in Fig. 3.27.

Code 3.14.A

```
1   clc
2   clear
3   close all
4   
5   %% Load data
6   Ex5 = xlsread ('Data','sheet5');    % Example 5
7   TrainTargets   = Ex5(1:767,1);      % Train Targets
8   TrainOutputs1  = Ex5(1:767,2);      % Train Outputs - Model 1
9   TrainOutputs2  = Ex5(1:767,3);      % Train Outputs - Model 2
10  TrainOutputs3  = Ex5(1:767,4);      % Train Outputs - Model 3
11  TrainOutputs4  = Ex5(1:767,5);      % Train Outputs - Model 4
12  TestTargets    = Ex5(768:end,1);    % Test Targets
13  TestOutputs1   = Ex5(768:end,2);    % Test Outputs  - Model 1
14  TestOutputs2   = Ex5(768:end,3);    % Test Outputs  - Model 2
15  TestOutputs3   = Ex5(768:end,4);    % Test Outputs  - Model 3
16  TestOutputs4   = Ex5(768:end,5);    % Test Outputs  - Model 4
17  
18  Train =
    [TrainTargets,TrainOutputs1,TrainOutputs2,TrainOutputs3,TrainOutputs4];
19  Test  =
    [TestTargets,TestOutputs1,TestOutputs2,TestOutputs3,TestOutputs4];
```

Code 3.14.B

```
1   %% Statistical indices of the Target and Models 1 to 4
2   % Training stage
3   Tr_LQ = quantile(Train,0.25);
4   Tr_Median=median(Train);
5   Tr_UQ=quantile(Train,0.75);
6   Tr_IQR = Tr_UQ - Tr_LQ;
7   Tr_Min = Tr_LQ - 1.5*Tr_IQR;
8   Tr_Max = Tr_UQ + 1.5*Tr_IQR;
9   Tr=[Tr_Min;Tr_LQ;Tr_Median;Tr_UQ;Tr_Max;];
10
11  % Testing stage
12  Ts_LQ = quantile(Test,0.25);
13  Ts_Median=median(Test);
14  Ts_UQ=quantile(Test,0.75);
15  Ts_IQR = Ts_UQ - Ts_LQ;
16  Ts_Min = Ts_LQ - 1.5*Ts_IQR;
17  Ts_Max = Ts_UQ + 1.5*Ts_IQR;
18  Ts=[Tr_Min;Ts_LQ;Ts_Median;Ts_UQ;Ts_Max;];
```

Code 3.14.C

```
1   %% Show the results of indices in the command window
2   Train1 = array2table([Tr]);
3   Train1.Properties.VariableNames = {'Target','M1','M2','M3','M4'};
4   Train1.Properties.RowNames = {'Minimum','Lower Quartile','Median','Upper
    Quartile','Maximum'}
5
6   Test1  = array2table([Ts]);
7   Test1.Properties.VariableNames = {'Target','M1','M2','M3','M4'};
8   Test1.Properties.RowNames = {'Minimum','Lower Quartile','Median','Upper
    Quartile','Maximum'}
```

Code 3.14.D

```
1  %% Boxplot
2  % Train
3  Train_All =
   [TrainTargets;TrainOutputs1;TrainOutputs2;TrainOutputs3;TrainOutputs4];
4  figure;
5  g = [zeros(length(TrainTargets), 1); ones(length(TrainOutputs1), 1);
   2*ones(length(TrainOutputs2), 1);3*ones(length(TrainOutputs3),
   1);4*ones(length(TrainOutputs4), 1)];
6  boxplot(Train_All,
   g,'Notch','off','labels',{'Target','M1','M2','M3','M4'})
7  ax = gca;
8  ax.FontSize = 16;
9  ax.FontWeight='normal';
10 ylabel('Samples')
11 title('Example 5 - Train','FontWeight','normal')
12 
13 % Test
14 Test_All =
   [TestTargets;TestOutputs1;TestOutputs2;TestOutputs3;TestOutputs4];
15 figure;
16 g = [zeros(length(TestTargets), 1); ones(length(TestOutputs1), 1);
   2*ones(length(TestOutputs2), 1);3*ones(length(TestOutputs3),
   1);4*ones(length(TestOutputs4), 1)];
17 boxplot(Test_All,
   g,'Notch','off','labels',{'Target','M1','M2','M3','M4'})
18 ax = gca;
19 ax.FontSize = 16;
20 ax.FontWeight='normal';
21 ylabel('Samples')
22 title('Example 5 - Test','FontWeight','normal')
23 
24 %% Clear all except mandatory information
25 clearvars -except Tr Ts
```

Fig. 3.28 presents the boxplot of the training and testing samples for all examples. For Example 1, in the training stage, the actual sample (i.e., Target) has one outlying data point which none of the models (i.e., Models 1–4) could have estimated it correctly. Model 2 performed better than others in the estimation of this sample. Estimating the outliers could be necessary in some cases, and it may not be vital in other cases. For example, predicting the peak flow, whose highest values often constitute an outlier, is very important in rainfall and runoff forecasting. The distribution of the estimated samples by different models is very close to the actual samples. Still, some differences should be considered, such as the Q1 of Model 2 that is lower than actual samples or higher values of the maximum for Models 2 and 4 and lower values of the maximum for Models 1 and 3. For the testing stage, the distribution of samples in M3 and M4 are very close to the Target, while there is some difference between the distribution of the samples in M2 and M3 with Target.

The boxplot results of Example 2 indicate a huge difference between the actual data and the estimated samples in both training and testing. This difference is especially noticeable in the distance between Q3 and the maximum, where this distance for M1–M4 is much greater than the value provided for Target. In addition, the IQR value in the four proposed models is very different from the Target IQR value.

For Example 3, the distribution of the estimated values compared to the actual ones is similar to Example 2, except that M2 in the training stage and M1 in both the training and test stages are very close to the actual data distribution. For M3 and M4 in both train and test stages, there are significant differences with Target.

In Example 4, the distribution of the estimated values in both training and testing of the M1−M4 are very close to the Target, indicating their performance is almost reasonable in estimating the desired target variable. However, there are differences in estimating the maximum value by M1−M4 with Target

The box plots provided for the estimated and actual values for Example 5 show that the IQRs for M2 and M3 in the training phase are smaller than in the Target, while this is not the case in the testing phase. Of all the models, the M3 is the most compatible with the Target, so the difference between the essential variables in the boxplot, including minimum, Q1, Q3, and maximum, is significantly similar to the Target.

According to the explanations of the performance of different models in estimating the target parameter in five different problems, it was observed that the use of boxplots could provide a good view of the distribution of estimated values by a given model compared to actual values. It should be noted that instead of using examples in drawing a boxplot, it can also use errors and compare the distribution of errors in different models with each other.

Data on boxplot can be found at Appendix 3A.

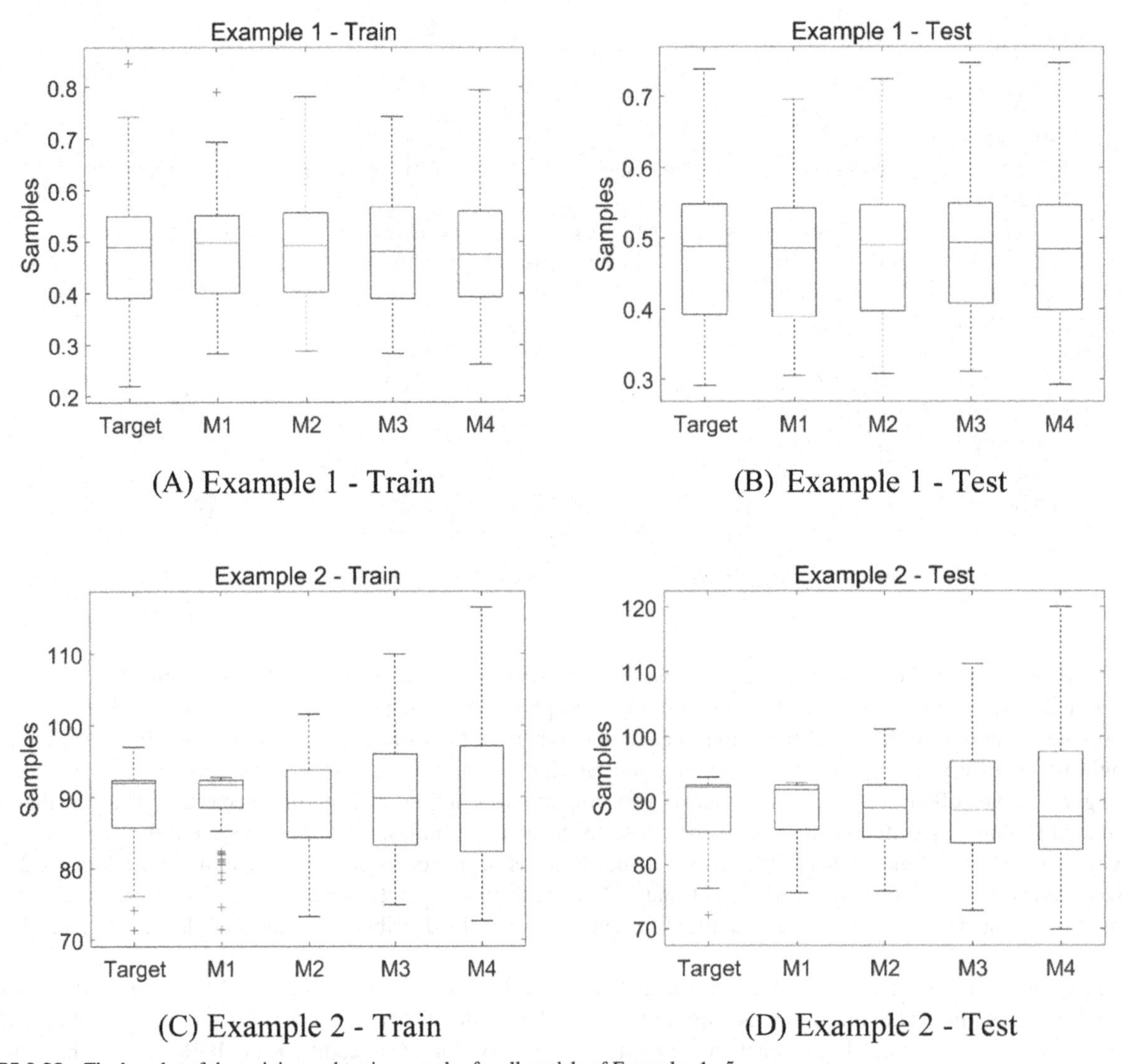

FIGURE 3.28 The boxplot of the training and testing samples for all models of Examples 1−5.

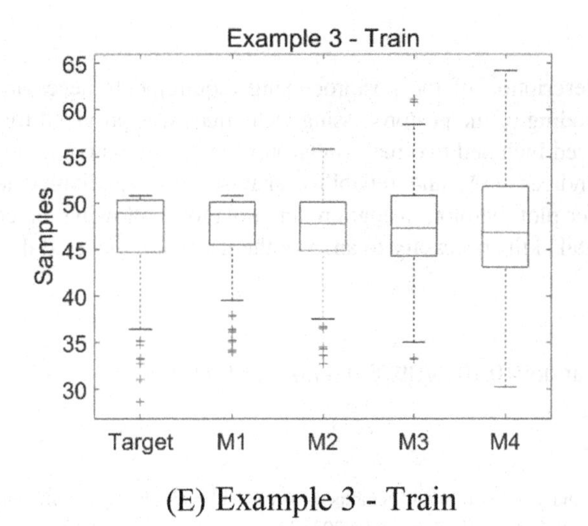

(E) Example 3 - Train

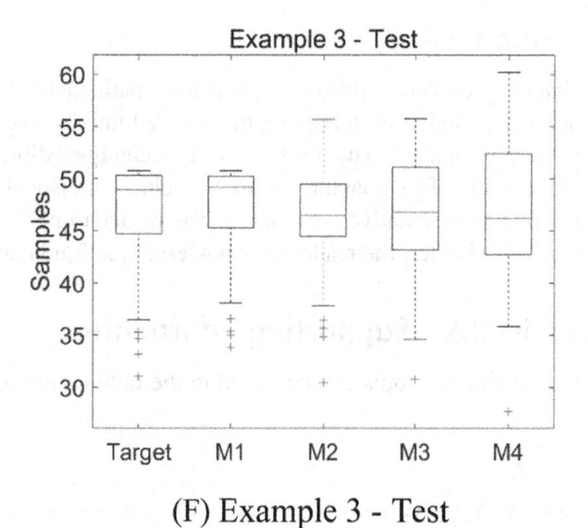

(F) Example 3 - Test

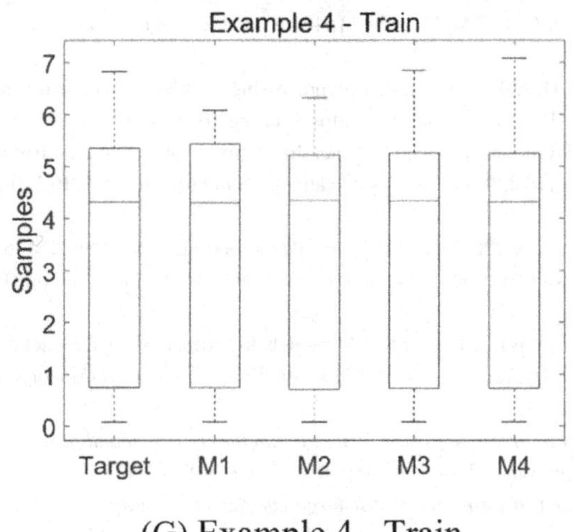

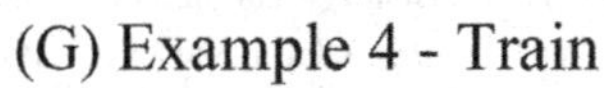

(G) Example 4 - Train

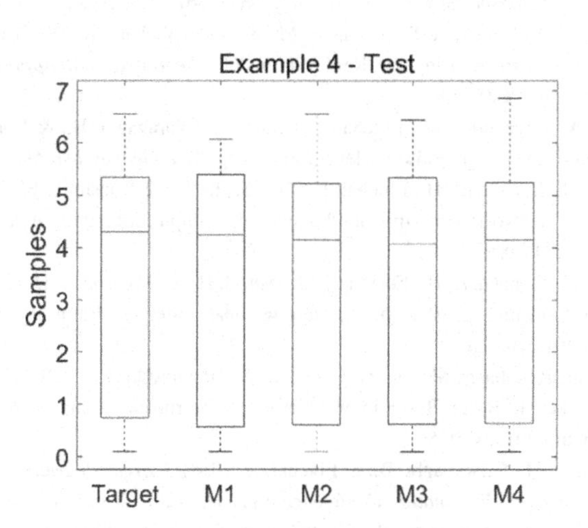

(H) Example 4 - Test

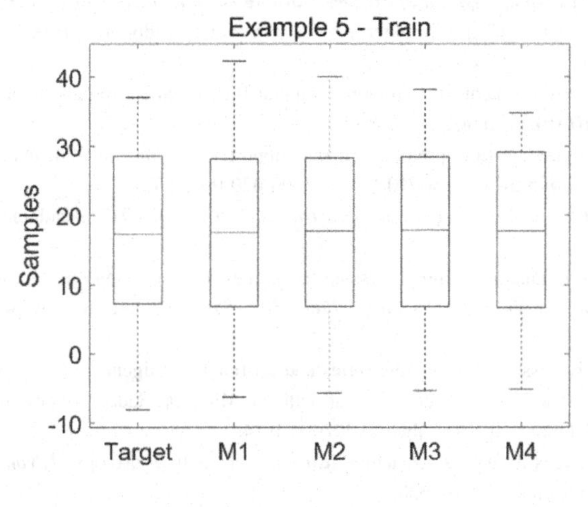

(I) Example 5 - Train

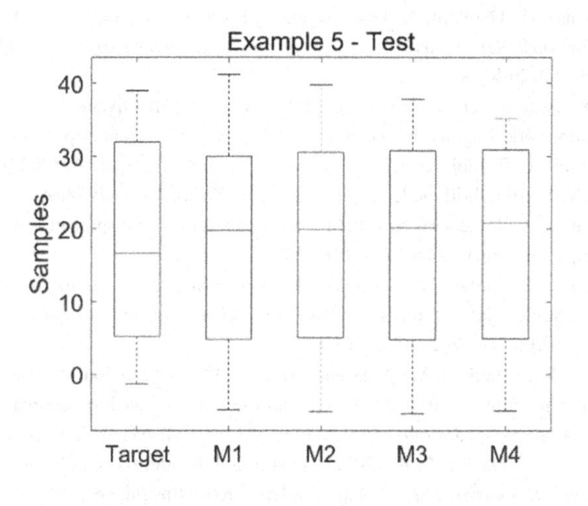

(J) Example 5 - Test

FIGURE 3.28 (Continued)

3.4 Summary

This chapter provided a full conceptual and mathematical description of the postprocessing requirements necessary in applying a ML model. Furthermore, the detailed line-by-line coding of the postprocessing techniques was provided for the MATLAB environment. The postprocessing techniques discussed included two main subgroups: (1) Quantitative tools, (2) Qualitative tools. The quantitative tools include statistical indices, UA, and reliability analysis. The qualitative tools included five representative approaches: Taylor diagram, scatter plot, barplot, histogram, and boxplot. Following the completion of this chapter, the reader now possesses the fundamental skills necessary to analyze the input and modeled data.

Appendix 3A Supporting information

Data on quantitative tools can be found in the online version at doi:10.1016/B978-0-443-15284-9.00002-1.

References

Ai, P., Song, Y., Xiong, C., Chen, B., & Yue, Z. (2022). A novel medium-and long-term runoff combined forecasting model based on different lag periods. *Journal of Hydroinformatics*, *24*(2), 367–387. Available from https://doi.org/10.2166/hydro.2022.116.

Anwar, S. A., Zakey, A. S., Robaa, S. M., & Abdel Wahab, M. M. (2019). The influence of two land-surface hydrology schemes on the regional climate of Africa using the RegCM4 model. *Theoretical and Applied Climatology*, *136*(3), 1535–1548. Available from https://doi.org/10.1007/s00704-018-2556-8.

Azari, A., Zeynoddin, M., Ebtehaj, I., Sattar, A., Gharabaghi, B., & Bonakdari, H. (2021). Integrated preprocessing techniques with linear stochastic approaches in groundwater level forecasting. *Acta Geophysica*, *69*(4), 1395–1411. Available from https://doi.org/10.1007/s11600-021-00617-2.

Azimi, H., Bonakdari, H., Ebtehaj, I., Gharabaghi, B., & Khoshbin, F. (2018). Evolutionary design of generalized group method of data handling-type neural network for estimating the hydraulic jump roller length. *Acta Mechanica*, *229*, 1197–1214. Available from https://doi.org/10.1007/s00707-017-2043-9.

Azimi, H., Bonakdari, H., Ebtehaj, I., Talesh, S. H. A., Michelson, D. G., & Jamali, A. (2017). Evolutionary Pareto optimization of an ANFIS network for modeling scour at pile groups in clear water condition. *Fuzzy Sets and Systems*, *319*, 50–69. Available from https://doi.org/10.1016/j.fss.2016.10.010.

Barzegar, R., Ghasri, M., Qi, Z., Quilty, J., & Adamowski, J. (2019). Using bootstrap ELM and LSSVM models to estimate river ice thickness in the Mackenzie River Basin in the Northwest Territories, Canada. *Journal of Hydrology*, *577*, 123903. Available from https://doi.org/10.1016/j.jhydrol.2019.06.075.

Bonakdari, H., Gharabaghi, B., & Ebtehaj, I. (2018). *Extreme learning machines in predicting the velocity distribution in compound narrow channels. Science and Information Conference* (pp. 119–128). Cham: Springer. Available from https://doi.org/10.1007/978-3-030-01177-2_9.

Bonakdari, H., Gharabaghi, B., Ebtehaj, I., & Sharifi, A. (2020). *A new approach to estimate the discharge coefficient in sharp-crested rectangular side orifices using gene expression programming. Science and information conference* (pp. 77–96). Cham: Springer. Available from https://doi.org/10.1007/978-3-030-52243-8_7.

Bonakdari, H., Gholami, A., Gharabaghi, B., Ebtehaj, I., & Akhtari, A. A. (2021). *An assessment of extreme learning machine model for estimation of flow variables in curved irrigation channels. Intelligent computing* (pp. 259–269). Cham: Springer. Available from https://doi.org/10.1007/978-3-030-80129-8_19.

Cui, Y., Fang, J., Li, Y., & Liu, H. (2022). Assessing effectiveness of a dual-barrier system for mitigating granular flow hazards through DEM-DNN framework. *Engineering Geology*, 106742. Available from https://doi.org/10.1016/j.enggeo.2022.106742.

Dehghani, R., Torabi Poudeh, H., Younesi, H., & Shahinejad, B. (2020). Daily streamflow prediction using support vector machine-artificial flora (SVM-AF) hybrid model. *Acta Geophysica*, *68*(6), 1763–1778. Available from https://doi.org/10.1007/s11600-020-00472-7.

Ebtehaj, I., & Bonakdari, H. (2016). Bed load sediment transport in sewers at limit of deposition. *Scientia Iranica*, *23*(3), 907–917. Available from https://doi.org/10.24200/sci.2016.2169.

Ebtehaj, I., Bonakdari, H., Zeynoddin, M., Gharabaghi, B., & Azari, A. (2020). Evaluation of preprocessing techniques for improving the accuracy of stochastic rainfall forecast models. *International Journal of Environmental Science and Technology*, *17*(1), 505–524. Available from https://doi.org/10.1007/s13762-019-02361-z.

Ebtehaj, I., Zeynoddin, M., & Bonakdari, H. (2020). Discussion of "comparative assessment of time series and artificial intelligence models to estimate monthly streamflow: A local and external data analysis approach" by Saeid Mehdizadeh, Farshad Fathian, Mir Jafar Sadegh Safari and Jan F. Adamowski. *Journal of Hydrology*, *583*, 124614. Available from https://doi.org/10.1016/j.jhydrol.2020.124614.

Gharib, A., & Davies, E. G. (2021). A workflow to address pitfalls and challenges in applying machine learning models to hydrology. *Advances in Water Resources*, *152*, 103920. Available from https://doi.org/10.1016/j.advwatres.2021.103920.

Gholami, A., Bonakdari, H., Ebtehaj, I., & Akhtari, A. A. (2017). Design of an adaptive neuro-fuzzy computing technique for predicting flow variables in a 90 sharp bend. *Journal of Hydroinformatics*, *19*(4), 572–585. Available from https://doi.org/10.2166/hydro.2017.200.

Khozani, Z. S., Bonakdari, H., & Ebtehaj, I. (2017). An analysis of shear stress distribution in circular channels with sediment deposition based on gene expression programming. *International Journal of Sediment Research*, *32*(4), 575–584. Available from https://doi.org/10.1016/j.ijsrc.2017.04.004.

Moeeni, H., Bonakdari, H., & Ebtehaj, I. (2017a). Integrated SARIMA with neuro-fuzzy systems and neural networks for monthly inflow prediction. *Water Resources Management*, *31*(7). Available from https://doi.org/10.1007/s11269-017-1632-7.

Moeeni, H., Bonakdari, H., & Ebtehaj, I. (2017b). Monthly reservoir inflow forecasting using a new hybrid SARIMA genetic programming approach. *Journal of Earth System Science*, *126*(2), 18. Available from https://doi.org/10.1007/s12040-017-0798-y.

Mohsenzadeh Karimi, S., Kisi, O., Porrajabali, M., Rouhani-Nia, F., & Shiri, J. (2020). Evaluation of the support vector machine, random forest and geo-statistical methodologies for predicting long-term air temperature. *ISH Journal of Hydraulic Engineering*, *26*(4), 376–386. Available from https://doi.org/10.1080/09715010.2018.1495583.

Mojtahedi, S. F. F., Ebtehaj, I., Hasanipanah, M., Bonakdari, H., & Amnieh, H. B. (2019). Proposing a novel hybrid intelligent model for the simulation of particle size distribution resulting from blasting. *Engineering with Computers*, *35*(1), 47–56. Available from https://doi.org/10.1007/s00366-018-0582-x.

Ray, A., Kumar, V., Kumar, A., Rai, R., Khandelwal, M., & Singh, T. N. (2020). Stability prediction of Himalayan residual soil slope using artificial neural network. *Natural Hazards*, *103*(3), 3523–3540. Available from https://doi.org/10.1007/s11069-020-04141-2.

Saberi-Movahed, F., Najafzadeh, M., & Mehrpooya, A. (2020). Receiving more accurate predictions for longitudinal dispersion coefficients in water pipelines: Training group method of data handling using extreme learning machine conceptions. *Water Resources Management*, *34*(2), 529–561. Available from https://doi.org/10.1007/s11269-019-02463-w.

Safari, M. J. S., Ebtehaj, I., Bonakdari, H., & Es-haghi, M. S. (2019). Sediment transport modeling in rigid boundary open channels using generalize structure of group method of data handling. *Journal of Hydrology*, *577*, 123951. Available from https://doi.org/10.1016/j.jhydrol.2019.123951.

Shukla, R., Kumar, P., Vishwakarma, D. K., Ali, R., Kumar, R., & Kuriqi, A. (2022). Modeling of stage-discharge using back propagation ANN-, ANFIS-, and WANN-based computing techniques. *Theoretical and Applied Climatology*, *147*(3), 867–889. Available from https://doi.org/10.1007/s00704-021-03863-y.

Siddik, M. A. Z. (2022). Application of machine learning approaches in predicting estuarine dissolved oxygen (DO) under a limited data environment. *Water Quality Research Journal*. Available from https://doi.org/10.2166/wqrj.2022.002.

Taylor, K. E. (2001). Summarizing multiple aspects of model performance in a single diagram. *Journal of Geophysical Research: Atmospheres*, *106* (D7), 7183–7192. Available from https://doi.org/10.1029/2000JD900719.

Zeynoddin, M., Bonakdari, H., Ebtehaj, I., Azari, A., & Gharabaghi, B. (2020). A generalized linear stochastic model for lake level prediction. *Science of The Total Environment*, *723*, 138015. Available from https://doi.org/10.1016/j.scitotenv.2020.138015.

Zeynoddin, M., Ebtehaj, I., & Bonakdari, H. (2020). Development of a linear based stochastic model for daily soil temperature prediction: One step forward to sustainable agriculture. *Computers and Electronics in Agriculture*, *176*, 105636. Available from https://doi.org/10.1016/j.compag.2020.105636.

Zeynolabedin, A., Ghiassi, R., Norooz, R., Najib, S., & Fadili, A. (2021). Evaluation of geoelectrical models efficiency for coastal seawater intrusion by applying uncertainty analysis. *Journal of Hydrology*, *603*, 127086. Available from https://doi.org/10.1016/j.jhydrol.2021.127086.

Zhang, Y., Gu, Z., Yang, S. X., & Gharabaghi, B. (2022). The discharge forecasting of multiple monitoring station for humber river by hybrid LSTM models. *Water*, *14*(11), 1794. Available from https://doi.org/10.3390/w14111794.

Chapter 4

Non-tuned single-layer feed-forward neural network learning machine—concept ☆

4.1 Machine learning application in applied science

Machine learning (ML) is a subset of artificial intelligence (AI) that provides modeling frameworks with the capacity to spontaneously learn and progress through their accumulated experience. An important feature of ML algorithms is that the learning undertaken by the model is performed without being explicitly programmed by a human user. The advancement of computing technologies has enabled the development of ML algorithms which can access information and utilize it to learn for themselves. Classical approaches such as artificial neural networks (ANNs) require many parameters to be fine-tuned prior to the commencement of modeling, with the results of the model being highly sensitive to the value of these tuned input parameters. In the case of the extreme learning machine (ELM), however, the only parameters that are required adjustment are the number of hidden layers and the type of activation function. The relationship between AI, ML, ANN, and ELM is summarized graphically in Fig. 4.1.

This chapter is presented for both beginner and advanced users (Fig. 4.2) and details the ELM algorithm in a simple and easy-to-understand way were discussed. In general, beginner users should have an understanding of the ELM concept, while ML users should gain a comprehensive understanding of the fundamental mathematical formulation of the ELM algorithm.

Based on the findings of a large number of studies in environmental science, the nature of the relationship between the dependent variable and the predictor(s) can be quite complex. For this reason, many studies employing classical multiple linear regression (MLR) tools have observed a low degree of precision, as these methods are often unable to describe such complicated relationships. An example of the inability of classical MLR tools to determine the relationship between dependent variables and predictor variables is sediment transport in sewers and open channels. Sediment transport in sewers can be problematic if the grade of the pipe is not sufficient to transport the suspended solids. This can lead to deposition and subsequent compaction of the solid materials leading to a reduced transport capacity of the pipe. The most important factors affecting sediment transport and deposition in open channels can be categorized into four groups, including the channel parameters and the fluid, sediment, and flow characteristics (Ebtehaj et al., 2020; Montes et al., 2021). The fluid characteristics include viscosity and specific gravity, while the volumetric sediment concentration, sediment particle size, and sediment-specific gravity comprise the sediment characteristics. The flow characteristics include channel geometry, such as the cross-sectional area of flow and the hydraulic radius, as well as the flow mean velocity and the gravitational acceleration constant. The bed friction factor and pipe diameter are considered to be channel parameters. In modeling analysis, these four groups of parameters have been investigated by considering seven dimensionless features. Reduction in dimensionality is commonly employed to increase predictive accuracy and the number of predictor variables.

The results of several studies regarding sediment deposition (Azamathulla et al., 2012; Roushangar & Ghasempour, 2017; Zounemat-Kermani et al., 2021) concluded that the MLR relationships did not perform well in the identified velocity ranges that limited sediment deposition. Furthermore, the application of ML models was found to have a greater predictive performance for the depth of deposited sediment when compared to classical models (Ebtehaj & Bonakdari, 2013, 2014, 2016; Tafarojnoruz & Sharafati, 2020). Another example of the power of ML techniques is in

☆. Sections 4.2.1 and 4.2.2 are provided for advance users.

Machine Learning in Earth, Environmental and Planetary Sciences. DOI: https://doi.org/10.1016/B978-0-443-15284-9.00001-X

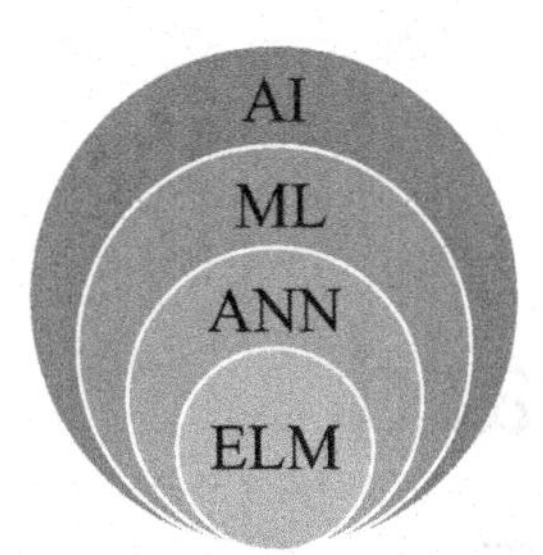

Artificial Intelligence (AI)
Any technique that helps machines perform tasks as human beings do

Machine learning (ML)
Adaptable algorithms that do not require explicit programming and that automate by observing examples

Artificial Neural network (ANN)
Machine learning based models inspired by the human brain

Extreme learning Machine (ELM)
A rapid single-layer feed-forward neural network

FIGURE 4.1 The relationship between artificial intelligence, machine learning, artificial neural network, and extreme learning machine.

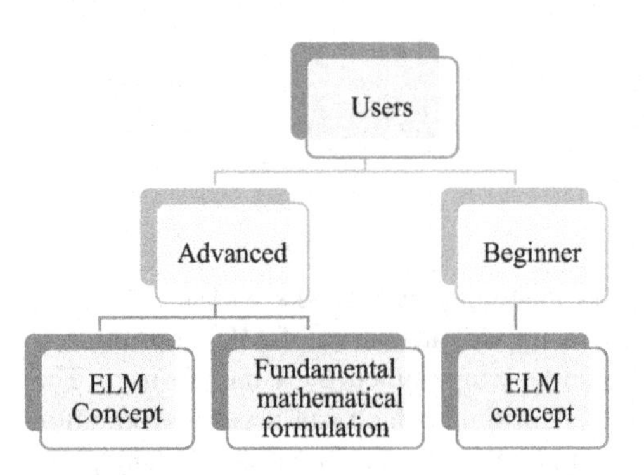

FIGURE 4.2 Chapter learning objectives for different users.

the prediction of the discharge coefficient for a side weir. Side weirs are commonly used to regulate flow and runoff in irrigation networks and wastewater networks, however, they have a complex flow regime near the outlet. Several research groups have employed ML-based studies in order to estimate the discharge coefficient of the weir (Azimi et al., 2017; Chen et al., 2022; Granata et al., 2019; Karbasi et al., 2021; Khoshbin et al., 2016). These studies demonstrated definitively that the application of ML techniques could improve the predicted discharge coefficient when compared to classical methods from the literature. The error distribution of ML-based results was found to be smaller for ML approaches (Khoshbin et al., 2016).

4.1.1 Background of the feed-forward neural network

Over the past decade, owing to their powerful nature, ML-based approaches have been identified as reliable nonlinear modeling tools that have a vast array of applications in environmental sciences. One of the best-known ML-based approaches is the feed-forward neural network (FFNN). The structure of the FFNN consists of three main layers, namely an input, hidden, and output layer. The number of neurons in the input layer is defined by the number of independent input variables that are considered in the specific application. The output layer generally consists of one neuron (the dependent variable), however, it may possess more than one neuron depending on how many dependent variables your application requires. Based on the above discussion, the size of the input and output layers are generally fixed, with the number of neurons in each layer dependent on the nature of the problem to which the model is applied. The number of neurons in the hidden layer, however, is a user-defined variable that can be adjusted to arrive at an optimal solution. Furthermore, the number of hidden layers can be adjusted in FFNNs. Training of the FFNN is the process that is undertaken to optimize the adjustable parameters of the network. Besides the user-defined size of the hidden layer, the adjustable parameters considered during model training are the input weights (InWs), the bias of the hidden neurons (BHN), and the output weights (OutWs). The InWs connect all neurons of the input layer (i.e., input parameters) to all the hidden neurons, while the OutWs connect all the hidden neurons to the output variable(s).

4.1.1.1 Feed-forward neural network with backpropagation algorithm

The backpropagation (BP) algorithm is the most well-known training algorithm in the optimization of FFNN parameters (i.e., weights and bias). The BP algorithm is a first-order gradient-based method which is applied to optimize the weights and biases of the FFNN (Albadra & Tiuna, 2017; Ng et al., 2012). BP is commonly employed to train classical neural networks by optimizing the learning algorithm and stabilizing the weight of the neurons. To do this, the BP algorithm calculates the reduction gradient of the cost function: this is why this algorithm is known as a gradient-based algorithm. Following the introduction of the BP algorithm for supervised learning, the use of this algorithm in training FFNN increased significantly, such that the FFNN has been applied to a wide range of research. Some of the research applications that consider the FFNN include: (1) approximation of the nonlinear mapping between input features and target parameters and (2) furnishing models for artificial and natural phenomena that present difficulty to classical multiple nonlinear regression tools.

4.1.1.1.1 Drawbacks of the classical feed-forward neural network

While the use of BP as a training algorithm for FFNN provides a relatively simple solution to the problem of optimization, there are, however, some critical drawbacks that limit its use in practice. Several of these drawbacks are shown in Fig. 4.3 and include the significant length of the training process, which is due to iterative adjustment of the FFNN parameters as well as slow convergence (Bonakdari & Ebtehaj, 2016; Bonakdari, Qasem et al. 2020; Bonakdari et al., 2019; Melo & Watada, 2016). It has also been reported that BP algorithms have low generalization ability, problems with local minima and maxima, as well as trouble overfitting the data (Abba et al., 2020; Bonakdari et al., 2019; Ebtehaj et al., 2018). Whenever nonlinearity is introduced into the problem, there is no guarantee of convergence for BP algorithms on the optimal solution.

4.1.1.1.2 Concepts of the existing drawbacks in feed-forward neural network

As mentioned earlier, slow convergence and overfitting are two critical drawbacks of the FFNN when employed with BP as the training algorithm. Indeed, the slow convergence rate significantly increases the modeling time, so that, in a problem with a large number of inputs, the modeling time may reach more than an hour. Overfitting produces a model that has been exclusively fine-tuned to the training data set and lacks the ability to be generalized to new data sets. For this reason, when the overfitted model is applied to data that did not play a role in calibrating the model, poor results are observed. In fact, the overfitted model often has so many parameters that it loses the ability to interpret any new data. For example, one parameter that can lead to the overfitting of FFNNs is the number of neurons in the hidden layer. Of course, if we continuously increase the number of hidden neurons (NHN), we will be able to achieve excellent performance during the training phase. However, there is a limit to the number of neurons that can be considered before overfitting occurs, and the model loses its ability to generalize to the testing set or any other unseen data. Indeed, the NHN should be considered in such a way that the model performs well during both the training and testing stages. The essence of overfitting shows that the calibrated model has actually extracted sampling variances from the training data set that will then form a part of the structure of the developed model (Burnham & Anderson, 2002).

4.1.2 Introduction to the extreme learning machine

From the above discussion, we can see that the development and training of ML-based algorithms still remain a challenge that is at the forefront of research. Scholars seek to implement an automated technique in which ML-based approaches can be universally developed without facing the drawbacks that currently exist with traditional methods (such as BP). To overcome the high number of tuned parameters and slow learning speed of the gradient-based algorithms, Huang et al. (2006) introduced a novel training algorithm for single-layer FFNN (SLFFNN), which is known as ELM.

4.1.2.1 The structure of the extreme learning machine model

The structure of the ELM model is composed of three layers, including the input layer, hidden layer, and output layer. In this sense, it is similar to the FFNN outlined previously. The size of the input and output layers is defined by the nature of the problem for which the model is being employed. The number of neurons in the input layer is equal to the number of input variables, while the output layer most often consists of a single neuron (output variable). It should be

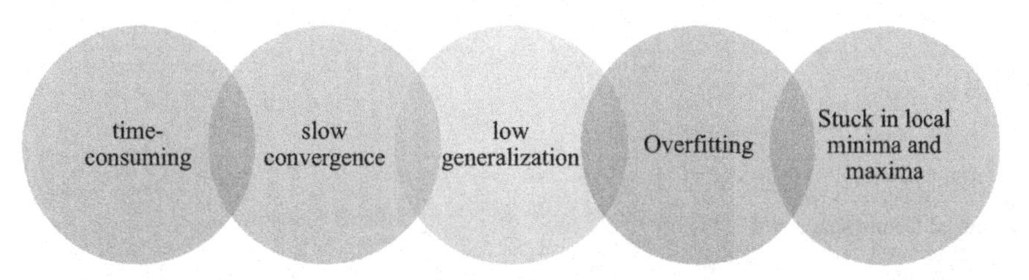

FIGURE 4.3 Main drawbacks of the classical feed-forward neural network.

noted that the output layer can include more than one neuron if the problem under consideration has more than one dependent variable. However, in most ELM problems, the ELM is defined with one output such that if there are several outputs (i.e., *n* outputs), the model will run n times to model all of the output variables separately. The number of neurons in the hidden layer is a parameter that is adjusted by the user before the commencement of the modeling process. The NHN to be considered should be tuned so as to balance high model performance and simplicity. The NHN should be considered as small as possible to simplify the model and prevent overfitting, however, the model should possess high performance in both the training and testing phases. A schematic of the ELM architecture is presented in Fig. 4.4. The structure shown in this figure is considered for a model with three input variables, four hidden neurons, and one output parameter. From the figure, we can see that the hidden neurons are defined as a "nontuned feature space." This phrase suggests that the hidden phase does not require any parameters to be tuned during model development. The NHN and the type of activation function employed are the only parameters that are allocated by the user in the ELM method, and the model does not require the definition of any other parameters.

4.1.2.2 Advantages and disadvantages of the extreme learning machine

Huang et al. (2006) demonstrated that by considering the infinitely differentiable activation function of the SLFFNN, the bias of the hidden layer as well as the InWs could be calculated. Through the training stage of the SLFFNN, three different matrices must be optimized, including the InWs, hidden neuron biases, and OutWs. The matrices of the InW and hidden neuron biases are randomly generated by the algorithm, and the ELM is then reduced to a linear system where the only unknown is the matrix of OutWs. The matrix of OutWs is calculated analytically by solving a linear system via an effortless generalized inverse process using the hidden layer output matrices. A critical advantage of this method is that it overcomes the tendency of traditional BP methods to become stuck in local minima in the solution space. Furthermore, this method only requires a single iteration for the learning algorithm to optimize the output matrix, resulting in a significant reduction in the processing time. Another advantage is that to model a nonlinear problem using the ELM, the only user-defined parameters are the number of hidden layers and the type of activation function. This means that the ELM approach requires minimum user intervention during the model construction process. A complete description of the advantages of the ELM is presented in Table 4.1.

The proposed ELM training process results not only in a high learning speed, which can be thousands of times quicker when compared to conventional methods but also better generalization performance. In contrast to classical training algorithms, the proposed ELM algorithm not only possess a propensity to achieve the lowest training error but also the lowest weights norm. A matrix norm is a number defined in terms of the entries of the matrix. The norm is a useful quantity which can give important information about a matrix. Simultaneous reduction of the model error and the weight norms increase the generalization of the model. For this reason, the proposed ELM algorithm has a tendency to achieve strong generalization performance for SLFFNN. Hence, the main advantages of the ELM learning process are threefold: High learning accuracy, low user invention, and fast learning speed (in seconds, milliseconds, and even microseconds) (Fig. 4.5).

4.2 Mathematical definition of extreme learning machine model

In this subsection, a comprehensive explanation of the SLFFNN with random hidden nodes is presented, including all the associated mathematical expressions. Following this, the minimum norm least-squares solution of the SLFFNN,

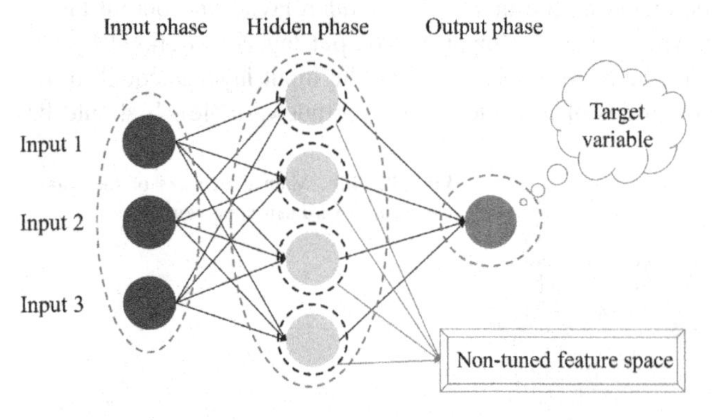

FIGURE 4.4 Schematic of the extreme learning machine model.

TABLE 4.1 Principle advantages and disadvantages of the extreme learning machine.

Pros	Cons
• Avoids becoming stuck in local minima	• Random allocation of the input weights and bias of hidden neurons
• Controllability	
• Fast learning rate	
• High accuracy	
• High generalization	
• Learning process requires only a single iteration	
• Low user intervention	
• Robustness	

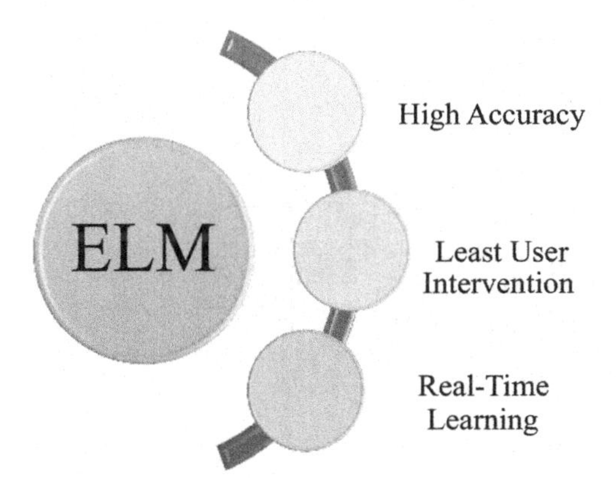

FIGURE 4.5 Advantages of extreme learning machine learning framework.

which is one of the main principles used in ELM, is presented. Finally, the proposed algorithm for SLFFNN training is presented, which is referred to as the ELM method. All of the subsections are divided into beginner and advanced, such that the beginner user can easily apply the developed code for nonlinear problem modeling using the ELM method (by employing the novel calculator seen in Fig. 4.6) without the requirement of reviewing all the advanced sections. Through the application of the provided code and its associated calculator (Fig. 4.6), beginners can easily use the developed model for unseen data. The advantage of the calculator for the beginner user is that only fundamental information regarding the MATLAB environment is required, and there is no need to have detailed information regarding the ELM and other MATLAB functions.

The schematic of the proposed calculator is provided in Fig. 4.6. Let us consider the sigmoid function as the optimum activation function. The main form of the Sigmoid activation function is $TV = \left[1/(1+exp(InW \times InV+BHN))\right]^T \times OutW$ where TV is the target variable, InW and OutW are the input and OutWs (respectively), InV is the matrix of input variables, and BHN is the bias of the hidden neurons. InV is known before the commencement of modeling from the data set under consideration. The three matrices (OutW, InW, and BHN) are then optimized through the training process, and from the developed calculator, the value of the target variable(s) is calculated for the unseen data set. The computed target variable can then be applied in practice to make decisions from the data regarding the problem that is being studied. A more detailed explanation of the proposed calculator is provided later.

4.2.1 Single-layer feed-forward neural network with random hidden nodes

If we consider the number of input parameters for a given problem to be equal to n and the total number of training samples to be equal to N, the vector of input parameters is $\mathbf{x}_i = [x_{i1}, x_{i2}, \ldots, x_{in}]^T \in \mathbf{R}^n, \quad (i = 1, \ldots, N)$. Similarly, for a problem with m output parameters and N training samples, the target parameter vector will be $\mathbf{t}_i = [t_{i1}, t_{i2}, \ldots, t_{im}]^T \in \mathbf{R}^m, (i = 1, \ldots, N)$.

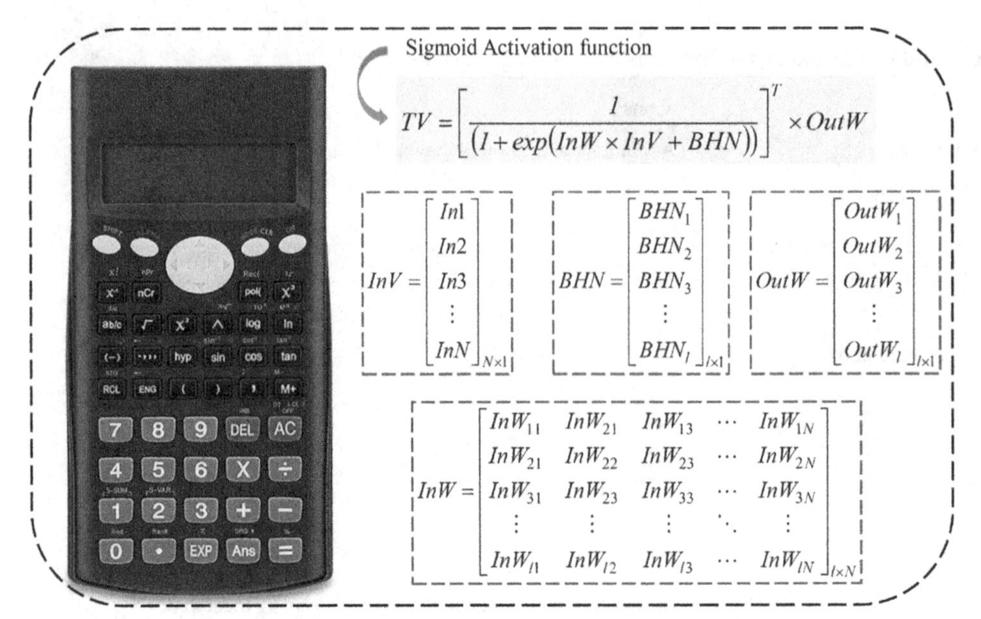

FIGURE 4.6 Schematic of the novel calculator tool.

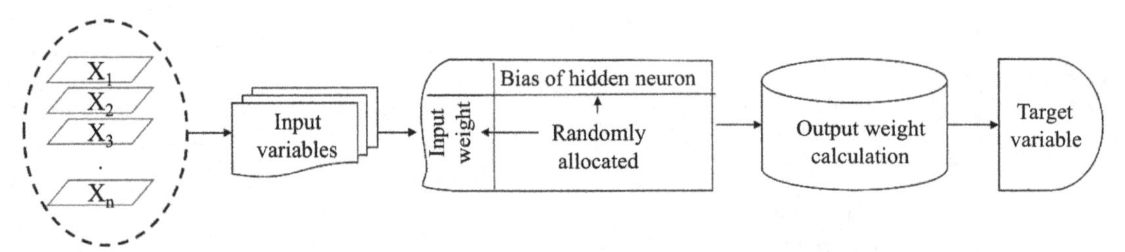

FIGURE 4.7 Extreme learning machine concept.

Fig. 4.7 provides a schematic of the ELM model. For a model with n inputs, the BHN (b_i) and the InW ($\mathbf{w}_i$) are randomly assigned. To find the weight vector linking the ith output node to the hidden nodes, $\beta_i = [\beta_{i1}, \beta_{i2}, \ldots, \beta_{im}]^T$, we need to consider an activation function $g(x)$. Therefore, the only unknown matrix in the SLFFNN that is optimized through the training phase is OutWs. If we consider l hidden neurons, then the relationship shown on the left-hand side of Eq. (4.1) is calculated to find the target values. The calculated output value for all training samples (N) is defined as $\mathbf{o}_j$ that will be calculated by the SLFFNN model. Due to the random allocation of the BHN (b_i) and the InW ($\mathbf{w}_i$), the left-hand side of Eq. (4.1) can be rewritten as shown below. Indeed, to find the target values, the OutWs ($\beta_i = [\beta_{i1}, \beta_{i2}, \ldots, \beta_{im}]^T$) are estimated through the training phase.

$$\sum_{i=1}^{l} \beta_i g_i(\mathbf{x}_j) = \sum_{i=1}^{l} \beta_i g_i(\mathbf{w}_i \cdot \mathbf{x}_j + b_i) = \mathbf{o}_j, \quad j = 1, \ldots, N \tag{4.1}$$

where l is the number of hidden layer neurons, $\beta_i = [\beta_{i1}, \beta_{i2}, \ldots, \beta_{im}]^T$ is the weight vector linking the ith output node to the hidden nodes, $g(x)$ is the activation function, $\mathbf{x}_i = [x_{i1}, x_{i2}, \ldots, x_{in}]^T \in \mathbf{R}^n$ is the input variables matrix, $\mathbf{t}_i = [t_{i1}, t_{i2}, \ldots, t_{im}]^T \in \mathbf{R}^m$ is the target output variable(s) matrix, b_i is the bias of the ith hidden node, and $\mathbf{w}_i \cdot \mathbf{x}_j$ indicates the inner product of the $\mathbf{w}_i$, and $\mathbf{x}_j$.

If Eq. (4.1) is capable of approximating all training samples (N) without any error, then the modeling error is zero, and the difference between the training target samples ($\mathbf{t}_i$) and the training output samples ($\mathbf{o}_i$) is given by

$$\sum_{j=1}^{l} \|\mathbf{o}_j - \mathbf{t}_j\| = 0 \tag{4.2}$$

Therefore the mathematical form of the SLFFNN [Eq. (4.1)] is equal to $\mathbf{t}_i$ when there is no error associated with the model. Indeed, there exist $b_i, \mathbf{w}_i, \beta_i$ such that

$$\sum_{i=1}^{l} \beta_i g_i(\mathbf{w}_i \cdot \mathbf{x}_j + b_i) = \mathbf{t}_j, \quad = 1, \ldots, N \tag{4.3}$$

The compact form of the N equations can be written in the following matrix form:

$$\mathbf{H}\beta = \mathbf{T} \tag{4.4}$$

where

$$\mathbf{T} = \begin{bmatrix} \mathbf{t}_1^{\mathrm{T}} \\ \vdots \\ \mathbf{t}_N^{\mathrm{T}} \end{bmatrix}_{N \times m} \tag{4.5}$$

$$\beta = \begin{bmatrix} \beta_1^{\mathrm{T}} \\ \vdots \\ \beta_l^{\mathrm{T}} \end{bmatrix}_{l \times m} \tag{4.6}$$

$$\mathbf{H} = \begin{bmatrix} g(\mathbf{w}_1 \cdot \mathbf{x}_1 + b_1) & \ldots & g(\mathbf{w}_l \cdot \mathbf{x}_1 + b_l) \\ \vdots & \ldots & \vdots \\ g(\mathbf{w}_1 \cdot \mathbf{x}_N + b_1) & \cdots & g(\mathbf{w}_l \cdot \mathbf{x}_N + b_l) \end{bmatrix}_{N \times l} \tag{4.7}$$

where $\mathbf{H}$ is the matrix of the hidden layer output related to the neural network. For an infinitely differentiable activation function $g(x)$, the maximum number of hidden nodes (l) is less than or equal to the number of training samples ($l \leq N$).

4.2.2 Minimum norm least-squares solution of single-layer feed-forward neural network

For an infinitely differentiable activation function, Huang et al. (2006) indicated that the bias of the hidden layer (b_i) and the InWs ($\mathbf{w}_i$) could be randomly allocated. Indeed, the allocated values of $\mathbf{w}_i$ and b_i remain unaltered after being assigned random values in the first stage of the learning process. Due to their fixed value, the $\mathbf{H}$ matrix can be easily calculated, and the only unknown parameter in Eq. (4.4) is β. Therefore, the least square (LS) solution of the linear system $\mathbf{H}\beta = \mathbf{T}$ is employed to find the weight vector linking the ith output node to the hidden nodes (β):

$$\left\| \mathbf{H}(\mathbf{w}_1, \ldots, \mathbf{w}_l, b_1, \ldots, b_l)\hat{\beta} - \mathbf{T} \right\| = \min_{\beta} \left\| \mathbf{H}(\mathbf{w}_1, \ldots, \mathbf{w}_l, b_1, \ldots, b_l)\beta - \mathbf{T} \right\| \tag{4.8}$$

If the number of hidden layer neurons is equal to the number of training samples ($N = l$), then H is an invertible and square matrix. Given that BHN (b_i) and the InW vector ($\mathbf{w}_i$) are randomly selected, the SLFFNN can approximate distinct training samples with zero error. In many cases, the number of hidden layer neurons is much smaller than the number of distinct training samples ($l << N$), and as a result, the approximated values of the target parameter are associated with an error. The lowest norm LS solution of $\mathbf{H}\beta = \mathbf{T}$ is shown in Eq. (4.9):

$$\hat{\beta} = \mathbf{H}^{+}\mathbf{T} \tag{4.9}$$

where $\mathbf{H}^{+}$ is the Moore–Penrose generalized inverse of $\mathbf{H}$ (Rätsch et al., 1998; Serre, 2002).

4.2.3 Least-square solution for finding output weights

Huang et al. (2006) reported the following important properties of the LS solution of the linear system:

(1) Eq. (4.9) is the LS solution of $\mathbf{H}\beta = \mathbf{T}$ as a general linear system. Indeed, the smallest modeling error is obtained by considering the Moore–Penrose generalized inverse:

$$\left\| \mathbf{H}\hat{\beta} - \mathbf{T} \right\| = \left\| \mathbf{H}\mathbf{H}^{+}\mathbf{T} - \mathbf{T} \right\| = \min_{\beta} \left\| \mathbf{H}\beta - \mathbf{T} \right\| \tag{4.10}$$

(2) The smallest norm amongst all existing LS solutions of the linear system ($\mathbf{H}\beta = \mathbf{T}$) is related to $\hat{\beta} = \mathbf{H}^{+}\mathbf{T}$ as a special solution:

$$\|\hat{\beta}\| = \|\mathbf{H}^{+}\mathbf{T}\| \leq \|\beta\|, \quad \forall \beta \in \{\beta : \|\mathbf{H}\beta - T\| \leq \|\mathrm{Hz} - T\|, \quad \forall \mathbf{z} \in \mathbf{R}^{l \times N}\} \tag{4.11}$$

(3) $\hat{\beta} = \mathbf{H}^{+}\mathbf{T}$ as the minimum norm LS solution of the linear system ($\mathbf{H}\beta = \mathbf{T}$) is unique.

4.2.4 Proposed learning algorithm for single-layer feed-forward neural network

Based on the detailed mathematical explanations in the previous subsections, a new algorithm for SLFFNN training, the ELM, is presented. A flowchart detailing the classical ELM method is presented in this section and can be seen in Fig. 4.8. The first step in the modeling process is data loading. Before the commencement of ELM modeling, the data

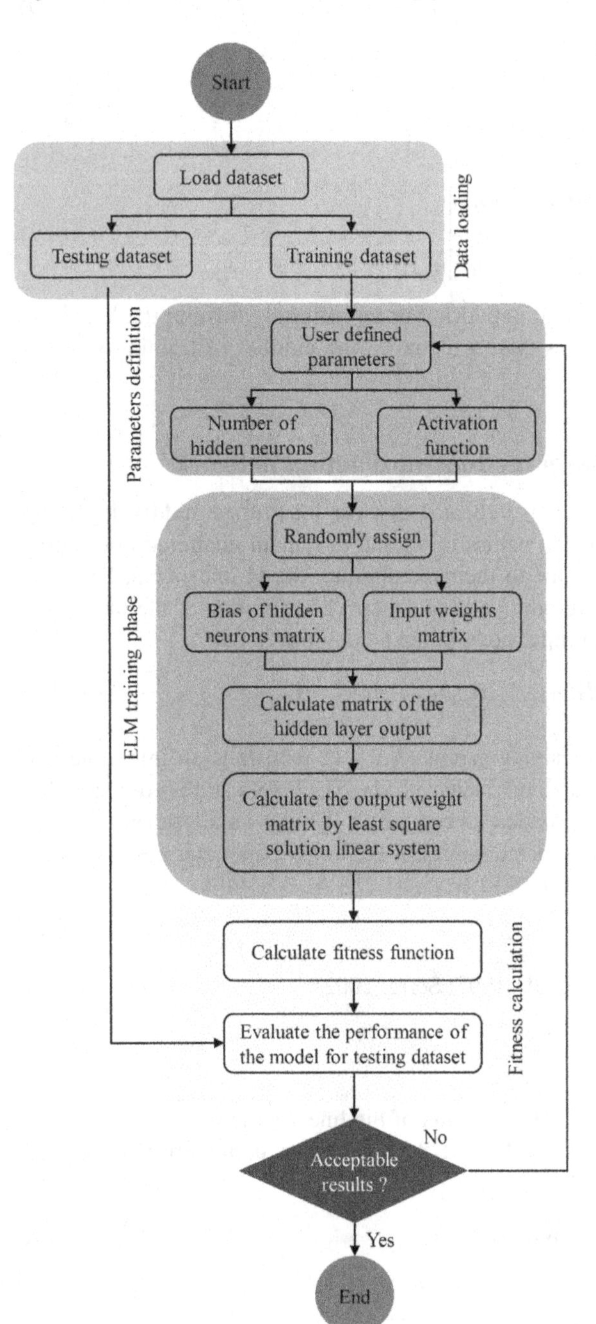

FIGURE 4.8 Flowchart of the classical ELM algorithm for SLFFNN training.

is first divided into two subsets, namely the training and testing subsets. Data classification into training and testing data sets can be done considering various proportions, such as 50%–50%, 60%–40%, 70%–30%, 80%–20%, and 90%–10% (Ebtehaj et al., 2020) for training and testing stages, respectively. In general, a data split of 70%–30% (Ebtehaj et al., 2016; Ebtehaj, Sammen, et al., 2021; Ebtehaj, Soltani, et al., 2021; Grégoire et al., 2022) or 80%–20% (Bonakdari, Ebtehaj, et al., 2020; Bonakdari et al., 2020; Mojtahedi et al., 2019) for training and testing data, respectively, are the most common proportions applied to many real-world problems. The second step is the definition of the model parameters. In this step, the activation function type and the number of hidden layers should be defined. The NHN is adjusted through a trial-and-error process. It should be noted that the NHN should be chosen so that the set of optimized parameters during the training process is less than the training samples. During the ELM training phase, three different matrices are optimized, including the InW, the BHN, and the OutW (the details are provided in Fig. 4.5). The BHN and InW matrices randomly assigned by the algorithm are structured as a matrix with one column and NHN rows, while the OutW matrix contains InV columns and NHN rows. On the other hand, the number of optimized values through the training stage must be less than the number of training samples. Therefore, the maximum number of allowable hidden layer neurons is obtained from the following equations (Ebtehaj, Soltani, et al., 2021; Zeynoddin et al., 2020):

$$NTrS > (InV + 2) \times NHN \tag{4.12}$$

$$NHN < \frac{NTrS}{(InV + 2)} \tag{4.13}$$

where *NTrS* denotes the number of training samples, *NIV* and *NHN* are the number of input variables and the number of hidden neurons, respectively.

The structure of SLFFNN is provided in Fig. 4.9. This model consists of three layers: the input layer, the hidden layer, and the output layer. The input layer is defined by the independent input variables. The input variables are obtained from the data regarding the problem being studied, and the number of these variables differs from 1 to *n*. The second layer, the hidden layer, is composed of a NHN. The size of the hidden layer is characterized by the NHN in the layer, a value which is defined by the user and can range from one to *l*. The third layer, the output layer, consists of the dependent output variables. The number of variables in this layer ranges from 1 to *m* depending on the nature of the problem under consideration, however, it should be noted that the number of output variables in most ELM applications is one. The w_{ij} in this figure denotes the InWs that connect the *i*th input neurons to the *j*th hidden neuron, while *b* is the BHN. The *b* and w_{ij} matrices are randomly allocated during the training phase of the ELM model. The β_{jk} denotes the weight vector that connects the *j*th hidden neuron to the *k*th output neuron. This matrix is estimated through the resolution of a linear system.

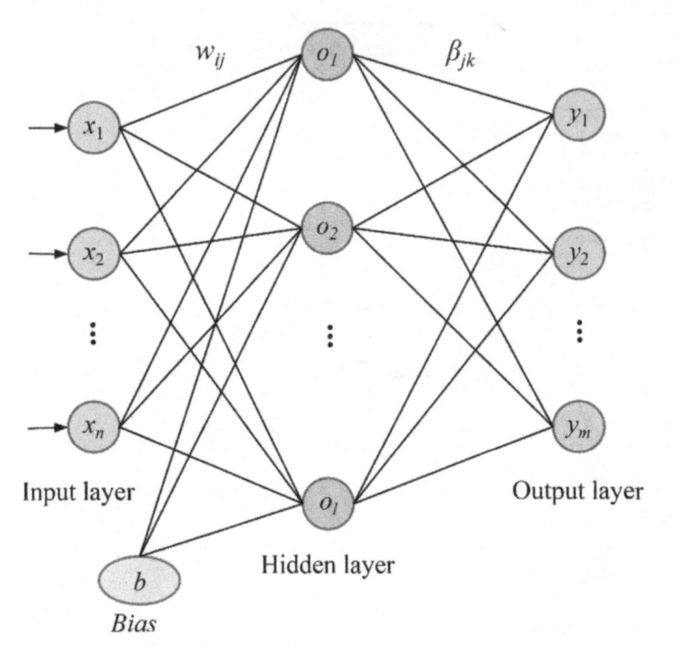

FIGURE 4.9 Single-layer feed-forward neural network structure.

4.2.5 Development of extreme learning machine-based modeling approach

After determining the NHN and the type of activation function to consider, we can then proceed to model using the ELM approach. First, the two vectors representing the BHN and the InW are randomly assigned, and then the hidden layer output matrix (**H**) is calculated using the activation function. After **H** has been determined, the OutW matrix is obtained using the LS solution of the linear system. Following this, the training and testing accuracy of the developed model is calculated. If the result is acceptable, the modeling stage is considered to be complete, otherwise, the NHN or type of activation function should be changed. This iterative loop is then repeated until the results of the model are acceptable based on the user-defined error criterion. A flowchart summarizing the required steps for the updated ELM process is shown in Fig. 4.10.

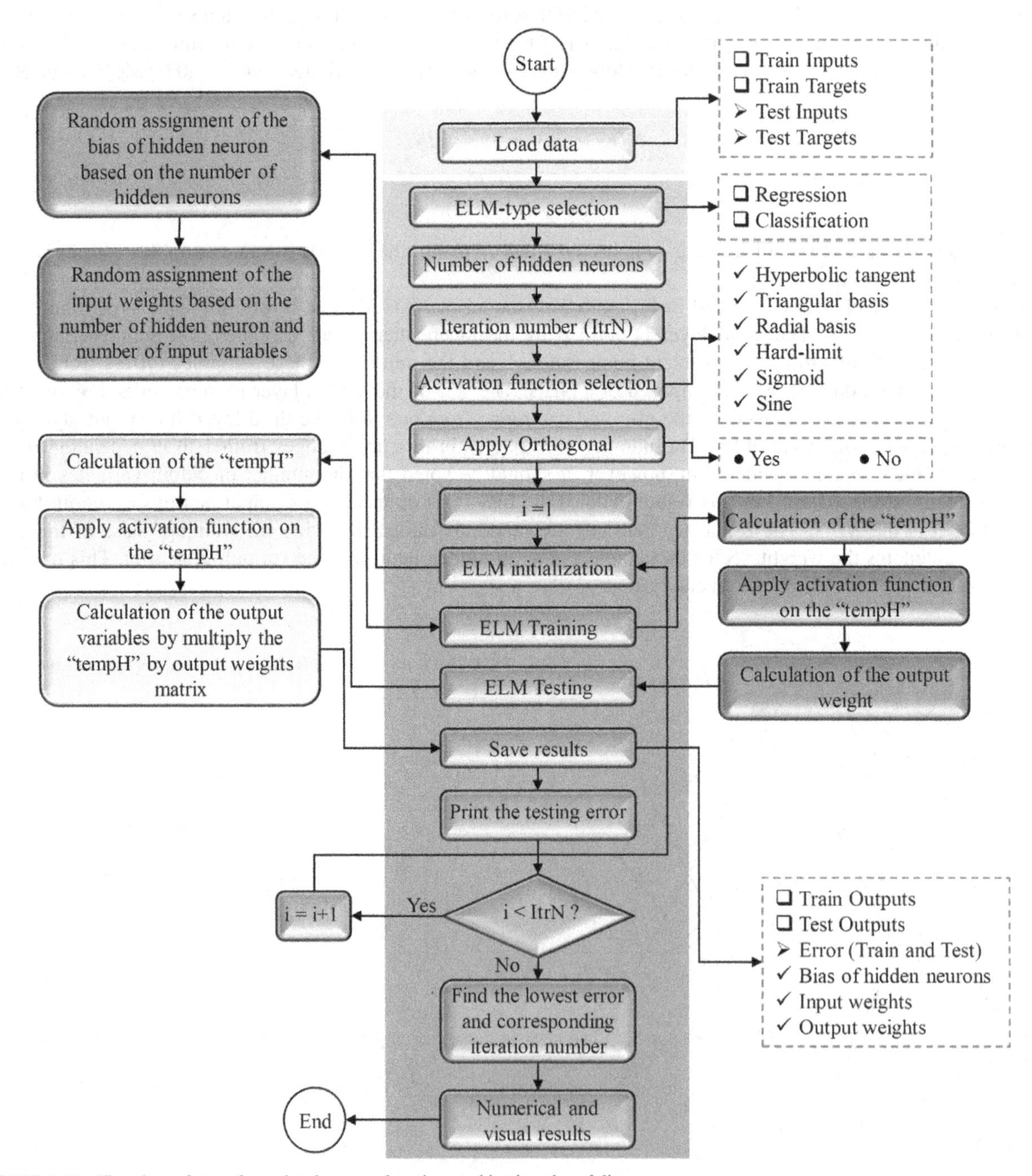

FIGURE 4.10 Flowchart of steps for updated extreme learning machine-based modeling.

From Fig. 4.6, the dimension of the BHN and the OutW are $l \times 1$, while for the InW, the dimension is $l \times N$ for InW. Therefore, the number of adjusted parameters (NAP) through the training phase can be calculated as follows:

$$NAP = D_{InW} + D_{BHN} + D_{OutW} = (l \times N) + (l \times 1) + (l \times 1) \tag{4.14}$$

$$NAP = (l \times N) + (l \times 1) + (l \times 1) \tag{4.15}$$

$$NAP = l \times (N + 2) \tag{4.16}$$

where NAP is the number of adjusted parameters through the training phase, l is the number of hidden neurons, and N is the number of independent input variables. Since the InW and BHN are randomly generated, the number of randomly adjusted parameters (NRAP) is determined as follows:

$$NRAP = D_{InW} + D_{BHN} = (l \times N) + (l \times 1) \tag{4.17}$$

$$NRAP = (l \times N) + (l \times 1) \tag{4.18}$$

$$NRAP = l \times (N + 1) \tag{4.19}$$

where NRAP is the number of randomly adjusted parameters. The ratio of the total number of tuned parameters to the randomly tuned parameters (R) is therefore:

$$R = \frac{NRAP}{NAP} = \frac{l \times (N+1)}{l \times (N+2)} = 1 - \frac{1}{N+2} \tag{4.20}$$

According to the above relationship, the greater the number of inputs to the problem, the greater the value of R. Therefore, the lowest value of R is obtained in the case where the problem has only one input, with R being equal to 0.6667. Indeed, at least 66% of all tuned parameters in the ELM algorithm are generated randomly. Therefore, any modeling approach considering the ELM algorithm should aim to reduce the impact of randomly allocated parameters on modeling performance.

As can be seen in Fig. 4.10, the flowchart has two important differences from the classical form of the ELM developed by Huang et al. (2006). These differences are aimed at decreasing the impact of the random selection of the InW and BHN matrices on modeling performance. The first development is the consideration of orthogonality for the randomly generated matrices, while the second is considering a simple iterative process to select a model with the highest generalizability. It should be noted that consideration of orthogonality is optionally applied at the discretion of the modeler, however, the user can quickly verify the model performance both with and without orthogonality. The orthogonality of two vectors is determined by their perpendicularity. Indeed, if the dot product of the two vectors is equal to zero, they are orthogonal. Consequently, if the user intends to use orthogonality, after randomly determining the InW and the BHN, the values of these matrices are changed in such a way that the sum of the dot product corresponding to all pairs of vectors becomes zero.

4.3 Activation function in the extreme learning machine model

In the ELM approach, activation functions defined by mathematical equations are applied to determine the output of the model. The activation function is the nonlinear transformation of the biases and weights of the input signals. The activation function acts on each neuron in the ELM structure and normalizes the output to a range between -1 and 1 or 0 and 1. The activation function in the ELM and its variants are necessary because they transform the input variables to a nonlinear feature space and help to enhance the performance of the ELM model in terms of prediction accuracy.

There are many types of activation functions that have been applied in ELM training, including the Hyperbolic tangent (tanh), Triangular basis (tribas), Radial basis (radbas), Hard-limit (hardlim), Sigmoid (Sig), and Sine (sin) functions, which are briefly defined below (Fig. 4.11):

- A sine function transforms the actual amount into the range of -1 to $+1$.
- Sigmoid is a continuous function that gradually changes between asymptotic values of -1 and $+1$ or 0 and 1.
- Hardlim transfer functions map the output to a value of 1 if the input threshold is met, otherwise, the output is 0.
- The radial basis function output is based on the distance of the point from the origin.
- A triangular basis function is very similar to the radial basis function so that it is simpler than the radial basis function.

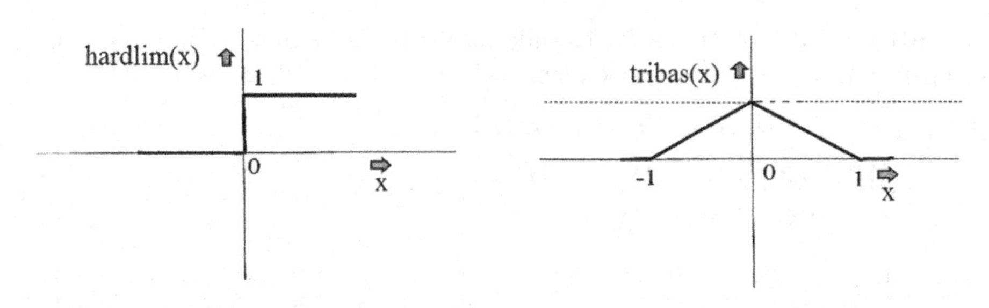

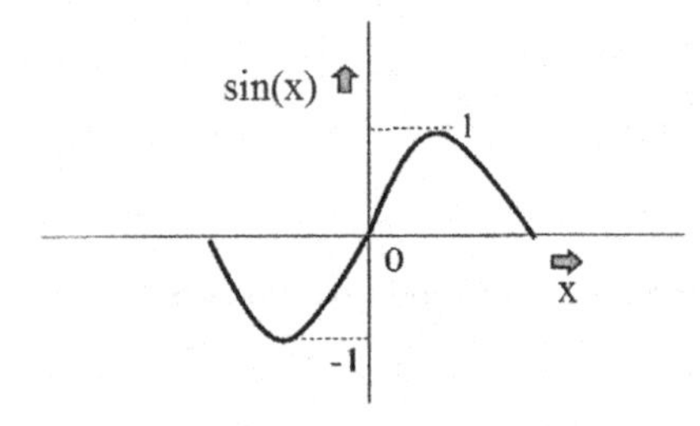

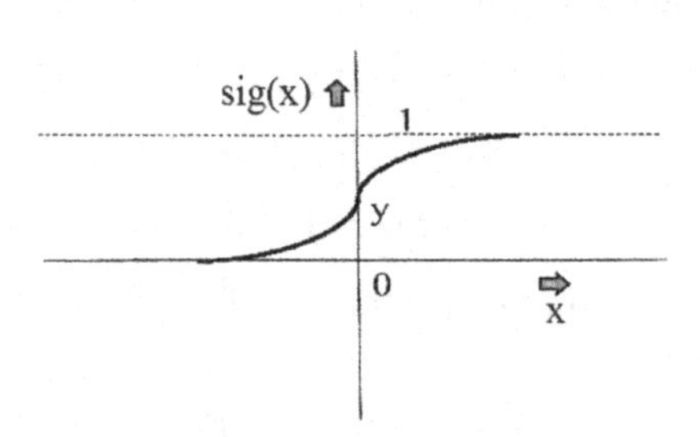

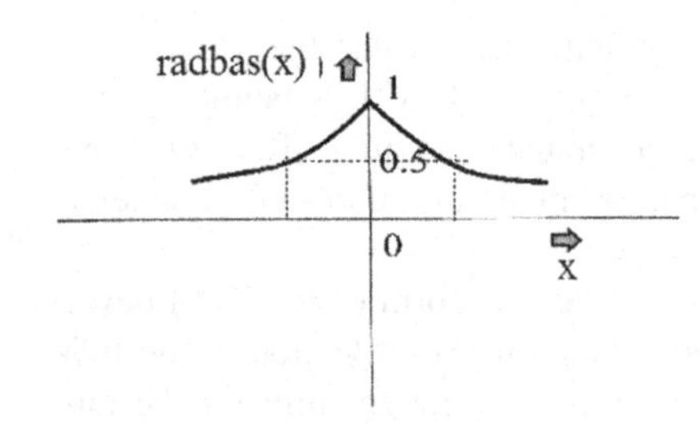

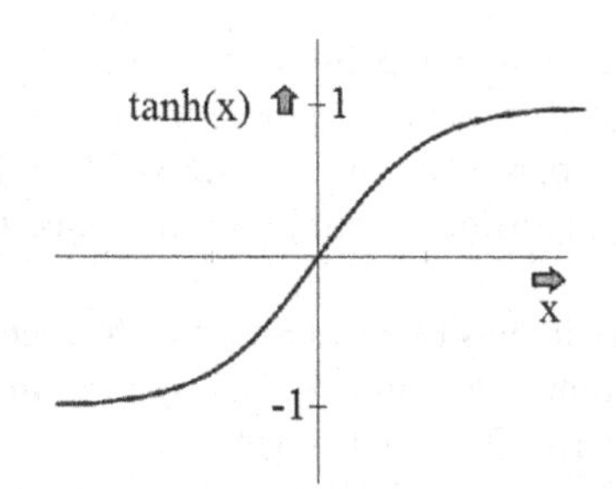

FIGURE 4.11 Different types of activation functions in extreme learning machine (Bonakdari, Moradi, et al., 2020).

- The hyperbolic tangent function equals 0 at the points $i\pi n$, and has singularities at the points $\pi i/2 + \pi in$, where n is an integer

The mathematical forms of the above-mentioned activation functions are as follows:

$$\text{sin } g(w, b, x) = \sin(w \cdot x + b) \tag{4.21}$$

$$\text{radbas } g(w, b, x) = \exp(-(w \cdot x + b)^2) \tag{4.22}$$

$$\text{tribas } g(w, b, x) = \begin{cases} 1 - |w \cdot x + b| & w \cdot x + b \geq 0 \\ 0 & otherwise \end{cases} \tag{4.23}$$

$$\text{hardlim } g(w, b, x) = \begin{cases} 1 & if \ w \cdot x + b \geq 0 \\ 0 & otherwise \end{cases} \tag{4.24}$$

$$\text{sig } g(w, b, x) = \frac{1}{1 + \exp(-(w, b, x))} \tag{4.25}$$

$$\text{tanh } g(w, b, x) = \tanh(w \cdot x + b) = \frac{\exp(2(w \cdot x + b)) - 1}{\exp(2(w \cdot x + b)) + 1} \tag{4.26}$$

4.4 Summary

In this chapter, the full mathematical and conceptual descriptions of the ELM were provided in detail. The ELM has two main user-defined parameters that must be optimized, including the NHN and the activation function type. Due to the random allocation of the BHN and InWs (InW) in the classical ELM, two new developments, including consideration of orthogonality and a simple iterative process, were introduced. Consideration of orthogonality is optional so that the user can verify model performance either with or without orthogonality. When a single iteration is considered with the rejection of orthogonality, the newly developed model is the same as the classical ELM model.

References

Azimi, H., Bonakdari, H., & Ebtehaj, I. (2017). A highly efficient gene expression programming model for predicting the discharge coefficient in a side weir along a trapezoidal canal. *Irrigation and Drainage*, *66*(4), 655–666. Available from https://doi.org/10.1002/ird.2127.

Azamathulla, H. M., Ghani, A. A., & Fei, S. Y. (2012). ANFIS-based approach for predicting sediment transport in clean sewer. *Applied Soft Computing*, *12*(3), 1227–1230. Available from https://doi.org/10.1016/j.asoc.2011.12.003.

Abba, S. I., Pham, Q. B., Usman, A. G., Linh, N. T. T., Aliyu, D. S., Nguyen, Q., & Bach, Q. V. (2020). Emerging evolutionary algorithm integrated with kernel principal component analysis for modeling the performance of a water treatment plant. *Journal of Water Process Engineering*, *33*, 101081. Available from https://doi.org/10.1016/j.jwpe.2019.101081.

Albadra, M. A. A., & Tiuna, S. (2017). Extreme learning machine: A review. *International Journal of Applied Engineering Research*, *12*(14), 4610–4623.

Burnham, K. P., & Anderson, D. R. (2002). *Model selection and multimodel inference (2nd ed.). Springer-Verlag.*

Bonakdari, H., Gholami, A., Mosavi, A., Kazemian-Kale-Kale, A., Ebtehaj, I., & Azimi, A. H. (2020). A novel comprehensive evaluation method for estimating the bank profile shape and dimensions of stable channels using the maximum entropy principle. *Entropy*, *22*(11), 1218. Available from https://doi.org/10.3390/e22111218.

Bonakdari, H., & Ebtehaj, I. (2016). A comparative study of extreme learning machines and support vector machines in prediction of sediment transport in open channels. *International Journal of Engineering*, *29*(11), 1499–1506.

Bonakdari, H., Ebtehaj, I., Gharabaghi, B., Sharifi, A., & Mosavi, A. (2020). *Prediction of discharge capacity of labyrinth weir with gene expression programming. Proceedings of SAI intelligent systems conference* (pp. 202–217). Cham: Springer. Available from https://doi.org/10.1007/978-3-030-55180-3_1.

Bonakdari, H., Ebtehaj, I., Samui, P., & Gharabaghi, B. (2019). Lake water-level fluctuations forecasting using minimax probability machine regression, relevance vector machine, gaussian process regression, and extreme learning machine. *Water Resources Management*, *33*(11), 3965–3984. Available from https://doi.org/10.1007/s11269-019-02346-0.

Bonakdari, H., Moradi, F., Ebtehaj, I., Gharabaghi, B., Sattar, A. A., Azimi, A. H., & Radecki-Pawlik, A. (2020). A non-tuned machine learning technique for abutment scour depth in clear water condition. *Water*, *12*, 301. Available from https://doi.org/10.3390/w12010301.

Bonakdari, H., Qasem, S. N., Ebtehaj, I., Zaji, A. H., Gharabaghi, B., & Moazamnia, M. (2020). An expert system for predicting the velocity field in narrow open channel flows using self-adaptive extreme learning machines. *Measurement*, *151*, 107202. Available from https://doi.org/10.1016/j.measurement.2019.107202.

Chen, W., Sharifrazi, D., Liang, G., Band, S. S., Chau, K. W., & Mosavi, A. (2022). Accurate discharge coefficient prediction of streamlined weirs by coupling linear regression and deep convolutional gated recurrent unit. *Engineering Applications of Computational Fluid Mechanics*, *16*(1), 965–976. Available from https://doi.org/10.1080/19942060.2022.2053786.

Ebtehaj, I., & Bonakdari, H. (2013). Evaluation of sediment transport in sewer using artificial neural network. *Engineering Applications of Computational Fluid Mechanics*, *7*(3), 382–392. Available from https://doi.org/10.1080/19942060.2013.11015479.

Ebtehaj, I., & Bonakdari, H. (2014). Performance evaluation of adaptive neural fuzzy inference system for sediment transport in sewers. *Water resources management*, *28*(13), 4765–4779. Available from https://doi.org/10.1007/s11269-014-0774-0.

Ebtehaj, I., & Bonakdari, H. (2016). Assessment of evolutionary algorithms in predicting non-deposition sediment transport. *Urban Water Journal*, *13* (5), 499–510. Available from https://doi.org/10.1080/1573062X.2014.994003.

Ebtehaj, I., Bonakdari, H., Moradi, F., Gharabaghi, B., & Khozani, Z. S. (2018). An integrated framework of extreme learning machines for predicting scour at pile groups in clear water condition. *Coastal Engineering*, *135*, 1–15. Available from https://doi.org/10.1016/j.coastaleng.2017.12.012.

Ebtehaj, I., Bonakdari, H., Safari, M. J. S., Gharabaghi, B., Zaji, A. H., Madavar, H. R., & Mehr, A. D. (2020). Combination of sensitivity and uncertainty analyses for sediment transport modeling in sewer pipes. *International Journal of Sediment Research*, *35*(2), 157–170. Available from https://doi.org/10.1016/j.ijsrc.2019.08.005.

Ebtehaj, I., Bonakdari, H., Zaji, A. H., Bong, C. H., & Ab Ghani, A. (2016). Design of a new hybrid artificial neural network method based on decision trees for calculating the Froude number in rigid rectangular channels. *Journal of Hydrology and Hydromechanics*, *64*(3), 252. Available from https://doi.org/10.1515/johh-2016-0031.

Ebtehaj, I., Sammen, S. S., Sidek, L. M., Malik, A., Sihag, P., Al-Janabi, A. M. S., & Bonakdari, H. (2021). Prediction of daily water level using new hybridized GS-GMDH and ANFIS-FCM models. *Engineering Applications of Computational Fluid Mechanics*, *15*(1), 1343–1361. Available from https://doi.org/10.1080/19942060.2021.1966837.

Ebtehaj, I., Soltani, K., Amiri, A., Faramarzi, M., Madramootoo, C. A., & Bonakdari, H. (2021). Prognostication of shortwave radiation using an improved no-tuned fast machine learning. *Sustainability*, *13*(14), 8009. Available from https://doi.org/10.3390/su13148009.

Granata, F., Di Nunno, F., Gargano, R., & de Marinis, G. (2019). Equivalent discharge coefficient of side weirs in circular channel—A lazy machine learning approach. *Water*, *11*(11), 2406. Available from https://doi.org/10.3390/w11112406.

Grégoire, G., Fortin, J., Ebtehaj, I., & Bonakdari, H. (2022). Novel hybrid statistical learning framework coupled with random forest and grasshopper optimization algorithm to forecast pesticide use on golf courses. *Agriculture*, *12*(7), 933. Available from https://doi.org/10.3390/agriculture12070933.

Huang, G. B., Zhu, Q. Y., & Siew, C. K. (2006). Extreme learning machine: Theory and applications. *Neurocomputing*, *70*(1–3), 489–501. Available from https://doi.org/10.1016/j.neucom.2005.12.126.

Karbasi, M., Jamei, M., Ahmadianfar, I., & Asadi, A. (2021). Toward the accurate estimation of elliptical side orifice discharge coefficient applying two rigorous kernel-based data-intelligence paradigms. *Scientific Reports*, *11*(1), 1–18. Available from https://doi.org/10.1038/s41598-021-99166-3.

Khoshbin, F., Bonakdari, H., Ashraf Talesh, S. H., Ebtehaj, I., Zaji, A. H., & Azimi, H. (2016). Adaptive neuro-fuzzy inference system multi-objective optimization using the genetic algorithm/singular value decomposition method for modelling the discharge coefficient in rectangular sharp-crested side weirs. *Engineering Optimization*, *48*(6), 933–948. Available from https://doi.org/10.1080/0305215X.2015.1071807.

Melo, H., & Watada, J. (2016). Gaussian-PSO with fuzzy reasoning based on structural learning for training a neural network. *Neurocomputing*, *172*, 405–412. Available from https://doi.org/10.1016/j.neucom.2015.03.104.

Mojtahedi, S. F. F., Ebtehaj, I., Hasanipanah, M., Bonakdari, H., & Amnieh, H. B. (2019). Proposing a novel hybrid intelligent model for the simulation of particle size distribution resulting from blasting. *Engineering with Computers*, *35*(1), 47–56. Available from https://doi.org/10.1007/s00366-018-0582-x.

Montes, C., Kapelan, Z., & Saldarriaga, J. (2021). Predicting non-deposition sediment transport in sewer pipes using random forest. *Water Research*, *189*, 116639. Available from https://doi.org/10.1016/j.watres.2020.116639.

Ng, S. C., Cheung, C. C., & Xu, S. (2012). *Magnified gradient function to improve first-order gradient-based learning algorithms. International symposium on* neural networks (pp. 448–457). Berlin, Heidelberg: Springer. Available from https://doi.org/10.1007/978-3-642-31346-2_51.

Rätsch, G., Onoda, T., & Müller, K. R. (1998). An improvement of AdaBoost to avoid overfitting. In *Proceedings of the fifth international conference on neural information processing (ICONIP'1998)*.

Roushangar, K., & Ghasempour, R. (2017). Prediction of non-cohesive sediment transport in circular channels in deposition and limit of deposition states using SVM. *Water Science and Technology: Water Supply*, *17*(2), 537–551. Available from https://doi.org/10.2166/ws.2016.153.

Serre, D. (2002). *Matrices: Theory and applications*. New York: Springer.

Tafarojnoruz, A., & Sharafati, A. (2020). New formulations for prediction of velocity at limit of deposition in storm sewers based on a stochastic technique. *Water Science and Technology*, *81*(12), 2634–2649. Available from https://doi.org/10.2166/wst.2020.321.

Zeynoddin, M., Bonakdari, H., Ebtehaj, I., Azari, A., & Gharabaghi, B. (2020). A generalized linear stochastic model for lake level prediction. *Science of the Total Environment*, *723*, 138015. Available from https://doi.org/10.1016/j.scitotenv.2020.138015.

Zounemat-Kermani, M., Mahdavi-Meymand, A., & Hinkelmann, R. (2021). Nature-inspired algorithms in sanitary engineering: modelling sediment transport in sewer pipes. *Soft Computing*, *25*(8), 6373–6390. Available from https://doi.org/10.1007/s00500-021-05628-1.

Chapter 5

Non-tuned single-layer feed-forward neural network learning machine—coding and implementation

5.1 Introduction

In this chapter the reader will be provided with detailed coding (including all written functions) of the extreme learning machine (ELM) in the MATLAB® environment, including initialization, training, and testing stages. Furthermore, the output of the ELM-based code, in terms of quantitative and qualitative results, is provided and interpreted for the reader for five real-world examples. To facilitate the application of the ELM for beginner users, a novel calculator is proposed, and its implementation as an approach to solving practical problems will be detailed. Finally, the effect of the user-defined ELM parameters, including the iteration number, number of hidden layer neurons, activation function types, and matrix orthogonality on the ELM model, is surveyed by analysis of different real-world problems. It should be noted that the iteration number and the use of orthogonal functions are two new parameters that are defined in addition to the developed version of the ELM described in the previous chapter.

The user-defined parameters of the newly developed ELM (compared to the classical ELM presented in the previous chapter) are shown in Fig. 5.1. As can be seen from the figure, among the four parameters provided, the activation function type and the use of orthogonal matrices are qualitative parameters, and optimal determination of them is not challenging. Numerous examples of possible activation functions were provided in the previous chapter. Considering the selection of orthogonality, the user determines whether or not to use this function in randomly determining the input weights and bias of the hidden neurons. In addition to these two parameters, the iteration number and number of hidden neurons must be defined numerically. Due to the ELM's very high-speed training phase, the iteration number can be considered large. Therefore, the only parameter that may be challenging to optimize is the number of hidden neurons.

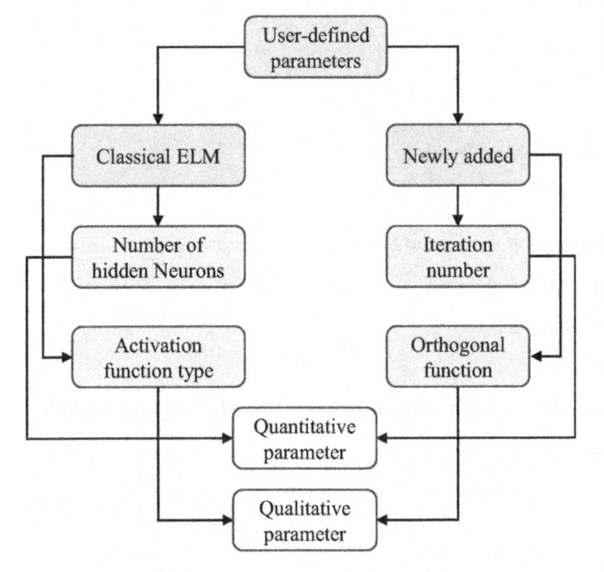

FIGURE 5.1 The user-defined parameters of the newly developed ELM. *ELM*, Extreme learning machine.

Machine Learning in Earth, Environmental and Planetary Sciences. DOI: https://doi.org/10.1016/B978-0-443-15284-9.00009-4

Both beginner and advanced users will find the presentation of ELM coding in this chapter to be simple and understandable. By the end of this chapter, beginners will understand how the user-defined parameters affect ELM results in solving real-world problems. The provided calculator will enable beginner users to employ ELM in the MATLAB environment without detailed knowledge regarding MATLAB coding or the mathematical formulation of the ELM. The advanced user will gain a thorough understanding of the coding process, the fundamentals of the ELM algorithm, as well as its application in the MATLAB environment. The provided MATLAB code can be employed by users for a wide range of data analysis in practice.

5.2 Extreme learning machine implementation in the MATLAB environment

5.2.1 MATLAB functions for extreme learning machine coding

This section describes in detail all steps required in modeling a real-world problem using ELM. To apply the developed ELM-based code in solving nonlinear problems, the "ELM" file in the provided package should be run. The concept of all functions contained within this ELM-based package is provided in Table 5.1. Moreover, the schematic relationship between the different functions is shown in Fig. 5.2. Details related to the inputs and outputs of each function along with their coding are presented in the following subsections.

It should be noted that beginner users can proceed by simply entering the user-defined variables (i.e., number of iterations, number of hidden neurons, activation function type, and whether or not to use the orthogonal function) into the provided calculator, and can therefore follow the remainder of this section without considering the detailed explanations of the developed MATLAB code (provided text in gray).

5.2.2 Steps of extreme learning machine coding

The first step in the modeling process is to load the data and divide it into training and testing subsets. To do this, an excel file is prepared with four different sheets containing the train inputs, train targets, test inputs, and test targets on sheet1 to sheet4, respectively (Fig. 5.3). The syntax for loading the data into excel is provided in Code 5.1. After preparation of the excel file, the `xlsread` function is employed to load the dataset. The schematic of Code 5.1 is provided in Fig. 5.4.

Code 5.1

```
1    %% Load data
2    TrainInputs   = xlsread('Example1','sheet1');
3    TrainTargets  = xlsread('Example1','sheet2');
4    TestInputs    = xlsread('Example1','sheet3');
5    TestTargets   = xlsread('Example1','sheet4');
```

TABLE 5.1 Functions considered in the developed extreme learning machine model.

Functions	Definitions
elm_initialization	Randomly assigns the input weights matrix and bias of hidden neurons matrix
elm_train	Training of the developed ELM model
elm_test	Testing of the developed ELM model
PlotResults	Provide the visual results
Regressor	Calculation of the beta variable for output matrix computing through the training stage
Statistical_Indices	Provide the numerical results

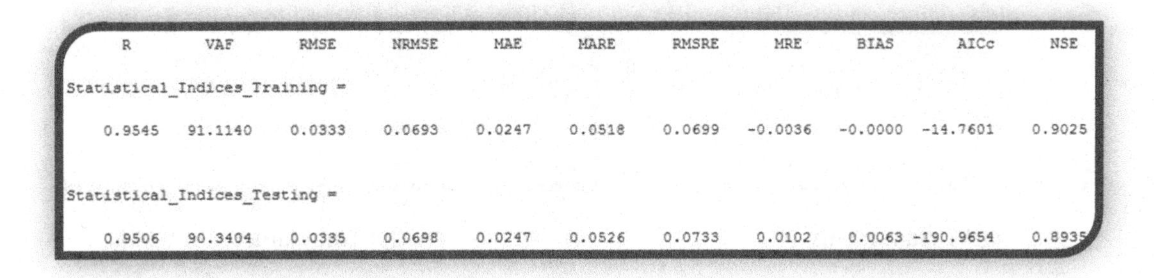

	R	VAF	RMSE	NRMSE	MAE	MARE	RMSRE	MRE	BIAS	AICc	NSE
Statistical_Indices_Training =	0.9545	91.1140	0.0333	0.0693	0.0247	0.0518	0.0699	-0.0036	-0.0000	-14.7601	0.9025
Statistical_Indices_Testing =	0.9506	90.3404	0.0335	0.0698	0.0247	0.0526	0.0733	0.0102	0.0063	-190.9654	0.8935

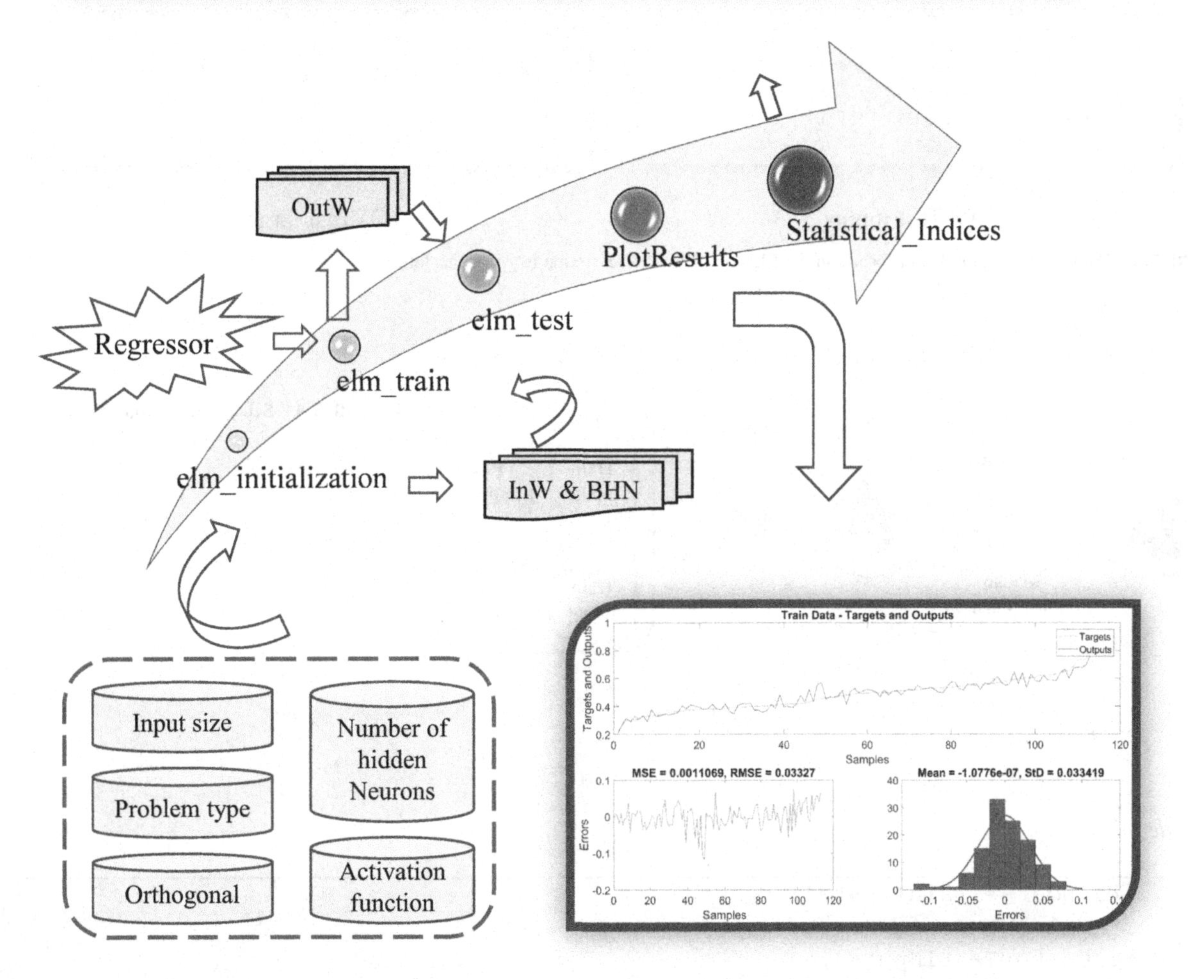

FIGURE 5.2 Schematic of the relationship between different functions in the ELM model. *ELM*, Extreme learning machine.

Here the built-in MATLAB function `xlsread` is used to load the input, where the first argument is the `filename` of the excel file and the second argument is the `sheet` containing the specific data for the variable you are defining. These should be adjusted on a case-by-case basis according to the chosen filename of the input files and the location of the data of interest within these files. If the number of inputs and the number of samples of the desired problem are denoted by m and n (respectively), the loading of data by `xlsread` will be in the form of $n \times m$. As a result, the number of columns in each of these matrices is equal to the number of inputs in the problem (i.e., one for train targets and test targets) and the number of rows is equal to the number of samples used in each training mode. To use the loaded data in the ELM model, the dimensions of the matrices must be changed to $m \times n$ by computing the transpose of the matrix as shown in Code 5.2. The schematic of this code is presented in Fig. 5.5.

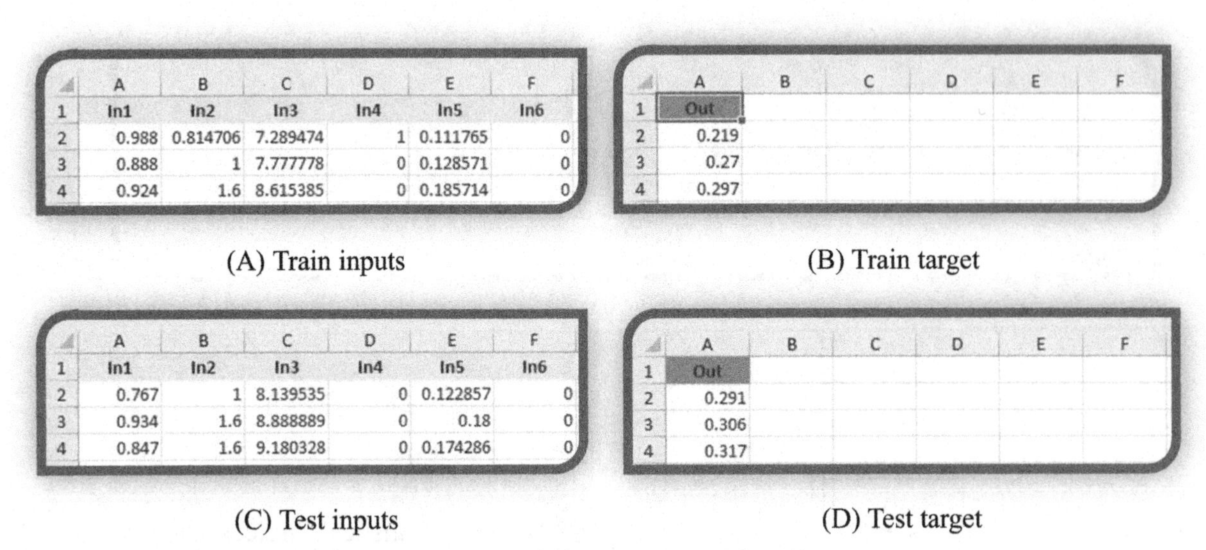

(A) Train inputs

	A	B	C	D	E	F
1	In1	In2	In3	In4	In5	In6
2	0.988	0.814706	7.289474	1	0.111765	0
3	0.888	1	7.777778	0	0.128571	0
4	0.924	1.6	8.615385	0	0.185714	0

(B) Train target

	A	B	C	D	E	F
1	Out					
2	0.219					
3	0.27					
4	0.297					

(C) Test inputs

	A	B	C	D	E	F
1	In1	In2	In3	In4	In5	In6
2	0.767	1	8.139535	0	0.122857	0
3	0.934	1.6	8.888889	0	0.18	0
4	0.847	1.6	9.180328	0	0.174286	0

(D) Test target

	A	B	C	D	E	F
1	Out					
2	0.291					
3	0.306					
4	0.317					

FIGURE 5.3 Data preparation in Excel for use in the ELM model. *ELM*, Extreme learning machine.

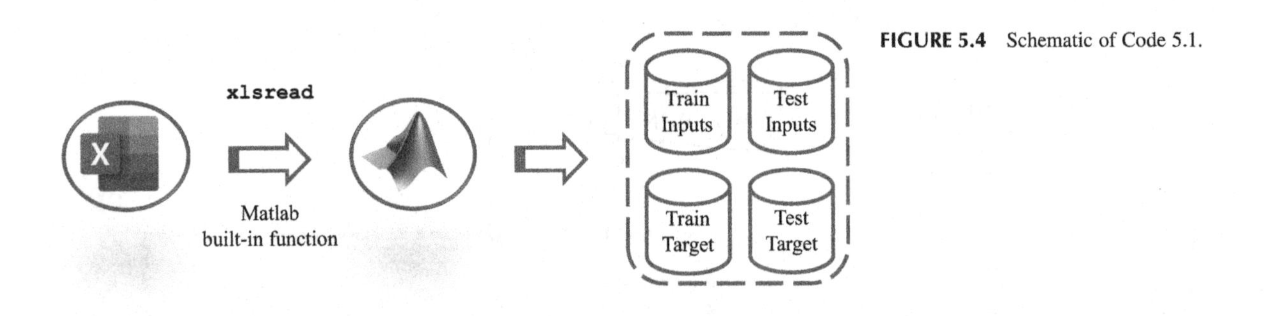

FIGURE 5.4 Schematic of Code 5.1.

Code 5.2

```
1    traindata=TrainInputs';
2    trainlabel=TrainTargets';
3    testdata=TestInputs';
4    testlabel=TestTargets';
```

The `inputsize` command, which calculates the input dimensions of the problem, is defined in Code 5.3 using the MATLAB `size` command. The concept of `size` command in MATLAB is defined graphically in Fig. 5.6.

Code 5.3

```
1    nn.inputsize = size(traindata,1);
```

The `size` function calculates the dimensions of a matrix (1: row; 2: column). Since we are only interested in the number of input features in the `inputsize` variable, we only need to use one dimension (row) of `traindata`. For this reason, the first argument of `size` function is the variable `traindata`, while the second argument is simply 1. Obviously, `testdata` can be used interchangeably with `traindata`. In addition, the `inputsize` is stored in the `nn` variable because of the simple call of this variable in the functions used at the later stages of the current program.

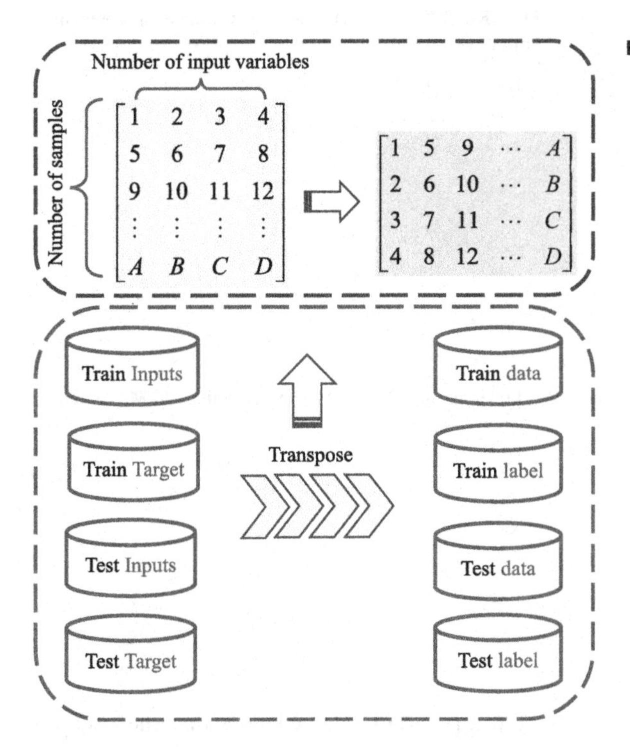

FIGURE 5.5 Schematic of Code 5.2.

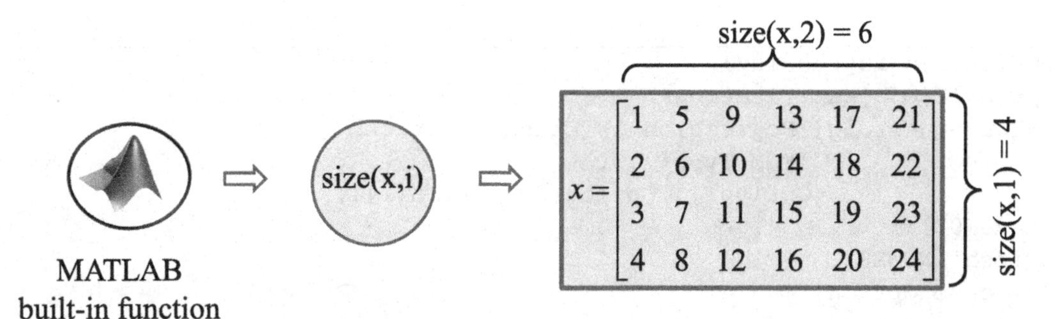

FIGURE 5.6 Size command in MATLAB.

The next step is related to problem type selection. Using the `questdlg` command, the user is asked which problem type is to be solved by the ELM. The schematic of this command is shown in Fig. 5.7. The general form of this command is as follows:

Code 5.4

```
1   button = questdlg(qstring,title,str1,str2,str3,default)
```

The `questdlg` command detailed in Code 5.5 should be applied to obtain the user-specified response as to whether the problem type is regression or classification. In lines 1–2, the variables `Option{1}` and `Option{2}` are specified as "Regression" and "Classification," respectively. In lines 3–5, the `questdlg` command is specified. The `qstring` argument is the question to be posed and is stored as a string of characters. In our case, this has been specified as "Regression or Classification?." The input argument `title` is used to provide a title to the dialog box, which in this example is "ELM Type." The `default` value has been set to `Option{1}`, regression analysis, which is most commonly employed in practical applications. Following the user selection of an answer (i.e., Regression or Classification), the

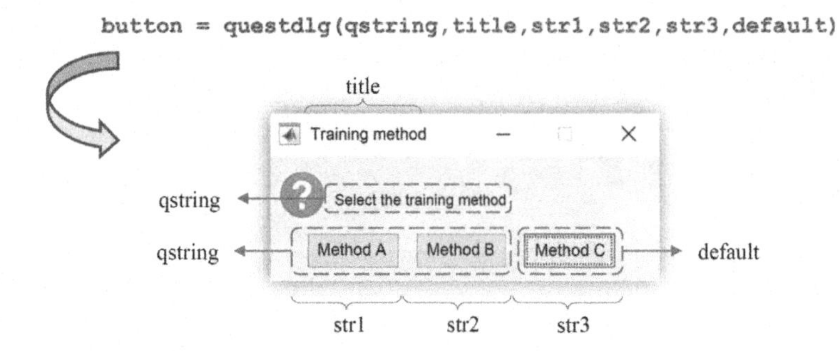

FIGURE 5.7 Schematic of the `questdlg` command with three options.

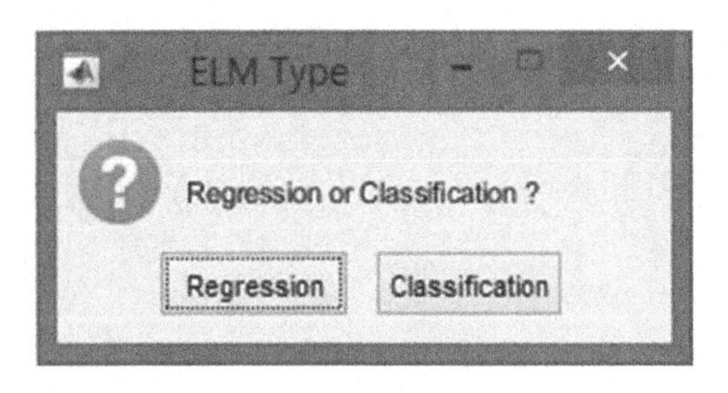

FIGURE 5.8 The `questdlg` prompt for ELM type selection. *ELM*, Extreme learning machine.

entry is then saved as the variable `nn.type` for use in other sections of the program. The window that the user is prompted with from the code is shown in Fig. 5.8.

Code 5.5

```
1   Option{1}='Regression';
2   Option{2}='Classification';
3   ANSWER=questdlg('Regression or Classification ?',...
4                   'ELM Type',...
5                   Option{1},Option{2},Option{1});
6   pause(0.1);
7   switch ANSWER
8       case Option{1}
9           nn.type = 'regression';
10      case Option{2}
11          nn.type = 'classification';
12  end
```

After determining the problem type, the "Number of Hidden Neurons," "Iteration Number," and "Activation Function Type" must be defined by the user. The "Number of Hidden Neurons" is chosen through a trial-and-error process, however, the simplest approach is to increase the value of the "Number of Hidden Neurons" to such an extent that further increase does not have a significant effect on the modeling results (i.e., point of diminishing returns). It should be noted that the maximum value of the "Number of Hidden Neurons" that can be considered during model development depends on the number of input variables and number of training samples and can be calculated according to Eq. (4.13).

Given that the input weights and bias of hidden neurons are randomly assigned in the ELM method, it is necessary to run the code N times to achieve a reliable model so that the influence of the random selection of the matrices is assessed. Following each run, the entries in the matrices are cleared and then re-populated. After determining the model performance over a given number of repetitions, which is generally considered as a very large value (more than 1000) due to the fast process time of the ELM, the answer to the best repetition is stored and used.

Code 5.6 details the required code to input the "Number of Hidden Neurons" and the "Iteration Number." The `inputdlg` command is used in line 6 to generate a dialog box where the user may enter the desired values. The first

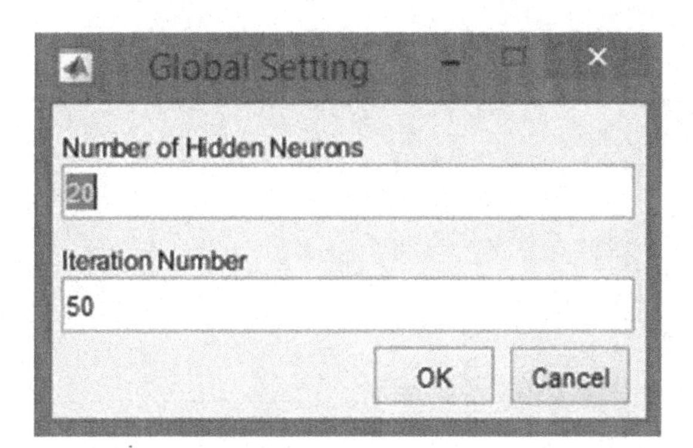

FIGURE 5.9 Dialog box used to assign the "Number of Hidden Neurons" and "Iteration Number" by the user.

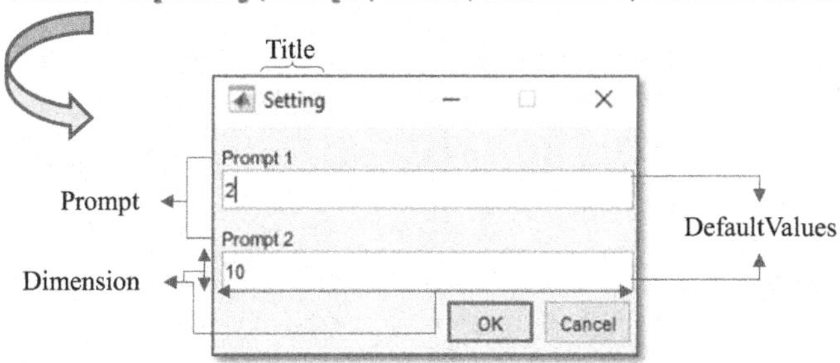

FIGURE 5.10 Schematic of the `inputdlg` command.

input argument, `Prompt`, specifies the character array describing the edit field for the user. In lines 2–3, it can be seen that these are "Number of Hidden Neurons" and "Iteration Number." In line 4, the `title` is specified as "Global Setting," the result of which can be seen at the top of the output dialog box in Fig. 5.9. The `[1 40]` input argument seen in line 6 indicates that each entry field is one line in height with a length of 40 characters. The `default` input is defined using the `DefaultValues` variable in line 5. The window that the user is prompted with from Code 5.6 is provided in Fig. 5.9. The schematic of the `inputdlg` command applied in Code 5.6 is provided in Fig. 5.10.

Code 5.6

```
1   %% Global Setting
2   Prompt={'Number of Hidden Neurons',...
3       'Iteration Number'};
4           Title='Global Setting';
5           DefaultValues={'20','50'};
6           PARAMS=inputdlg(Prompt,Title,[1 40],DefaultValues);
7           pause(0.1);
8           nn.hiddensize=str2double(PARAMS{1});
9           N=str2double(PARAMS{2});
```

Since more than three activation function types are considered in this algorithm, it is not possible to use the `questdlg` command. The `questdlg` command can only be programmed to contain three responses. For this reason, the `spaceList` command must be employed to select the activation function type, as seen in Code 5.7. The schematic of this code is shown in Fig. 5.11.

Six activation functions (which were reviewed in earlier chapters) are considered in Code 5.7 and are indexed using the numbers one to six in the order of hyperbolic tangent (Azimi & Shiri, 2021; Bonakdari et al., 2020), triangular basis (Bonakdari & Ebtehaj, 2016a; Sattar et al., 2019), radial basis (Bonakdari et al., 2019; Yaseen et al., 2018), hard-limit (Rohim et al., 2022; Tao et al., 2019), sigmoid (Azimi et al., 2017; Bonakdari & Ebtehaj, 2016b; Ebtehaj et al., 2016, 2020a), and sine (Sin) (Ebtehaj et al., 2018; Nou et al., 2022). Once the user selects the type of activation function to consider, the index value is returned by the function to the program, and the `nn.activefunction` variable is defined. The prompt that the user is presented from the `spaceList` command in Code 5.7 is shown in Fig. 5.12. It is clear that

```
idx = listdlg('ListString', {'Model 1','Model 2','Model 3','Model 4'}, ...
    'SelectionMode', 'Single', 'PromptString', 'Select item', ...
    'Initialvalue', 3, 'Name', 'Method Selection','ListSize',Dimension);
```

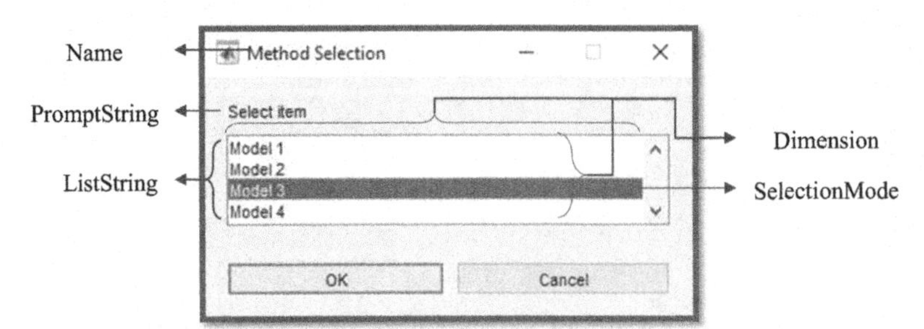

FIGURE 5.11 Schematic of Code 5.7.

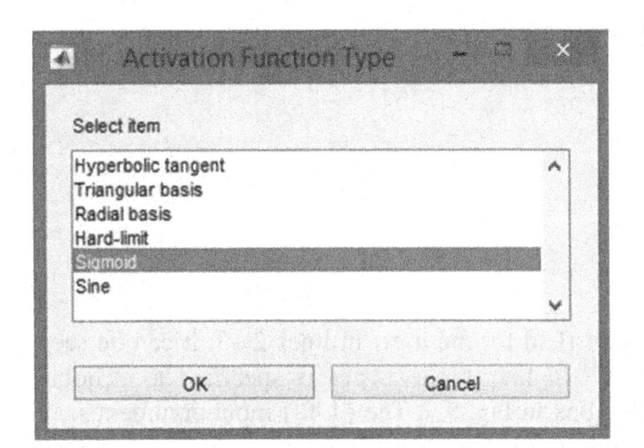

FIGURE 5.12 Question dialog box for selection of activation function type.

the sigmoid function, highlighted in blue, is selected as the default activation function type. The default activation function is specified in line 3 of Code 5.7 by setting the `Initialvalue` input argument to an index of 5, corresponding to the sigmoid activation function.

Code 5.7

```
1   spaceList = {'Hyperbolic tangent','Triangular basis','Radial
    basis','Hard-limit','Sigmoid','Sine'};
2           [idx, ~] = listdlg('ListString', spaceList,...
3   'SelectionMode', 'Single', 'PromptString', 'Select item', 'Initialvalue',
    5,'Name', 'Activation Function Type','ListSize',[300 100]);
4           if idx == 1
5               nn.activefunction='tanh';
6           elseif idx == 2
7               nn.activefunction='tribas';
8           elseif idx == 3
9               nn.activefunction='radbas';
10          elseif idx == 4
11              nn.activefunction='hardlim';
12          elseif idx == 5
13              nn.activefunction='sig';
14          elseif idx == 6
15              nn.activefunction='sin';
16          end
```

The last user-defined input that should be determined before proceeding with the ELM modeling is whether or not to consider an orthogonal matrix using the `orth` function. The `orth` function calculates the orthonormal basis for the matrix. The choice of orthogonality is posed to the user following the selection of activation function type using the `questdlg` command. The detailed MATLAB coding is shown in Code 5.8, while the question dialog box that the user is prompted with is shown in Fig. 5.13. The schematic of Code 5.8 is presented in Fig. 5.14. The `questdlg` syntax has already been discussed earlier in this chapter. In line 4 of Code 5.8, it is seen that the `qstring` input argument is set to "Orthogonal?" while the dialog box title is set to "T&F." In line 6, the `default` input argument is specified as `Option {1}`, which corresponds to "True." For this reason, in Fig. 5.13, it can be seen that the "True" button is selected.

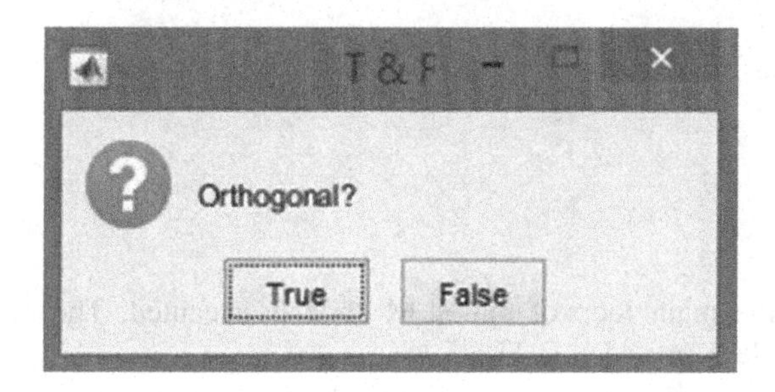

FIGURE 5.13 Question dialog box used to determine whether or not orthogonal matrices are considered during ELM initialization. *ELM*, Extreme learning machine.

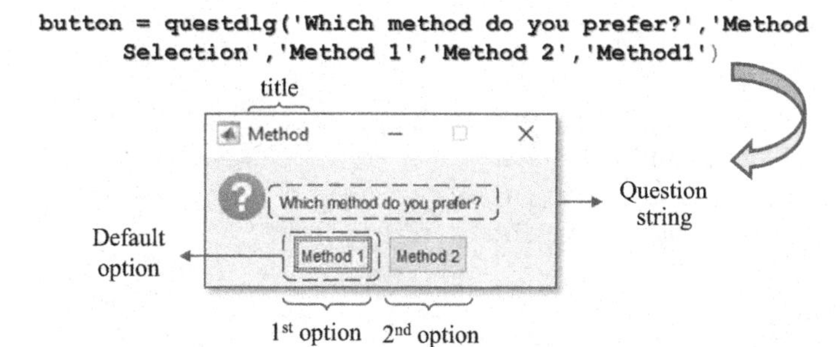

FIGURE 5.14 Schematic of a `questdlg` command with two options.

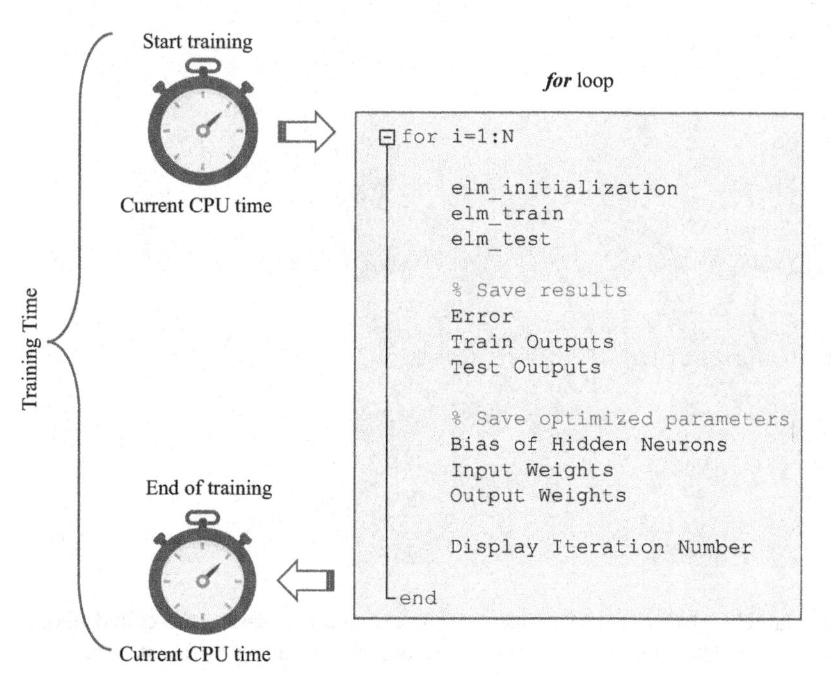

FIGURE 5.15 Schematic of Code 5.9.

Code 5.8

```
1   %% Orthogonal
2   Option{1}='True';
3   Option{2}='False';
4   ANSWER=questdlg('Orthogonal?',...
5                   'T & F',...
6                   Option{1},Option{2},Option{1});
7   pause(0.1);
8   switch ANSWER
9       case Option{1}
10          nn.orthogonal = 'true';
11      case Option{2}
12          nn.orthogonal = 'false';
13  end
```

Following the initiation of the user-defined ELM parameters, the main loop of the ELM code is executed. The detailed coding is provided in Code 5.9, and the schematic of this code is provided in Fig. 5.15.

Code 5.9

```
1   %% ELM
2   nn.method = 'ELM';
3   start_time_train=cputime;
4   for i=1:N
5       nn                  = elm_initialization(nn);
6       [nn, acc_train]     = elm_train(traindata, trainlabel, nn);
7       [nn1, acc_test]     = elm_test(testdata, testlabel, nn);
8
9       if strcmp(nn.type, 'regression')
10
11          Error(i)=acc_test;
12          TrainOutputs(:,i)= nn.Y_hat';
13          TestOutputs(:,i) = nn1.Y_hat';
14
15          A = nn1.W;
16          B = nn1.b;
17          C = (nn1.beta)';
18
19          BHN(:,i) = B;
20          OutW(:,i) = C;
21          InW(:,:,i) = A;
22
23          disp(['Iteration ' num2str(i) ': Best Cost = '
    num2str(Error(i))]);
24
25      end
26  end
27
28  end_time_train=cputime;
```

It is observed that three general functions, namely `elm_initialization`, `elm_train` and `elm_test`, are called upon, and several variables (i.e., `BHN`, `OutW`, `InW`, `Error`, `TrainOutputs`, and `TestOutputs`) are stored during different iterations.

The reason for storing these variables throughout the numerous iterations is that after selecting the best model (according to the modeling error criterion), the `BHN` (bias of hidden neurons), `OutW` (output weight) and `InW` (input weight) matrices must be saved so that they can be applied in future application to unseen data. To calculate training time, the `CPU time` value is saved before the start of the training in line 3 (`start_time_train = cputime`) and after it is finished in line 28 (`end_time_train = cputime`). The difference between the two `cputime` values will yield the training time.

Over the next several pages, the three functions `elm_initialization`, `elm_train`, and `elm_test`, which form the core of the ELM, are explained in detail.

5.2.3 Details of the `elm_initialization` function

Each function that is called in MATLAB has its own input, which is fed from the main program, and output parameters, which it returns to the main program after being called upon. In fact, only a small number of total input parameters are passed to the specific functions, where they are required to calculate the output to be passed back to the main program. The input argument of the `elm_initialization` function is `nn`, which is a composite consisting of `Inputsize`, `type`, `hiddensize`, `activefunction`, `orthogonal`, and `method`. The `Inputsize` is the number of input variables, `type` is the ELM type (i.e., regression or classification), `hiddensize` is the number of hidden neurons, `activefunction` is the type of activation function that was selected by the user, `orthogonal` is whether or not the user decided to consider an orthogonal matrix, and `method` is ELM. An example of the `nn` composite is given in Fig. 5.16.

The output argument of `elm_initialization` is also termed `nn`, and randomly assigns the bias of hidden neurons and the input weights in the ELM model. The bias of the hidden neurons and the input weights are allocated in this section and termed `nn.b` and `nn.W`, respectively. As shown in Code 5.10 below, random generation of these matrices is performed by using the `rand` command. It should be noted that the dimensions of `nn.b` and `nn.W` are (`nn.hiddensize`,1) and (`nn.hiddensize`, `nn.inputsize`), respectively. After the random allocation of the bias of hidden neurons and the input weights, the input condition regarding orthogonality is checked. If it is true, the orthogonal of `nn.W` and `nn.b` is calculated using the `orth` command. Otherwise, the randomly assigned `nn.W` and `nn.b` matrices are reported as the output of the `elm_initialization` function. The `orth(K)` function returns an orthonormal basis for the range of `K`. The range or column space of a matrix `K` is the collection of all linear combinations of the columns of `K`. Two nonzero vectors are known as orthogonal when their dot product is zero. An example detailing the application of the `orth` function is presented in Fig. 5.17. Moreover, the schematic of Code 5.10 is shown in Fig. 5.18.

```
nn:
  struct with fields:

       inputsize: 6
            type: 'regression'
      hiddensize: 5
  activefunction: 'sig'
      orthogonal: 'true'
          method: 'ELM'
```

FIGURE 5.16 Example of the nn composite.

$$x = \begin{bmatrix} 1 & -1 & 0 \\ 0 & -2 & 1 \\ 1 & 0 & -1 \end{bmatrix} \xrightarrow{x = \text{orth}(x)} x_{orth} = \begin{bmatrix} -0.12 & -0.8097 & 0.5744 \\ 0.9018 & 0.1531 & 0.4042 \\ -0.4153 & 0.5665 & 0.7118 \end{bmatrix}$$

(columns A, B, C) → Verification

$A \cdot B$	$B \cdot C$	$A \cdot C$
0.0972	−0.4651	−0.0689
0.1381	0.0619	0.3645
−0.2353	0.4032	−0.2956
+ ✓0.0000	✓0.0000	✓0.0000

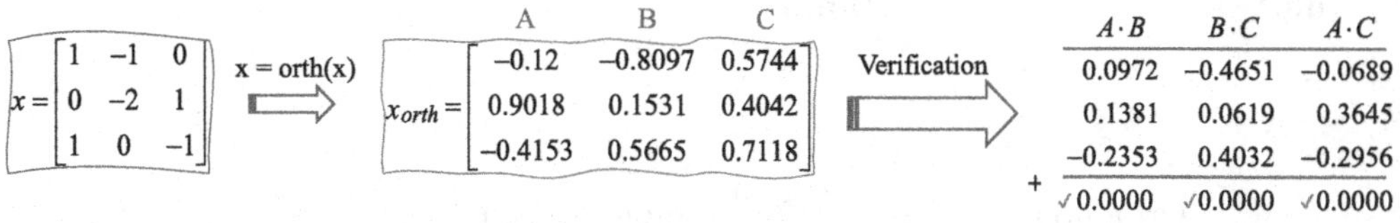

FIGURE 5.17 An example of the orthogonalization of a matrix.

Code 5.10

```
1    function nn = elm_initialization(nn)
2    nn.b = 2*rand(nn.hiddensize,1)-1;
3    nn.W = 2*rand(nn.hiddensize, nn.inputsize)-1;
4
5    if nn.orthogonal
6        if nn.hiddensize > nn.inputsize
7            nn.W = orth(nn.W);
8        else
9            nn.W = orth(nn.W')';
10       end
11       nn.b=orth(nn.b);
12
13   end
14
15   end
```

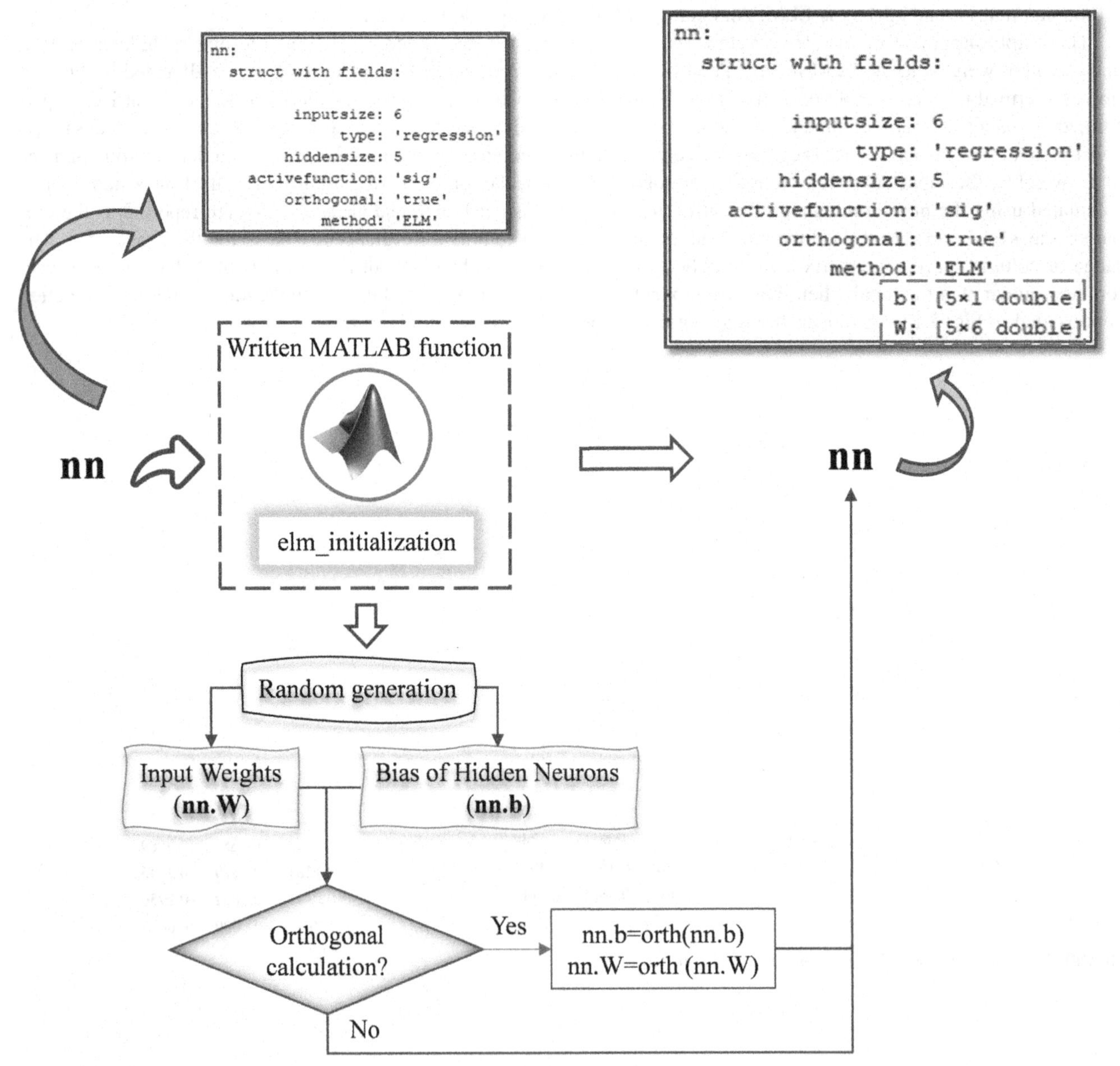

FIGURE 5.18 Schematic of Code 5.10.

5.2.4 Details of the elm_train function

After determining the activation function type, the number of hidden neurons, and randomly assigning the bias of hidden neurons and the input weights, the ELM method uses the `elm_train` function to find the output weight matrix. To determine the output weight matrix, the algorithm must resolve the nonlinear system outlined in Chapter 4, Section 4.3.2. The input arguments of this function are `nn` (similar to what was seen for the `elm_initialization` function), and X and Y, which are the training inputs and targets, respectively.

With the value of input weights and bias of hidden neurons stored as `nn.W` and `nn.b`, respectively, the value of the variable `tempH` can be calculated using the MATLAB command `repmat` as shown in Code 5.11. The schematic of this code is outlined in Fig. 5.19.

Code 5.11

```
1    tempH = nn.W*X + repmat(nn.b,1,ndata);
```

It should be noted that in Code 5.11, `repmat (K,D)` returns an array containing `D` copies of the `K` in the row and column dimensions.

In the above relation, if the hidden layer size is 5,`nn.b` will be a 5 × 1 matrix (5 rows and a column). If we were to consider a case where the number of training samples is equal to 10, the dimensions of the new matrix created using `repmat (nn.b, 1, ndata)` will be equal to 5 × 10. Fig. 5.20 shows the results of a problem with 5 neurons in the hidden layer as well as the results of the `repmat (nn.b, 1, ndata)` function for a problem with 10 training samples. It is clear that all of the columns in Fig. 5.20 are identical and have been copied a number of times equal to the number of training samples (i.e., 10). The matrix `nn.W` will be defined with a dimension of NHN × NIV (NHN is the number of the

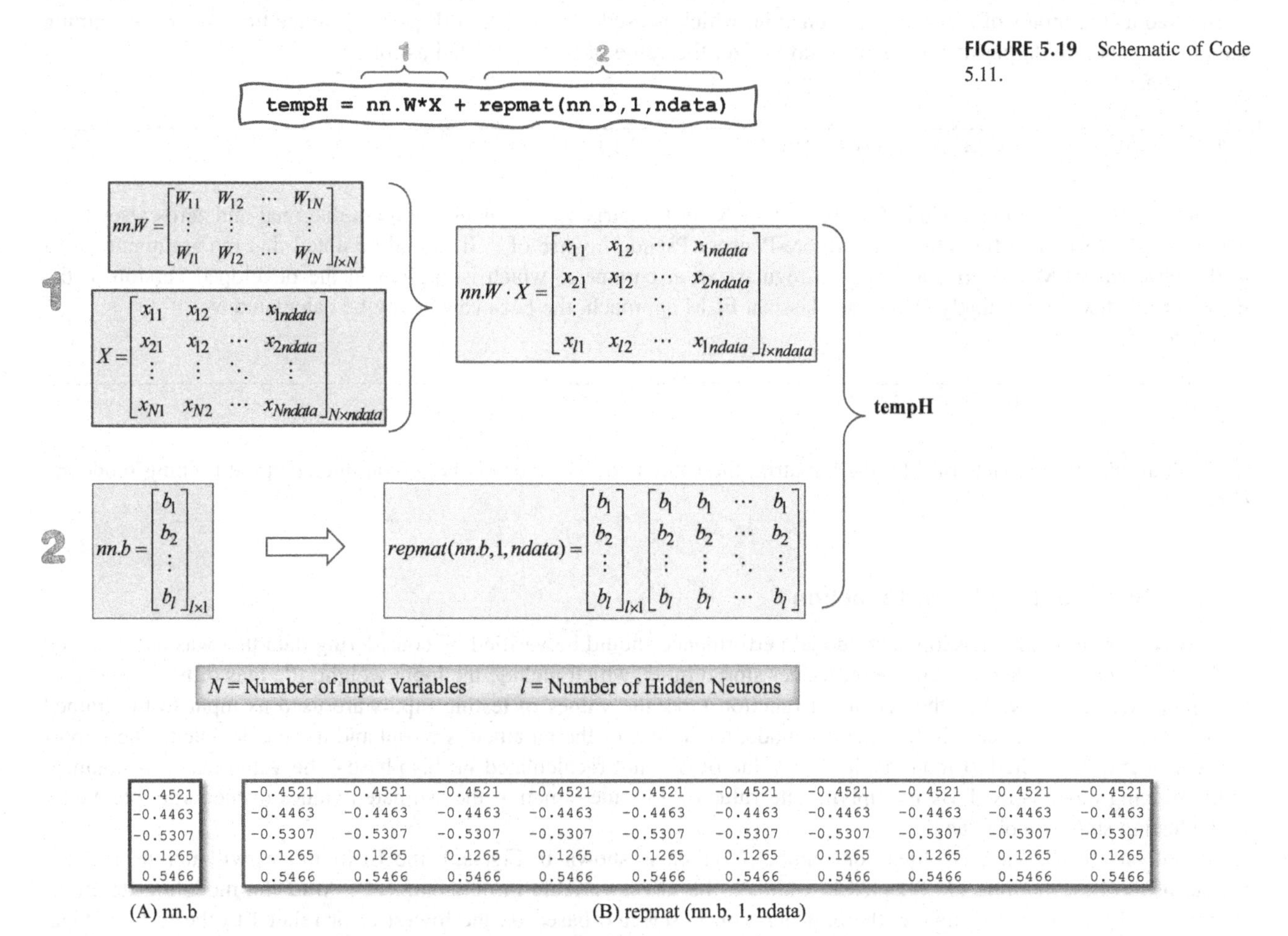

FIGURE 5.19 Schematic of Code 5.11.

(A) nn.b

nn.b
-0.4521
-0.4463
-0.5307
0.1265
0.5466

(B) repmat (nn.b, 1, ndata)

1	2	3	4	5	6	7	8	9	10
-0.4521	-0.4521	-0.4521	-0.4521	-0.4521	-0.4521	-0.4521	-0.4521	-0.4521	-0.4521
-0.4463	-0.4463	-0.4463	-0.4463	-0.4463	-0.4463	-0.4463	-0.4463	-0.4463	-0.4463
-0.5307	-0.5307	-0.5307	-0.5307	-0.5307	-0.5307	-0.5307	-0.5307	-0.5307	-0.5307
0.1265	0.1265	0.1265	0.1265	0.1265	0.1265	0.1265	0.1265	0.1265	0.1265
0.5466	0.5466	0.5466	0.5466	0.5466	0.5466	0.5466	0.5466	0.5466	0.5466

FIGURE 5.20 The output of the `repmat` command for a problem with 5 hidden neurons and 10 training samples.

hidden neuron and NIV is the number of input variables), and the dimension of tempH is therefore NHN × TS. It should be noted that X is the matrix of the input variables and has a dimension of NIV × TS, where TS is the training samples.

After calculating the variable tempH, the user-defined activation function is applied to this matrix, and the hidden layer output matrix (H) [Eq. (4.7)] is obtained using Code 5.12.

Code 5.12

```
1    switch lower(nn.activefunction)
2        case{'s','sig','sigmoid'}
3            H = 1 ./ (1 + exp(-tempH));
4        case{'t','tanh'}
5            H = tanh(tempH);
6        case {'sin','sine'}
7            H = sin(tempH);
8        case {'hardlim'}
9            H = double(hardlim(tempH));
10       case {'tribas'}
11           H = tribas(tempH);
12       case {'radbas'}
13           H = radbas(tempH);
14   end
```

The values of the H matrix calculated in Code 5.12 and the target variable (Y—the matrix of training targets) are considered as the inputs of theregressor function, which is used to calculate β [Eq. (4.9)]. Since the number of training samples is generally larger than the hidden layer size, the value of β is calculated as follows.

Code 5.13

```
1    beta = pinv(H'*H+lamda*eye(size(H,2)))*H'*Y;
```

The eye (K,L) function in Code 5.13 returns an K-by-L matrix with ones on the main diagonal and zeros elsewhere, while the pinv(h) function returns the Moore-Penrose Pseudo inverse of H. It should be noted that the argument lamda in the classical ELM is zero. Lambda is a regularization parameter which is applied in the developed version of the ELM for this text. Accordingly, using the classical ELM approach, the beta can finally be calculated as:

Code 5.14

```
1    beta = pinv(H'*H)*H'*Y;
```

By multiplying the value of β by the H matrix, the estimated values of the target parameter in the training mode are obtained.

5.2.5 Details of the elm_test function

After model training and development, model performance should be verified by considering data that was not involved in the training stage. Therefore, using the values stored in nn, which include the input weight, the bias of hidden neuron, the output weight, as well as the activation function type, the values of testing inputs are used as input to the trained model. Then, as was the case in the training mode, the values of the parameters tempH and H are calculated. The important consideration in the test mode is that the value of β is not recalculated in this phase—the value that was obtained in the training mode is used. By multiplying the value of β by the H matrix, the estimated values of the target parameter in the testing stage are obtained.

Based on the user-defined "Iteration number," which is shown in Fig. 5.9, the ELM model will iterate, and the results of the best iteration are selected to estimate the target variable in new data sets. After the modeling iterations (training and testing) are complete, the best iteration is selected based on the lowest error related to the testing stage. Therefore, the matrices of the input weights, the bias of hidden neurons, and output weights are saved for only the

```
Command Window
Iteration 44: Best Cost = 0.037263
Iteration 45: Best Cost = 0.036979
Iteration 46: Best Cost = 0.039333
Iteration 47: Best Cost = 0.040458
Iteration 48: Best Cost = 0.036077
Iteration 49: Best Cost = 0.044367
Iteration 50: Best Cost = 0.035672

Training Time = 1.4531

Min Error is 0.033452 at position 14

      R         VAF        RMSE       NRMSE       MAE        MARE       RMSRE       MRE        BIAS        AICc        NSE

Statistical_Indices_Training =

    0.9545    91.1140    0.0333    0.0693    0.0247    0.0518    0.0699   -0.0036   -0.0000  -14.7601    0.9025


Statistical_Indices_Testing =

    0.9506    90.3404    0.0335    0.0698    0.0247    0.0526    0.0733    0.0102    0.0063 -190.9654    0.8935

fx >>
```

FIGURE 5.21 The output of the ELM modeling in the MATLAB command window.

selected best iteration. The output of the ELM modeling is evaluated in terms of quality and computational time. An example of the results output to the user in the MATLAB Command Window is provided in Fig. 5.21.

It can be seen in Fig. 5.21 that the training time for all iterations is presented in the Command Window after the final iteration (i.e., 50 for the current example). In this case, the training time for the model was found to be 1.4531 seconds. Following the display of the training time, the position of the best iteration as well as its associated minimum error, is provided. The iteration number associated with the minimum error is provided to the user to indicate the location of the saved best input weights, bias of hidden neurons, and output weights matrices. For the user's convenience, eleven commonly used statistical indices are calculated and displayed to provide an indication of the performance of the ELM model at the best iteration (i.e., which was at position 14 for the current example). These statistical indices include the coefficient of determination (R), the Variance Accounted For (VAF), The Root Mean Square Error (RMSE), the Normalized Root Mean Square Error (NRMSE), the Mean Absolute Error (MAE), the Mean Absolute Percentage Error (MARE), the Root Mean Squared Relative Error (RMSRE), the Mean Relative Error (MRE), the bias, the Akaike Information Criteria (AIC), and the Nash Sutcliffe Efficiency (NSE). The qualitative output of the ELM code is discussed in the next section. Finally, the results of the ELM implementation are saved using Code 5.15.

Code 5.15

```
1    save('Results')
```

5.3 Extreme learning machine modeling output

Fig. 5.22 demonstrates the MATLAB modeling results that are presented to the user during the training stage. This figure includes three subfigures, which are now discussed in detail. The first figure (shown at the top of Fig. 5.22) displays a graph containing the estimated values by the ELM model (Outputs) and the actual values (Targets) for all training samples. In this figure, the abscissa corresponds to the index of each training sample, while the ordinate displays the predicted and observed values of Outputs and Targets, respectively. Therefore, the range of the ordinate axis is given as [min (Outputs, Targets), max (Outputs, Targets)]. As shown in the figure, the training samples are arranged in ascending order. In addition, it is clear (from the comparison of the blue and red plots) that the model performance both over- and underestimates the target values, a trend that is observed throughout the range of all studied samples. Indeed, it cannot be said with any degree of certainty that the model has a tendency to underestimate or overestimate the value of the target data.

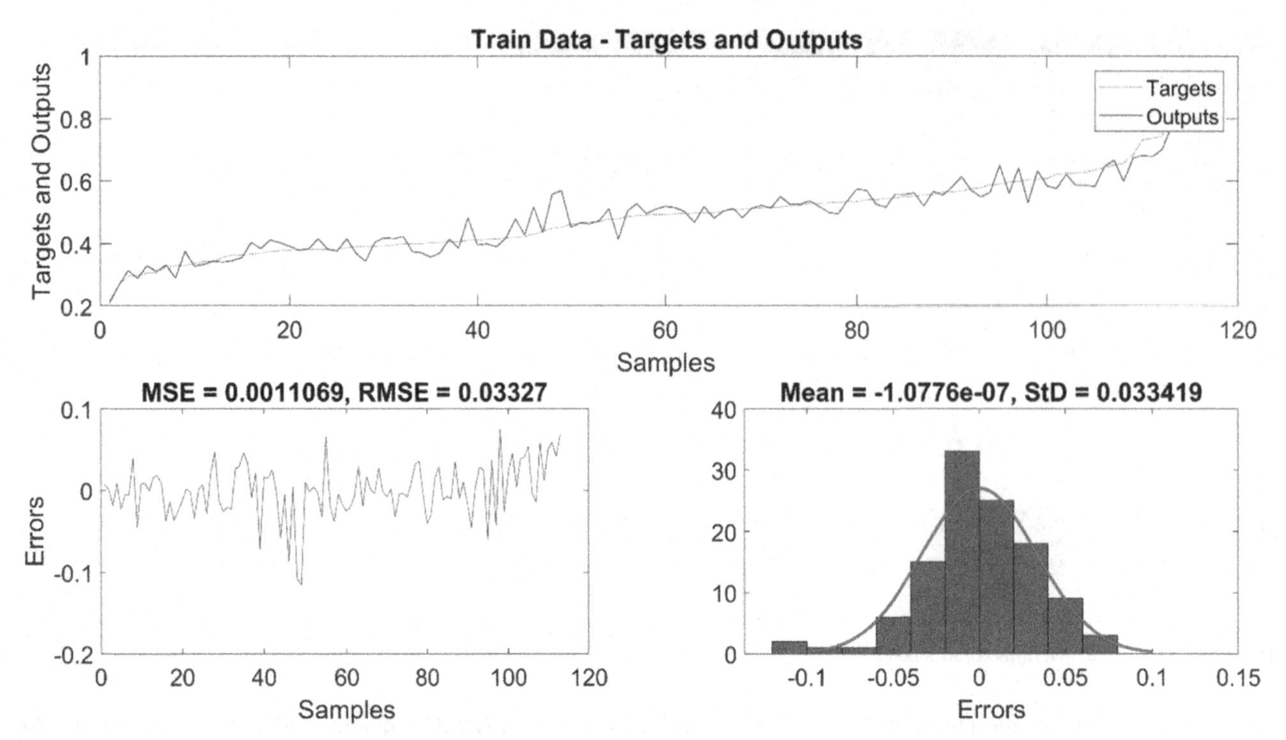

FIGURE 5.22 The quantitative results of the developed ELM model during the training phase. *ELM*, Extreme learning machine.

In addition to this, there are two other subfigures in Fig. 5.22 related to the modeling error. The graph shown in the bottom left of Fig. 5.22 outlines the difference between the Targets and Outputs for each sample. For the current example, the error values are approximately constant in the range of ± 0.1. Displayed at the top of the chart are the values of the Mean Square Error (MSE) and Root Mean Square Error (RMSE) for the training stage. The chart shown on the bottom right of Fig. 5.22 shows an error histogram for the training stage and considers bins spanning from $[-0.15, 0.15]$. Indeed, this plot compares the histogram of the errors to the normal distribution. The results of this figure show that most of the estimated samples have an error (|Targets-Outputs|) less than 0.05, and the errors are approximately normally distributed. For the user's convenience, the values of the mean and standard deviation (StD) are also computed and displayed above the plot.

Fig. 5.23 shows the results that are displayed to the user following the completion of the ELM testing stage. Similar to Fig. 5.22, three plots are presented that show the estimated (Outputs) and actual values (Targets), the errors related to each of the testing samples, and their histogram compared with the normal distribution. The distribution of testing stage errors indicates that the target parameter estimation error is in the range of $(-0.12, 0.05)$. In fact, on average, test results show a tendency to underestimate, although many examples can be seen where the predicted output of the model is, in fact, an overestimation. Similar to the previous figure, the error histogram follows a normal distribution, with most errors being concentrated within the range of $(-0.05, 0.05)$.

Fig. 5.24 presents the scatter plot of the actual and predicted values of the developed ELM model during the training and testing stages. The abscissa displays the observed values (Targets) while the ordinate axis displays the predicted values (Outputs). For the purpose of easy comparison, the perfect linear relationship between the predicted and observed values is plotted as a dotted line. In addition, the Pearson correlation coefficient (R) for model performance during both the training and testing stages is provided above the plot.

ELM modeling results show that the R-values for the training and testing phase are 0.9545 and 0.9506, respectively. In the current case, the R-values of both the testing and training phases are close to each other, and their difference is considered insignificant. In ML approaches, there is often an observable difference between the computed statistical indices during the training and testing phases, however, this difference should be reduced as much as possible. In fact, because the testing data is excluded from the model development, the performance of the model in the testing phase will almost certainly be lower than in the training phase, whose data was involved in model training and development. From the results of the modeling process, it can be observed that the target values in both the training and testing phases are almost the same, such that the maximum difference between the actual and predicted values in the range is 0.4–0.5.

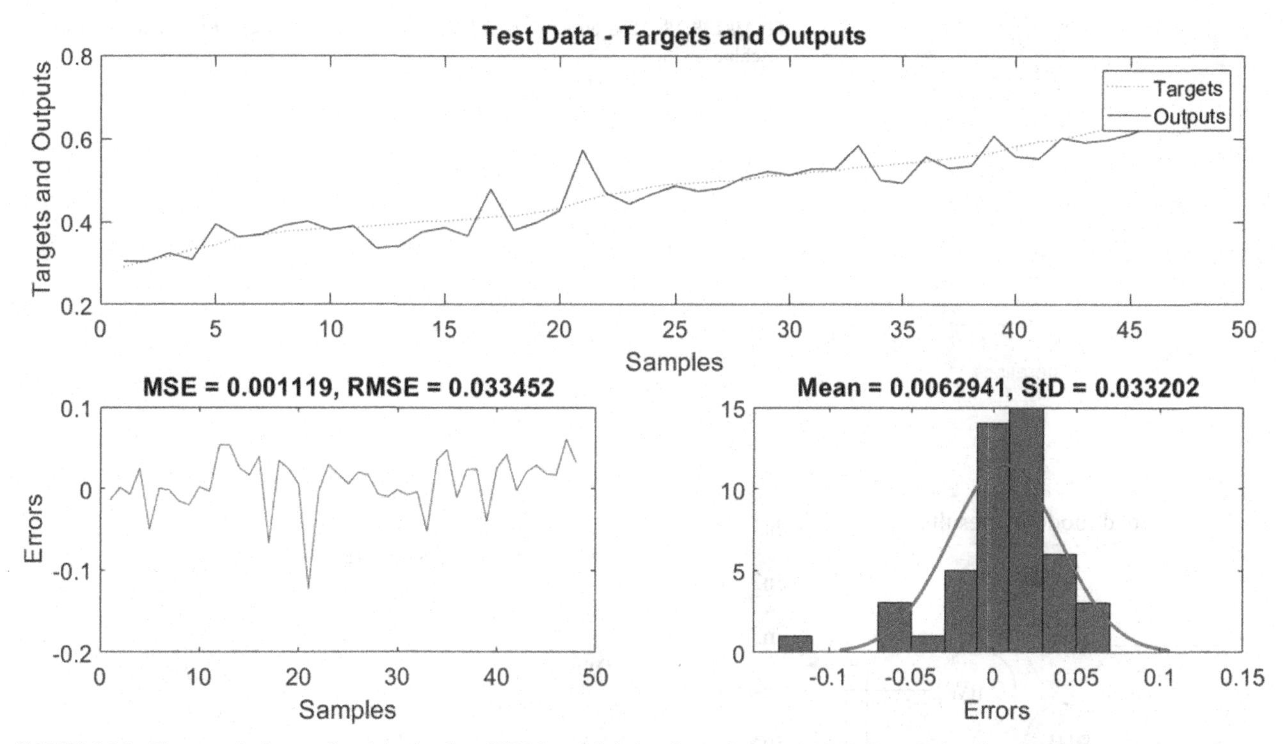

FIGURE 5.23 The quantitative results of the developed ELM model during the testing stage. *ELM*, Extreme learning machine.

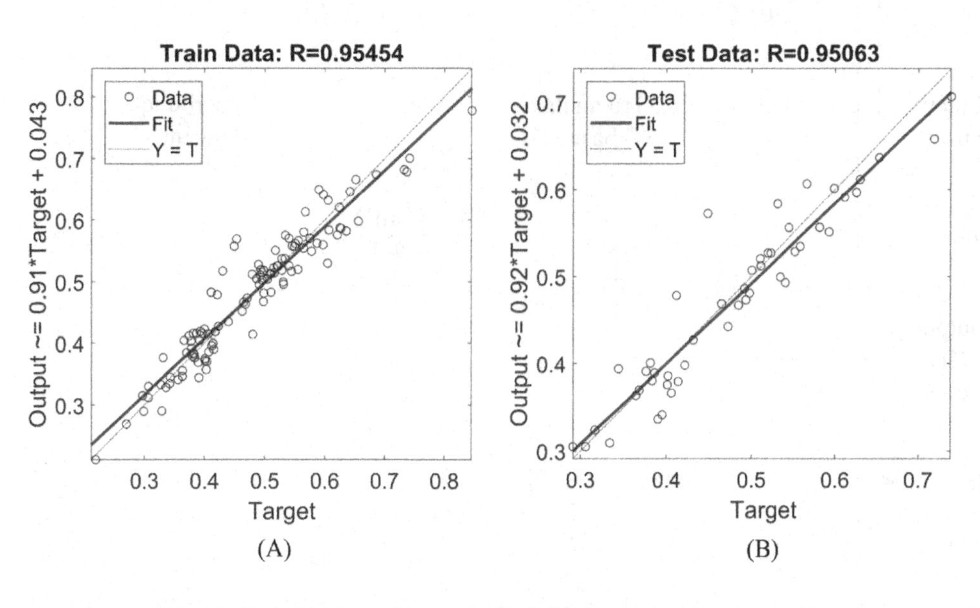

FIGURE 5.24 Scatter plot of the observed and predicted values of the developed extreme learning machine ELM model during the: (A) training stage and (B) testing stage.

Fig. 5.25 indicates the error of each iteration in ELM modeling. The ELM method is not an iteration-based method, however, it must be run several times to achieve acceptable results with previously unseen data. This is because the bias of hidden neurons and matrix of input weights are randomly assigned [recall from Eq. (4.20) that more than 66% of all tuned parameters through the training phase were randomly assigned], the values of which have a significant effect on the output weight matrix and hence the model results. Therefore, the number of iterations is a model input that must be determined prior to the commencement of the modeling process. In the current problem, the iteration number was considered to be 50. Fig. 5.25 demonstrates that the iteration with the lowest model error when considering unseen data was obtained at iteration 14. Fig. 5.21 also displays, through the MATLAB command window, that the best position is located at iteration 14. The maximum error of the model is approximately 0.05 (as interpreted from Fig. 5.25), which is about 50% higher than the optimal error value (from Fig. 5.21, we can see that the lowest error is

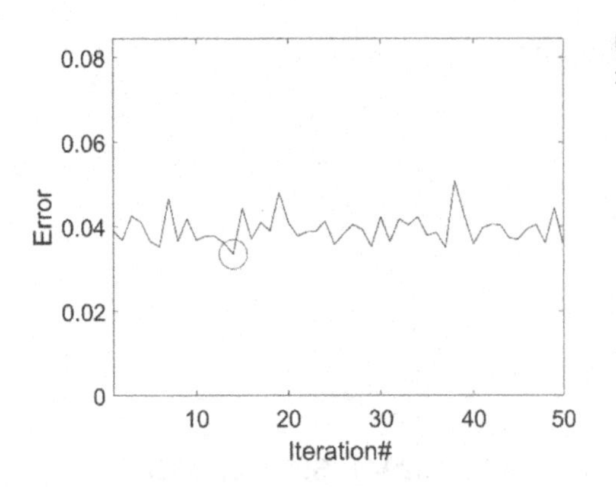

FIGURE 5.25 The error of the ELM model at each iteration. *ELM*, Extreme learning machine.

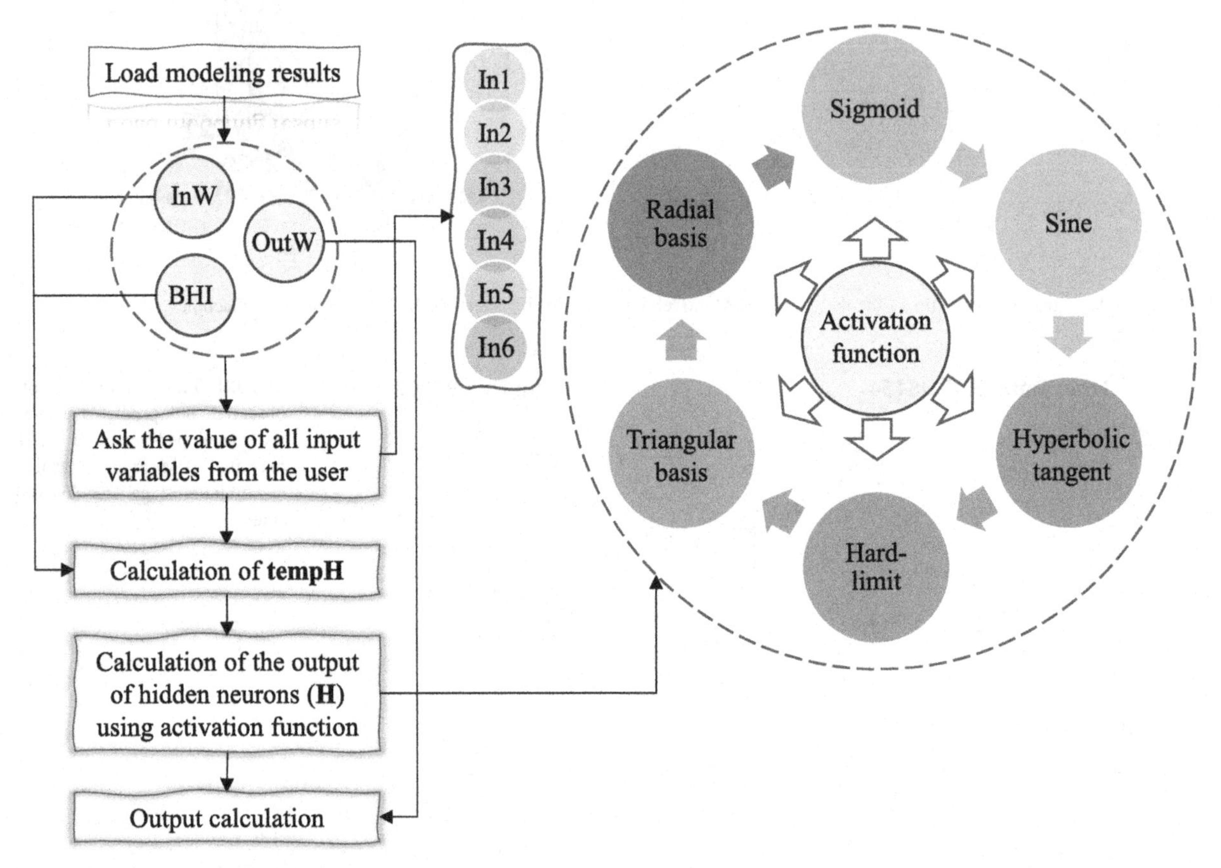

FIGURE 5.26 Schematic of Code 5.16.

0.033). This example demonstrates the importance of selecting an appropriate number of iterations in your model, so that the optimal result is obtained by the end of the modeling procedure. From Fig. 5.21, we can observe that by considering 50 iterations, the training time of the ELM model was only 1.45 seconds. With such a quick computational time and, therefore, high modeling speed, it is recommended that at least 1000 iterations be considered to achieve the optimal model.

5.4 Calculator for extreme learning machine model

As an alternative to the direct implementation of the ELM MATLAB code, a proposed calculator for beginner users is shown in Code 5.16, and the schematic of this code is shown in Fig. 5.26. To apply this calculator, it is necessary that the results from the previous subsection are loaded by the program. As can be seen in the provided program syntax, the model loads the `results` variable automatically in line 6. Therefore, once the previous model results have been read into the calculator, the only input required for the model is the numeric values of the various inputs (Fig. 5.27). The `inputdlg` command is used in line 12 to prompt the reader to input the value of up to 6 variables. Following this, the algorithm uses the `ActivationFunction` stored within the `Results` variable and computes the corresponding `Output`. It should be noted that this code could be easily modified to accommodate problems with a lower or higher number of input variables. To do so, four general changes are needed: (1) increase or decrease the number of inputs specified in line 8; (2) increase or decrease the number of default values in line 10 to match the number of input variables; (3) increase or decrease the number of variables contained in lines 14–19 according to the number of input variables (simply copy the format of line 14–19 and specify In6, In7, ..., etc. if the number of variables is increased); increase or decrease the number of input variables contained within the array specified in line 21.

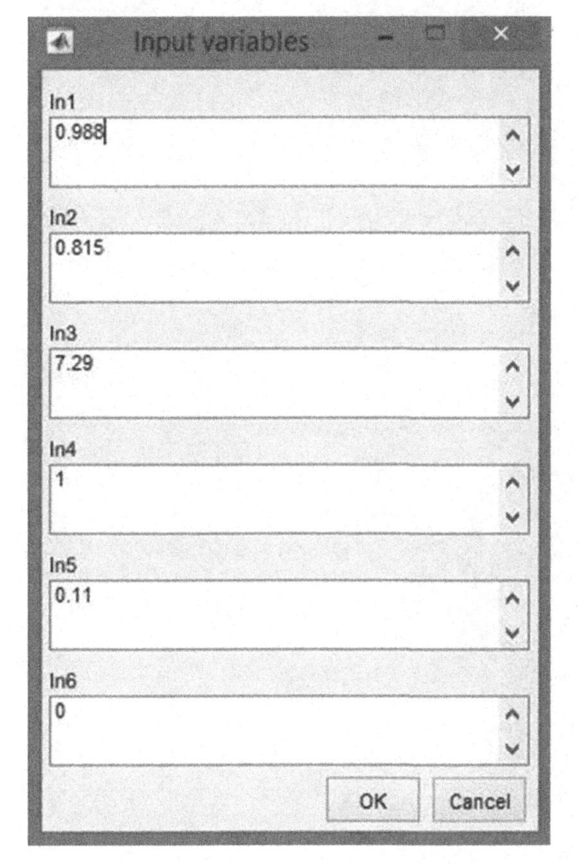

FIGURE 5.27 The dialog box of the proposed ELM calculator. *ELM*, Extreme learning machine.

Code 5.16

```
1	clear all
2	close all
3	clc
4	
5	%% Load data
6	load('Results')
7	%% Output Calculation
8	Prompt={'In1','In2','In3','In4','In5','In6'};
9	Title='Input variables';
10	DefaultValues={'0.5','0.5','0.5','0.5','0.5','0.5'};
11	
12	PARAMS=inputdlg(Prompt,Title,[2 40],DefaultValues);
13	
14	In1=str2num(PARAMS{1});
15	In2=str2num(PARAMS{2});
16	In3=str2num(PARAMS{3});
17	In4=str2num(PARAMS{4});
18	In5=str2num(PARAMS{5});
19	In6=str2num(PARAMS{6});
20	
21	InV=[In1 In2 In3 In4 In5 In6];
22	
23	tempH = (InW*InV')+BHN;
24	
25	switch lower(ActivationFunction)
26	    case{'s','sig','sigmoid'}
27	        H = 1 ./ (1 + exp(-tempH));
28	        Outputs = H'*OutW
29	    case{'t','tanh'}
30	        H = tanh(tempH);
31	        Outputs = H'*OutW
32	    case {'sin','sine'}
33	        %%%%%%%% Sine
34	        H = sin(tempH);
35	        Outputs = H'*OutW
36	    case {'hardlim'}
37	        %%%%%%%% Hard Limit
38	        H = double(hardlim(tempH));
39	        Outputs = H'*OutW
40	    case {'tribas'}
41	%%%%%%%% Triangular basis function
42	        H = tribas(tempH)
43	        Outputs = H'*OutW
44	    case {'radbas'}
45	        %%%%%%%% Radial basis function
46	        H = radbas(tempH);
47	        Outputs = H'*OutW
48	end
```

5.5 Effect of the extreme learning machine parameters

As discussed in the previous subsections, to apply the ELM to solve real-world problems, the values of two user-defined inputs must be specified: the number of hidden layer neurons and the type of activation function. In addition to these two inputs, the user must define the number of iterations that will be considered in the development of the optimal model. As discussed in Section 5.3, the number of iterations should be considered in excess of 1000 to overcome the random generation of the input weight and bias of hidden neuron matrices. To find the optimal model that provides good predictive performance for unseen data, there is a need to run the ELM model through an iterative process such that the model with the minimum testing error is selected as the best one. The effect of ELM parameters on model performance will be discussed over the next few subsections using the example datasets introduced at the beginning of the text. The ELM parameters to be discussed are summarized in Fig. 5.28.

5.5.1 Iteration number

In this subsection, the effect of iteration number on ELM model performance is evaluated, with the objective of identifying the optimal case (in terms of modeling error). For this purpose, the five examples defined in Chapter 1 are employed. To isolate the effect of iteration number on achieving optimal results for all examples, the other ELM input parameters are held constant. This includes a constant number of hidden neurons, the use of the sigmoid activation function, and the selection of orthogonality of the randomly allocated matrices (Input weights and bias of hidden neurons).

5.5.1.1 The effect of iteration number on Example 1

Table 5.2 presents the results of the ELM modeling when different numbers of iterations were considered in solving Example 1. It is clear that, as the number of iterations of the model increases, so too does the training time of the model. Although the training time of the model increases, the training time, even at the maximum iteration number (i.e., 1000), still remains less than 40 seconds. Therefore, it can be concluded that the ELM exhibits a high degree of speed during the training phase regardless of the number of iterations considered. As the number of iterations increases, the results obtained for the minimum error are also improved, so that the best performance of the model is found after 1000 iterations. An interesting observation regarding the results is that the best position is not necessarily always found

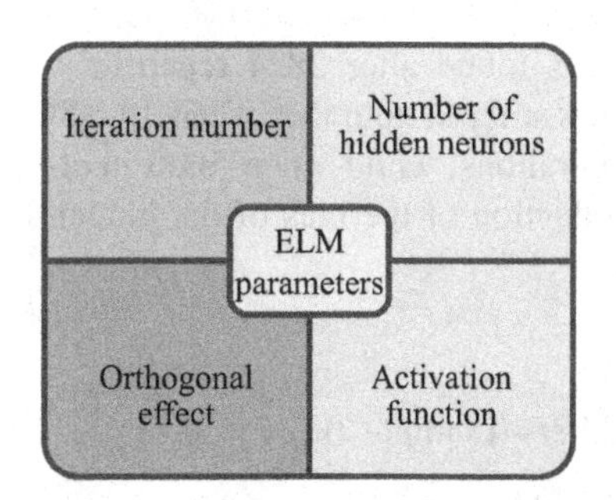

FIGURE 5.28 The ELM parameters. *ELM*, Extreme learning machine.

TABLE 5.2 Results of extreme learning machine Modeling with different iteration numbers (Example 1).

Example no.	Iteration no.	Training time	Best position	Min error	Max error
Example 1	50	1.875	23	0.0361	0.0506
Example 1	100	4.0781	30	0.0346	0.0498
Example 1	200	8.25	132	0.0337	0.0526
Example 1	300	11.3594	161	0.0354	0.0514
Example 1	400	16.1719	57	0.0328	0.0539
Example 1	500	19.2031	438	0.0322	0.0525
Example 1	1000	39.1563	436	0.0316	0.0567

at the maximum number of iterations. For example, the best position when 100 iterations (or 400 iterations) were considered was found at iteration 30 (or iteration 57 in the case of 400 iterations). This is due to the random nature of the assignment of the bias of hidden neurons and the input weights matrices in the ELM method. Therefore, as the number of iterations increases, the probability of obtaining the optimal answer also increases, however, it cannot be said that the modeling error will always be lowest at the maximum iteration number. The range of the minimum error in the model extends from 0.0361 to 0.0316. In fact, the increase in the number of iterations had reduced the error value in the optimal model by over 12%. The maximum error values presented in this table show that if only a single iteration is considered during model development, the amount of modeling error will increase to 0.0567, which is 60% more than the value obtained for the minimum error of the optimal results (position 436 at 1000 iterations).

5.5.1.2 The effect of iteration number on Example 2

Table 5.3 indicates the results of the ELM modeling with different iteration numbers for Example 2. It should be noted that the data set considered in Example 2 is smaller than that considered in Example 1. For this reason, the training time is reduced by 2 seconds compared to Example 1 after 1000 iterations. The minimum error found after 50 iterations is the highest value compared to other iteration numbers (i.e., 100, 200, . . ., 1000). Modeling errors are expected to decrease with an increasing number of iterations, however, this is not always the case in practice due to the random allocations of the input weight and bias of hidden neuron matrices. For example, the minimum error value recorded following 400 iterations (Min Error = 1.6521) is less than when 500 iterations are considered (Min Error = 1.6936), although this amount of error is considered negligible. However, a further increase in the number of iterations to 1000 reduced the value of the recorded minimum error, making it smaller in magnitude compared to all other cases. Given that the minimum error value recorded after 50 iterations (Min Error = 2.6484) is very large compared to other iterations (e.g., for 400 iterations, Min Error = 1.6521), the reduction in minimum error after 5,000 and 10,000 iterations is also verified, and their effect on obtaining the minimum error is examined. The results of the minimum error for these high values of iterations demonstrate that the probability of obtaining an improved result is increased as the number of iterations increases. It was observed that the minimum error recorded after 5000 and 10,000 iterations rank second and first, respectively, in terms of the recorded error. In fact, the developed ELM code allows us to run the model for multiple iterations and store the results of the best iteration during the training and testing process. In addition, this code stores ELM matrices (i.e., bias of hidden neurons, input weights, and output weights) related to the best iteration to apply in future work.

Fig. 5.29 indicates that when 5000 iterations are considered, the optimum result is found after 2824 repetitions, while for 10,000 iterations, the optimum result is found after 4817 repetitions. It is interesting to note that when 10,000 iterations are considered, the optimal result is obtained by repeating less than 5000 iterations, while when 5000 iterations are considered, this result is not obtained. The reason for this is the random production of the bias of the hidden neuron and input weight matrices, which have a significant impact on the results.

TABLE 5.3 Results of extreme learning machine modeling with different iteration numbers (Example 2)

Example no.	Iteration no.	Training time	Best position	Min error	Max error
Example 2	50	2.3281	24	2.6484	16.4316
Example 2	100	3.5625	29	1.9938	18.5595
Example 2	200	7.4531	73	1.8149	36.0738
Example 2	300	11.2344	258	1.80172	62.4808
Example 2	400	14.6406	49	1.6521	72.4584
Example 2	500	18.7813	311	1.6936	37.2052
Example 2	1000	37.75	49	1.6485	91.7188
Example 2	5000	188.9688	2824	1.2624	64.2539
Example 2	10000	343.9688	4817	0.94838	90.8259

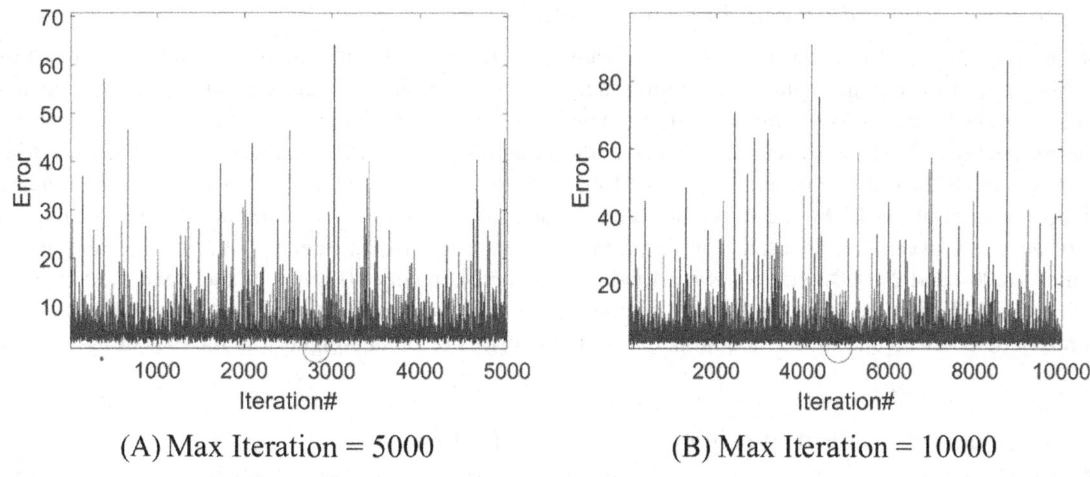

FIGURE 5.29 The error of each iteration for ELM-based modeling of Example 2. *ELM*, Extreme learning machine. (A) = 5000 iterations, (B) = 10,000 iteration.

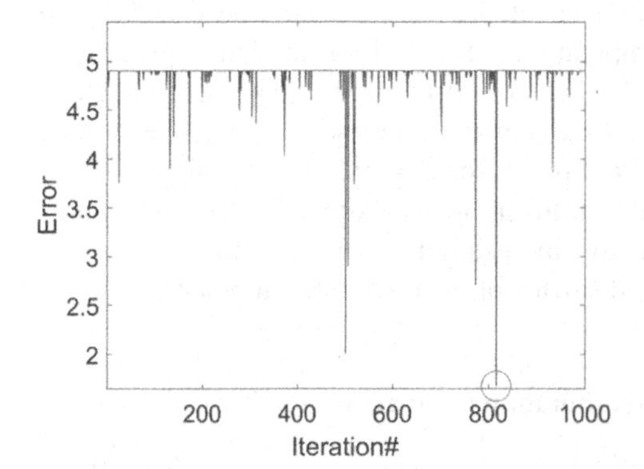

FIGURE 5.30 Error value for ELM-based modeling of Example 3 with 1000 iterations. *ELM*, Extreme learning machine.

Fig. 5.29 shows that the range of responses obtained for ELM can be very wide. For example, the maximum recorded error when 5,000 iterations are considered is greater than 60: this is more than 40 times the optimal response obtained in this mode. In addition, for the case of 10,000 iterations, the maximum error is recorded to be greater than 80, which is at least 80 times greater than the minimum error value that was obtained at this iteration number.

Therefore, it is concluded that according to the type of problem, a number of different iterations should be considered in this method. As noted, for Example 1, 1000 iterations were sufficient, and a further increase in the number of iterations did not reduce the modeling error. However, in Example 2, a greater number of iterations were required, with the error obtained after 5000 iterations being more than 25% lower than that observed after 1000 iterations. The remainder of this section will investigate the performance of the model when applied to different examples.

Based on the experience gained from Example 2, at the commencement of modeling for Example 3, we initially elect to consider an iteration number equal to 1000. Based on the model results for this Example, the number of iterations may be altered based on a trial-and-error approach to arrive at the optimal solution. Fig. 5.30 presents the error of the ELM-based modeling of Example 3. The training time for 1000 iterations was 32.3 seconds, and the best position was obtained following 816 iterations with an error of 1.6726. From the plot of the error shown in Fig. 5.30, it can be seen that most of the iterations had an error of around 4.5 in magnitude.

It should be noted that the average of the modeling error is 4.8706, which is almost three times the minimum error found at iteration 816. Based on the results seen after 1,000 iterations, it is concluded that similar to Example 2, a wide range of minimum errors is observed for the ELM applied to Example 3. The number of iterations is therefore increased to examine whether this can achieve a further reduction in the model error.

5.5.1.3 The effect of iteration number on Example 3

Table 5.4 provides the results of the ELM-based modeling of Example 3 with different numbers of iterations. The results of this table show that increasing the iteration number from 1000 to 5000 and 10,000 significantly increases the accuracy of the model (similar to Example 2). It is noteworthy that by increasing the iteration number from 10,000 to 20,000, the minimum error obtained with the higher iteration number is actually increased when compared to the case of 10,000 iterations. In fact, it can generally be stated that due to the random allocation of bias of hidden neurons and input weights matrices of the ELM, the minimum error is rarely obtained after 1000 iterations, and we may not reach that level of accuracy over many thousands of iterations. Another point to consider is that the maximum error values obtained after 10,000 and 20,000 iterations were found to be approximately 10 times greater than the values obtained when 1,000 and 5,000 iterations were considered. Similar to the minimum error, these values obtained for the maximum error will rarely be obtained due to the random nature of the matrices of input weights and the bias of hidden neurons.

5.5.1.4 The effect of iteration number on Examples 4 and 5

Table 5.5 provides the results of ELM modeling with the different number of iterations for Examples 4 and 5. Based on the experience gained from previous examples, we begin modeling Example 4 using 500 iterations where the minimum error was found to be about 0.5. As the iteration number increases to 1000 and 2000, it is observed that the minimum error values present very small differences compared to when 500 iterations were considered. In fact, it can be concluded that for this example, 500 iterations are sufficient to obtain a good result. However, it may be beneficial to investigate the model performance with fewer iterations to reduce the computational effort of our modeling procedure. If only 50 iterations were considered, the minimum error value will be 0.577, which is 14% higher than the case where 500 iterations are considered. Therefore, due to the already low training time and the increase in predictive performance, 500 iterations are an acceptable value for Example 4. Similar to Example 4, for Example 5, the effects of 500, 1000, and 2000 iterations were examined. It was observed that the minimum, mean, and maximum values are almost the same for all three iteration numbers. Therefore, it was concluded that, for this example, there was no advantage to increasing the number of iterations and that 500 iterations can be considered sufficient for model development.

TABLE 5.4 The results of extreme learning machine Modeling with different iteration numbers (Example 3).

Example no.	Iteration no.	Training time	Best position	Min error	Max error
Example 3	1000	32.2969	816	1.6726	4.9101
Example 3	5000	82.0469	1835	1.144	4.9323
Example 3	10000	186.5	7481	0.82518	43.2209
Example 3	20000	383.3906	15617	0.95722	45.5383

TABLE 5.5 The results of extreme learning machine modeling with different iteration numbers (Examples 4 & 5).

Example no.	Iteration no.	Training time	Best position	Min error	Mean error	Max error
Example 4	500	10.375	319	0.50267	1.02	1.8468
Example 4	1000	18.5625	866	0.51257	1.0412	2.0236
Example 4	2000	32.9688	1538	0.49978	1.0298	2.5088
Example 5	500	41.1563	261	3.2303	3.3804	4.1243
Example 5	1000	81.3594	9	3.2155	3.3852	4.2971
Example 5	2000	150.3594	1043	3.2061	3.3783	4.251

5.5.2 Number of hidden neurons

This subsection investigates the effect of varying the number of neurons in the hidden layer on ELM performance. Table 5.6 shows the maximum number of allowable hidden neurons for each example, which was determined by the application of Eq. (4.13) and proposed by Ebtehaj et al. (2021a). To apply this equation, the number of training samples and the number of input variables are required. Unless otherwise stated, a training percentage of 70 is considered throughout this text. Therefore, the number of training samples can be directly computed from the total number of samples contained in each Example data set. Using the number of training samples and the number of inputs, the maximum allowable number of hidden neurons is calculated.

By considering 1000 iterations (see the previous subsection) and the sigmoid activation function, the performance of the ELM model subject to different numbers of hidden neurons is studied.

5.5.2.1 The effect of hidden layer neurons on Example 1

Fig. 5.31 presents the statistical indices for the ELM-based modeling of Example 1 with different numbers of hidden neurons. From Fig. 5.31A, we can observe that strong correlations, as determined by the values of the correlation coefficient (R), are observed between the predicted and observed values. In addition, the value account for (VAF) and the Nash−Sutcliffe efficiency (NSE) values for models with more than one and five hidden neurons (respectively) are greater than 0.8. Recall that the best performance of the model is obtained when R = NSE = VAF = 1. For all three indices, the best model performance for Example 1 is achieved when using 14 hidden neurons, which is the maximum number of neurons that can be considered in the analysis. Another five statistical indices, including the RMSE, the NRMSE, the MARE, the RMSRE, and the MRE, are presented in Fig. 5.31B. All of these statistical indices indicate optimal model performance when they have a value of 0. The lowest value of all five of these indices, similar to the previous indices (R, VAF, NSE), is observed whenever the model considers the maximum permissible number of hidden neurons.

Recall that NHN and InV are the numbers of hidden neurons and input variables, respectively, and that the number of tuned parameters through ELM-based modeling is equal to (InV + 2) × NHN [Eq. (4.12) and Ebtehaj et al., 2021a].

TABLE 5.6 Maximum number of allowable hidden neurons for each example.

Example no.	Samples no.	Training (%)	Training samples	Inputs no.	Maximum allowable hidden neurons
Example 1	161	70	113	6	14
Example 2	100	70	70	3	14
Example 3	240	70	168	4	28
Example 4	250	70	175	5	25
Example 5	1095	70	767	2	191

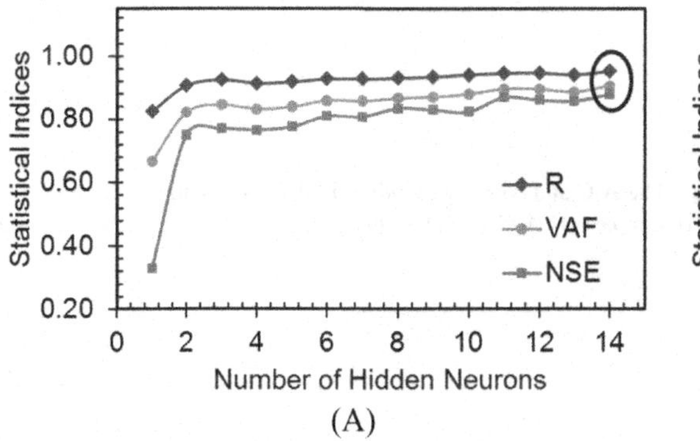

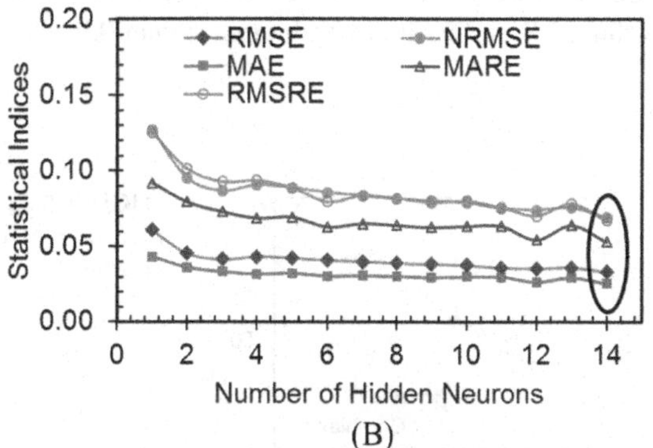

FIGURE 5.31 Statistical indices of the ELM-based model for Example 1 with a different number of hidden neurons. *ELM*, Extreme learning machine. (A) correlation-based indices, (B) relative and absolute indices.

The number of InV in Example 1 is 6. Therefore, increasing the NHN value leads to an eightfold increase in the number of parameters that must be defined during the training process. From Fig. 5.31, we can observe that as the number of hidden neurons increases, the difference in the statistical indices is low. For example, the difference between relative error in models with 9 (MARE = 6.28%) and 14 (MARE = 5.33%) hidden neurons is less than 1%.

Therefore, higher values of R, VAF, NSE, and lower values of the RMSE, NRMSE, MARE, RMSRE, and MRE cannot be the only selection criteria when developing an optimal model. It is suggested that, when selecting the most optimum model, the complexity of the model should be examined and the superior model should be selected according to both accuracy and complexity, simultaneously. For this purpose, the use of the Akaike Information Criterion (AIC) index (Azimi & Shiri, 2021a, 2021b; Azimi et al., 2019, 2021; Ebtehaj et al., 2014; Ebtehaj et al., 2021b; Safari et al., 2019) is recommended, which is defined as follows:

$$AIC = N\ln(\sigma_{\varepsilon}^{2}) + 2k \tag{5.1}$$

where k is the number of tuned variables through the training phase, σ_{ε} is the standard deviation of the difference between the target and output values, and N is the number of training samples. The first term of this relationship denotes the accuracy of the model, while the second one denotes its complexity. Other types of AIC index are available in the literature (Azari et al., 2021; Ebtehaj et al., 2019, 2020b; Khalid & Sarwat, 2021; Lin et al., 2022, Mehdizadeh et al., 2020, Zeynoddin et al., 2019, 2020; Żymełka & Szega, 2021).

Fig. 5.32 presents the AIC index and the complexity of the ELM model when developed using different numbers of hidden neurons. As mentioned earlier, according to Eq. (4.12), increasing the number of hidden neurons by a single neuron increases the complexity by eight units. Therefore, the complexity and number of hidden neurons are directly related to each other (complexity = 8 × NHN). For the AIC index, the lowest value is obtained at NHN = 2, and as the number of hidden neurons increases, so too does the AIC value. In fact, for NHN greater than 2, complexity outweighs accuracy, and virtually governs the selection of the optimal model.

Suppose the error value obtained in the model with two hidden neurons is not acceptable for the nature of the problem under investigation. In that case, it is possible to select a model with a greater number of hidden neurons to arrive at a more accurate solution. For example, suppose a relative error of less than 6% (models 12 and 14) and an NSE value of more than 0.85 (models with 11−14 hidden neurons) indicate a desirable model. Based on this description of model suitability, the models that meet the above conditions are those that consider 12 and 14 hidden neurons. Therefore, the model presented with 12 hidden neurons, which has a lower complexity and still meets the desired error criterion, is selected as the optimal model.

5.5.2.2 The effect of hidden layer neurons on Example 2

Fig. 5.33 shows the statistical indices of the ELM-based modeling of Example 2 subject to different numbers of hidden neurons. The number of iterations and activation function type for all models were fixed as 10,000 and sigmoid, respectively. Similar to Example 1, the weakest ELM model performance is achieved when a single hidden neuron is considered to the point where the NSE value in this model is negative. For NHN values of 9 and 11−14, the value of the NSE is greater than 0.9. For a NHN greater than or equal to 6, the MARE is found to be slightly less than 2%. The values of the AIC index that take into account the accuracy and complexity of the model simultaneously [Eq. (5.1)] lead to the model with the lowest AIC for those with a NSE greater than 0.9. The relative error of this model is found to be 1.1%, and the value of R and NSE are both greater than 0.9.

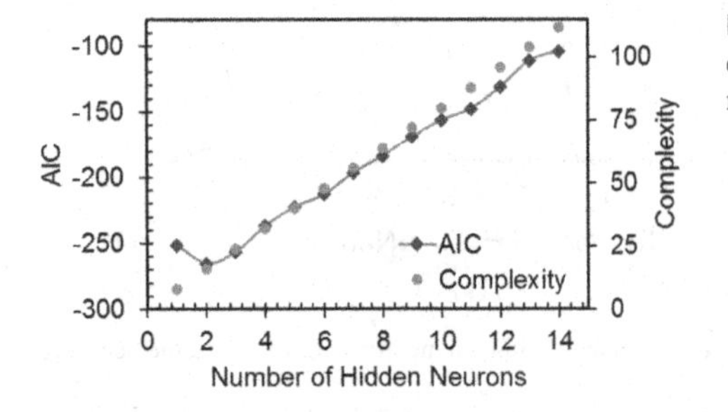

FIGURE 5.32 The AIC and complexity of the ELM model with different numbers of hidden neurons (Example 1). *AIC*, Akaike Information Criterion; *ELM*, extreme learning machine.

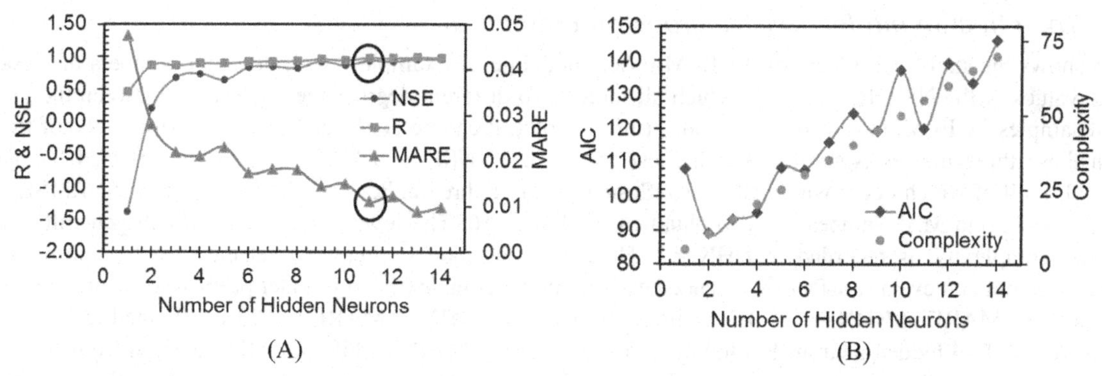

FIGURE 5.33 Statistical indices of the ELM model with different numbers of hidden neurons (Example 2). *ELM*, Extreme learning machine. (A) correlation—based indices, (B) relative and absolute indices.

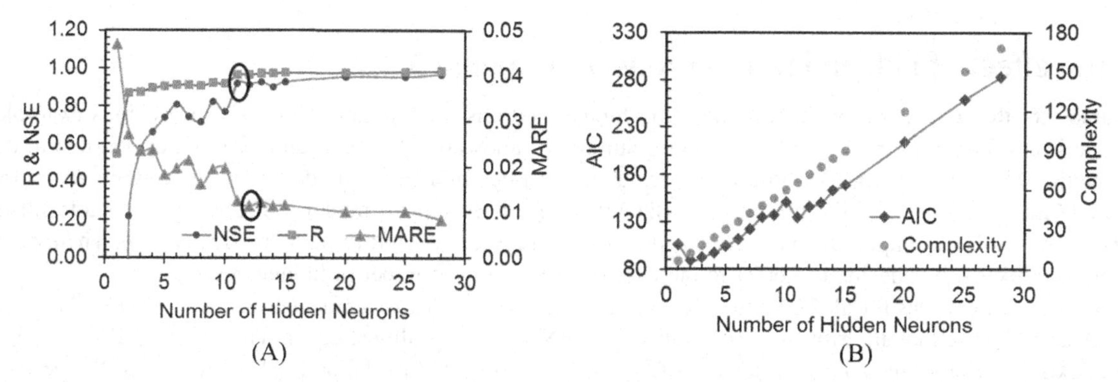

FIGURE 5.34 Statistical indices of the ELM model with a different number of hidden neurons (Example 3). *ELM*, Extreme learning machine. (A) R, NSE, and MARE indices, (B) AIC and complexity.

5.5.2.3 The effect of hidden layer neurons on Example 3

Fig. 5.34 presents the statistical indices of the ELM-based modeling of Example 3 with different numbers of hidden neurons. In this example, similar to Example 2, the NSE index of the ELM with a single hidden neuron is negative, which is disregarded during further analysis due to the discrepancy with other index values. There is also a clear difference between the Pearson correlation coefficient in the model with a single hidden neuron and models with a greater number of neurons. As the number of neurons in the hidden layer increases to 15, it is observed that the values of R and NSE reach 0.97 and 0.92, respectively. It is also observed that the models with 11−15 neurons have practically the same performance, with the values of these two indices remaining almost constant in this range of hidden neurons. In addition, the maximum recorded relative error associated with the model with one hidden neuron is about 4%, which was found to decrease significantly as the number of hidden layer neurons increases. The use of 11 hidden neurons in ELM modeling leads to a model with a relative error of about 1.2%. Increasing the number of neurons from 11 to 15 reduces the relative error by less than 0.1%. Given that the maximum number of neurons that can be considered in this example is 28, the performance of the model for the cases that use 20, 25, and 28 hidden neurons are also examined. Increasing the number of hidden neurons raises the NSE from 0.92 (NHN = 15) to 0.96 (NHN = 28). As mentioned earlier, the relationship between complexity and NHN is a linear relationship in the form of "complexity = (InV + 2) × NHN." Example 3 has four inputs, so this relationship for this problem will be as "complexity = 6 × NHN" [Eq. (4.12)]. In fact, increasing the number of hidden neurons from 15 to 28 increases the complexity value by 78 units. According to Fig. 5.34, the AIC Index varies from 80 to 280. Indeed, an increase of 78 units of complexity leads to an increase of 156 units of AIC [Eq. (5.1)], while the total change in AIC is about 200 units. Therefore, although the use of 28 hidden neurons provides the best results in terms of relative error, the model with 11 hidden neurons is selected as it best balances optimal model performance and model simplicity. Models with a lower number of hidden neurons tend to have increased MARE values and reduced R values, as seen in Fig. 5.34.

5.5.2.4 *The effect of hidden layer neurons on Example 4*

Fig. 5.35 shows the statistical indices of the ELM-based modeling of Example 4 with different numbers of hidden neurons. The values of the NSE, R, and VAF, which all indicate to different degrees the correlation between the target and predicted samples by ELM, have a similar trend. From this plot, it can be concluded that the performance of the model based on these three indices is similar. The highest values of R, NSE, and VAF indicators are 0.98, 0.97, and 0.98 (respectively), all of which occur with NHN = 16. Similar to these three indices, the lowest values for NRMSE, RMSE, RMSRE, MARE, and MAE indices are also obtained in NHN = 16. The lowest AIC value, which generally increases with model complexity, is recorded in NHN = 2. However, this model cannot be selected as the optimum model, because it does not achieve a satisfactory balance between model complexity and model predictive accuracy. For example, the value of MARE at NHN = 2 is 0.9, while with increasing NHN, a MARE = 0.23 is obtained at NHN = 16. In fact, an increase of 14 hidden neurons has led to a 75% decrease in MARE, while the AIC has risen from − 36 to 112. Therefore, it is not possible to say which one is the optimal model, and it is up to the user to choose the optimal model. The user should define a criterion for selecting the optimal model based on the required accuracy of the prediction. The required accuracy can be governed by the constraints of the particular problem under consideration. According to the authors' experience, choosing ELM with 16 hidden neurons is the optimal choice.

5.6 The effect of hidden layer neurons on Example 5

From Table 5.6, the maximum number of allowable hidden neurons for Example 5 is 193. In previous examples, the NHN value is considered from one to the maximum number of allowable. In this example, it is not possible to develop a model with 193 hidden neurons because considering such a large number of hidden neurons leads to an impractical model. In addition, the reproducibility and generalizability of the developed model will not be acceptable for unseen data. Therefore, in this example, the number of neurons is initially considered to be 5, 10, 15, and 20. Then, according to the obtained results, it is decided whether to increase or decrease the number of hidden neurons.

The number of iterations for all ELM models with different NHN is set to a constant value of 10,000. The results of Fig. 5.36A demonstrate that all error indices for all four NHN values are almost equal. Therefore, NHN = 5, which has less complexity than other models (Complexity = (2 + 2) × 5), is selected as the optimal model. Next, the performance

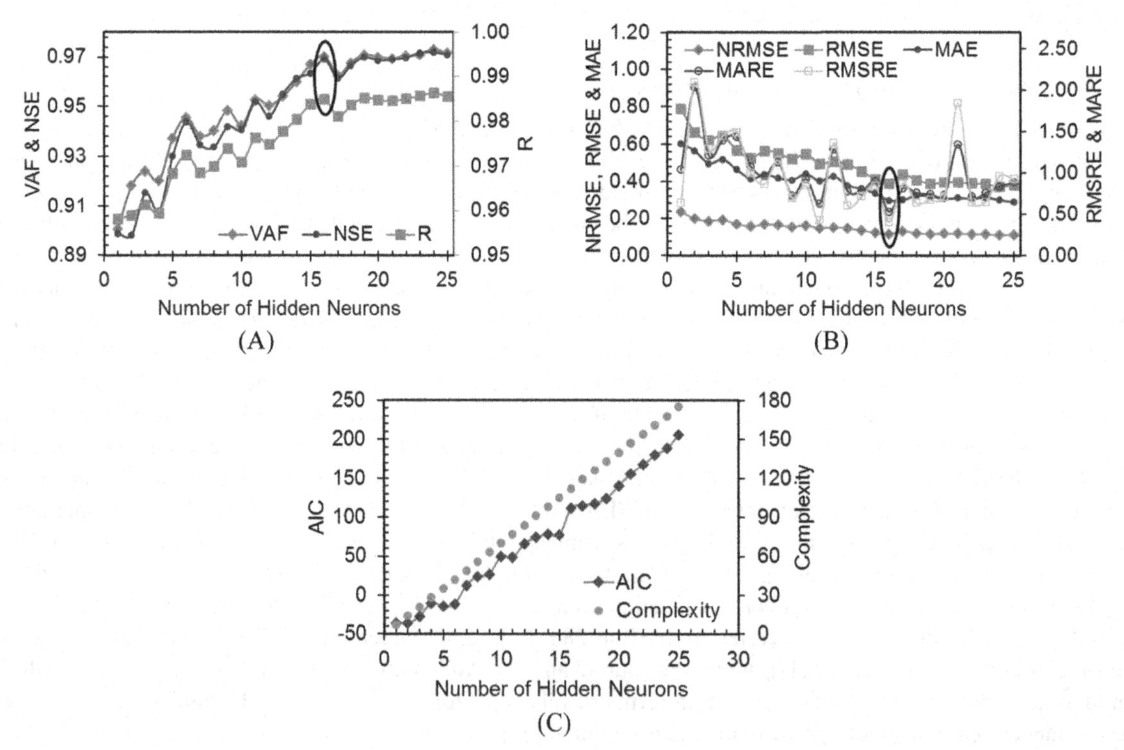

FIGURE 5.35 Statistical indices of the ELM model with a different number of hidden neurons (Example 4). *ELM*, Extreme learning machine. (A) Correlation-based indices, (B) relative and absolute indices, (C) AIC and complexity.

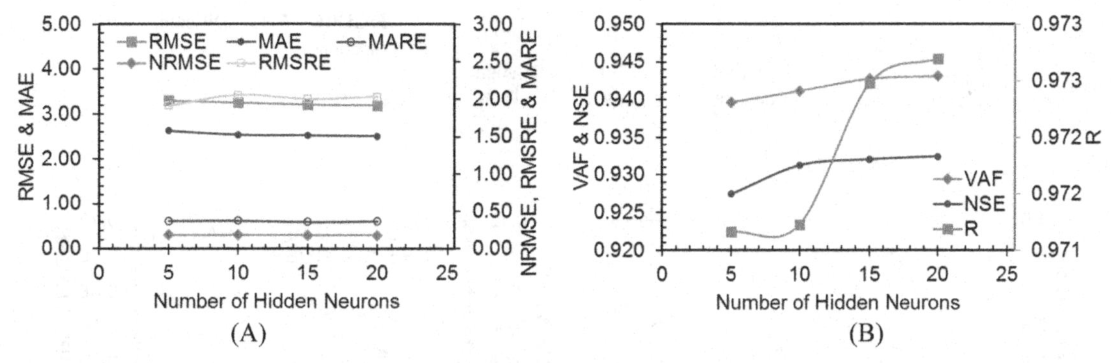

FIGURE 5.36 Statistical indices of the ELM model with a different number of hidden neurons (Example 5). *ELM*, Extreme learning machine. (A) Relative and absolute indices, (A) correlation-based indices.

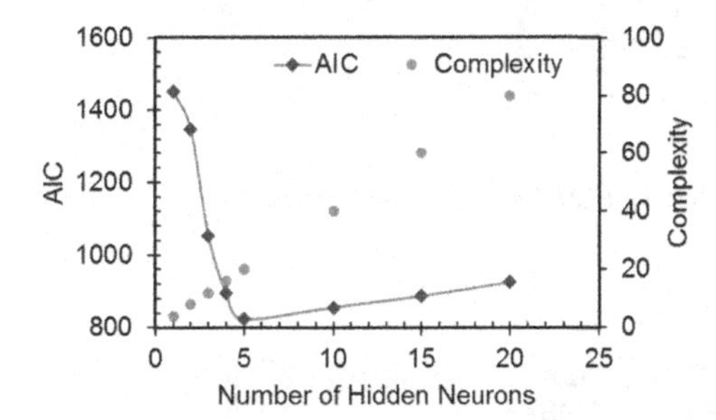

FIGURE 5.37 The AIC and the complexity of the ELM model with different numbers of hidden neurons (Example 5). *ELM*, Extreme learning machine.

of the ELM model with different NHN (NHN = 1, 2, 3, 4) and through 10,000 iterations is also evaluated, and its results are compared with NHN = 5.

Fig. 5.37 indicates the AIC and the complexity of the ELM model with the different number of hidden neurons for Example 5. The number of inputs to consider in Example 5 is 2. Therefore, the complexity and NHN have a linear relationship given by "Complexity = 4 × NHN." The results of this figure show that the lowest AIC value is related to NHN = 5. Considering the different NHNs in the range [1, 5], the relationship between NHN and AIC is an indirect one, so that the lowest value of AIC is related to NHN = 5. Although complexity is directly related to NHN, the low accuracy of different models with NHN >5 has led to the fact that the predominant parameter between accuracy and complexity is accuracy, but for NHN more than 5, the predominant parameter will be complexity. As the NHN increases, the accuracy of the model was not observed to change significantly, and therefore the complexity of the model becomes the governing decision criterion. Among the five examples presented, the only case where the optimal model can be clearly chosen is Example 5.

5.6.1 Activation function

In this subsection, the performance of the six different activation functions, including hyperbolic tangent (tanh) (Maimaitiyiming et al., 2019; Ratnawati et al., 2020), triangular basis (Tribas) (Sattar et al., 2019; Owolabi and Abd Rahman, 2021), radial basis (Radbas) (Samal et al., 2022; Tripathi et al., 2020), hard-limit (Hardlim) (Suchithra & Pai, 2020; Zeynoddin et al., 2018), sigmoid (Sig) (Calabrò et al., 2021; Ebtehaj et al., 2017), and sine (Sin) (Owolabi & Abd Rahman, 2021), are examined for the five examples considered in the previous section (Fig. 5.38). It should be noted that the iteration number and number of hidden neurons for each example are selected based on the optimum values identified in the previous subsections. Additionally, recall that all of the models in the previous subsections were developed considering the sigmoidal activation function.

5.6.1.1 The effect of activation function on Example 1

Fig. 5.39 presents the statistical indices of the ELM model with different activation functions for Example 1. The optimum values of the number of iterations and number of hidden neurons for this example were previously found to be 1000 and 12, respectively.

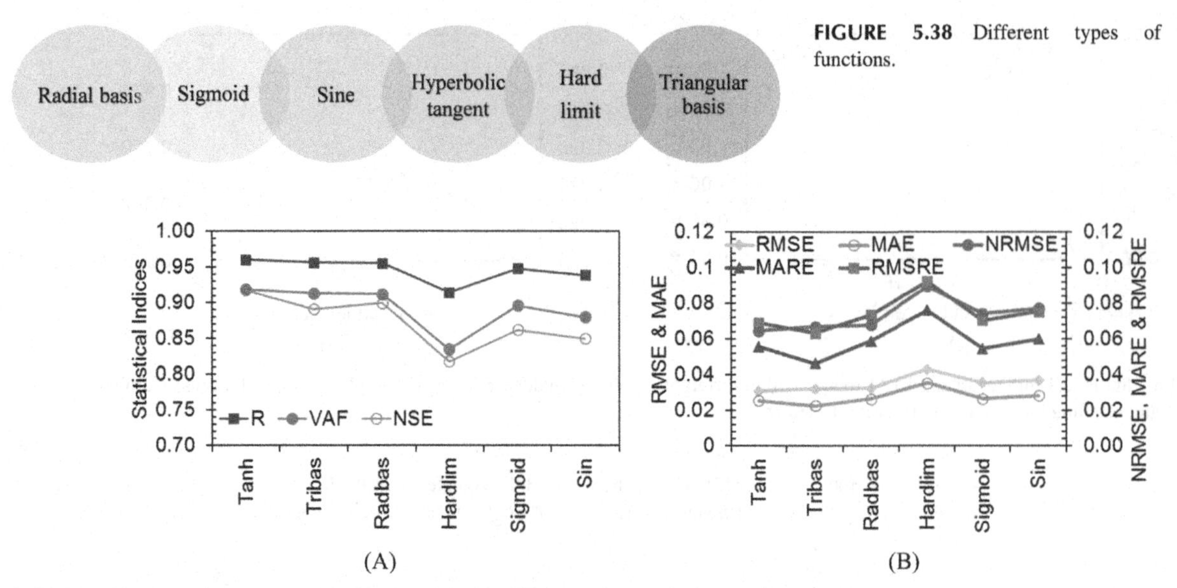

FIGURE 5.38 Different types of activation functions.

FIGURE 5.39 Statistical indices of the ELM model with different activation functions (Example 1). *ELM*, Extreme learning machine. (A) Correlation-based indices, (B) relative and absolute indices.

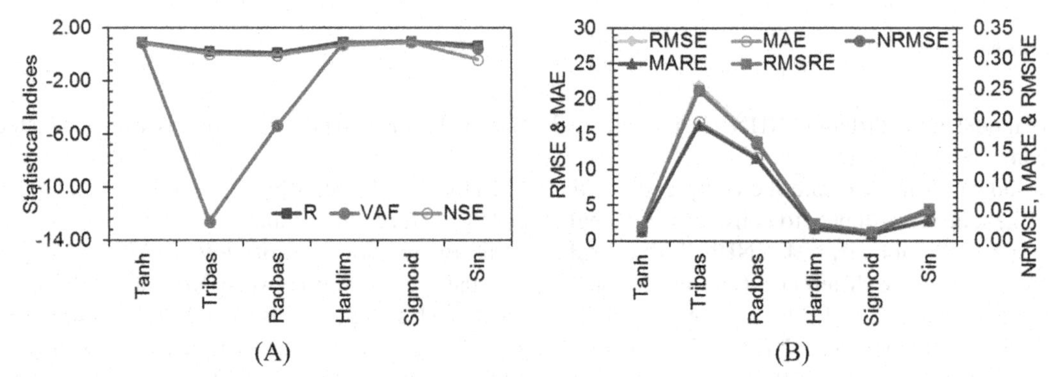

FIGURE 5.40 Statistical indices of the ELM model with different activation functions (Example 2). *ELM*, Extreme learning machine. (A) Correlation-based indices, (B) relative and absolute indices.

For this example, the performance of different activation functions was determined by considering the same statistical indices that were used to evaluate the impact of hidden layer size. From the plot in Fig. 5.39A, it can be seen that the values of R, VAF, and NSE do not differ significantly, and the maximum difference is less than 5%. The lowest values of R, VAF, and NSE are related to the Hardlim activation function, while the only activation function with a NSE value greater than 0.9 is the Tanh activation function. The best performance of activation functions in terms of RMSE and MAE as absolute indices is related to the Tanh and Tribas activation functions, respectively. Additionally, it was found that the lowest values of the MARE, RMSRE, and NRMSE (as three relative indices) are related to the Tribas, Tribas, and Tanh activation functions, respectively.

According to the explanations given in this example, both Tanh and Tribas activation functions perform well in estimating the value of the target variable. In addition to these two functions, Radbas, Sigmoid, Sin, and Hardlim functions were found to be the next best performing functions.

5.6.1.2 The effect of activation function on Example 2

Fig. 5.40 presents the statistical indices of the ELM model with different activation functions for Example 2. The optimum value of the number of iterations and number of hidden neurons for this example was previously found to be 10,000 and 11, respectively. For this example, the results are different from Example 1. Although the Tanh activation function in this example also performs relatively well, the best results were observed when the sigmoid activation function was applied. More specifically, the sigmoid activation function was found to have the highest values of R, VAF, and NSE indices and the lowest values in the other indices (RMSE, MARE, MAE, RMSRE, NRMSE). The Tribas activation function, which was considered to be one of the most effective activation functions for Example 1, was consistently found to be the most poorly performing activation function in the case of Example 2. The amount of VAF obtained from Tribas and Radbas is a negative value that can be reported as zero, according to the Okada (2017) proposal. It is noteworthy that the Hardlim activation function, which had the weakest performance in Example 1, ranks second in this example.

5.6.1.3 The effect of activation function on Example 3

Fig. 5.41 presents the statistical indices of the ELM model with different activation functions for Example 3. The optimum value of the iteration number and number of hidden neurons for this example was previously found to be 10,000 and 11, respectively. It is clear from the plot that the sigmoid activation function consistently yields a superior modeling performance as determined in terms of the R, VAF, and NSE. The VAF for the Tribas and Radbas are negative, similar to Example 2, and can be considered as zero (Okada, 2017). Moreover, the NSE of the Hardlim and Sin activation functions are also negative. The Tanh activation function for the current example is ranked second in terms of model performance with $R = 0.81$, MARE $= 0.05$, and NSE $= 0.024$, while these indices for Sigmoid as the best activation function are $R = 0.96$, MARE $= 0.01$ and NSE $= 0.92$. Examining all the indices for the six activation functions, it is concluded that only Sigmoid offers acceptable performance, and other functions are not suitable for estimating the target variable of Example 3.

5.6.1.4 The effect of activation function on Example 4

Fig. 5.42 presents the statistical indices of the ELM model with different activation functions for Example 4. The optimum value of the iteration number and number of hidden neurons for this example were previously found to be 2000

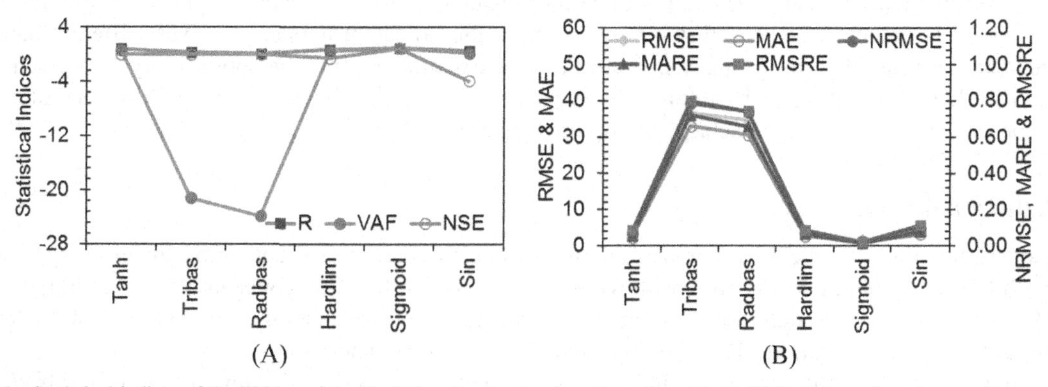

FIGURE 5.41 Statistical indices of the ELM model with different activation functions (Example 3). *ELM*, Extreme learning machine. (A) Correlation-based indices, (B) relative and absolute indices.

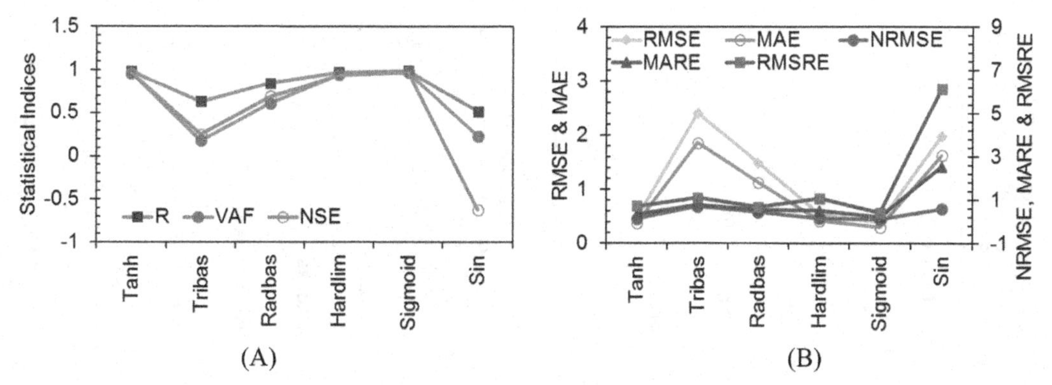

FIGURE 5.42 Statistical indices of the ELM model with different activation functions (Example 4). *ELM*, Extreme learning machine. (A) Correlation-based indices, (B) relative and absolute indices.

and 16, respectively. For this example, correlation-based indices indicated that the Tanh, Sigmoid, and Hardlim functions performed well, and their performance as activation functions is very similar. Following these three top-performing indices, the Radbas activation function is the next highest-performing ($R = 0.84$; VAF = 0.6; NSE = 0.69). If we want to find the best function using these three indicators, the best solution is to consider the average values of the indices for each activation function. Therefore, Sigmoid, Tanh, Hardlim, Radbas, Tribas, and Sin functions are ranked 1–6 in order of model performance.

Examination of other indices shows that the Sigmoid, Tanh, and Hardlim are still in the first to third ranks, however, the fourth highest-ranked function differs when these five error-based indices are considered. The Sin activation function ranks fourth in the RMSE, NRMSE, and MAE indices, while in the two relative MARE and RMSRE indices, the fourth rank belongs to the Radbas activation function. Finally, Tribas ranks fifth in all indices.

5.6.1.5 *The effect of activation function on Example 5*

Fig. 5.43 presents the statistical indices of the ELM model with different activation functions for Example 5. The optimum value of the iteration number and number of hidden neurons for this example were previously found to be 2000 and 5, respectively. Similar to Examples 1, 3, and 4, the Tanh and Sigmoid activation functions have a good performance in Example 5. The use of these two functions leads to a significantly higher value of statistical indices when compared to the other functions, so much so that the use of the Tribas, Radbas, and Sin functions in modeling using ELM leads to negative values of the NSE index. It is noteworthy that after the Sigmoid and Tanh activation functions, which are in the first and second ranks, the Hardlim activation function is third-ranked, with only a slightly lower performance level.

Based on the explanations provided, it can be said that Sigmoid and Tanh functions have more acceptable performance in different examples (different inputs with different numbers of samples) than other functions. After these two functions, the Hardlim activation function also performed relatively well, however, using this function or the remaining three activation functions (i.e., Sin, Radbas, and Tribas), generally did not provide acceptable results. It should be noted that the fact that an activation function provided acceptable (or unacceptable) results for any of the above examples do not confirm or deny that it may be applicable to future applications with different datasets. The nature of the datasets considered in the analysis will to a large extent determine the applicability of a given activation function and its relative performance. Therefore, it is suggested that in other problems, the performance of all six activation functions should be examined to achieve the optimal answer.

5.6.2 Orthogonal effect

In the previous subsections, the iteration number, number of hidden neurons, and activation function type were examined for five different examples. In the current subsection, the effect of the orthogonalization on the initialization step of the ELM is investigated. This step results in the orthogonalization of the randomly allocated bias of hidden neurons and input weights matrices using a MATLAB-built-in function (i.e., `orth` function).

Fig. 5.44 presents the statistical indices of the ELM model with or without orthogonalization of the InW and BHN during the initialization step of the ELM. The correlation indices (R, VAF, and NSE) and AIC indicate that for all examples (Examples 1–5), initialization with orthogonalization of the BHN and InW matrices results in a higher level

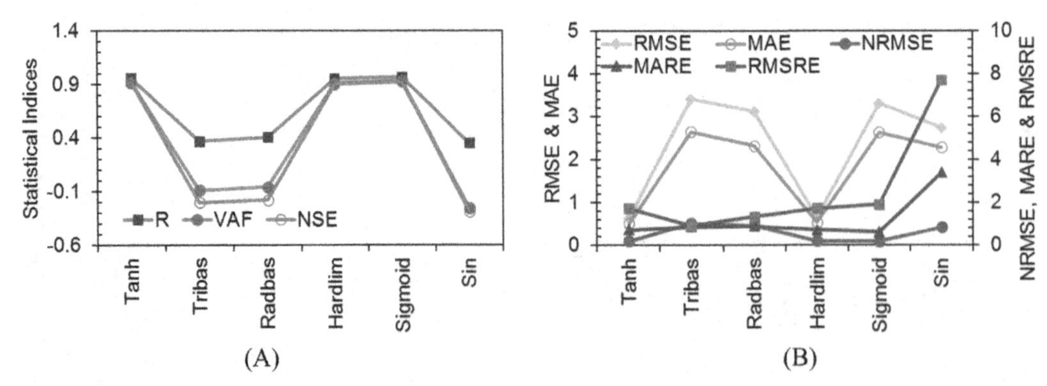

FIGURE 5.43 Statistical indices of the ELM model with different activation functions (Example 5). *ELM*, Extreme learning machine. (A) Correlation-based indices, (B) relative and absolute indices.

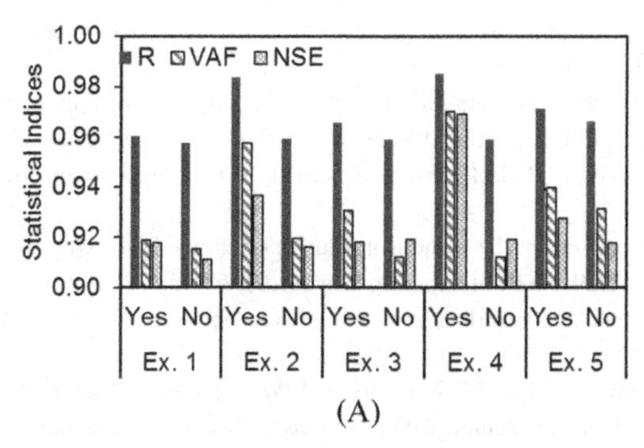

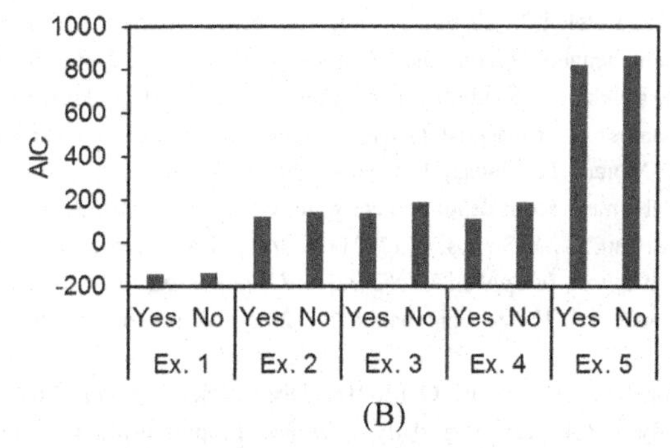

FIGURE 5.44 The statistical indices of the ELM model with or without orthogonal effect. *ELM*, Extreme learning machine. (A) Correlation-based indices, (B) AIC index.

of accuracy in comparison to the case where they are nonorthogonal. It should be noted that the mentioned difference between the consideration of the orthogonality of the BHN and InW matrices is low. For example, the maximum decrease of the Pearson correlation coefficient is related to Example 2, where a change from 0.9836 to 0.9591 is seen.

5.7 Summary

In this chapter the detailed coding of the ELM in the MATLAB environment was presented line-by-line. Furthermore, a novel MATLAB-based calculator was introduced for application to unseen datasets (see Appendix 5.A Supporting information). The activation function type and the number of hidden neurons are two primary user-defined parameters of the classical ELM. This study develops the classical ELM model for a large number of iterations to overcome the weakness of the ELM in the random assignment of two main matrices (i.e., Input weight & bias of hidden neurons), where at least 66% of the parameters set in the training phase are randomly allocated. Moreover, a Matlab built-in function (i.e., `orth`) was employed to orthogonalize these matrices to reduce the effect of their random generation. Finally, the effect of the mentioned parameters, including iteration number, number of hidden neurons, activation function type, and orthogonality, was examined for five different examples with different numbers of input variables (ranging from 2 to 6) and different numbers of training samples (ranging from 70 to 767). After reading this chapter, the reader has the ability to model using ELM, and for the advanced user, can use their experience in ELM coding to develop models for use in practical work.

Appendix 5.A Supporting information

Data on ELM implemention can be found in the online version at https://doi.org/10.1016/B978-0-443-15284-9.00009-4.

References

Azari, A., Zeynoddin, M., Ebtehaj, I., Sattar, A. M., Gharabaghi, B., & Bonakdari, H. (2021). Integrated preprocessing techniques with linear stochastic approaches in groundwater level forecasting. *Acta Geophysica*, *69*(4), 1395–1411.

Azimi, H., & Shiri, H. (2021). Assessment of ice-seabed interaction process in clay using extreme learning machine. *International Journal of Offshore and Polar Engineering*, *31*(04), 411–420. Available from https://doi.org/10.17736/ijope.2021.mt31.

Azimi, H., & Shiri, H. (2021a). Sensitivity analysis of parameters influencing the ice–seabed interaction in sand by using extreme learning machine. *Natural Hazards*, *106*(3), 2307–2335. Available from https://doi.org/10.1007/s11069-021-04544-9.

Azimi, H., & Shiri, H. (2021b). Assessment of ice-seabed interaction process in clay using extreme learning machine. *International Journal of Offshore and Polar Engineering*, *31*(04), 411–420. Available from https://doi.org/10.17736/ijope.2021.mt31.

Azimi, H., Bonakdari, H., & Ebtehaj, I. (2017). Sensitivity analysis of the factors affecting the discharge capacity of side weirs in trapezoidal channels using extreme learning machines. *Flow Measurement and Instrumentation*, *54*, 216–223.

Azimi, H., Bonakdari, H., & Ebtehaj, I. (2019). Gene expression programming-based approach for predicting the roller length of a hydraulic jump on a rough bed. *ISH Journal of Hydraulic Engineering*, *27*(sup1), 77–87.

Azimi, H., Shiri, H., & Malta, E. R. (2021). A non-tuned machine learning method to simulate ice-seabed interaction process in clay. *Journal of Pipeline Science and Engineering*, *1*(4), 379–394. Available from https://doi.org/10.1016/j.jpse.2021.08.005.

Bonakdari, H., & Ebtehaj, I. (2016a). Predicting velocity at limit of deposition in storm channels using two data mining techniques. Hydraulic structures and water system management. In: *6th IAHR international symposium on hydraulic structures*, Portland, OR, 27–30 June (pp. 72–79). Available from https://doi.org/10.15142/T3100628160853 (ISBN 978-1-884575-75-4).

Bonakdari, H., & Ebtehaj, I. (2016b). A comparative study of extreme learning machines and support vector machines in prediction of sediment transport in open channels. *International Journal of Engineering*, *29*(11), 1499–1506.

Bonakdari, H., Ebtehaj, I., Samui, P., & Gharabaghi, B. (2019). Lake water-level fluctuations forecasting using minimax probability machine regression, relevance vector machine, Gaussian process regression, and extreme learning machine. *Water Resources Management*, *33*(11), 3965–3984.

Bonakdari, H., Moradi, F., Ebtehaj, I., Gharabaghi, B., Sattar, A. A., Azimi, A. H., & Radecki-Pawlik, A. (2020). A non-tuned machine learning technique for abutment scour depth in clear water condition. *Water*, *12*(1), 301.

Calabrò, F., Fabiani, G., & Siettos, C. (2021). Extreme learning machine collocation for the numerical solution of elliptic PDEs with sharp gradients. *Computer Methods in Applied Mechanics and Engineering*, *387*, 114188. Available from https://doi.org/10.1016/j.cma.2021.114188.

Ebtehaj, I., Bonakdari, H., & Gharabaghi, B. (2019). A reliable linear method for modeling lake level fluctuations. *Journal of Hydrology*, *570*, 236–250.

Ebtehaj, I., Bonakdari, H., & Kisi, O. (2021b). Discussion of "ANFIS Modeling with ICA, BBO, TLBO, and IWO optimization algorithms and sensitivity analysis for predicting daily reference evapotranspiration" by Maryam Zeinolabedini Rezaabad, Sadegh Ghazanfari, and Maryam Salajegheh. *Journal of Hydrologic Engineering*, *26*(12), 07021006.

Ebtehaj, I., Bonakdari, H., & Shamshirband, S. (2016). Extreme learning machine assessment for estimating sediment transport in open channels. *Engineering with Computers*, *32*(4), 691–704.

Ebtehaj, I., Bonakdari, H., & Sharifi, A. (2014). Design criteria for sediment transport in sewers based on self-cleansing concept. *Journal of Zhejiang University Science A*, *15*(11), 914–924. Available from https://doi.org/10.1631/jzus.A1300135.

Ebtehaj, I., Bonakdari, H., Moradi, F., Gharabaghi, B., & Khozani, Z. S. (2018). An integrated framework of extreme learning machines for predicting scour at pile groups in clear water condition. *Coastal Engineering*, *135*, 1–15.

Ebtehaj, I., Bonakdari, H., Safari, M. J. S., Gharabaghi, B., Zaji, A. H., Madavar, H. R., & Mehr, A. D. (2020a). Combination of sensitivity and uncertainty analyses for sediment transport modeling in sewer pipes. *International Journal of Sediment Research*, *35*(2), 157–170. Available from https://doi.org/10.1016/j.ijsrc.2019.08.005.

Ebtehaj, I., Bonakdari, H., Zeynoddin, M., Gharabaghi, B., & Azari, A. (2020b). Evaluation of preprocessing techniques for improving the accuracy of stochastic rainfall forecast models. *International Journal of Environmental Science and Technology*, *17*(1), 505–524. Available from https://doi.org/10.1007/s13762-019-02361-z.

Ebtehaj, I., Sattar, A. M., Bonakdari, H., & Zaji, A. H. (2017). Prediction of scour depth around bridge piers using self-adaptive extreme learning machine. *Journal of Hydroinformatics*, *19*(2), 207–224. Available from https://doi.org/10.2166/hydro.2016.025.

Ebtehaj, I., Soltani, K., Amiri, A., Faramarzi, M., Madramootoo, C. A., & Bonakdari, H. (2021a). Prognostication of shortwave radiation using an improved No-Tuned fast machine learning. *Sustainability*, *13*(14), 8009.

Khalid, A., & Sarwat, A. I. (2021). Unified univariate-neural network models for lithium-ion battery state-of-charge forecasting using minimized akaike information criterion algorithm. *Ieee Access*, *9*, 39154–39170. Available from https://doi.org/10.1109/ACCESS.2021.3061478.

Lin, H., Gharehbaghi, A., Zhang, Q., Band, S. S., Pai, H. T., Chau, K. W., & Mosavi, A. (2022). Time series-based groundwater level forecasting using gated recurrent unit deep neural networks. *Engineering Applications of Computational Fluid Mechanics*, *16*(1), 1655–1672. Available from https://doi.org/10.1080/19942060.2022.2104928.

Maimaitiyiming, M., Sagan, V., Sidike, P., & Kwasniewski, M. T. (2019). Dual activation function-based extreme learning machine (ELM) for estimating grapevine berry yield and quality. *Remote Sensing*, *11*(7), 740. Available from https://doi.org/10.3390/rs11070740.

Mehdizadeh, S., Ahmadi, F., & Kozekalani Sales, A. (2020). Modelling daily soil temperature at different depths via the classical and hybrid models. *Meteorological Applications*, *27*(4), e1941. Available from https://doi.org/10.1002/met.1941.

Nou, M. R. G., Foroudi, A., Latif, S. D., & Parsaie, A. (2022). Prognostication of scour around twin and three piers using efficient outlier robust extreme learning machine. *Environmental Science and Pollution Research*, 1–14. Available from https://doi.org/10.1007/s11356-022-20681-5.

Okada, K. (2017). Negative estimate of variance-accounted-for effect size: How often it is obtained, and what happens if it is treated as zero. *Behavior research methods*, *49*(3), 979–987.

Owolabi, T. O., & Abd Rahman, M. A. (2021). Prediction of band gap energy of doped graphitic carbon nitride using genetic algorithm-based support vector regression and extreme learning machine. *Symmetry*, *13*(3), 411. Available from https://doi.org/10.3390/sym13030411.

Ratnawati, D. E., Marjono, W., & Anam, S. (2020). *Comparison of activation function on extreme learning machine (ELM) performance for classifying the active compound, AIP conference proceedings* (2264, p. 140001). AIP Publishing LLC September. Available from https://doi.org/10.1063/5.0023872.

Rohim, M. A. S., Nazmi, N., Bahiuddin, I., Mazlan, S. A., Norhaniza, R., Yamamoto, S. I., & Abdul Aziz, S. A. (2022). Prediction for magnetostriction magnetorheological foam using machine learning method. *Journal of Applied Polymer Science*, *139*(34), e52798. Available from https://doi.org/10.1002/app.52798.

Safari, M. J. S., Ebtehaj, I., Bonakdari, H., & Es-haghi, M. S. (2019). Sediment transport modeling in rigid boundary open channels using generalize structure of group method of data handling. *Journal of Hydrology*, *577*, 123951.

Samal, D., Dash, P. K., & Bisoi, R. (2022). Modified added activation function based exponential robust random vector functional link network with expanded version for nonlinear system identification. *Applied Intelligence*, *52*(5), 5657–5683. Available from https://doi.org/10.1007/s10489-021-02664-0.

Sattar, A., Ertuğrul, Ö. F., Gharabaghi, B., McBean, E. A., & Cao, J. (2019). Extreme learning machine model for water network management. *Neural Computing and Applications*, *31*(1), 157–169. Available from https://doi.org/10.1007/s00521-017-2987-7.

Suchithra, M. S., & Pai, M. L. (2020). Improving the prediction accuracy of soil nutrient classification by optimizing extreme learning machine parameters. *Information processing in Agriculture*, *7*(1), 72–82. Available from https://doi.org/10.1016/j.inpa.2019.05.003.

Tao, H., Ebtehaj, I., Bonakdari, H., Heddam, S., Voyant, C., Al-Ansari, N., & Yaseen, Z. M. (2019). Designing a new data intelligence model for global solar radiation prediction: Application of multivariate modeling scheme. *Energies*, *12*(7), 1365.

Tripathi, D., Edla, D. R., Kuppili, V., & Bablani, A. (2020). Evolutionary extreme learning machine with novel activation function for credit scoring. *Engineering Applications of Artificial Intelligence*, *96*, 103980. Available from https://doi.org/10.1016/j.engappai.2020.103980.

Yaseen, Z. M., Deo, R. C., Ebtehaj, I., & Bonakdari, H. (2018). *Hybrid data intelligent models and applications for water level prediction. Handbook of research on predictive modeling and optimization methods in science and engineering* (pp. 121–139). IGI Global. Available from https://doi.org/10.4018/978-1-5225-4766-2.ch006.

Zeynoddin, M., Bonakdari, H., Azari, A., Ebtehaj, I., Gharabaghi, B., & Madavar, H. R. (2018). Novel hybrid linear stochastic with non-linear extreme learning machine methods for forecasting monthly rainfall a tropical climate. *Journal of Environmental Management*, *222*, 190–206. Available from https://doi.org/10.1016/j.jenvman.2018.05.072.

Zeynoddin, M., Bonakdari, H., Ebtehaj, I., Azari, A., & Gharabaghi, B. (2020). A generalized linear stochastic model for lake level prediction. *Science of The Total Environment*, *723*, 138015.

Zeynoddin, M., Bonakdari, H., Ebtehaj, I., Esmaeilbeiki, F., Gharabaghi, B., & Haghi, D. Z. (2019). A reliable linear stochastic daily soil temperature forecast model. *Soil and Tillage Research*, *189*, 73–87. Available from https://doi.org/10.1016/j.still.2018.12.023.

Żymełka, P., & Szega, M. (2021). Short-term scheduling of gas-fired CHP plant with thermal storage using optimization algorithm and forecasting models. *Energy Conversion and Management*, *231*, 113860. Available from https://doi.org/10.1016/j.enconman.2021.113860.

Chapter 6

Outlier-based models of the non-tuned neural network—concept

6.1 Background of extreme learning machines

In recent years, the number of available datasets for statistical modeling and machine learning (ML) approaches has been increasing, resulting in easy access to large quantities of information regarding the phenomena of interest in a given study. Depending on the sampling methodologies and the chosen instrumentation, errors in the recorded data may exist. Outlying data points may generally be described as points that are situated unusually far from the bulk of the collected data and are often aberrant observations. Care should be taken when handling outlying data points because it should be ascertained whether the outlying point represents a significant phenomenon in our data or whether it represents an instrumental or systematic error in data collection. Traditional ML algorithms have shown a tendency to favor outlying data points during their training and development stage in order to capture the full variability of the data set. However, this can lead to a significantly reduced accuracy and precision of the model, and as a result, it is required that modified learning techniques are implemented to be able to handle variability in the data set under consideration.

6.1.1 Most critical advantages of the extreme learning machine

The extreme learning machine (ELM), an ML technique introduced by Huang et al. (2004, 2006), is a powerful and effective learning algorithm that offers a significantly more efficient performance compared to classical ML techniques (Huang, 2014). Some of the benefits of ELM are described below:

(1) Similar to most of the neural network-based approaches, the ELM has a wide generalization and approximation ability due to the large number of training cases that are used to develop its underlying functions (Ebtehaj & Bonakdari, 2022). However, the ELM also provides a very efficient and rapid training process (Dong et al., 2020) which is accomplished through the resolution of a simple linear system.

(2) Many numerical studies have indicated that the ELM has better generalization and scalability performance in comparison with the traditional neural networks, including the backpropagation (BP) neural network (Behbahani et al., 2018; Bonakdari et al., 2019a; Ebtehaj et al., 2016) and support vector machine (Bonakdari & Ebtehaj, 2016; Ebtehaj et al., 2017; Zhong et al., 2014).

(3) The number of user-adjusted (tunable) parameters in the ELM is minimal such that in addition to the activation function type (which only contains several options), only the number of neurons in the hidden layer must be determined before proceeding with the modeling phase.

(4) Perhaps the greatest advantage of the ELM is that its learning speed is extremely fast, overcoming a major challenge of traditional neural networks. These methods commonly employed BP training algorithms and displayed high computational costs. Recall from Chapter 4 that two of the three matrices in the ELM model [input weight and bias of hidden neurons (BHNs)] are randomly generated prior to consideration of the training samples. Therefore only the remaining matrix [i.e., matrix of output weights (OutW)] is analytically determined by solving a linear system (Fig. 6.1).

6.1.2 The main limitations of the extreme learning machine

As a result of the aforementioned advantages, the ELM technique has attracted the attention of many scholars in various fields of science and engineering as a powerful technique capable of providing an accurate prediction of output variables, while having a relatively fast training phase and low computational requirements. The fields of application of the

Machine Learning in Earth, Environmental and Planetary Sciences. DOI: https://doi.org/10.1016/B978-0-443-15284-9.00008-2

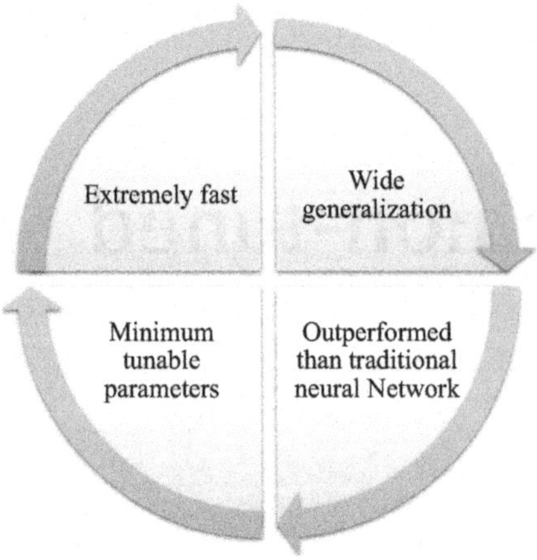

FIGURE 6.1 Most essential benefits of the extreme learning machine.

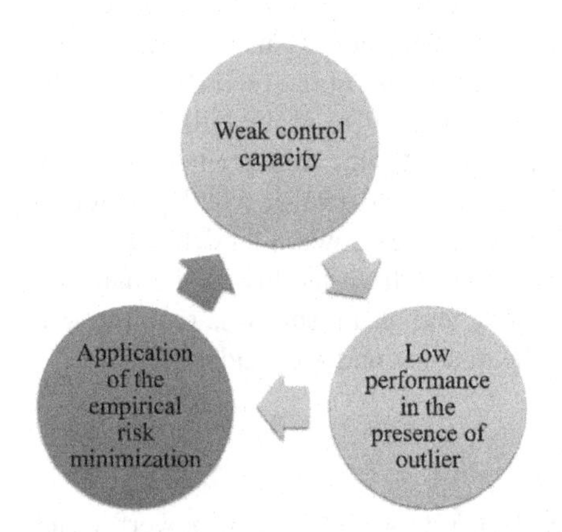

FIGURE 6.2 The main disadvantages of the original extreme learning machine.

ELM to real-world problems are rapidly expanding but are known to include discharge capacity (Azimi et al., 2017), radiation (Ebtehaj et al., 2021; Hai et al., 2020); river morphology (Gholami et al., 2020; Yousif et al., 2019); wastewater (Lotfi et al., 2019; Zhao et al., 2021), scour depth (Ebtehaj et al., 2018, 2019a, 2019b), velocity field in sewers (Bonakdari et al., 2019b; 2020), rainfall (Zeynoddin et al., 2018) and lake water-level (Bonakdari et al., 2019c). While the advantages of the ELM model have driven its development and application in engineering and science practice, it is worth mentioning several disadvantages of the model that must be considered when being employed. Several of these disadvantages are presented below:

(1) The original ELM model was developed based on empirical risk minimization (ERM). ERM is a principle in statistical theory (Vapnik, 1995) that defines a family of learning algorithms in order to provide theoretical bounds on their performance. The main idea in ERM is that prior to modeling, one cannot say exactly which algorithm from the family will perform better because we do not know the true distribution of samples that the model will encounter in practice (i.e., true "risk"). However, the performance of the algorithm can be computed based on the consideration of a set of training samples with a known distribution (i.e., "empirical" risk). By employing ERM in the development of ELM models, overfitting is commonly encountered, which limits the applicability of the model to new and unseen data sets.

(2) In the original ELM, which is a single-layer feed-forward neural network (SLFFNN), there are three matrices, including the input weights (InW), the BHNs, and the OutWs. The first two matrices (i.e., InW and BHN) are randomly adjusted, therefore, only the OutW matrix, which links the hidden layer neurons to the output variable(s), is analytically calculated. Indeed, we have low control over the solution because we only resolve one matrix by computing the minimum norm least-squares solutions.

(3) One of the most significant drawbacks of the ELM methodology is that the original ELM is less robust when heteroskedasticity or significant amounts of outlying data points are observed in the data set.

A summary of the main disadvantages of the classical ELM is provided in graphical format in Fig. 6.2.

6.2 Extreme learning machine in the presence of outliers

From the above discussion of the advantages and disadvantages of ELM, it is clear that a critical drawback of this method is the reduction of model performance in the presence of outlying data in the training samples. The presence of outliers in real-world applications, especially in the field of environmental science/engineering, is abundant. For example, significant dispersion of the data was observed in a recent study considering the electrocoagulation (EC) of synthetic wastewater (Akhbari et al., 2017; Bonakdari et al., 2017). In this study, color removal efficiency (CR%) and energy consumption (EnC) were investigated, and the maximum value of the target parameters differed significantly from the average. Other examples of data sets containing significant variation in environmental sciences include rainfall (Ebtehaj et al., 2020a; Salih et al., 2020; Yaseen et al., 2018, 2019a) and runoff (Ebtehaj et al., 2020b; Soltani et al., 2021; Walton et al., 2019; Yaseen et al., 2017, 2019b) forecasting. In the case of the tropical dry forest, rainfall is highly seasonal in nature; therefore, significant rainfall upwards of 5 times the monthly average may be experienced during the wet season. Often when we are considering rainfall and runoff variables during analysis, our goal is to predict the maximum values such that we can make appropriate decisions as engineering professionals regarding flood prevention and water resource management. If the classical ELM is used to predict these variables, it may exhibit a poor performance due to the wide variability of the data and the presence of outlying data points. This can result in inaccurate flood prediction from the model, for instance. As a result, it is critical that the classical ELM is developed further to be able to exhibit strong reliability and generalizability in the presence of outlying data points.

6.2.1 Proposed solutions to overcome the extreme learning machine weakness in the presence of outliers

To the best of the authors' knowledge, few studies have been presented in which the original ELM is developed to provide improved model performance in the presence of outliers in the training data set. One of the first works in this regard was a study by Deng et al. (2009), which proposed a combination of weighted least squares and regularized ELM to overcome the reduction of ELM performance in problems with outliers. The results from this study demonstrated that the predictive modeling performance was significantly improved when the developed weighted regularized ELM (WRELM) method was employed in comparison to the original ELM. In addition, a regularized ELM (RELM) was introduced by Huang et al. (2011), which considers the norm of the OutW as a regularization term.

The two modified ELM methods (i.e., RELM and WRELM) were developed to enhance the ELM in the presence of outliers during model training, resulting in a more generalizable model with higher predictive performance. However, in both models, the value of the training loss function is determined using $\ell 2$-norm (sum of squares). Wang (2012) showed that the use of $\ell 2$-norm in the presence of outliers will magnify their effect in the case of large deviations and can cause the ELM to become unstable when outliers are present in the training set (Zhang & Luo, 2015).

6.2.2 Evolution of extreme learning machine -based approaches in the presence of outliers

Zhang and Luo (2015) introduced the outlier robust ELM (ORELM) to overcome the limitation of the previous versions of the developed ELM-based model (i.e., RELM and WRELM), which considered the $\ell 2$-norm. The core of the proposed ORELM is the application of the $\ell 1$-norm instead of the $\ell 2$-norm for the training loss function. The reason behind the selection of the $\ell 1$-norm (as opposed to the $\ell 2$-norm) is twofold. Firstly, the $\ell 1$-norm loss function has greater robustness than the $\ell 2$-norm and has been widely applied to tackle outliers (Chuang et al., 2002; Daszykowski et al., 2007; McQuarrie & Tsai, 1998; Shi & Guoying, 1994). Second, it has been shown that even though outliers typically occupy a small region of the space of the training set, sparsity can be achieved when considering the $\ell 1$-norm (Candes et al., 2008; Candès et al., 2011; Donoho & Tsaig, 2008; Donoho, 2006a, 2006b; Wright et al., 2009). A sparse matrix or sparse array is a matrix in which most of the elements are zero. The number of zero-valued elements divided by the total number of elements (e.g., $m \times n$ for an $m \times n$ matrix; m and n are the number of rows and columns, respectively) is called the sparsity of the matrix. Using those definitions, a matrix will be sparse when its sparsity is greater than 0.5. An example of a sparse matrix is provided in Fig. 6.3.

The main differences between the original ELM and the modified ELM approaches (RELM, WRELM, ORELM) that have been developed to increase the performance in the presence of the outliers are shown in Table 6.1. From Table 6.1, we can observe that all the developed methods are similar to the original ELM in that the structure is that of an SLFFNN. Therefore, the neurons considered in each of the developed models are contained within a single hidden layer in the model architecture. In addition, as the choice of activation function forms an integral component of all

$$A=\begin{bmatrix} 0 & 0.2 & 0 & 0.3 \\ 0.5 & 0 & 0 & 0 \\ 0 & 0 & 0 & 0.4 \\ 0.6 & 0 & 0.9 & 0 \end{bmatrix}_{4\times 4} \quad \frac{\textit{Number of zero elements}}{m\times n} = \frac{10}{16} = 0.625 > 0.5$$

FIGURE 6.3 An example of a sparse matrix.

TABLE 6.1 Main characteristics of the developed extreme learning machine-based models in the presence of outliers.

Model	Network type	Activation function	Training loss function		Regularization parameter (C)
			ℓ1-norm	ℓ2-norm	
ELM	SLFFNN	✓	-	✓	-
RELM	SLFFNN	✓	-	✓	✓
WRELM	SLFFNN	✓	-	✓	✓
ORELM	SLFFNN	✓	✓	-	✓

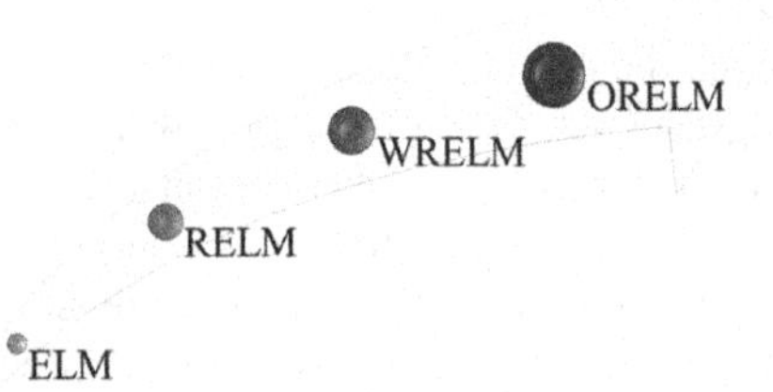

FIGURE 6.4 Evolution of extreme learning machine-based approaches in the presence of outliers.

ELM-based methods, the application in the original ELM and its variants is practically identical. The training loss function provided in RELM and WRELM is ℓ2-norm, while ORELM uses ℓ1-norm according to its successful application in outlier-based problems. In addition, it should be noted that the regularization parameter is used in all ELM-based developed models. The evolution and development of ELM-based approaches to overcome the outlier limitation of the original ELM is presented in Fig. 6.4. Fig. 6.5 demonstrates the conceptual diagram of the developed ELM-based models for application in the presence of outliers. As illustrated in this figure, the input weights and BHNs matrices are randomly generated while the OutWs are analytically calculated. For the developed ELM models (i.e., RELM, WRELM, ORELM), the regularization parameter is required during the analytical computation of the OutW and is defined as the norm of the OutWs matrix. One can also observe that the training loss function in RELM and WRELM is defined based on the ℓ2-norm, while the function in the ORELM is determined considering the ℓ1-norm.

6.3 Mathematical definition of extreme learning machine-based models

In the previous chapter, the original ELM model was described in complete detail, including a mathematical derivation of the model and an in-depth consideration of the impact of different user-defined input variables. In this section, a brief mathematical description of the original ELM model is presented, as this is the basis for the development of the variant models capable of handling outliers in the training data. Following this introductory presentation of the ELM model, the necessary modifications leading to the development of the regularized ELM (RELM), the weighted regularized ELM (WRELM), and the outlier robust ELM (ORELM) are presented with relevant details.

The discussion of this topic will again be divided into two categories based on user proficiency: beginner and advanced. The sections presented for the advanced user consider information regarding MATLAB coding details and the models' mathematical formulae. In the case of the beginner sections, a calculator is presented such that the variant ELM methodology can be employed in real-world applications without in-depth knowledge regarding the mathematical

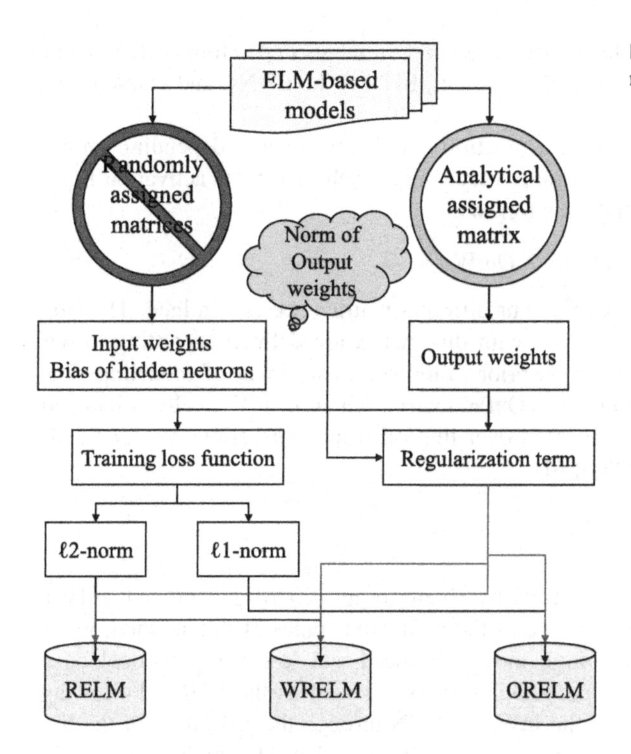

FIGURE 6.5 The conceptual diagram of the developed extreme learning machine-based models in the presence of outliers.

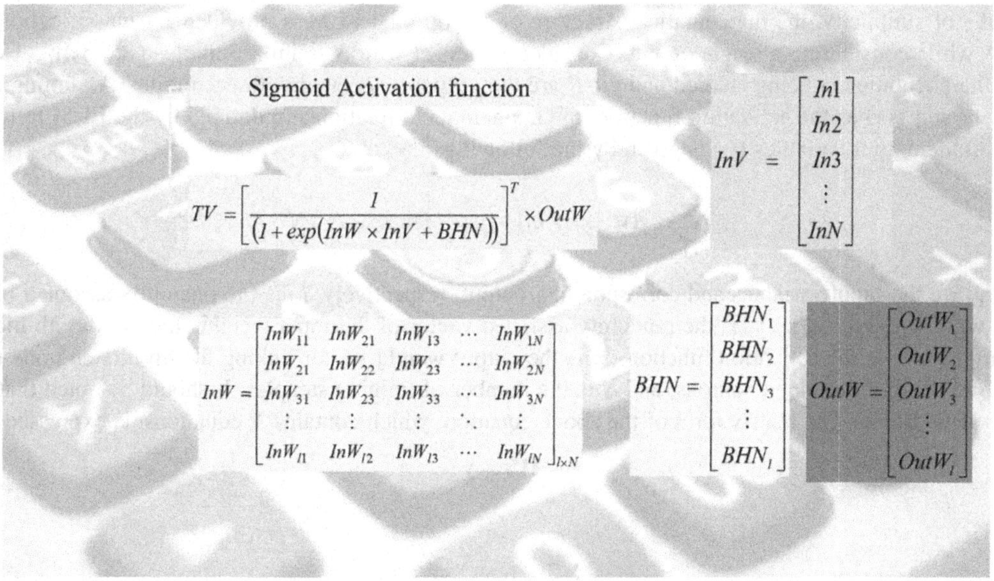

FIGURE 6.6 A schematic of the extreme learning machine calculator for practical engineering application.

derivation of the model. By employing the novel calculator derived in the current study (Fig. 6.6), new data samples which did not form a part of the training or testing data set may be evaluated by the model in order to estimate the target variable(s). To use the developed calculator in this study, the only requirement is that the user has access to a computer that runs the provided MATLAB code—detailed knowledge of MATLAB and mathematical formulation is not required.

An outline of the proposed calculator for use in practical engineering is presented in Fig. 6.6. Assuming that the activation function is of Sigmoidal format, the general relation is presented as follows:

$$\mathrm{TV} = \left[1/(1+\exp(InW \times InV + BHN))\right]^T \times OutW \tag{6.1}$$

where TV is the target variable(s), which may be a single variable or include several variables depending on the nature of the phenomena being studied, InV is the input variable, InW is the input weight, BHN is the BHNs, and OutW is the output weight.

It should be noted that activation functions other than the sigmoidal function may be considered depending on the nature of the data set. Consideration of an alternative activation function is very simple such that if the activation function is the hyperbolic tangent, for example, the general relationship is as follows:

$$\mathrm{TV} = [\tanh(InW \times InV + BHN)]^T \times OutW \tag{6.2}$$

To employ the provided calculator, it is required that the values for four different matrices be determined. The first one to consider is the InV matrix. The independent variables associated with this matrix are defined according to our data set. Recall that the InW and BHN matrices are randomly assigned prior to the commencement of modeling. The fourth matrix that must be determined through the ELM algorithm is the OutW matrix, which is analytically calculated [see Eq. (6.4)]. More details of the proposed calculator will be presented over the following subsections for all ELM-based models handling outlier data (i.e., RELM, WRELM, and ORELM).

6.3.1 Original extreme learning machine[1]

The original ELM was introduced as a novel algorithm for the first time by Huang et al. (2006) and proved to be a simple method to generate an SLFFNN. As discussed in Section 6.2, one of the main advantages of this method, when compared to classical methods (classical neural network), is the high modeling speed and low computational cost. Indeed, the ELM algorithm randomly assigns the values of the input weights (InW) as well as the BHNs. Following this, the OutW vector is determined analytically using the values of the InW and BHN through the resolution of the linear system presented in Eq. (6.4). Using the OutW matrix and the matrix of hidden layer output (H), the input variables can be related to the output or target variable.

For the sake of simplicity, the mathematical structure of the original ELM is provided for classification and regression problems while considering a single output. Consider a data set with N training samples ($\{(x_i, y_i)\}_{i=1}^{N}$) where $x_i \in R^n$ are inputs to the phenomenon being studied and $y_i \in R$ are the output associated with each input. Assuming L neurons in the hidden layer and $g(x)$ as the activation function, the mathematical relationship defined by the ELM to map the input variables (x_i) to the output variables (y_i) is given by the following:

$$\sum_{i=1}^{L} \beta_i g(\mathbf{w}_i \cdot \mathbf{x}_j + b_i) = y_j, \quad j = 1, 2, \ldots, N \tag{6.3}$$

where $\mathbf{x}_i$ and y_i are the input variables and corresponding output, respectively, b_i is the randomly assigned bias of the ith hidden node, $\mathbf{w}_i = [w_{i1}, w_{i2}, \ldots, w_{in}]^T$ is the randomly assigned vector of the input weights linking the ith hidden node to the input neurons, $g(x)$ is the activation function, β_i is the output weight vector linking the ith hidden node to the output variable, L is the number of hidden neurons, and N is the number of training samples. It should be noted that $\mathbf{w}_i \cdot \mathbf{x}_j$ is the inner product of $\mathbf{w}_i$ and $\mathbf{x}_j$. The matrix form of the above equation, which contains N equations, is expressed as follows:

$$\mathbf{H}\beta = \mathbf{y} \tag{6.4}$$

where

$$\mathbf{y} = [y_1, \ldots, y_N]^T \tag{6.5}$$

$$\beta = [\beta_1, \ldots, \beta_L]^T \tag{6.6}$$

$$\mathbf{H}(\mathbf{w}_1, \ldots, \mathbf{w}_L, \mathbf{x}_1, \ldots, \mathbf{x}_N, b_1, \ldots, b_L,) = \begin{bmatrix} g(\mathbf{w}_1 \cdot \mathbf{x}_1 + b_1) & \cdots & g(\mathbf{w}_L \cdot \mathbf{x}_1 + b_L) \\ \vdots & \ddots & \vdots \\ g(\mathbf{w}_1 \cdot \mathbf{x}_N + b_1) & \cdots & g(\mathbf{w}_L \cdot \mathbf{x}_N + b_L) \end{bmatrix}_{N \times L} \tag{6.7}$$

As previously mentioned, the InW and BHN matrices are randomly assigned. The structure of these two matrices is defined by the number of hidden neurons, and they are populated without the use of training samples. Therefore, the only matrix that must be calculated through the training process is the OutW matrix. This matrix is obtained by solving the linear system defined in Eq. (6.4), where the only unknown parameter is the OutW (i.e., β).

1. This section is provided for advanced users.

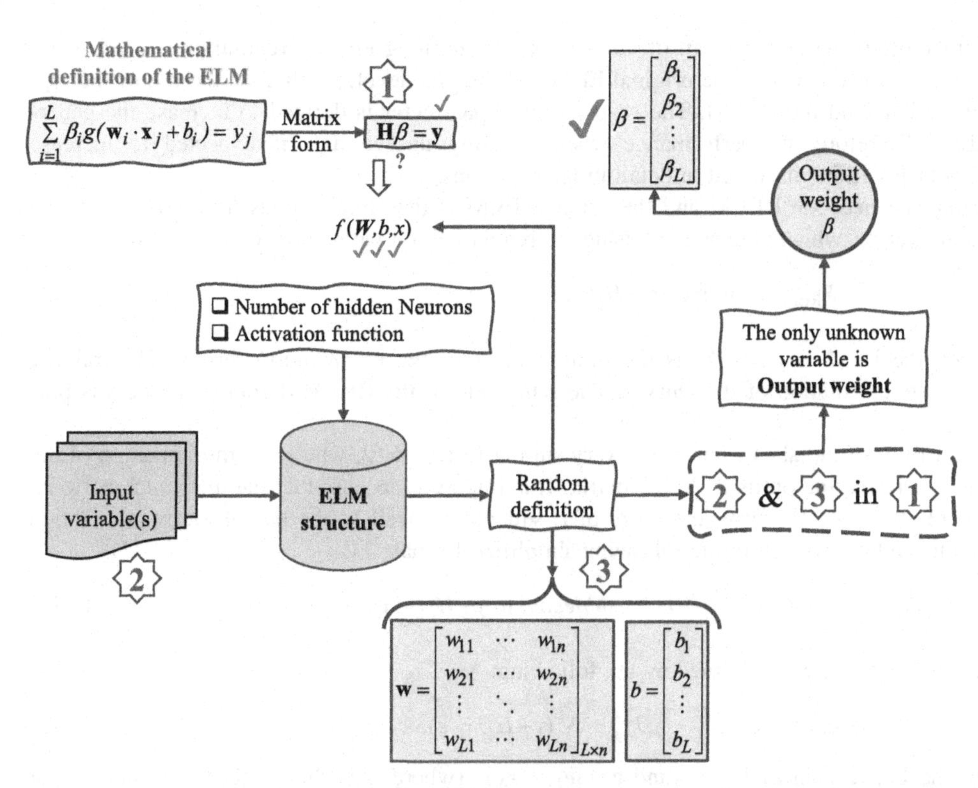

FIGURE 6.7 Schematic flowchart of the extreme learning machine.

Given that in most cases the matrix **H** is a nonsquare matrix, Eq. (6.4) cannot simply be used to obtain the value of β (Huang et al., 2006). The simplest way to deal with this problem is to calculate the optimal least square of $\hat{\beta}$ using the minimization of loss function as follows:

$$E_{ELM} = \min \| y - H\beta \| \tag{6.8}$$

Therefore, the optimal solution of the ELM in the sense of minimum ℓ2-norm is as follows:

$$\hat{\beta} = \mathbf{H}^{+}\mathbf{y} \tag{6.9}$$

where $\mathbf{H}^{+}$ is the Moore–Penrose generalized inverse of the **H** matrix (Rao & Mitra, 1971).

Since the number of hidden layer neurons (*L*) is generally smaller than the number of training samples (*N*), the above equation can be written as follows:

$$\hat{\beta} = (\mathbf{H}^{T}\mathbf{H})^{-1}\mathbf{H}^{T}\mathbf{y} \tag{6.10}$$

The schematic of the ELM flowchart is presented in Fig. 6.7.

6.3.2 Regularized extreme learning machine[2]

As mentioned in the previous section, when modeling using ELM, the values of the input weights, the BHNs, and the OutWs will be adjusted according to the nature of the problem under consideration. The first two matrices (InW, BHN) are randomly populated by the algorithm, while the OutWs are analytically determined during the modeling process. Indeed, the most important part of the ELM is calculating this matrix using training error minimization. However, the determination of the optimal OutW matrix through training error minimization can sometimes lead to overfitting, which will yield poor generalizability of the model when dealing with unseen data. Bartlett (1997) indicated that the greatest generalizability in a feed-forward neural network (FFNN) is obtained when the training error and norm of OutWs simultaneously have their lowest value. Indeed, the best generalizability in ELM is achieved when there is the best trade-off between the norm of OutWs and the training error.

2. This section is provided for advanced users.

To determine the optimal trade-off between the norm of OutWs and the training error, a regularization parameter (C) is introduced. Since the only difference between the original ELM and the one developed here is the use of the regularization parameter, this method is called the RELM. The regularization parameter is defined to increase the generalizability of the original ELM. Therefore, its performance when handling unseen data is expected to increase compared to the original ELM, which will be discussed in detail in later sections.

As mentioned, the only difference between RELM and the original ELM is the simultaneous minimization of the norm of the OutWs and the training error, which is expressed using the regularization parameter (C) as follows:

$$E_{\mathrm{RELM}} = \min_{\beta} C\|y - H\beta\|_2^2 + \|\beta\|_2^2 \tag{6.11}$$

The main difference between the ELM and RELM is the minimization of the OutW norm through the training phase, which has as its objective the calculation of the OutWs. The schematic of the RELM definition process is provided in Fig. 6.8.

It can be seen that the general form of the above relation is very similar to Eq. (6.8), where the minimization of the loss function is used in order to determine the optimal OutW matrix. It is important to note that the difference between Eq. (6.8) and the one shown here is the second term, the norm of β squared, as well as the use of the regularization parameter, C. The above equation can be rewritten in the following simplified format:

$$E_{RELM} = \min_{\beta} C\|\mathbf{e}\|_2^2 + \|\beta\|_2^2 \quad \textit{subjected to } y - H\beta = \mathbf{e} \tag{6.12}$$

The corresponding Lagrangian function can be defined by the following:

$$L(\beta, \mathbf{e}, \lambda) = C\|\mathbf{e}\|_2^2 + \|\beta\|_2^2 + \lambda^T(y - H\beta - e) \tag{6.13}$$

where λ is a column-vector of the Lagrangian multiplier and $\mathbf{e} = [e_1, .., e_N]^T$ (where N is the number of training error variables). The optimal condition of the above-mentioned Lagrangian equation is gained by the resolution of the partial derivatives (Tabak & Kuo, 1971).

$$\begin{cases} \dfrac{\partial L}{\partial \beta} = 0 & \Rightarrow \quad 2\beta - \mathbf{H}^T\lambda = 0 \\ \dfrac{\partial L}{\partial \mathbf{e}} = 0 & \Rightarrow \quad 2C\mathbf{e} - \lambda = 0 \\ \dfrac{\partial L}{\partial \lambda} = 0 & \Rightarrow \quad y - H\beta - \mathbf{e} = 0 \end{cases} \tag{6.14}$$

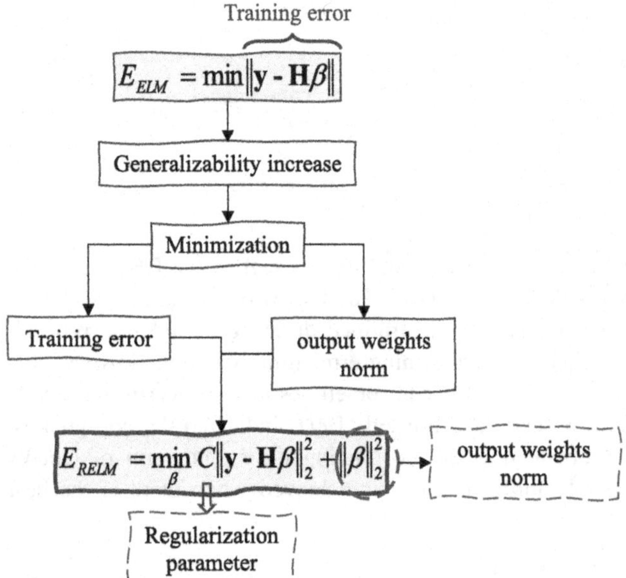

FIGURE 6.8 Schematic of the regularized extreme learning machine.

Consequently, the solution of the OutWs (β) is as follows:

$$\hat{\beta} = \left(\mathbf{H}^T\mathbf{H} + \frac{\mathbf{I}}{C}\right)^{-1}\mathbf{H}^T\mathbf{y} \tag{6.15}$$

where **I** is the identity matrix.

The above equation is provided for the case where the number of hidden neurons is less than the number of training samples. If the number of hidden neurons is greater than the number of training samples, the solution of the OutWs (β) is calculated as follows:

$$\hat{\beta} = \mathbf{H}^T\left(\mathbf{H}\mathbf{H}^T + \frac{\mathbf{I}}{C}\right)^{-1}\mathbf{y} \tag{6.16}$$

where all parameters have previously been defined.

6.3.3 Weighted regularized extreme learning machine[3]

To decrease the effect of outlying data points on the development of the ELM network, the WRELM was introduced based on structural risk minimization (Deng et al., 2009). The training process followed by the WRELM includes three steps:

1. The RELM is first employed with the training data to initialize the model network. Due to the presence of outliers in the training samples, the OutWs matrix obtained through the training process may not be stable. Therefore, the OutWs must be updated as the results obtained using this model are often unreliable in nature.
2. The RELM training error obtained following step 1 is used to introduce a weighting process. In this process, training samples that have a high error in step 1 are given small weights so that the detrimental effects of outliers are significantly reduced in the developed model.
3. The OutW matrix is revised and updated according to the application of RELM with information regarding the weighting from the previous step.

When compared to the RELM, the distinguishing feature of the WRELM is the implementation of a weighting system to better deal with outlying data points. To arrive at the WRELM model, the RELM model shown in Eq. (6.12) can be modified by considering the RELM error as an e_i variable and applying a weighting factor, w_i. Therefore, the $\|\mathbf{e}\|_2^2$ term in the RELM model [Eq. (6.12)] is extended as $\|\mathbf{we}\|_2^2$ where $\mathbf{w} = diag(w_1, \ldots, w_N)$. In order to most accurately estimate the value of **w**, the standard deviation of the RELM error variables ($\hat{s}$) should be considered and is defined as follows:

$$\hat{s} = \frac{IQR}{2 \times 0.6745} \tag{6.17}$$

where IQR is the interquartile range. The IQR is defined as the difference between the 75th and 25th percentile. Following the computation of the standard deviation, $\hat{s}$, Eq. (6.18) may be used to arrive at an estimation for **w** (Deng et al., 2009; Suykens et al., 2002).

$$w_i = \begin{cases} 1, & \left|\frac{e_i}{\hat{s}}\right| \le c_1 \\ \frac{c_2 - \left|\frac{e_i}{\hat{s}}\right|}{c_2 - c_1} & c_1 \le \left|\frac{e_i}{\hat{s}}\right| \le c_2 \\ 10^{-4} & Otherwise \end{cases} \tag{6.18}$$

Several alternative methods for the computation of the weighting factor, according to numerous authors, are presented in Table 6.2. It should be noted that the value of r is calculated as follows:

$$r = \frac{e}{tune \times \hat{s}} \tag{6.19}$$

where the *tune* is a tuning constant that is suggested to be higher than 1.

3. This section is provided for advanced users.

TABLE 6.2 Robust weight functions.

Robust weight function	Equation	Default tuning constant
Bisquare	$w = (abs(r) < 1 \times (1 - r^2)^2$	4.685
Huber	$w = 1/\max(1, abs(r))$	1.345
Andrews	$w_i = (abs(r) < pi) \times \sin(r)/r$	1.339
Fair	$w_i = 1/(1 + abs(r))$	1.400
Cauchy	$w_i = 1/(1 + r^2)$	2.385
Logistic	$w_i = \tanh(r)/r$	1.205
Talwar	$w_i = 1 \times abs(r) < 1$	2.795
Welsch	$w_i = exp\,(-(r^2))$	2.985
$\ell 1$ -norm	$w_i = 1/\max(0.0001, abs(e));\quad w = w_i / \sum_{i=1}^{N} w_i$	—

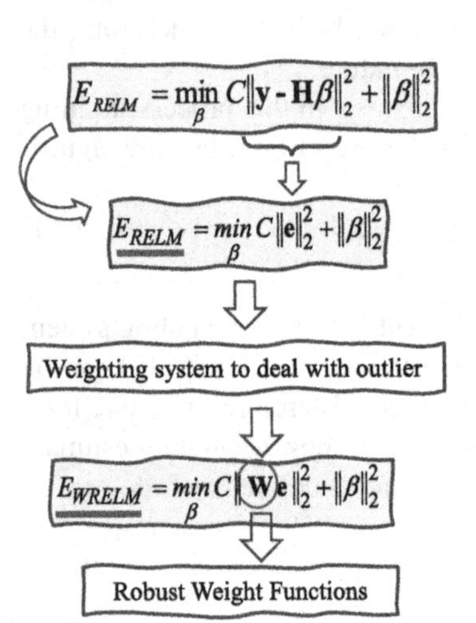

FIGURE 6.9 Schematic definition of the loss function in weighted regularized extreme learning machine.

By computing the standard deviation, $\hat{s}$, it is determined how much the calculated error distribution deviates from the Gaussian distribution. Given that in the Gaussian distribution, a small number of residuals are more than $2.5\hat{s}$, the constant coefficients in Eq. (6.18) can be considered as $c_1 = 2.5$ and $c_2 = 3$ (Deng et al., 2009).

According to the provided explanations, the mathematical form of the WRELM loss function can be defined as (Fig. 6.9):

$$E_{WRELM} = \min_{\beta} C\|\mathbf{We}\|_2^2 + \|\beta\|_2^2 \qquad \textit{subjected to } y - H\beta = \mathbf{e} \tag{6.20}$$

Using the same structure as seen in Eq. (6.15), the β solution could be found as follows:

$$\hat{\beta} = \left(\mathbf{H}^T\mathbf{W}^2\mathbf{H} + \frac{\mathbf{I}}{C}\right)^{-1} \mathbf{H}^T\mathbf{W}^2\mathbf{y} \tag{6.21}$$

The main goal of the WRELM method is to focus on eliminating the impact of outliers. The application of the WRELM has shown promise as an acceptable solution to increasing the accuracy of ELM when dealing with outlying data points. However, this model also has drawbacks that limit its performance when applied in practice. One of the most significant drawbacks of this method is the estimation of weights based on RELM results. Weight estimates in the WRELM are based on results obtained from solving the problem using the RELM and can significantly impact the final results of the model. Therefore, in order to achieve acceptable results, the WRELM must have a good RELM solution. Accordingly, this is a significant challenge. The second major drawback of the WRELM method is related to the removal of outliers during the preprocessing stage. If outliers are effectively removed by the modeler during the pre-processing stage to an extent that their effect is reduced, then some good samples which are not outlying might be assigned a small weight. This can have a significant impact on model training and may reduce its accuracy when applied practically. The impact of this drawback is most dramatic whenever the sample training error obtained from the RELM is relatively large in nature. Logically, an alternative to overcome the limitations of this model is to use iteratively reweighted methods, although this process also significantly increases the computational cost, which counteracts the main advantage of ELM-based methods based on fast learning. In the next section, a new method is proposed to overcome the drawbacks of the RELM and WRELM methods, which were originally introduced to improve the ELM model performance in the presence of outliers.

6.3.4 Outlier robust extreme learning machine[4]

It is clear that outlying data points make up only a small fraction of the totality of the training samples. To enhance the performance of the ELM-based model in the presence of outliers, this characteristic for training error (**e**) could be considered as sparsity. As mentioned in the previous sections, the sparsity can be reflected better by ℓ_0-norm than ℓ_2-norm. In fact, the modeling purpose in ORELM is to search for OutW (β) with small ℓ_2-norm so that the training error is sparse.

$$E_{ORELM} = C\min_{\beta} \|\mathbf{e}\|_0 + \|\beta\|_2^2 \quad \textit{subjected to} \quad \mathbf{y} - \mathbf{H}\beta = \mathbf{e} \tag{6.22}$$

The above-mentioned equation is a nonconvex programming problem. One of the easiest ways to resolve this equation for β is to reform it into a tractable convex relaxation scheme so that the sparsity characteristic will remain. Due to robust principal component analysis and compressive sensing, the sparse term could be attained with the ℓ_1-norm. Replacing the ℓ_1-norm with the ℓ_0-norm in Eq. (6.22) not only results in overall minimization but also guarantees the sparsity features. Consequently, the convex relaxation scheme of the previous equation is considered as follows (Fig. 6.10):

$$E_{ORELM} = \min_{\beta} \|\mathbf{e}\|_1 + \frac{1}{C}\|\beta\|_2^2 \quad \textit{subjected to} \quad \mathbf{y} - \mathbf{H}\beta = \mathbf{e} \tag{6.23}$$

Eq. (6.23) is a constrained convex optimization problem. It fits a well-defined range of the augmented Lagrange multiplier (ALM) approach. The augmented Lagrange is defined as

$$L(\beta, \mathbf{e}, \lambda) = \|\mathbf{e}\|_1 + \frac{1}{C}\|\beta\|_2^2 + \lambda^T(y - H\beta - e) + \frac{\mu}{2}\|y - H\beta - \mathbf{e}\|_2^2 \tag{6.24}$$

where μ denotes the penalty parameter and λ is the vector of the Lagrange multiplier ($\lambda \in R^n$). The μ is determined using $\mu = 2N/\|\mathbf{y}\|_1$, where N is the number of training samples (Yang & Zhang, 2011). The ALM algorithm is applied to estimate the Lagrange multiplier (λ) and optimal solution (**e**, β) through the iterative process of the augmented Lagrangian function minimization as follows:

$$\begin{cases} (\mathbf{e}_{k+1}, \beta_{k+1}) = \arg\min L_\mu(\mathbf{e}, \beta, \lambda_k) & (a) \\ \lambda_{k+1} = \lambda_k + \mu(y - H\beta_{k+1} - \mathbf{e}_{k+1}) & (b) \end{cases} \tag{6.25}$$

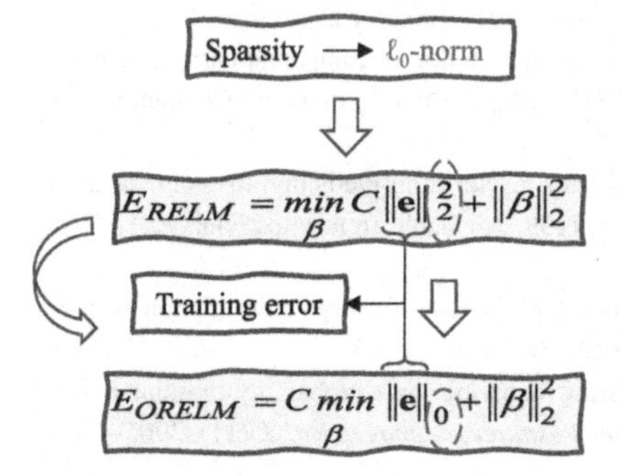

FIGURE 6.10 Schematic of the outlier robust extreme learning machine loss function.

4. This section is provided for advanced users.

In order to minimize the first relationship [Eq. (6.25(a))] through the ALM iterative process, methods are used that serially update the **e** and β parameters. Indeed, the use of ALM leads to solving the following equation to produce a new iteration:

$$\begin{cases} \beta_{k+1} = \arg\min_{\beta} L_{\mu}(\mathbf{e_k}, \beta, \lambda_k) & (a) \\ \mathbf{e}_{k+1} = \arg\min_{\mathbf{e}} L_{\mu}(\mathbf{e}, \beta_{k+1}, \lambda_k) & \\ \lambda_{k+1} = \lambda_k + \mu(y - H\beta_{k+1} - \mathbf{e}_{k+1}) & (b) \end{cases} \quad (6.26)$$

The explicit solution of the β_{k+1} and $\mathbf{e}_{k+1}$ are as follows (Zhang et al., 2012):

$$\beta_{k+1} = (\mathbf{H}^T\mathbf{H} + 2/C\mu\mathbf{I})^{-1}\mathbf{H}^T(\mathbf{y} - \mathbf{e_k} + \lambda_k/\mu) \quad (6.27)$$

$$\begin{aligned} \mathbf{e}_{k+1} &= shrink(\mathbf{y} - \mathbf{H}\beta_{k+1} + \lambda_k/\mu, 1/\mu) \\ &\cong \max\{|\mathbf{y} - \mathbf{H}\beta_{k+1} + \lambda_k/\mu| - 1/\mu, 0\} \circ sign(\mathbf{y} - \mathbf{H}\beta_{k+1} + \lambda_k/\mu) \end{aligned} \quad (6.28)$$

where " ∘ " denotes the element-wise multiplication.

It is clear that most of the computation time is related to solving the inverse of $(\mathbf{H}^T\mathbf{H} + 2/C\mu\mathbf{I})^{-1}\mathbf{H}^T$. The existence of a matrix similar to this matrix in each iteration that can be precalculated before the iterations begin leads to a fast calculation in this model.

6.4 Summary

In the current chapter, a detailed description of the limitations of the classical ELM model in the presence of outlier data points was presented. Due to this limitation, the mathematical formulation of three modified ELM-based models capable of handling outlying data points was provided in detail (RELM, WRELM, and ORELM). After completing this chapter, the reader will understand the main concepts related to the ELM as well as its weaknesses in dealing with outliers. Moreover, the reader will become familiar with the methods available to overcome the problem of outliers in ELM as well as the differences in the three main approaches used to deal with this problem. A complete understanding of this chapter will greatly aid the reader in understanding the coding process of these three ELM-based methods, which will be presented in detail in the next chapter. It should be noted that since the coding used in this book is presented in the simplest possible way, there is no significant dependence between different chapters. Therefore, the reader can proceed to the next chapter without being required to have read all the coding material outlined in previous chapters.

References

Akhbari, A., Bonakdari, H., & Ebtehaj, I. (2017). Evolutionary prediction of electrocoagulation efficiency and energy consumption probing. *Desalination and Water Treatment, 64*, 54–63. Available from https://doi.org/10.5004/dwt.2017.20235.

Azimi, H., Bonakdari, H., & Ebtehaj, I. (2017). Sensitivity analysis of the factors affecting the discharge capacity of side weirs in trapezoidal channels using extreme learning machines. *Flow Measurement and Instrumentation, 54*, 216–223. Available from https://doi.org/10.1016/j.flowmeasinst.2017.02.005.

Bartlett, P. L. (1997). For valid generalization the size of the weights is more important than the size of the network. *In Advances in neural information processing systems*, 134–140.

Behbahani, H., Amiri, A. M., Imaninasab, R., & Alizamir, M. (2018). Forecasting accident frequency of an urban road network: A comparison of four artificial neural network techniques. *Journal of Forecasting, 37*(7), 767–780. Available from https://doi.org/10.1002/for.2542.

Bonakdari, H., Gharabaghi, B., & Ebtehaj, I. (2019a). Extreme learning machines in predicting the velocity distribution in compound narrow channels. In K. Arai, S. Kapoor, R. Bhatia, et al. (Eds.), *Advances in intelligent systems and computing* (857). Springer, Cham, Intelligent Computing. SAI 2018. Available from https://doi.org/10.1007/978-3-030-01177-2_9.

Bonakdari, H., & Ebtehaj, I. (2016). A comparative study of extreme learning machines and support vector machines in prediction of sediment transport in open channels. *International Journal of Engineering-Transactions B: Applications, 29*(11), 1499. Available from https://doi.org/10.5829/idosi.ije.2016.29.11b.00.

Bonakdari, H., Ebtehaj, I., & Akhbari, A. (2017). Multi-objective evolutionary polynomial regression-based prediction of energy consumption probing. *Water Science and Technology, 75*(12), 2791–2799. Available from https://doi.org/10.2166/wst.2017.158.

Bonakdari, H., Ebtehaj, I., Samui, P., & Gharabaghi, B. (2019c). Lake water-level fluctuations forecasting using minimax probability machine regression, relevance vector machine, Gaussian process regression, and extreme learning machine. *Water Resources Management, 33*(11), 3965–3984. Available from https://doi.org/10.1007/s11269-019-02346-0.

Bonakdari, H., Qasem, S. N., Ebtehaj, I., Zaji, A. H., Gharabaghi, B., & Moazamnia, M. (2019b). An expert system for predicting the velocity field in narrow open channel flows using self-adaptive extreme learning machines. *Measurement, 151*, 107202. Available from https://doi.org/10.1016/j.measurement.2019.107202.

Bonakdari, H., Zaji, A. H., Gharabaghi, B., Ebtehaj, I., & Moazamnia, M. (2020). More accurate prediction of the complex velocity field in sewers based on uncertainty analysis using extreme learning machine technique. *ISH Journal of Hydraulic Engineering*, *26*(4), 409–420. Available from https://doi.org/10.1016/j.measurement.2019.107202.

Candès, E. J., Li, X., Ma, Y., & Wright, J. (2011). Robust principal component analysis. *Journal of the ACM (JACM)*, *58*(3), 1–37. Available from https://doi.org/10.1145/1970392.1970395.

Candes, E. J., Wakin, M. B., & Boyd, S. P. (2008). Enhancing sparsity by reweighted $\ell 1$ minimization. *Journal of Fourier analysis and applications*, *14*(5–6), 877–905. Available from https://doi.org/10.1007/s00041-008-9045-x.

Chuang, C. C., Su, S. F., Jeng, J. T., & Hsiao, C. C. (2002). Robust support vector regression networks for function approximation with outliers. *IEEE Transactions on Neural Networks*, *13*(6), 1322–1330. Available from https://doi.org/10.1109/TNN.2002.804227.

Daszykowski, M., Kaczmarek, K., Vander Heyden, Y., & Walczak, B. (2007). Robust statistics in data analysis—A review: Basic concepts. *Chemometrics and intelligent laboratory systems*, *85*(2), 203–219. Available from https://doi.org/10.1016/j.chemolab.2006.06.016.

Deng, W., Zheng, Q., & Chen, L. (2009, March). Regularized extreme learning machine. In *2009 IEEE symposium on computational intelligence and data mining* (pp. 389–395). IEEE. https://doi.org/10.1109/CIDM.2009.4938676

Dong, J., Wu, L., Liu, X., Li, Z., Gao, Y., Zhang, Y., & Yang, Q. (2020). Estimation of daily dew point temperature by using bat algorithm optimization based extreme learning machine. *Applied Thermal Engineering*, *165*, 114569. Available from https://doi.org/10.1016/j.applthermaleng.2019.114569.

Donoho, D. L. (2006a). Compressed sensing. *IEEE Transactions on information theory*, *52*(4), 1289–1306. Available from https://doi.org/10.1109/TIT.2006.871582.

Donoho, D. L. (2006b). For most large underdetermined systems of linear equations the minimal ℓ_1 -norm solution is also the sparsest solution. *Communications on Pure and Applied Mathematics: A Journal Issued by the Courant Institute of Mathematical Sciences*, *59*(6), 797–829. Available from https://doi.org/10.1002/cpa.20132.

Donoho, D. L., & Tsaig, Y. (2008). Fast solution of l-norm minimization problems when the solution may be sparse. *IEEE Transactions on Information Theory*, *54*(11), 4789–4812. Available from https://doi.org/10.1109/TIT.2008.929958.

Ebtehaj, I., & Bonakdari, H. (2022). A reliable hybrid outlier robust non-tuned rapid machine learning model for multi-step ahead flood forecasting in Quebec, Canada. *Journal of Hydrology*, 128592. Available from https://doi.org/10.1016/j.jhydrol.2022.128592.

Ebtehaj, I., Bonakdari, H., & Shamshirband, S. (2016). Extreme learning machine assessment for estimating sediment transport in open channels. *Engineering with Computers*, *32*(4), 691–704. Available from https://doi.org/10.1007/s00366-016-0446-1.

Ebtehaj, I., Bonakdari, H., & Gharabaghi, B. (2019a). Closure to "An integrated framework of extreme learning machines for predicting scour at pile groups in clear water condition" by: I. Ebtehaj, H. Bonakdari, F. Moradi, B. Gharabaghi, Z. Sheikh Khozani. *Coastal Engineering*, *147*, 135–137. Available from https://doi.org/10.1016/j.coastaleng.2019.02.011.

Ebtehaj, I., Bonakdari, H., Moradi, F., Gharabaghi, B., & Khozani, Z. S. (2018). An integrated framework of extreme learning machines for predicting scour at pile groups in clear water condition. *Coastal Engineering*, *135*, 1–15. Available from https://doi.org/10.1016/j.coastaleng.2017.12.012.

Ebtehaj, I., Bonakdari, H., Zaji, A. H., & Sharafi, H. (2019b). Sensitivity analysis of parameters affecting scour depth around bridge piers based on the non-tuned, rapid extreme learning machine method. *Neural Computing and Applications*, *31*(12), 9145–9156. Available from https://doi.org/10.1007/s00521-018-3696-6.

Ebtehaj, I., Bonakdari, H., Zeynoddin, M., Gharabaghi, B., & Azari, A. (2020a). Evaluation of preprocessing techniques for improving the accuracy of stochastic rainfall forecast models. *International Journal of Environmental Science and Technology*, *17*(1), 505–524. Available from https://doi.org/10.1007/s13762-019-02361-z.

Ebtehaj, I., Sattar, A. M., Bonakdari, H., & Zaji, A. H. (2017). Prediction of scour depth around bridge piers using self-adaptive extreme learning machine. *Journal of Hydroinformatics*, *19*(2), 207–224. Available from https://doi.org/10.2166/hydro.2016.025.

Ebtehaj, I., Soltani, K., Amiri, A., Faramarzi, M., Madramootoo, C. A., & Bonakdari, H. (2021). Prognostication of shortwave radiation using an improved No-Tuned fast machine learning. *Sustainability*, *13*(14), 8009. Available from https://doi.org/10.3390/su13148009.

Ebtehaj, I., Zeynoddin, M., & Bonakdari, H. (2020b). Discussion of "Comparative assessment of time series and artificial intelligence models to estimate monthly streamflow: A local and external data analysis approach" by Saeid Mehdizadeh, Farshad Fathian, Mir Jafar Sadegh Safari and Jan F. Adamowski. *Journal of Hydrology*, *583*, 124614. Available from https://doi.org/10.1016/j.jhydrol.2020.124614.

Gholami, A., Bonakdari, H., Ebtehaj, I., & Khodashenas, S. R. (2020). Reliability and sensitivity analysis of robust learning machine in prediction of bank profile morphology of threshold sand rivers. *Measurement*, *153*, 107411. Available from https://doi.org/10.1016/j.measurement.2019.107411.

Hai, T., Sharafati, A., Mohammed, A., Salih, S. Q., Deo, R. C., Al-Ansari, N., & Yaseen, Z. M. (2020). Global solar radiation estimation and climatic variability analysis using extreme learning machine based predictive model. *IEEE Access*, *8*, 12026–12042. Available from https://doi.org/10.1109/ACCESS.2020.2965303.

Huang, G. B. (2014). An insight into extreme learning machines: Random neurons, random features and kernels. *Cognitive Computation*, *6*(3), 376–390. Available from https://doi.org/10.1007/s12559-014-9255-2.

Huang, G. B., Zhou, H., Ding, X., & Zhang, R. (2011). Extreme learning machine for regression and multiclass classification. *IEEE Transactions on Systems, Man, and Cybernetics, Part B (Cybernetics)*, *42*(2), 513–529. Available from https://doi.org/10.1109/TSMCB.2011.2168604.

Huang, G.B., Zhu, Q.Y., & Siew, C.K. (2004, July). Extreme learning machine: A new learning scheme of feedforward neural networks. In *2004 IEEE international joint conference on neural networks (IEEE Cat. No. 04CH37541)* (Vol. 2, pp. 985–990). https://doi.org/10.1109/IJCNN.2004.1380068

Huang, G. B., Zhu, Q. Y., & Siew, C. K. (2006). Extreme learning machine: theory and applications. *Neurocomputing*, *70*(1–3), 489–501. Available from https://doi.org/10.1016/j.neucom.2005.12.126.

Lotfi, K., Bonakdari, H., Ebtehaj, I., Mjalli, F. S., Zeynoddin, M., Delatolla, R., & Gharabaghi, B. (2019). Predicting wastewater treatment plant quality parameters using a novel hybrid linear-nonlinear methodology. *Journal of Environmental Management*, *240*, 463–474. Available from https://doi.org/10.1016/j.jenvman.2019.03.137.

McQuarrie, A. D., & Tsai, C. L. (1998). *Regression and time series model selection*. World Scientific. Available from https://doi.org/10.1142/3573.

Rao, C.R., & Mitra, S.K. (1971). *Generalized inverse of matrices and its applications*. John Wiley & Sons. Inc., New York.

Salih, S. Q., Sharafati, A., Ebtehaj, I., Sanikhani, H., Siddique, R., Deo, R. C., Bonakdari, H., Shahid, S., & Yaseen, Z. M. (2020). Integrative stochastic model standardization with genetic algorithm for rainfall pattern forecasting in tropical and semi-arid environments. *Hydrological Sciences Journal*. Available from https://doi.org/10.1080/02626667.2020.1734813.

Shi, P., & Guoying, L. (1994). On the rates of convergence of "minimum l1-norm" estimates in a partly linear model. *Communications in Statistics-Theory and Methods*, *23*(1), 175–196. Available from https://doi.org/10.1080/03610929408831246.

Soltani, K., Ebtehaj, I., Amiri, A., Azari, A., Gharabaghi, B., & Bonakdari, H. (2021). Mapping the spatial and temporal variability of flood susceptibility using remotely sensed normalized difference vegetation index and the forecasted changes in the future. *Science of The Total Environment*, *770*145288. Available from https://doi.org/10.1016/j.scitotenv.2021.145288.

Suykens, J. A., De Brabanter, J., Lukas, L., & Vandewalle, J. (2002). Weighted least squares support vector machines: robustness and sparse approximation. *Neurocomputing*, *48*(1–4), 85–105. Available from https://doi.org/10.1016/S0925-2312(01)00644-0.

Tabak, D., & Kuo, B. C. (1971). *Optimal control by mathematical programming* (pp. 19–20). Englewood Cliffs, NJ: Prentice-Hall, *ISBN 0-13-638106-5*.

Vapnik, V. N. (1995). *The nature of statistical learning theory*. New York: Springer.

Walton, R., Binns, A., Bonakdari, H., Ebtehaj, I., & Gharabaghi, B. (2019). Estimating 2-year flood flows using the generalized structure of the group method of data handling. *Journal of Hydrology*, *575*, 671–689. Available from https://doi.org/10.1016/j.jhydrol.2019.05.068.

Wang, H. (2012). Block principal component analysis with L1-norm for image analysis. *Pattern recognition letters*, *33*(5), 537–542. Available from https://doi.org/10.1016/j.patrec.2011.11.029.

Wright, J., Ganesh, A., Rao, S., Peng, Y., & Ma, Y. (2009). Robust principal component analysis: Exact recovery of corrupted low-rank matrices via convex optimization. *In Advances in neural information processing systems*, 2080–2088.

Yang, J., & Zhang, Y. (2011). Alternating approximation algorithms for ℓ1-problems in compress sensing. *SIAM Journal of Scientific Computing*, *33* (1), 250–278. Available from https://doi.org/10.1137/090777761.

Yaseen, Z. M., Ebtehaj, I., Kim, S., Sanikhani, H., Asadi, H., Ghareb, M. I., & Shahid, S. (2019a). Novel hybrid data-intelligence model for forecasting monthly rainfall with uncertainty analysis. *Water*, *11*(3), 502. Available from https://doi.org/10.3390/w11030502.

Yaseen, Z. M., Mohtar, W. H. M. W., Ameen, A. M. S., Ebtehaj, I., Razali, S. F. M., Bonakdari, H., & Shahid, S. (2019b). Implementation of univariate paradigm for streamflow simulation using hybrid data-driven model: Case study in tropical region. *IEEE Access*, *7*, 74471–74481. Available from https://doi.org/10.1109/ACCESS.2019.2920916.

Yaseen, Z. M., Ebtehaj, I., Bonakdari, H., Deo, R. C., Mehr, A. D., Mohtar, W. H. M. W., Diop, L., El-shafie, A., & Singh, V. P. (2017). Novel approach for streamflow forecasting using a hybrid ANFIS-FFA model. *Journal of Hydrology*, *554*, 263–276. Available from https://doi.org/10.1016/j.jhydrol.2017.09.007.

Yaseen, Z. M., Ghareb, M. I., Ebtehaj, I., Bonakdari, H., Siddique, R., Heddam, S., Yusif, A., & Deo, R. (2018). Rainfall pattern forecasting using novel hybrid intelligent model based ANFIS-FFA. *Water Resources Management*, *32*(1), 105–122. Available from https://doi.org/10.1007/s11269-017-1797-0.

Yousif, A. A., Sulaiman, S. O., Diop, L., Ehteram, M., Shahid, S., Al-Ansari, N., & Yaseen, Z. M. (2019). Open channel sluice gate scouring parameters prediction: Different scenarios of dimensional and non-dimensional input parameters. *Water*, *11*(2), 353. Available from https://doi.org/10.3390/w11020353.

Zeynoddin, M., Bonakdari, H., Azari, A., Ebtehaj, I., Gharabaghi, B., & Riahi Madavar, H. (2018). Novel hybrid linear stochastic with non-linear extreme learning machine methods for forecasting monthly rainfall a tropical climate. *Journal of Environmental Management*, *222*, 190–206. Available from https://doi.org/10.1016/j.jenvman.2018.05.072.

Zhang, K., & Luo, M. (2015). Outlier-robust extreme learning machine for regression problems. *Neurocomputing*, *151*, 1519–1527. Available from https://doi.org/10.1016/j.neucom.2014.09.022.

Zhang, L., Yang, M., Feng, X., Ma, Y., & Zhang, D. (2012). *Collaborative representation based classification for face recognition. arXiv preprint arXiv*, *1204*, 2358. Available from https://doi.org/10.48550/arXiv.1204.2358.

Zhao, F., Liu, M., Wang, K., Wang, T., & Jiang, X. (2021). A soft measurement approach of wastewater treatment process by lion swarm optimizer-based extreme learning machine. *Measurement*, *179*, 109322. Available from https://doi.org/10.1016/j.measurement.2021.109322.

Zhong, H., Miao, C., Shen, Z., & Feng, Y. (2014). Comparing the learning effectiveness of BP, ELM, I-ELM, and SVM for corporate credit ratings. *Neurocomputing*, *128*, 285–295. Available from https://doi.org/10.1016/j.neucom.2013.02.054.

Chapter 7

Outlier-based models of the non-tuned neural network—coding and implementation

7.1 Developed extreme learning machine-based approaches in the presence of outliers

The previous chapter highlighted several ELM-based methods which have been developed to overcome the limitations of the original ELM model in the presence of outlying data. In the current section, a flowchart detailing all required steps for the developed models is provided in Fig. 7.1.

The first step when performing any modeling task is to load the data. The data can be contained in a number of different file formats, however, most often is within the Excel environment. Once the data has been loaded, it must be split into two separate categories, namely the training and testing data. In order to categorize the data into these two categories, it is necessary to define the percentage of training and testing data. We want to ensure that a sufficient number of training samples are available to provide the model with an adequate amount of experience, allowing it to properly handle future cases involving previously unseen cases. It is clear that training data should comprise at least half of the total data (i.e., 50% of all samples), however, the fundamental question is, What percentage of the total data should we consider during the training stage? Data segmentation into training and testing subsets has been previously divided using different percentages including 50%–50% (Khozani et al., 2017; Zhang et al., 2018; Azimi et al., 2017, 2019), 60%–40% (Jahani & Rayegani, 2020; Ebtehaj et al., 2020), 70%–30% (Qasem et al., 2017b; Gholami et al., 2019; Ebtehaj et al., 2021a), 75%–25% (Gholami et al., 2017; Ebtehaj et al., 2018), 80%–20% (Mojtahedi et al., 2019; Chen et al., 2020; Halpern-Wight et al., 2020), 90%–10% (Erbir & Ünver, 2021). It is clear that when a higher percentage of training cases is considered, the model will possess greater experience compared to the case where fewer training cases are considered (i.e., when 50% is used for training data). On the other hand, there should be a sufficient number of testing data cases to determine the model performance. The model performance should demonstrate acceptable generalizability considering a large range of data, specifically data that the model does not have experience with. Considering these two requirements, it could be concluded that 70%–30% (Qasem et al., 2017a; Ebtehaj & Bonakdari, 2017; Sihag et al., 2019), 75%–25% (Adams & Dymond, 2018; Elbadawi et al., 2020; Diop et al., 2020), and 80%–20% (Zhang et al., 2021; Kumar et al., 2021; Essam et al., 2022) are the most common percentage splits that are applicable for use in real-world problems. The second step in applying ELM-based models is to define the initial parameters required to produce the model structures. The user must first define whether orthogonal calculations of the matrices will be considered once the input weight and bias of hidden neuron matrices have been randomly generated. Following this, it should be defined whether the type of problem being considered as one of regression or classification. The basic parameters of the model are then defined, including the activation function type, the number of hidden neurons, the regularization parameter, and the iteration number. In the present study, six activation functions will be considered: Sigmoid, hyperbolic tangent, triangular basis function, radial basis function, sine, and hard limit. It should be noted that the details of each of these functions were provided in detail in Chapter 4.

Calculation of the regularization parameter will be considered in subsequent sections, however, it should be noted that its value must be determined in accordance with the type of practical problem that is under investigation. The choice regarding the number of hidden neurons should be made based on the complexity of the problem being studied. Generally, we want the simplest model possible that provides accurate predictive output or classification. However, as the complexity of the problem increases, it will be required to consider a greater number of neurons to arrive at a model capable of good results. It should be noted that increasing the number of hidden neurons can also lead to a reduction in

Machine Learning in Earth, Environmental and Planetary Sciences. DOI: https://doi.org/10.1016/B978-0-443-15284-9.00005-7

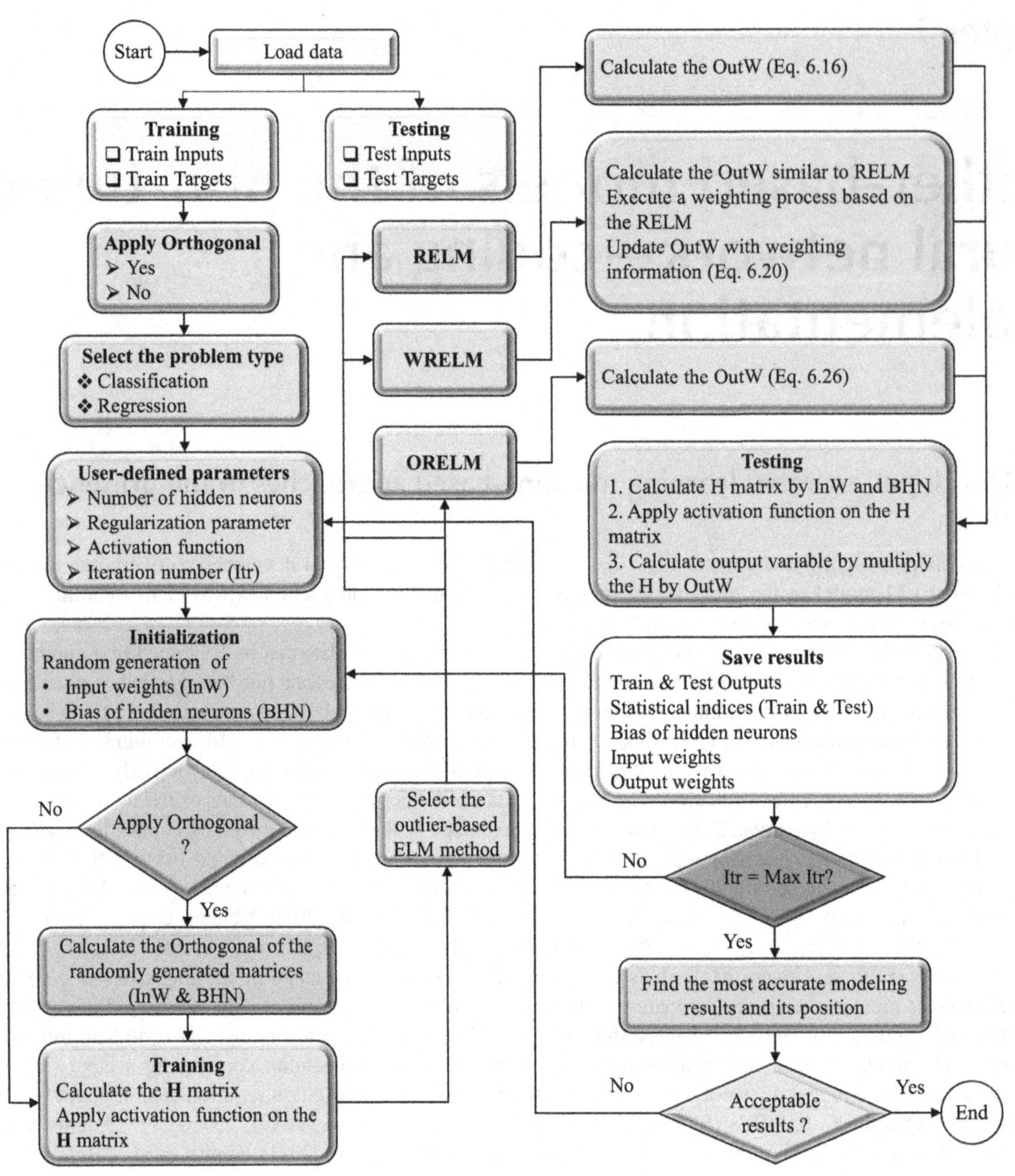

FIGURE 7.1 Flowchart of the developed extreme learning machine-based models in the presence of outliers.

the generalizability of the model. As the number of neurons is increased, the model becomes "fine-tuned" to the training set and loses the ability to perform well when considering previously unseen data. Therefore, when selecting the number of hidden neurons, a trade-off should be made considering the complexity, accuracy, and generalizability of the model. One of the well-known approaches to consider a trade-off between complexity and accuracy (or generalizability) is using the Akaike Information Criteria (Ebtehaj & Bonakdari, 2016; Walton et al., 2019; Ebtehaj et al., 2019; Montes et al., 2020; Nguyen et al., 2021) and its correction versions which were provided in detail in Chapter 3. The maximum value of the number of hidden neurons (NHNs) with respect to the number of training samples (NTrSs) and the number of input variables (NIVs) is defined as follows (Zeynoddin et al., 2020; Ebtehaj et al., 2021b).

$$\mathrm{NHN} < \frac{\mathrm{NTrS}}{(\mathrm{NIV} + 2)} \tag{7.1}$$

The final parameter to be defined prior to beginning modeling is the number of iterations to consider. Given that, in the original ELM and all developed ELM-based approaches [i.e., regularized ELM (RELM), weighted regularized ELM (WRELM), and outlier robust ELM (ORELM)], the two matrices of input weights and bias of hidden neurons are randomly assigned, it is expected that this will have a significant impact on the model performance for testing samples. After randomly initializing these two matrices, ELM-based methods calculate the only remaining matrix (matrix of output weights) through the resolution of a linear system. Since the resolution of the matrix of output weights is by direct resolution of a simple linear system, it can be easily seen that these methods are not based on an iterative approach. Therefore, in this study, each ELM-based approach is modeled considering an iterative process, and the model with the best performance when considering unseen data is selected. For this superior model, the parameters are saved to be used in future modeling applications. This iterative approach is feasible because of the very high speeds of ELM models when compared with other algorithms.

The third step consists of the initialization of the ELM model in the programming environment. This step is twofold; the matrix of input weights and the bias of the hidden neurons should be randomly initialized, after which they should be made orthogonal (orthogonal was defined as true by the user).

Step four, which is the model training, is the most important step in the developed ELM-based techniques. In this step, the hidden layer output (**H**) is first calculated by applying the activation function, which was defined in step 2 (i.e., Apply orthogonal function, select the problem type, and user-defined parameters). Following this, the matrix of output weights is calculated depending on the ELM method selected. Eqs. (6.16), (6.21), and (6.27) are applied to the **H** matrix to calculate the output weights matrix in the RELM, WREM, and ORELM, respectively.

In the fifth step, using the randomly generated matrices (i.e., input weights and bias of hidden neurons), the analytically calculated one (i.e., output weights), and the activation function type, the performance of the developed model in the testing stage is evaluated.

In the sixth and final step, the defined matrices from ELM-based modeling (i.e., input weights, the bias of hidden neurons, and output weights), the training and testing output calculated by the desired model, and the corresponding statistical indices for each stage are saved for future use and evaluation.

Steps 3–6 are repeated until the maximum number of iterations (Itr) is reached. Once the maximum number of iterations has been reached, the superior model is selected based on the statistical error provided from the testing stage. If the model performance is considered acceptable, the modeling is finished. In the case that the model performance is unacceptable, the modeling will be repeated by resetting the modeling parameters in step 2. Acceptability can be determined by a user-defined error threshold or constraints from the problem under evaluation.

7.2 Implementation of the developed extreme learning machine-based models in the MATLAB[1]

This section presents the MATLAB-based code required to implement the developed ELM-based approaches in complete detail. To apply the MATLAB code for the developed models (i.e., RELM, WRELM, and ORELM), the "ELMsModel" file located in the provided package should be run. This package includes five Excel-based files (Examples 1–5, see in Appendix 7A), two m-files (i.e., "ELMsModel" and calculator), and eight functions. The main application of each file provided in the package of the ELM-based approach in the presence of outliers is presented in Table 7.1.

7.2.1 Detailed coding of the main MATLAB code

7.2.1.1 Loading of data for all outlier-based extreme learning machine methods

To start modeling a real-world problem using developed ELM-based models in the current chapter, the "ELMsModel" m-file is opened and employed to run the main code. The full details of the MATLAB-based coding of this m-file are provided below. It should be noted that beginner users can follow the modeling process by only considering the explanation regarding the user-defined variables (iteration number, number of hidden neurons, activation function type, and regularization parameter) without following the detailed development of the MATLAB code (which is provided text in the gray boxes).

1. This section is provided for advanced users.

TABLE 7.1 Required m-files for developed extreme learning machine-based models.

Function	Definition
"ELMsModel"	Main m-file for all developed models
"Calculator"	Calculator of the developed models for unseen data
"elm_initialization"	Random generation of the input weights and bias of hidden neurons
"elm_train"	Training ELM-based approaches
"Elm_test"	Testing ELM-based approaches
"regressor"	Calculation of the output weights for RELM and WRELM
"regressor_alm"	Calculation of the output weights for ORELM
"weight_fun"	Considering the weight information for WRELM
"Statistical_Indices"	Calculation of statistical indices for training and testing stages
"PlotResults"	Plotting of visual results for developed models

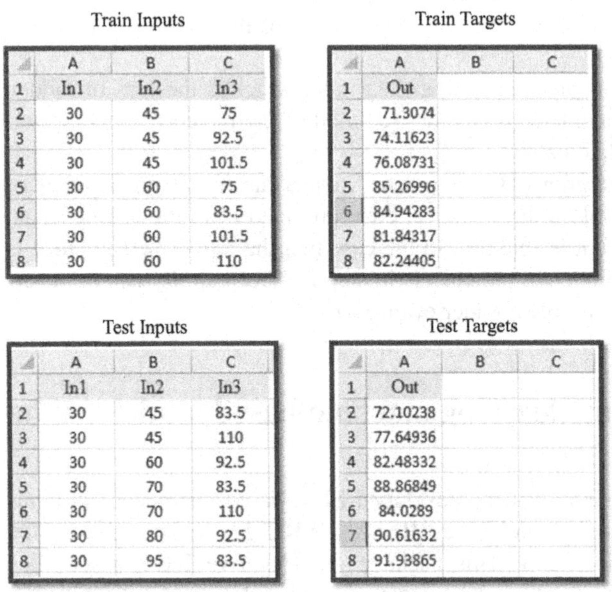

Train Inputs

	A	B	C
1	In1	In2	In3
2	30	45	75
3	30	45	92.5
4	30	45	101.5
5	30	60	75
6	30	60	83.5
7	30	60	101.5
8	30	60	110

Train Targets

	A	B	C
1	Out		
2	71.3074		
3	74.11623		
4	76.08731		
5	85.26996		
6	84.94283		
7	81.84317		
8	82.24405		

Test Inputs

	A	B	C
1	In1	In2	In3
2	30	45	83.5
3	30	45	110
4	30	60	92.5
5	30	70	83.5
6	30	70	110
7	30	80	92.5
8	30	95	83.5

Test Targets

	A	B	C
1	Out		
2	72.10238		
3	77.64936		
4	82.48332		
5	88.86849		
6	84.0289		
7	90.61632		
8	91.93865		

FIGURE 7.2 Data preparation and organization for loading into the developed extreme learning machine-based models.

The first step in the modeling process is to load the data contained in Excel into the MATLAB environment. The associated coding details for this step are presented in Code 7.1 below. To load the data from Excel using this MATLAB code, the data should be organized into four sheets. The first and second sheets contain the training inputs and targets, respectively, while the third and fourth sheets contain the testing inputs and targets, respectively. An Example of the required format for data loading into the developed code is provided in Fig. 7.2.

The data contained within these four sheets is read into the MATLAB environment using the built-in `xlsread` function in lines 2–5 of Code 7.1. From the provided code, it can be seen that the train inputs, train targets, test inputs, and test targets are called from sheets 1 to 4, respectively. The first argument of the `xlsread` command is the filename of the Excel sheet, while the second argument is the sheet name that contains the data of interest. Note that if the file name that contains the data is changed, "`Example1`" must be altered to the name of the file. Given that the training and testing data in ELM-based approaches are loaded in columns while the calculations performed by this code are in rows, the transpose of each of the loaded matrices must be calculated. To achieve this, MATLAB's transpose operator (`'`) is employed in lines 7 and 8 for the training and testing data, respectively. The schematic representation of Code 7.1 is provided in Fig. 7.3.

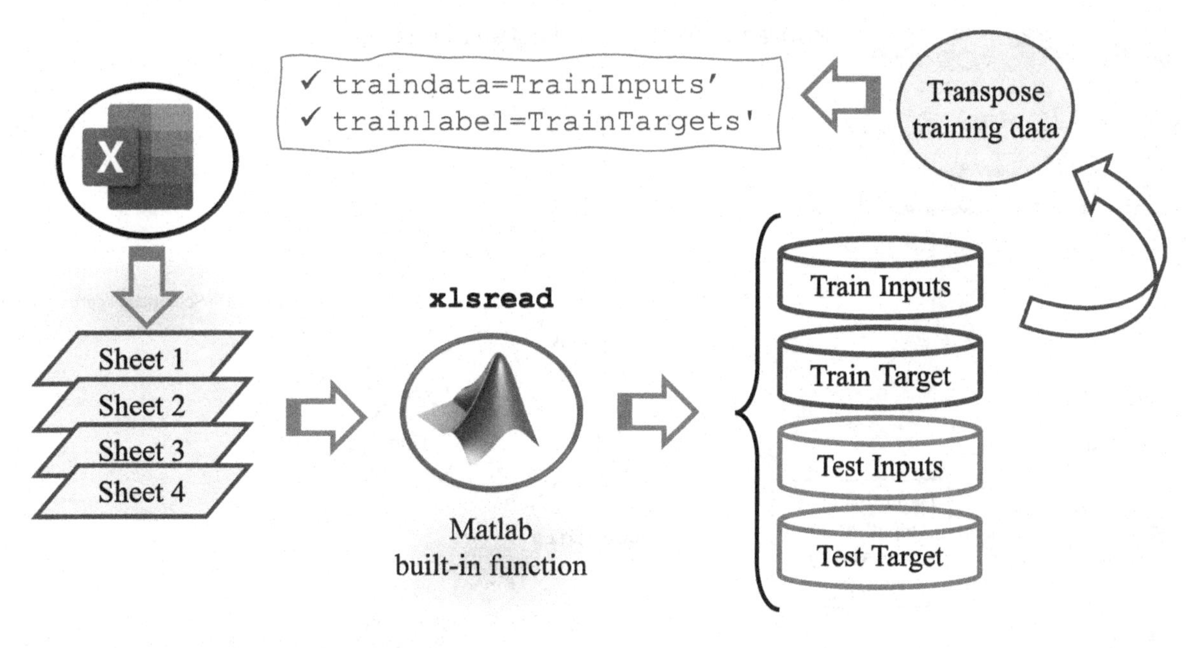

FIGURE 7.3 Schematic representation of Code 7.1.

Code 7.1

```
1    %% Load data
2    TrainInputs  = xlsread('Example1','sheet1');
3    TrainTargets = xlsread('Example1','sheet2');
4    TestInputs   = xlsread('Example1','sheet3');
5    TestTargets  = xlsread('Example1','sheet4');
6
7    traindata=TrainInputs';
8    trainlabel=TrainTargets';
```

To calculate the dimension of the input, the `inputsize` variable is defined. It is calculated as follows using MATLAB's built-in `size` function as shown in Code 7.2.

Code 7.2

```
1    nn.inputsize = size(traindata,1);
```

7.2.1.2 Definition of the outlier-based model type

The next step is selecting the problem type (i.e., whether the problem is regression-based or classification-based). Code 7.3 outlines the MATLAB code related to this input, while Fig. 7.4 displays the resulting question dialog box that the user is prompted with. It is clear that the default ELM type is regression-based problems. The `questdlg` function in line 4 is used to create a question dialog box – this is what the user is prompted with to input the required parameter. The main format of this command is `questdlg(qstring,title,str1,str2,str3,default)`. Therefore, we need to define `qstring`, `title`, the response strings (`str1`, `str2`, `str3`), and the `default` response to apply this command. The `qstring` and `title` input arguments are specified directly in the `questdlg` command in line 4 as "Regression or Classification?" and "ELM Type," respectively. The implementation of these input arguments can be clearly seen in Fig. 7.4, where the title at the top of the window is "ELM Type" while the question string shown next to the blue icon is "Regression or Classification?." Next, the user should define which of the two problem types will be considered. The response strings

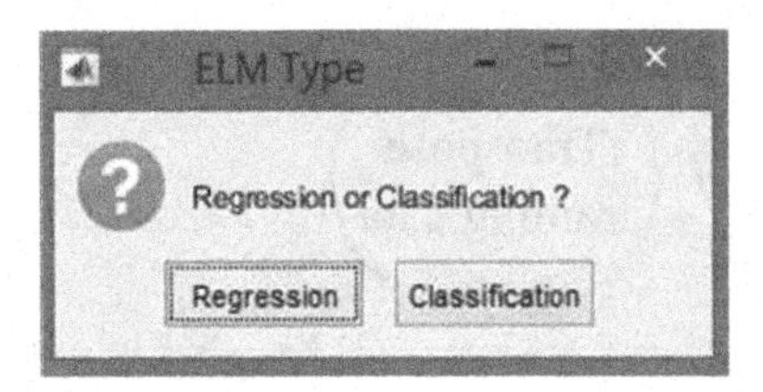

FIGURE 7.4 Output question dialog box for ELM type selection.

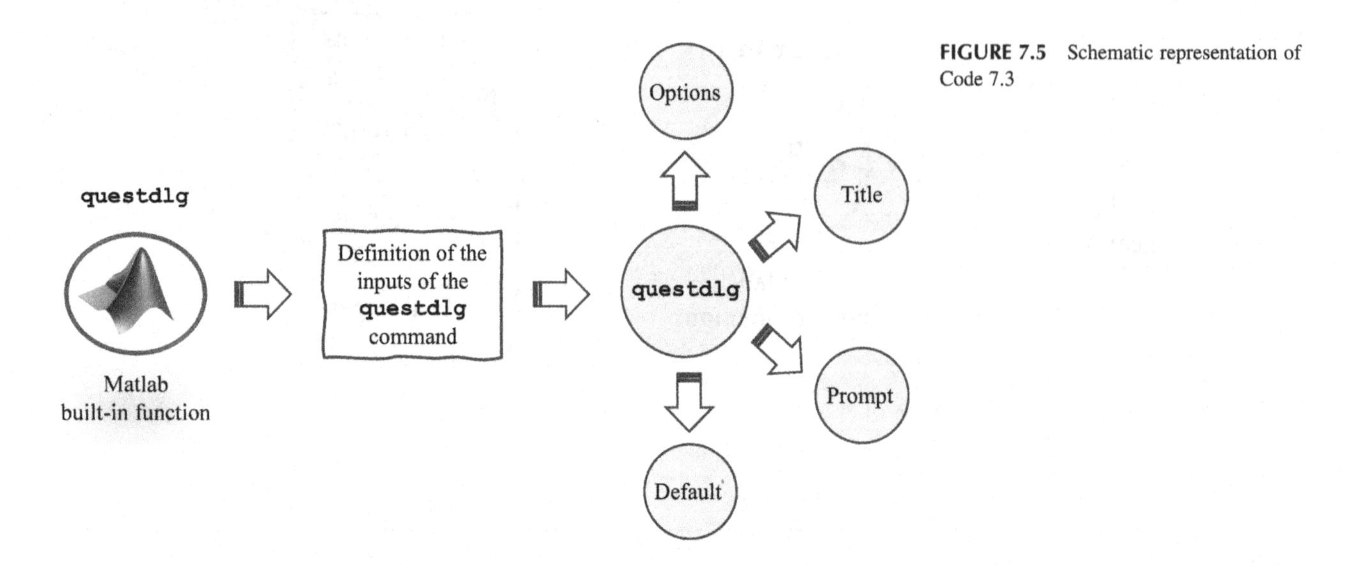

FIGURE 7.5 Schematic representation of Code 7.3

in the problem type, therefore, include only two options as `Option{1}` and `Option{2}`which are specified as "Regression" and "Classification," respectively (lines 2–3). It should be noted that the maximum number of allowable respond strings in the `questdlg` command is three. If we consider only two response strings, the third string is considered as the default response. The default response for the defined `questdlg` command is `Option{1}`, as can be seen by `Option{1}` being included as the argument for the default value in line 6. The schematic representation of Code 7.3 is provided in Fig. 7.5.

Code 7.3

```
1   %% Problem type
2   Option{1}='Regression';
3   Option{2}='Classification';
4   ANSWER=questdlg('Regression or Classification ?',...
5                   'ELM Type',...
6                   Option{1},Option{2},Option{1});
7   pause(0.1);
8   switch ANSWER
9       case Option{1}
10          nn.type = 'regression';
11      case Option{2}
12          nn.type = 'classification';
13  end
```

7.2.1.3 Definition of main user-defined input for outlier-based methods

To read the number of hidden neurons (NHN) and the number of iterations, the `inputdlg` command is applied in Code 7.4. The resulting dialog box from its execution is shown in Fig. 7.6. The syntax used for the `inputdlg` arguments is the following: `inputdlg(Prompt,Title,1,DefaultValues)`. The `prompt` argument is used to create a user-edited field in

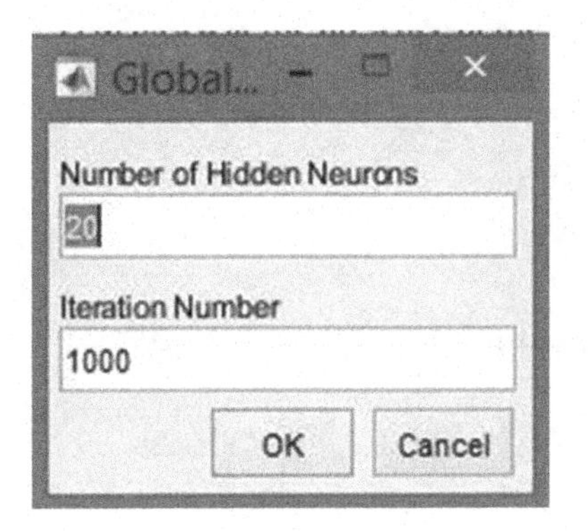

FIGURE 7.6 The output dialog box for the assignment of the NHN and the iteration number.

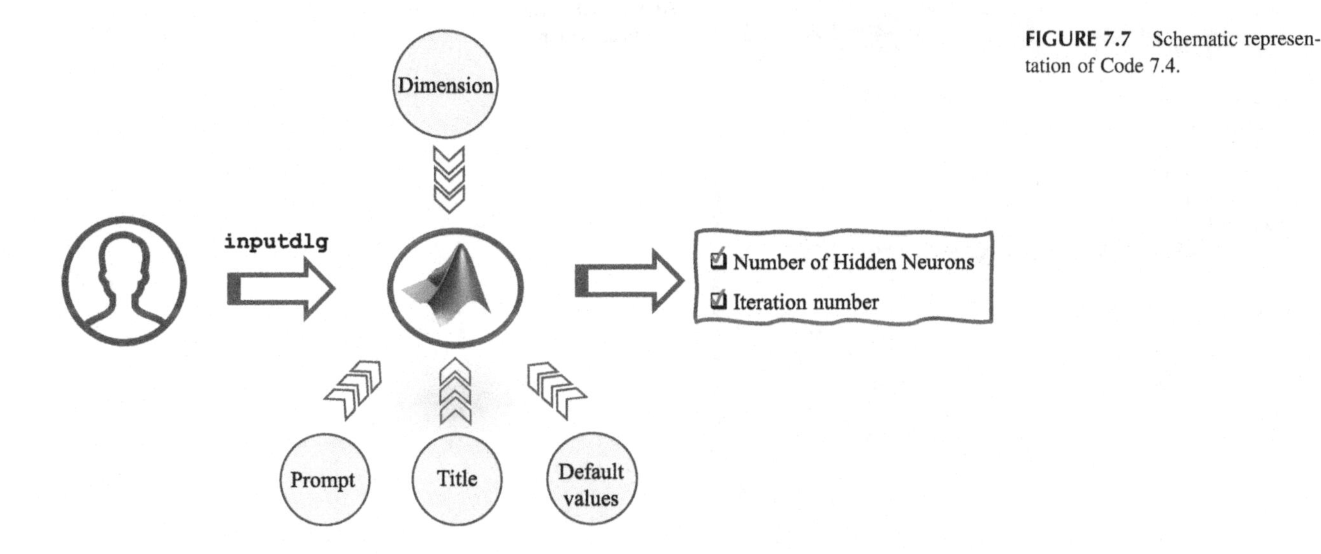

FIGURE 7.7 Schematic representation of Code 7.4.

the dialog box. In this case, there are two prompt arguments in lines 2–3 (i.e., the Number of Hidden Neurons and the Iteration Number), which return the user-defined input to the program. The `title` argument is used to provide a title to the dialog box and is designated as "Global Setting" in line 4. The `DefaultValues` argument specifies the default value for each of the fields defined by the prompt argument. This includes the default values of the first (i.e., Number of Hidden Neurons) and second (i.e., Iteration Number) parameters. It should be noted that the number "1" in the input arguments for `inputdlg` specifies the dimension of the input allowed from the edit field. In this case, a scalar value indicates that a single line of text can be considered. The `dims` argument can also take on the form of a column vector and an $m \times 2$ array. In the case of the column vector, the entries limit the height of each of the fields in the dialog box. For the "$m \times 2$ array," the first column specifies the height allowed for the field entry, while the second column value specifies the permissible width of the field in character units. The result of this command is shown in Fig. 7.6.

The NHN is a user-defined value and might be determined through professional experience or a trial-and-error process. In this case, the NHN could be considered to range from 1 to the maximum number of allowable NHN, which is provided in Eq. (7.1). It should be noted that, although increasing the value of the NHN may provide improved results during model training, the model performance during the testing stage may suffer due to a lack of generalizability of the model. In this regard, the NHN should be iterated to find a balance between the training and testing results, leading to a generalizable model that performs well on unseen data. One of the most important features of the ELM is the remarkable learning speed of this model. As a result of this speed, modeling using this method is usually done in a small number of seconds. Therefore, repeating the modeling process using the ELM approach will not be time-consuming. As a result, it is suggested that the number of iterations is considered to be at least 1000. The schematic representation of Code 7.4 is provided in Fig. 7.7.

Code 7.4

```
1    %% Global Setting
2    Prompt={'Number of Hidden Neurons',...
3        'Iteration Number'};
4            Title='Global Setting';
5            DefaultValues={'20','50'};
6            PARAMS=inputdlg(Prompt,Title,1,DefaultValues);
7            pause(0.1);
8            nn.hiddensize=str2double(PARAMS{1});
9            N=str2double(PARAMS{2});
```

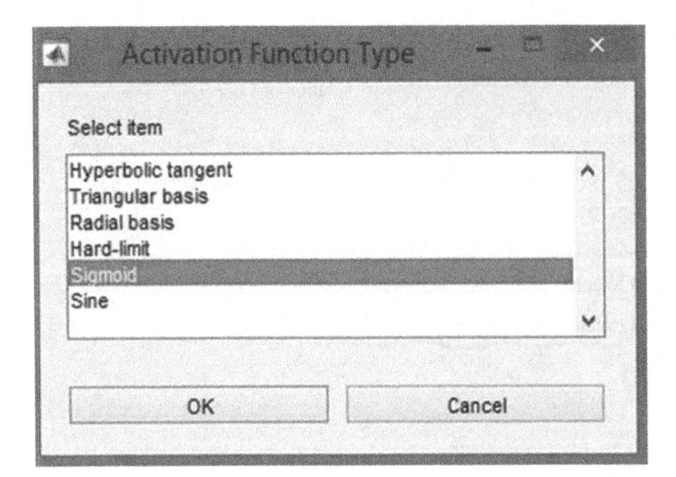

FIGURE 7.8 The output of the question box for the selection of activation function type.

After the selection of the ELM type as well as the definition of the NHN and number of iterations using the global settings dialog box, the activation function type should be selected. Code 7.5 details the MATLAB coding required to select the activation function, while Fig. 7.8 demonstrates the resulting dialog box.

Using the `questdlg` command, the maximum number of parameters permissible is limited to three. The number of activation functions considered in the current code is six. Therefore, it is necessary to employ the `listdlg` command in order to accommodate the number of activation functions. The required input arguments for the `listdlg` MATLAB command are in the form of `listdlg("ListString,"list,Name,Value)` and are discussed below. The `list` argument of the `listdlg` command is used to present all of the list items in the dialog box. The `ListString` argument is a cell array of character vectors that specify the list items. Typically, each element in the cell array corresponds to an item in the list. For our purposes, the `spaceList` argument includes all six activation functions consisting of Hyperbolic tangent, Triangular basis, Radial basis, Hard-limit, Sigmoid, and Sine. These activation functions must be defined in the `spaceList` prior to the use of the `listdlg` command. The`SelectionMode` argument is used to define how many selections the user may make from the list. This parameter may be used to allow single and multiple selections, with possible inputs being "single" and "multiple" (default). The `PromptString` argument is used to define the prompt that appears above the list box, which in this case is defined as "Select item." The `Initialvalue` argument is a vector of indices of the list box items that are initially selected. In the current command, the index 5 is selected for the Sigmoid activation function. The `Name` argument in the dialog box defines the title of the dialog box and is defined as "Activation Function Type." Finally, the `ListSize` argument is the List box size in pixels, specified as a two-element vector of the form `[width height]`. The specified value in the current command is as `[300 100]`. The schematic representation of Code 7.5 is provided in Fig. 7.9.

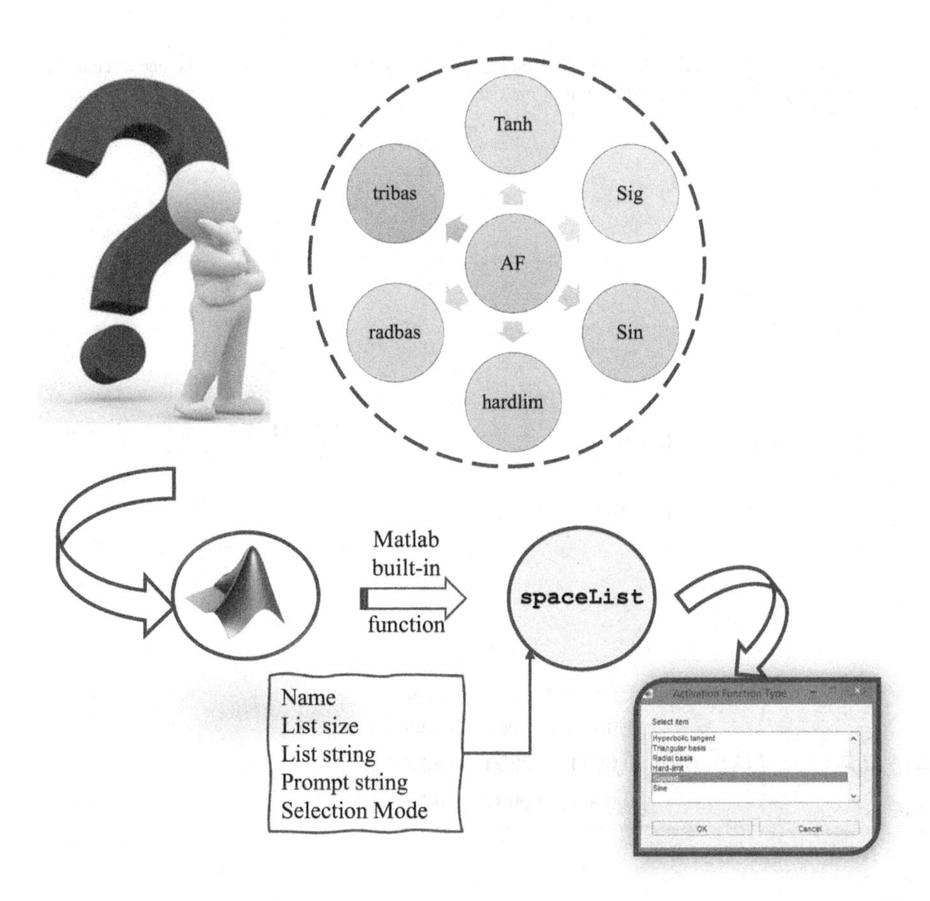

FIGURE 7.9 Schematic representation of Code 7.5.

Code 7.5

```
1   %% Activation function selection
2   spaceList = {'Hyperbolic tangent','Triangular basis','Radial
    basis','Hard-limit','Sigmoid','Sine'};
3           [idx, ~] = listdlg('ListString', spaceList,...
                'SelectionMode', 'Single', 'PromptString', 'Select item',
4   'Initialvalue', 5,'Name', 'Activation Function Type','ListSize',[300
    100]);
5           if idx == 1
6               nn.activefunction='tanh';
7           elseif idx == 2
8               nn.activefunction='tribas';
9           elseif idx == 3
10              nn.activefunction='radbas';
11          elseif idx == 4
12              nn.activefunction='hardlim';
13          elseif idx == 5
14              nn.activefunction='sig';
15          elseif idx == 6
16              nn.activefunction='sin' ;
17          end
```

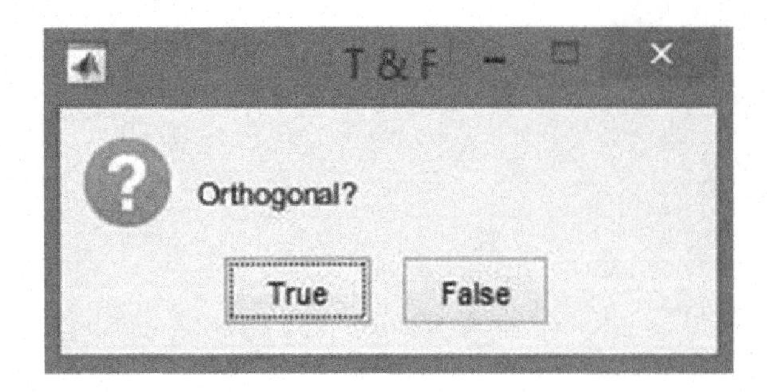

FIGURE 7.10 The output dialog box for the selection of whether to consider orthogonalization of matrices.

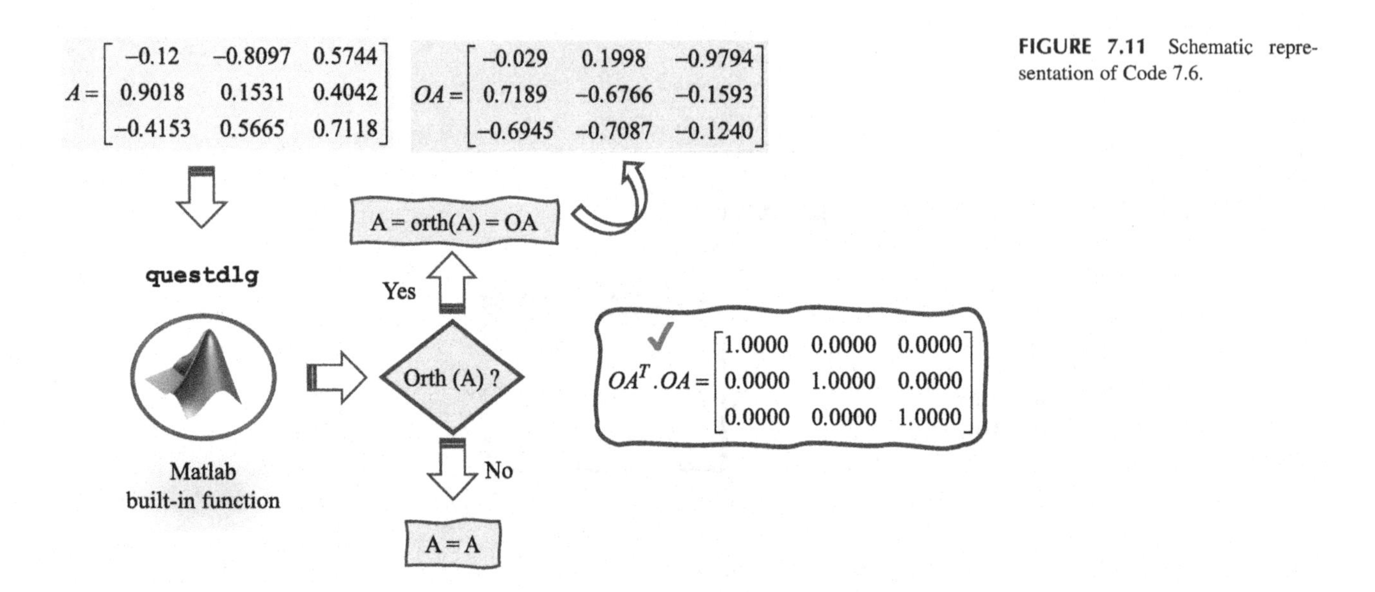

FIGURE 7.11 Schematic representation of Code 7.6.

The next step is for the user to input whether or not they want to consider orthogonalization of the randomly generated matrices of the bias of hidden neurons and input weight. Similar to the dialog box for ELM type selection, the `questdlg` command is applied in Code 7.6 to allow the user to select whether or not to use this parameter. Fig. 7.10 demonstrates the resulting dialog box that the user is presented with. The schematic representation of Code 7.6 is presented in Fig. 7.11.

Code 7.6

```
1    %% Orthogonal
2    Option{1}='True';
3    Option{2}='False';
4    ANSWER=questdlg('Orthogonal?',...
5                    'T & F',...
6                    Option{1},Option{2},Option{1});
7    pause(0.1);
8    switch ANSWER
9        case Option{1}
10           nn.orthogonal = 'true';
11       case Option{2}
12           nn.orthogonal = 'false';
13   end
```

7.2.1.4 Selection of the desired method and definition of its parameters

Following the selection of the initial parameters, including ELM type, the NHN, iteration number, activation function type, and orthogonality, the ELM-based model should be selected to begin the training stage. The `questdlg` command is again applied to allow the user to select whether to consider the original ELM or the developed ones (Fig. 7.12A). If the Classical ELM is selected, modeling is commenced using the original ELM; otherwise, the value of the regularization parameter (C) (Fig. 7.12B) must be specified. The regularization parameter is a positive number that is defined in the range of (0, ∞). Once the regularization parameter has been defined, the user must specify which developed outlier-based ELM model (RELM, WRELM, and ORELM) will be considered for modeling (Fig. 7.12C).

7.2.2 Details of the initialization process

After selecting the modeling method, the development of the model is begun following the respective specific algorithm (see Fig. 7.1). To model a real-world problem, the `elm_initialization`, `elm_train` and `elm_test` functions are run to present a model for target value estimation. The `elm_initialization` function is run to randomly produce the input weights and bias of hidden neurons, while the `elm_train` function is run to calculate the output weights analytically. The performance of the developed ELM-based model with the presence of outliers is evaluated using the `elm_test` function.

The main code of the `elm_initialization` function is presented in Code 7.7. It should be noted that this function is identical to the original ELM and all developed ELM-based methods with the presence of outliers (i.e., RELM, WRELM, ORELM). The input of this function is the number of input variables, which is characterized by the input size and the size of the number of hidden neurons. These two values are both saved in the variable `nn`. From lines 5 and

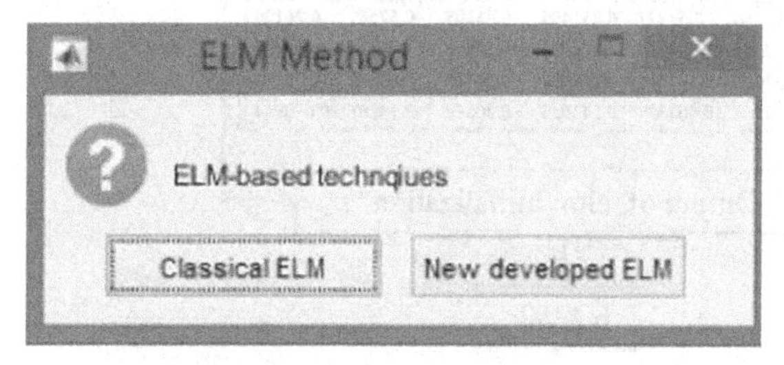

(A) ELM or outlier-based ELM

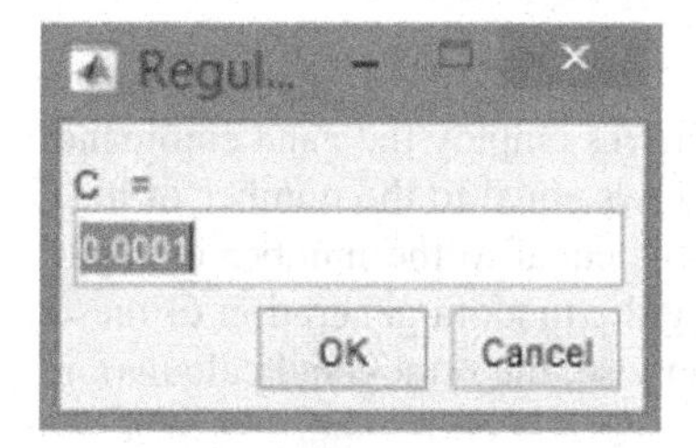

(B) Regularization parameter

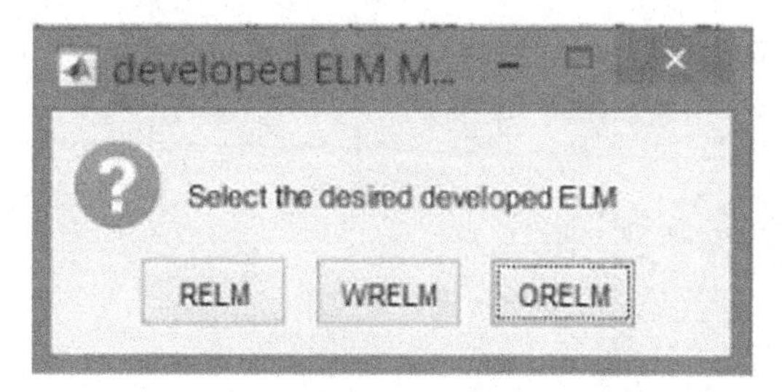

(C) Select the outlier-based ELM

FIGURE 7.12 Selection of extreme learning machine (ELM)-based model type. (A) ELM or outlier-based ELM, (B) Regularization parameter, and (C) Select the outlier-based ELM.

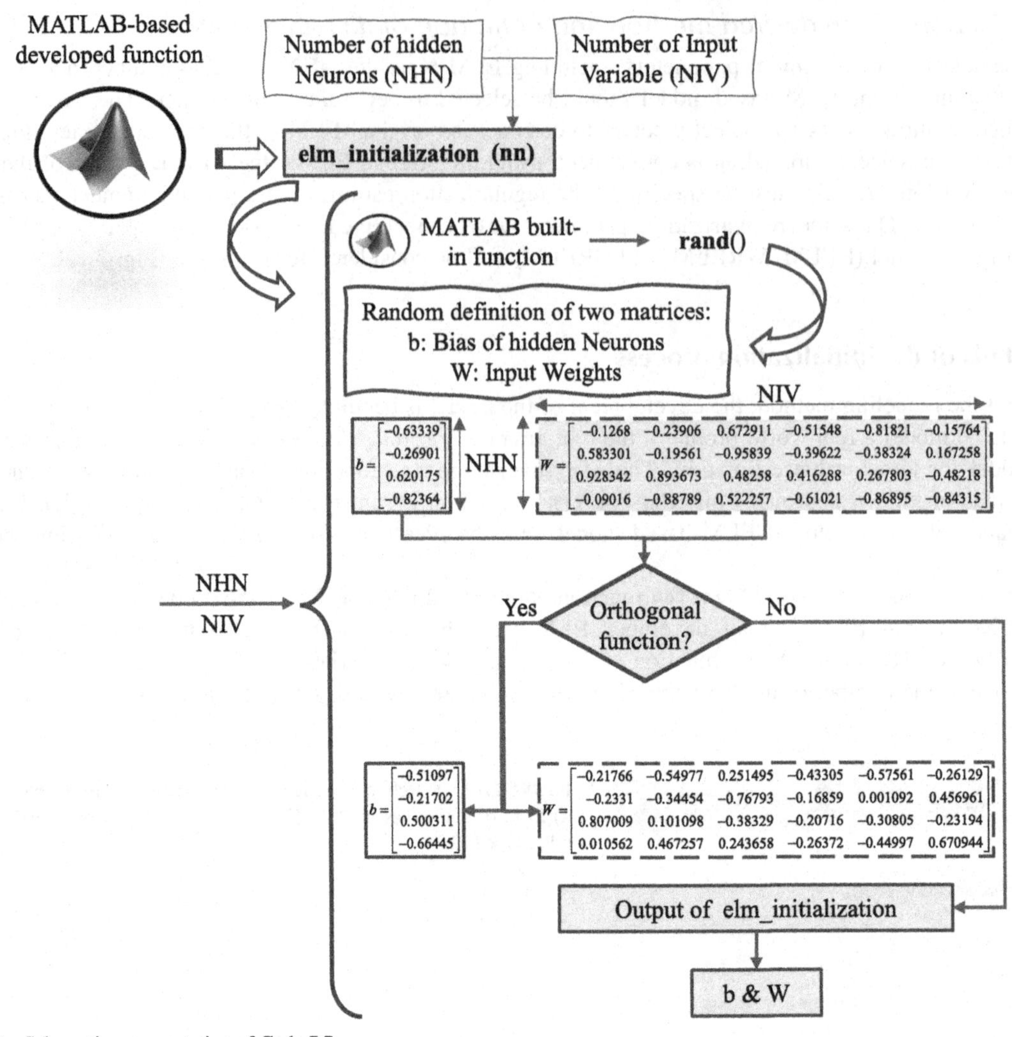

FIGURE 7.13 Schematic representation of Code 7.7.

6 below, we can observe that both the bias of hidden neurons and the input weights matrices employ the `rand` command to populate. The bias of hidden neurons is a matrix with one column and a number of rows equal to the number of hidden layer neurons. On the other hand, the input weight vector has a number of columns equal to the number of input variables and a number of rows equal to the number of hidden layer neurons. Following the random generation of these two matrices, the orthogonal of each one, if desired by the user, is also calculated. Otherwise, the orthogonal calculation is ignored. The schematic representation of Code 7.7 is provided in Fig. 7.13.

Code 7.7

```
1   function nn = elm_initialization(nn)
2
3   % biases and input weights
4
5   nn.b = 2*rand(nn.hiddensize,1)-1;
6   nn.W = 2*rand(nn.hiddensize, nn.inputsize)-1;
7
8       if nn.orthogonal
9           if nn.hiddensize > nn.inputsize
10              nn.W = orth(nn.W);
11          else
12              nn.W = orth(nn.W')';
13          end
14          nn.b=orth(nn.b);
15      end
16
17  end
```

7.2.3 Details of the training process

After the population of the two randomly generated ELM-based matrices (i.e., the bias of hidden neurons and input weights), the training process is initiated to calculate the output weights using the defined methodology for each ELM-based method. The function elm_train (shown in Code 7.8) is used to train and develop the model using data from a real-world problem. The inputs to this function are the training input variables (i.e., X) and their corresponding outputs (i.e., *Y*) as well as nn (Line 1). As previously mentioned, the variables *nn* comprises the regularization parameters, activation function type, orthogonalization decision, ELM type, and the number of hidden neurons. The first step in the training stage is the calculation of the hidden layer output without considering the activation function type – this is stored in the variable tempH in line 3. Following this, the hidden layer output (H) is calculated by considering the variable tempH as the input to the user-defined activation function type. Up until this point, the training methodology is identical for all of the ELM-based models that we have presented in this text.

The next step is to determine the output weights, the calculation of which is dependent on the type of ELM selected (classical or RELM, WRELM, ORELM). The calculation of the output weights according to the RELM model is slightly different from the classical ELM method, as the norm of the output weights and training error are simultaneously minimized through the user-defined regularization parameter C. The schematic representation of Code 7.8 is presented in Fig. 7.14.

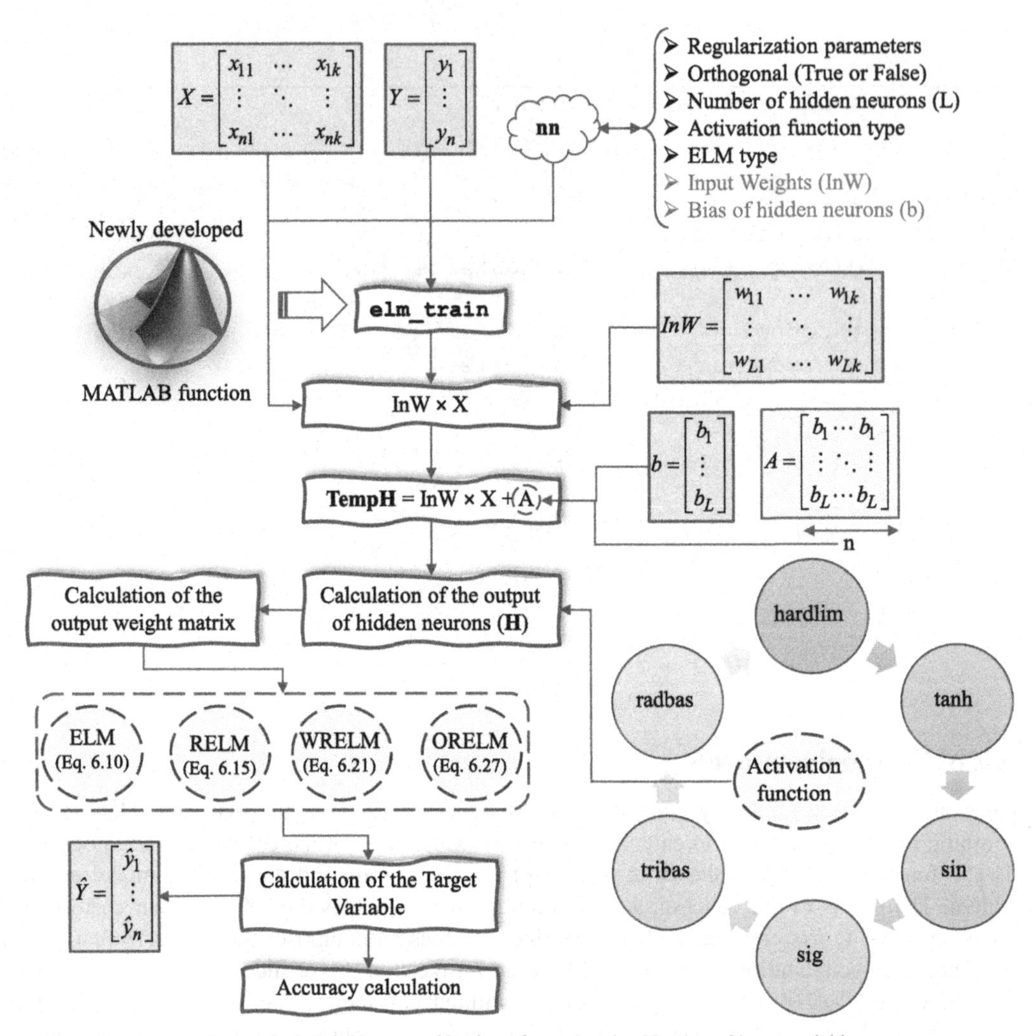

FIGURE 7.14 Schematic representation of Code 7.8. Note: n = Number of samples; k = Number of input variables.

Code 7.8

```
1   function [nn, acc_train] = elm_train(X, Y, nn)
2   ndata = size(X,2);
3   tempH = nn.W*X + repmat(nn.b,1,ndata);
4   
5   switch lower(nn.activefunction)
6       case{'s','sig','sigmoid'}
7           H = 1 ./ (1 + exp(-tempH));
8       case{'t','tanh'}
9           H = tanh(tempH);
10      case {'sin','sine'}
11          %%%%%%%% Sine
12          H = sin(tempH);
13      case {'hardlim'}
14          %%%%%%%% Hard Limit
15          H = double(hardlim(tempH));
16      case {'tribas'}
17          %%%%%%%% Triangular basis function
18          H = tribas(tempH);
19      case {'radbas'}
20          %%%%%%%% Radial basis function
21          H = radbas(tempH);
22          %%%%%%%% More activation functions can be added here
23  end
24  
25  clear tempH;
26  
27  switch(nn.method)
28      case 'ELM'
29          [beta] = regressor(H', Y', 0);
30      case 'RELM'
31          [beta] = regressor(H', Y', nn.C);
32      case 'WRELM'
33          [beta] = regressor(H', Y', nn.C);
34          e = beta'*H - Y;
35          s = median(abs(e))/0.6745;
36          w = weight_fun(e, nn.wfun, s);
37          [beta] = regressor(repmat(sqrt(w'),1,size(H,1)).*H',
    repmat(sqrt(w'),1,size(Y,1)).*Y', nn.C);
38      case 'ORELM'
39          [beta] = regressor_alm(H', Y', nn.C, 20);
40  end
41  
42  nn.beta  = beta';
43  Y_hat    = nn.beta*H;
44  nn.Y_hat=Y_hat;
45  
46  if ismember(nn.type,{'c','classification','Classification'})
47      [~,label_actual]  = max(Y_hat,[],1);
48      [~,label_desired] = max(Y,[],1);
49      acc_train = sum(label_actual==label_desired)/ndata;
50  else
51     acc_train = sqrt(mse(Y-Y_hat));
52  
53  end
54  nn.trainlabel  = Y_hat;
55  nn.acc_train   = acc_train;
```

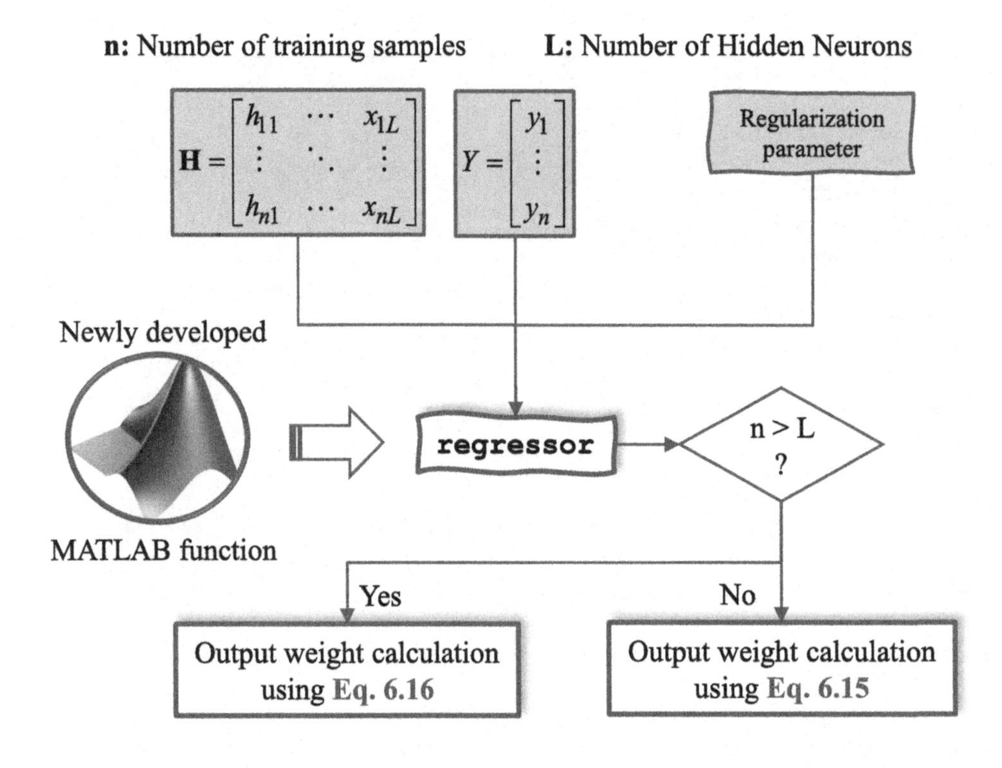

FIGURE 7.15 Schematic representation of Code 7.9.

The calculation of the ELM and RELM output weights is done using the regressor function, which is detailed in Code 7.9. The inputs to this function include the output of the hidden layer neurons (H), the training output (Y), and the regularization parameter (C), which is denoted in this function as lambda in line 1. It can be seen that in this function, there is no command to select the method. In fact, according to the value of the regularization parameter, the type of ELM approach to be considered is determined. In the classical ELM method, the value of the regularization parameter is considered equal to zero, while the value of this parameter is positive for RELM. Initially, the number of training samples, defined as ndata in line 3, is calculated using the MATLAB function size (H, 1). It is necessary to calculate this parameter to compare the number of data points with the number of hidden layer neurons (i.e., size (H, 2)). If the selected method is ELM, the output weights are calculated Moore–Penrose generalized inverse of the output weight matrix (H) using the pinv command. It should be noted that if the selected method was RELM, a new term is added to the pinv command, which is a combination of the regularization parameter and the eye command (which generates the identity matrix). Generally, the Regressor function performs the calculations of Eqs. (6.15) and (6.16) in the RELM subsection. The schematic representation of Code 7.9 is provided in Fig. 7.15.

Code 7.9

```
1   function [beta] = regressor(H, Y, lamda)
2
3   ndata = size(H,1);
4   if ndata < size(H,2)
5       beta = H'*pinv(H*H'+lamda*eye(ndata))*Y;
6   else
7       beta = pinv(H'*H+lamda*eye(size(H,2)))*H'*Y;
8   end
```

To calculate the output weights according to the WRELM approach, the calculated output weight from the RELM is considered to be the initial value. The initial matrix defined from the RELM solution is then applied to determine the e, s, and w values as defined in Eqs. (6.20), (6.17), and (6.18), respectively. After calculating these parameters, the regressor function is applied again to update output weight (β) using the computed weight information, as can be seen in line 4 of Code 7.10. It should be noted that the weight information is calculated using the weight_fun function, which will be subsequently described in detail in Code 7.11.

Code 7.10

```
1            e = beta'*H - Y;
2            s = median(abs(e))/0.6745;
3            w = weight_fun(e, nn.wfun, s);
4            [beta] = regressor(repmat(sqrt(w'),1,size(H,1)).*H',
      repmat(sqrt(w'),1,size(Y,1)).*Y', nn.C);
```

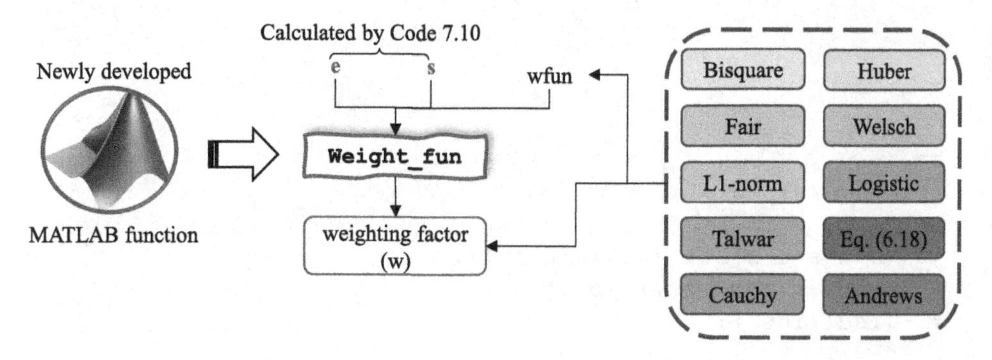

FIGURE 7.16 Schematic representation of Code 7.11.

7.2.3.1 *Weight function for weighted regularized extreme learning machine*

The weight_fun function is presented in Code 7.11. The values of *s* and *e* should be determined prior to calling this function since they are considered as input arguments and are required in order to calculate *w*. The weight_fun function provides 10 different methods to calculate the value of the weight function, the first of which is the program default model [Eq. (6.18)]. In order to effectively model a real-world problem, all 10 of the provided methods should be used to verify the weight calculation. It should be noted that the wfun, which is the input of the weight_fun function, should be initiated (from 1 to 10) in the main training code to check each method on weight information calculation. The schematic representation of Code 7.11 is presented in Fig. 7.16.

Code 7.11

```
1    function [w] = weight_fun(e, wfun, s)
2
3    if nargin < 2
4        wfun = '1';
5        s     = median(abs(e(:)))/0.6745;
6    end
7
8    w  = zeros(size(e));
9
10   switch(wfun)
11
12       case{'1','default'}
13
14           a=2.5;   b=3;
15           ind1     = (abs(e/s)<=a);
16           w(ind1)  = 1;
17           ind2     = ((a<abs(e/s)) & (abs(e/s)<b));
18           w(ind2)  = (b-abs(e(ind2)/s))/(b-a);
19           ind3     = ~(ind1|ind2);
20           w(ind3)  = 1e-4;
21
22       case{'2','bisquare','b'}
23           tune = 4.685;
24           r    = e/(tune*s);
25           w    = (abs(r)<1) .* (1 - r.^2).^2;
26
27       case{'3','huber','h'}
28           tune = 1.345;
29           r    = e/(tune*s);
30           w    = 1./max(1,abs(r));
31
32       case{'4','lp','l'}
33           p = 1; % L1-norm
34           w = 1./max(0.0001,abs(e).^(2-p));
35          w = w/sum(w);
36               case{'5','andrews'}
37
38                   tune  = 1.339;
39                   r     = e/(tune*s);
40                   w     = (abs(r)<pi) .* sin(r) ./ r;
```

```
41
42            case{'6','fair'}
43                tune = 1.400;
44                r    = e/(tune*s);
45                w    = 1 ./ (1 + abs(r));
46
47            case{'7','cauchy'}
48                tune = 2.385;
49                r    = e/(tune*s);
50                w    = 1 ./ (1 + r.^2);
51
52            case{'8','logistic'}
53                tune = 1.205;
54                r    = e/(tune*s);
55                w    = tanh(r) ./ r;
56
57            case{'9','talwar'}
58                tune = 2.795;
59                r    = e/(tune*s);
60                w    = 1 * (abs(r)<1);
61
62            case{'10','welsch'}
63                tune = 2.985;
64                r    = e/(tune*s);
65                w    = exp(-(r.^2));
66
67    end
```

7.2.3.2 The regressor_alm function for outlier robust extreme learning machine

The above discussion presented a detailed explanation of the output weight calculation method for the ELM, RELM, and WRELM methods. The output weight for all of these methods was determined using the regressor function. Next, we will shift our focus to the calculation of the output weight for the ORELM method. To calculate the output weight analytically in the ORELM method, the regressor_alm function is employed. The input of this function includes the hidden layer output (H), training targets, regularization parameters, and the maximum number of iterations for the ORELM modeling (which are denoted as A, b, C, and maxIter, respectively in Code 7.12). The default value of maxIter is 20, which can be optimized to find the best model results. Based on the values of the number of hidden neurons (n) and the number of training samples (m), the pinv command is applied to calculate the Moore-Penrose generalized inverse of the output weight matrix. Then, during an iterative process, the output weight value is modified using Eq. (26.6) schematic representation of Code 7.12 is presented in Fig. 7.17.

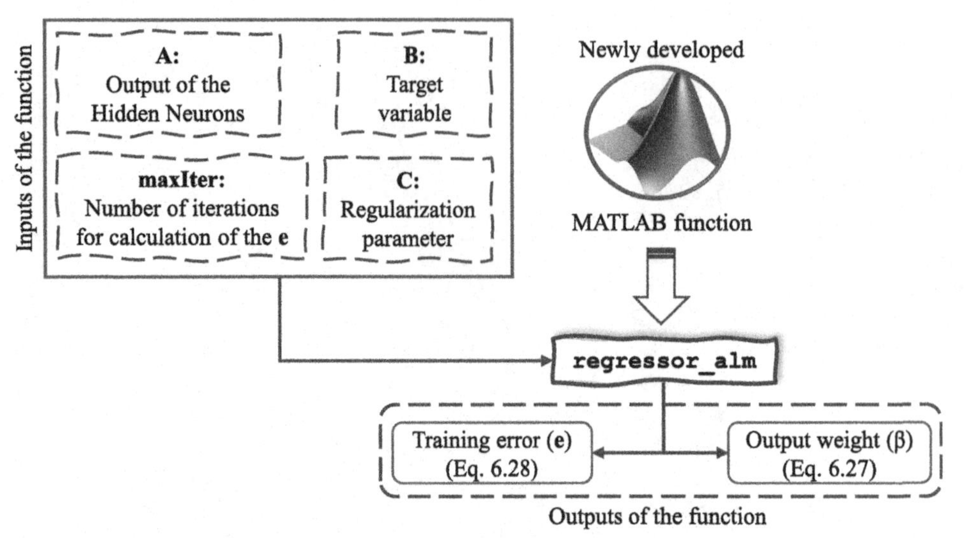

FIGURE 7.17 Schematic representation of Code 7.12.

Code 7.12

```
1     function [x, e] = regressor_alm(A, b, C, maxIter)
2
3     C = 1/C;
4
5     [m,n]  = size(A) ;
6     kappa  = 1/C;
7     nIter  = 0 ;
8     mu     = 2*m/norm(b,1);
9     lambda = zeros(m,1);
10    e      = zeros(m,1);
11    converged_main = 0;
12    muInv  = 1/mu ;
13    if n<m
14        Proj_M = pinv(A'*A+2*kappa*muInv*eye(n))*A';
15    else
16        Proj_M = A'*pinv(A*A'+2*kappa*muInv*eye(m));
17    end
18
19    while ~converged_main
20        lambdaScaled = muInv*lambda ;
21        nIter  = nIter + 1 ;
22        x      = Proj_M*(b-e+lambdaScaled );
23        temp   = b + lambdaScaled - A*x ;
24        e      = shrink(temp,muInv) ;
25        lambda = lambda + mu*(b - A*x - e) ;
26        if nIter >= maxIter
27            converged_main = 1 ;
28        end
29    end
30
31    function Y = shrink(X, alpha)
32
33    Y = sign(X).*max(abs(X)-alpha,0);
```

7.2.4 Extreme learning machine-based model testing phase

Once the training and development of the ELM-based models are complete (including the classical ELM, RELM, WRELM, and ORELM), the performance of the model must be verified using previously unseen data. Indeed, the generalizability of the developed model is determined using the elm_test function, which is detailed in Code 7.13. The input to the elm_test function includes the test inputs, the corresponding test targets, and the nn parameter. The nn parameter includes the number of hidden neurons, the activation function type, the bias of hidden neurons (b), the input weights (w), and the calculated output weights (β). To verify the performance of the developed model against previously unseen data, the first step is the computation of the tempH variable in line 4 using the input weights, the bias of hidden neurons, and the number of input variables. Using tempH and the selected activation function type, the output of the hidden neurons (H) matrix can be calculated. Once the H matrix has been determined, it is multiplied by the output weight to estimate the value of the target variables. Finally, the performance of the estimated value based on the problem type (i.e., Regression or classification) is evaluated using various statistical indices. The schematic representaion of Code 7.13 is presented in Fig. 7.18.

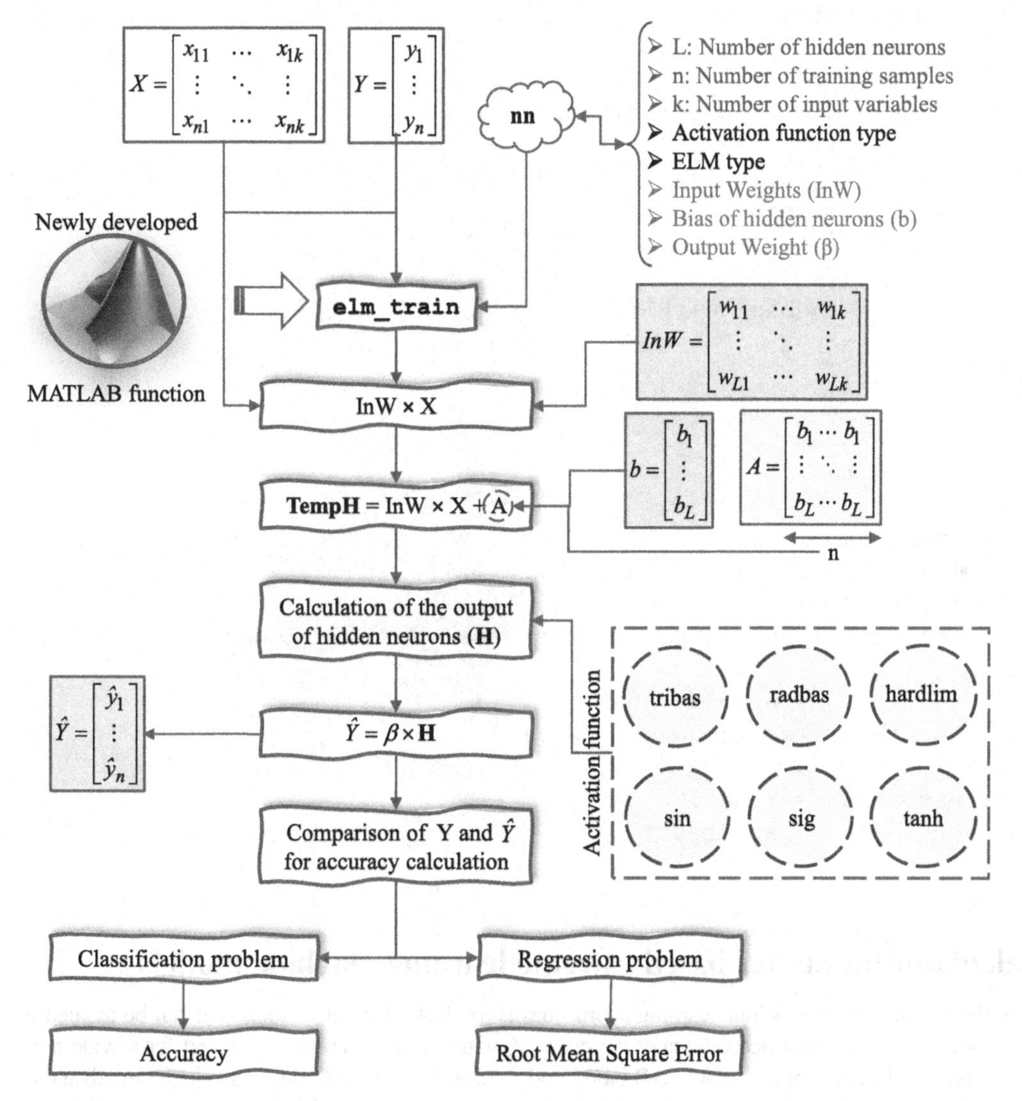

FIGURE 7.18 Schematic representation of Code 7.13.

Code 7.13

```
1   function [nn, acc_test] = elm_test(X,Y,nn)
2
3   ndata         = size(X, 2);
4   tempH         = nn.W*X + repmat(nn.b,1,ndata);
5
6   switch lower(nn.activefunction)
7       case{'s','sig','sigmoid'}
8           H = 1 ./ (1 + exp(-tempH));
9       case{'t','tanh'}
10          H = tanh(tempH);
11      case {'sin','sine'}
12          %%%%%%%% Sine
13          H = sin(tempH);
14      case {'hardlim'}
15          %%%%%%%% Hard Limit
16          H = double(hardlim(tempH));
17      case {'tribas'}
18          %%%%%%%% Triangular basis function
19          H = tribas(tempH);
20      case {'radbas'}
21          %%%%%%%% Radial basis function
22          H = radbas(tempH);
23          %%%%%%%% More activation functions can be added here
24  end
25
26  Y_hat     = nn.beta*H;
27  nn.Y_hat=Y_hat;
28
29  if ismember(nn.type,{'c','classification','Classification'})
30      [~,label_actual]  = max(Y_hat,[],1);
31      [~,label_desired] = max(Y,[],1);
32      acc_test = sum(label_actual==label_desired)/ndata;
33  else
34      normfro    = norm(Y-Y_hat,'fro');
35      acc_test = sqrt(mse(Y-Y_hat));
36  end
37  nn.testlabel    = Y_hat;
38  nn.acc_test     = acc_test;
```

7.3 Calculator for outlier-based extreme learning machine models

A significant drawback of most machine learning approaches is the lack of intuitive tools that can be applied to solve practical tasks. In the current study, a simplified calculator is presented, which can be rapidly employed for a wide number of practical situations. The provided calculator is a MATLAB-based code and is detailed in Code 7.14. The main advantage of this code is the fact that in-depth knowledge about the various ELM-based approaches or the MATLAB environment is not required for its application. In order to execute the developed calculator tool, the only knowledge that is required is the ability to execute a MATLAB file in its environment. Upon execution of the calculator program code, the user is faced with two different dialog windows. The first one is related to the number of input variables (Fig. 7.19), while the second one concerns the value of each variable (Fig. 7.20). The second window is designed to consider one to ten input variables. The extension of the provided calculation for more than 10 input variables is also a trivial exercise.

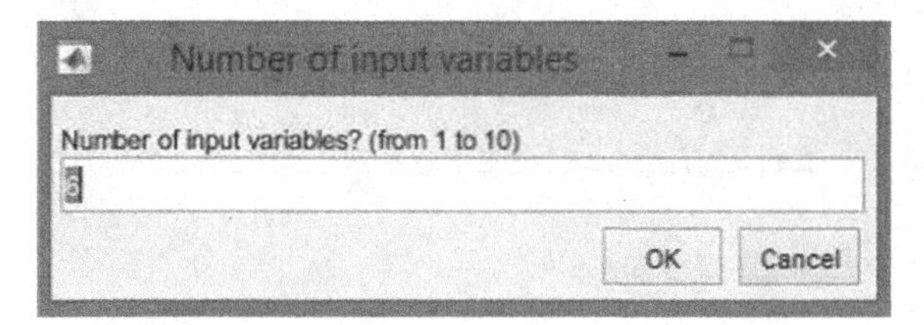

FIGURE 7.19 Output dialog box requesting the number of input variables for the developed calculator.

Code 7.14

```
1   close all
2   clear
3   clc
4
5   %% Load data
6   load('Results')
7
8   %% Output Calculation
9   Prompt={'Number of input variables? (from 1 to 10)'};
10  Title='Number of input variables';
11  DefaultValues={'6'};
12  PARAMS=inputdlg(Prompt,Title,[1 56],DefaultValues);
13  NIV=str2double(PARAMS{1});
14
15  if NIV==1
16      Prompt={'In1'};
17      Title='Input variables';
18      DefaultValues={'0.988'};
19      PARAMS=inputdlg(Prompt,Title,[1 45],DefaultValues);
20      In1=str2double(PARAMS{1});
21      InV=[In1];
22
23  elseif NIV==2
24      Prompt={'In1','In2'};
25      Title='Input variables';
26      DefaultValues={'0.988','0.815'};
27      PARAMS=inputdlg(Prompt,Title,[1 45],DefaultValues);
28      In1=str2double(PARAMS{1});
29      In2=str2double(PARAMS{2});
30      InV=[In1 In2];
31
32  elseif NIV==3
33      Prompt={'In1','In2','In3'};
34      Title='Input variables';
35      DefaultValues={'0.988','0.815','7.29'};
36      PARAMS=inputdlg(Prompt,Title,[1 45],DefaultValues);
37      In1=str2double(PARAMS{1});
38      In2=str2double(PARAMS{2});
39      In3=str2double(PARAMS{3});
40      InV=[In1 In2 In3];
```

```
41
42   elseif NIV==4
43       Prompt={'In1','In2','In3','In4'};
44       Title='Input variables';
45       DefaultValues={'0.988','0.815','7.29','1'};
46       PARAMS=inputdlg(Prompt,Title,[1 45],DefaultValues);
47       In1=str2double(PARAMS{1});
48       In2=str2double(PARAMS{2})
49       In3=str2double(PARAMS{3});
50       In4=str2double(PARAMS{4});
51       InV=[In1 In2 In3 In4];
52
53   elseif NIV==5
54       Prompt={'In1','In2','In3','In4','In5'};
55       Title='Input variables';
56       DefaultValues={'0.988','0.815','7.29','1','0.11'};
57       PARAMS=inputdlg(Prompt,Title,[1 45],DefaultValues);
58       In1=str2double(PARAMS{1});
59       In2=str2double(PARAMS{2});
60       In3=str2double(PARAMS{3});
61       In4=str2double(PARAMS{4});
62       In5=str2double(PARAMS{5});
63       InV=[In1 In2 In3 In4 In5];
64
65   elseif NIV==6
66       Prompt={'In1','In2','In3','In4','In5','In6'};
67       Title='Input variables';
68       DefaultValues={'0.988','0.815','7.29','1','0.11','0'};
69       PARAMS=inputdlg(Prompt,Title,[1 45],DefaultValues);
70       In1=str2double(PARAMS{1});
71       In2=str2double(PARAMS{2});
72       In3=str2double(PARAMS{3});
73       In4=str2double(PARAMS{4});
74       In5=str2double(PARAMS{5});
75       In6=str2double(PARAMS{6});
76       InV=[In1 In2 In3 In4 In5 In6];
77
78   elseif NIV==7
79       Prompt={'In1','In2','In3','In4','In5','In6','In7'};
80       Title='Input variables';
```

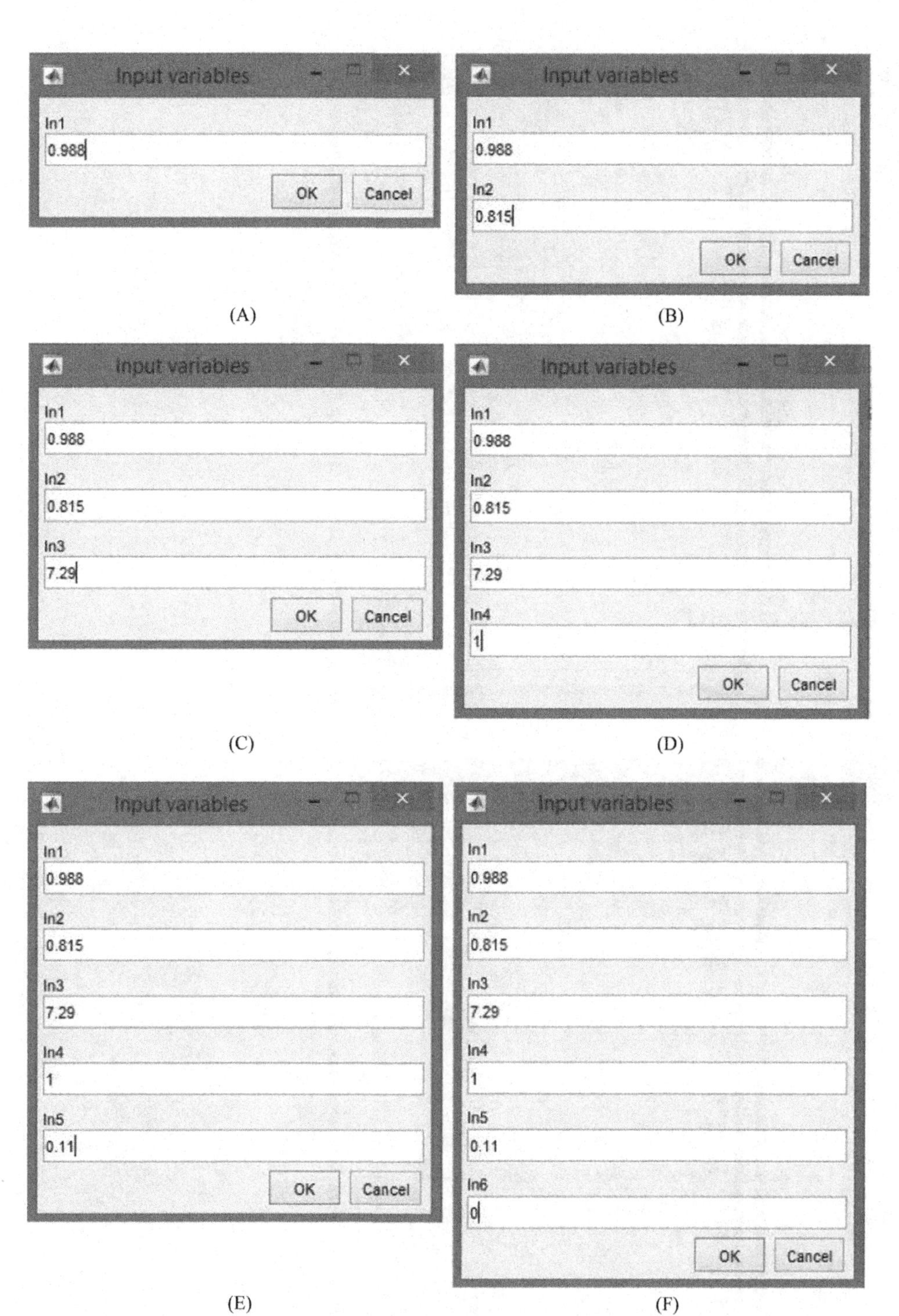

FIGURE 7.20 The output of the dialog window for input variable definition: **(a)** to **(j)** represent the observed window for 1 to 10 inputs, respectively.

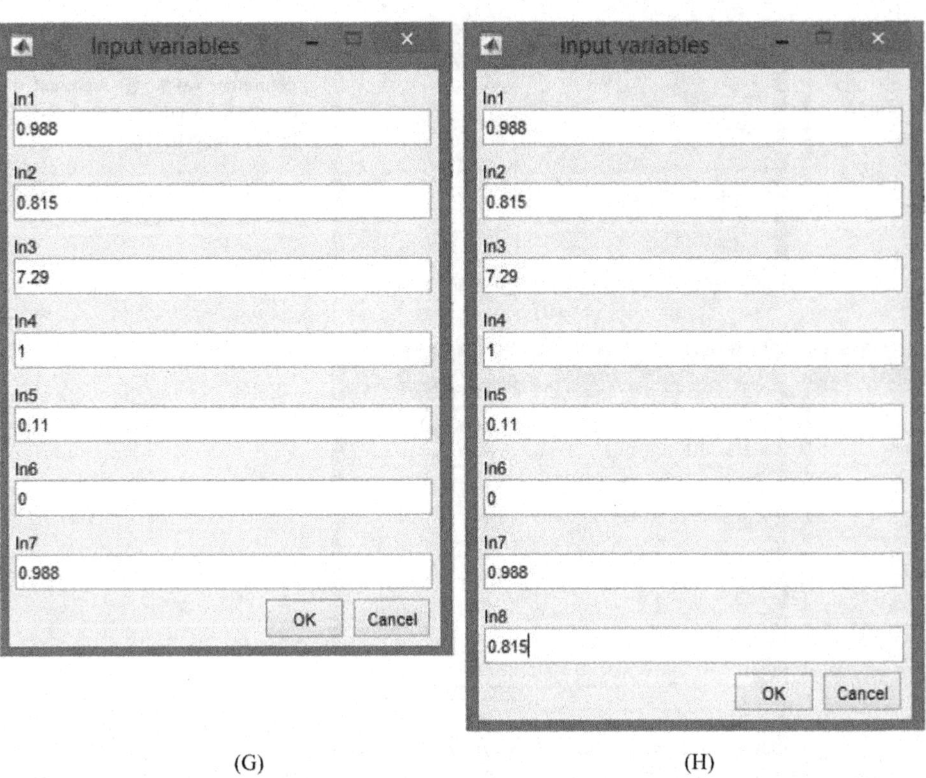

(G) (H)

Input variables

In1
0.988
In2
0.815
In3
7.29
In4
1
In5
0.11
In6
0
In7
0.988
In8
0.815
In9
7.29

OK Cancel

(I)

Input variables

In1
0.988
In2
0.815
In3
7.29
In4
1
In5
0.11
In6
0
In7
0.988
In8
0.815
In9
7.29
In10
1

OK Cancel

(J)

FIGURE 7.20 (Continued).

```
81          DefaultValues={'0.988','0.815','7.29','1','0.11','0','0.988'};
82          PARAMS=inputdlg(Prompt,Title,[1 45],DefaultValues);
83
84          In1=str2double(PARAMS{1});
85          In2=str2double(PARAMS{2});
86          In3=str2double(PARAMS{3});
87          In4=str2double(PARAMS{4});
88          In5=str2double(PARAMS{5});
89          In6=str2double(PARAMS{6});
90          In7=str2double(PARAMS{7});
91          InV=[In1 In2 In3 In4 In5 In6 In7];
92
93      elseif NIV==8
94          Prompt={'In1','In2','In3','In4','In5','In6','In7','In8'};
95          Title='Input variables';
96
        DefaultValues={'0.988','0.815','7.29','1','0.11','0','0.988','0.815'};
97          PARAMS=inputdlg(Prompt,Title,[1 45],DefaultValues);
98          In1=str2double(PARAMS{1});
99          In2=str2double(PARAMS{2});
100         In3=str2double(PARAMS{3});
101         In4=str2double(PARAMS{4});
102         In5=str2double(PARAMS{5});
103         In6=str2double(PARAMS{6});
104         In7=str2double(PARAMS{7});
105         In8=str2double(PARAMS{8});
106         InV=[In1 In2 In3 In4 In5 In6 In7 In8];
107
108     elseif NIV==9
109         Prompt={'In1','In2','In3','In4','In5','In6','In7','In8','In9'};
110         Title='Input variables';

111     DefaultValues={'0.988','0.815','7.29','1','0.11','0','0.988','0.815','7.2
        9'};
112         PARAMS=inputdlg(Prompt,Title,[1 45],DefaultValues);
113         In1=str2double(PARAMS{1});
114         In2=str2double(PARAMS{2});
115         In3=str2double(PARAMS{3});
116         In4=str2double(PARAMS{4});
117         In5=str2double(PARAMS{5});
118         In6=str2double(PARAMS{6});
119         In7=str2double(PARAMS{7});
120         In8=str2double(PARAMS{8});
121         In9=str2double(PARAMS{9});
122         InV=[In1 In2 In3 In4 In5 In6 In7 In8 In9];
123
124     elseif NIV==10
125
        Prompt={'In1','In2','In3','In4','In5','In6','In7','In8','In9','In10'};
```

```
126         Title='Input variables';
127     DefaultValues={'0.988','0.815','7.29','1','0.11','0','0.988','0.815','7.2
        9','1'};
128         PARAMS=inputdlg(Prompt,Title,[1 45],DefaultValues);
129         In1=str2double(PARAMS{1});
130         In2=str2double(PARAMS{2});
131         In3=str2double(PARAMS{3});
132         In4=str2double(PARAMS{4});
133         In5=str2double(PARAMS{5});
134         In6=str2double(PARAMS{6});
135         In7=str2double(PARAMS{7});
136         In8=str2double(PARAMS{8});
137         In9=str2double(PARAMS{9});
138         In10=str2double(PARAMS{10});
139         InV=[In1 In2 In3 In4 In5 In6 In7 In8 In9 In10];
140     else
141         disp("The number of input variables is not in the range of [1 10]");
142         return
143     end
144
145
146     tempH = (InW*InV')+BHI;
147
148     switch lower(ActivationFunction)
149         case{'s','sig','sigmoid'}
150             H = 1 ./ (1 + exp(-tempH));
151             Outputs = H'*OutW
152         case{'t','tanh'}
153             H = tanh(tempH);
154             Outputs = H'*OutW
155         case {'sin','sine'}
156             %%%%%%%% Sine
157             H = sin(tempH);
158             Outputs = H'*OutW
159         case {'hardlim'}
160             %%%%%%%% Hard Limit
161             H = double(hardlim(tempH));
162             Outputs = H'*OutW
163         case {'tribas'}
164             %%%%%%%% Triangular basis function
165             H = tribas(tempH)
166             Outputs = H'*OutW
167         case {'radbas'}
168             %%%%%%%% Radial basis function
169             H = radbas(tempH);
170             Outputs = H'*OutW
171     end
172
```

TABLE 7.2 Optimum values of user-defined Input for the original extreme learning machine and all examples.

Iteration	Iteration No.	NHN	Activation function	Orthogonal effect
Example 1	1000	12	Tanh	Yes
Example 2	10000	11	Sigmoid	Yes
Example 3	10000	11	Sigmoid	Yes
Example 4	500	16	Sigmoid	Yes
Example 5	2000	5	Sigmoid	Yes

7.4 Evaluating the effects of user-defined parameters on the modeling results of the extreme learning machine-based models

In this section, the effect of the user-defined parameters is investigated, including the regularization parameter for all developed ELM-based models (i.e., RELM, WRELM, ORELM), the weight function for the WRELM, and the maximum number of iterations for the ORELM. The other user-defined values are fixed based on the optimum values determined using the original ELM in Chapter 5. The values of these optimized user-defined inputs are presented in Table 7.2 for the convenience of the reader.

7.4.1 Regularization parameter

This section evaluates the performance of the RELM, WRELM, and ORELM models for different regularization parameters using Examples 1 to 5. It should be noted that the main parameters of the original ELM for all models are considered as provided in Table 7.2.

7.4.1.1 Optimum value of the regularization parameter for Example 1

The performance of the RELM, WRELM, and ORELM, as determined through a variety of statistical indices, is evaluated for a number of different regularization parameters for Example 1 (shown in Fig. 7.21). The regularization parameters considered for all models are 0.0001, 0.0005, 0.001, 0.005, 0.01, 0.05, 0.1, 0.2, 0.3, 0.4, and 0.5. It should be noted that there is no defined approach to the selection of the regularization parameter, and its value should be determined through a trial and error-approach. Recall that for the classical ELM model, the regularization parameter is 0, while for the developed ELM-based models its value should be selected as less than 1. From Fig. 7.10, we can observe for the RELM that as the value of the regularization parameter increases, there is a general trend for the variance accounted for (VAF), the Nash Sutcliffe error (NSE), and the Pearson correlation coefficient (R) to decrease. The lowest value for C does not have the best performance in these three indices, so that the best performance of correlation-based indices (R, VAF, NSE) is obtained at $C = 0.005$, $C = 0.0001$, and $C = 0.001$ for RELM, WRELM, and ORELM, respectively. It should be noted that although the global trend suggests that increasing regularization parameter value leads to a decrease in correlation accuracy, there are several instances where the increase in C leads to an increase in correlation accuracy on a local scale. The trends of changes for root mean square error (RMSE), normalized root mean square error (NRMSE), mean absolute error (MAE), mean absolute percentage error (MARE), and root mean squared relative error (RMSRE) indices are quite similar to the trend of correlation coefficients previously discussed. In this case, the overall trend shows a decrease in model accuracy as the regularization parameter increases, and the best performance is related to $C = 0.005$. The general trend of changes presented in the figure for the WRELM and ORELM methods is similar to RELM, such that the relationship between the regularization parameter and model accuracy is inversely related. It is noteworthy, however, that the optimal value of the regularization parameter in each method is not the same, so the best performance of the WRELM and ORELM is found at $C = 0.001$, which is a different value from RELM. The overall best for the RELM is found at $C = 0.005$. The difference between the $C = 0.005$ and $C = 0.005$ for RELM is insignificant. It should be noted that in the WRELM method, the value of $C = 0.05$ also has a performance very close to the optimal value.

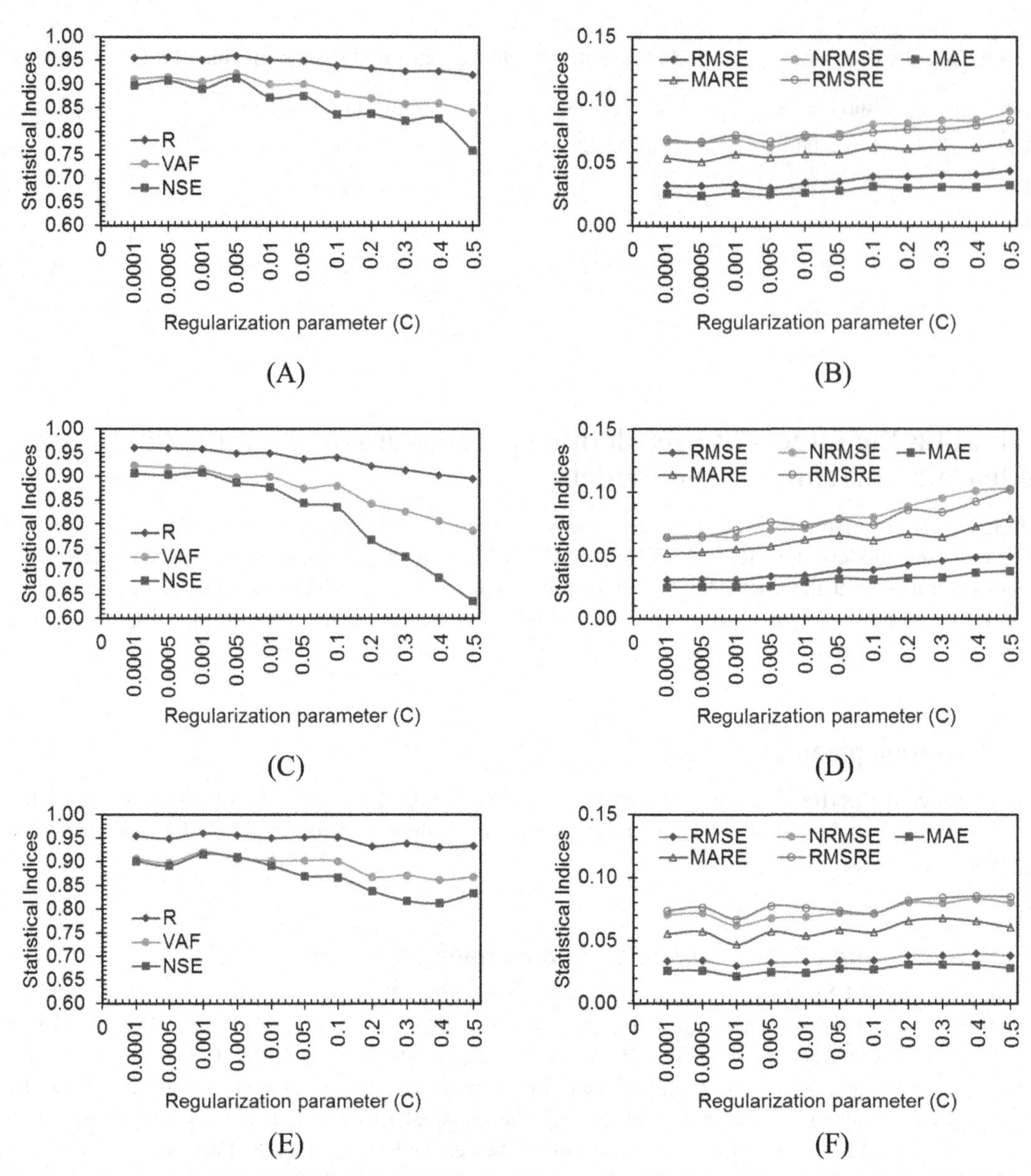

FIGURE 7.21 Performance of the RELM, WRELM, and ORELM with different regularization parameters for Example 1. (A) RELM—Correlation-based indices, (B) RELM—other indices, (C) WRELM—Correlation-based indices, (D) WRELM—other indices, (E) ORELM—Correlation-based indices, (F) ORELM—other indices.

7.4.1.2 Optimum value of the regularization parameters for Example 2

Fig. 7.22 demonstrates the performance of the RELM, WRELM, and ORELM when varying values of the regularization parameter are considered for Example 2. As shown in the figure, changes in modeling accuracy are not significantly related to changes in the regularization parameters. This can be stated due to the relatively stable values of the correlation coefficients and statistical indices as the value of the regularization parameter are varied. A zig-zag pattern can be observed in all models with only a small variation in model performance. Therefore, it is not possible to say exactly which value of C is related to the range of decrease or increase of modeling accuracy in RELM, WRELM, and ORELM methods. Considering the values of all indices, in general, it can be said that the optimal value of C in RELM, WRELM, and ORELM methods is equal to 0.2, 0.05, and 0.05, respectively. It can be seen that the optimal value of C, as in Example 1, is the same in both WRELM and ORELM methods.

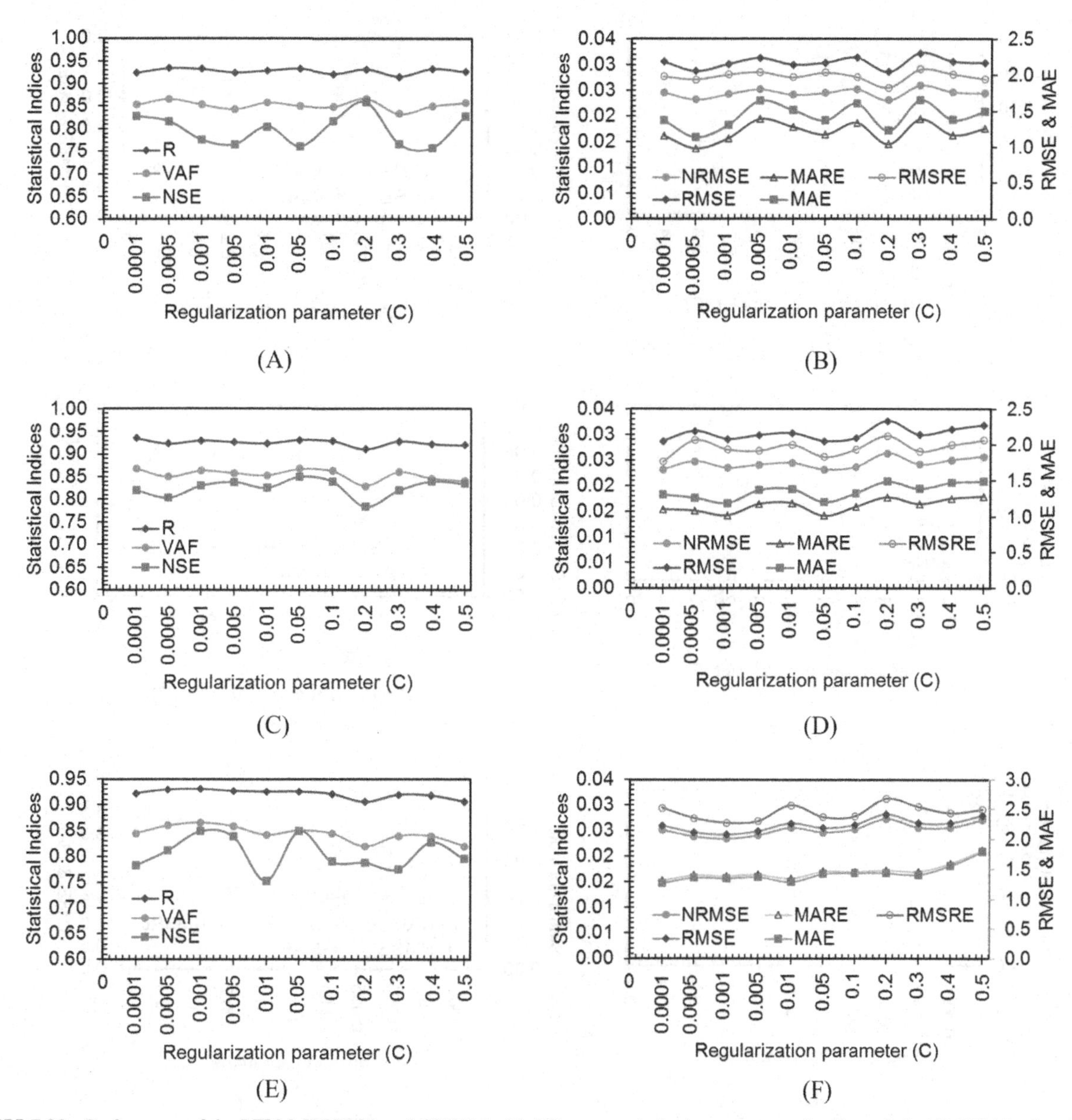

FIGURE 7.22 Performance of the RELM, WRELM, and ORELM with different regularization parameters for Example 2. (A) RELM—Correlation-based indices, (B) RELM—other indices, (C) WRELM—Correlation-based indices, (D) WRELM—other indices, (E) ORELM—Correlation-based indices, and (F) ORELM—other indices.

7.4.1.3 Optimum value of the regularization parameters for Example 3

Fig. 7.23 shows the performance of the RELM, WRELM, and ORELM versus the regularization parameter for Example 3. Similar to Example 2, the trend changes of the NRMSE, RMSE, MARE, MAE, and RMSRE is a zigzag trend that indicates the sensitivity of all models to the regularization parameter. A noteworthy point in all three ELM-based methods is that the NSE index decreases significantly with changes in the regularization parameter so that in the ORELM method, the value of this index is less than 0.2. In fact, changes in C lead to unsteady trends in model performance, suggesting the need to choose its value with high sensitivity. In the case of an incorrect selection of this parameter, whose value will change according to the nature of the data, the user's ability to estimate nonlinear problems using this model may be compromised.

7.4.1.4 Optimum value of the regularization parameters for Example 4

Fig. 7.24 indicates the performance of the RELM, WRELM, and ORELM with different regularization parameters for Example 4. The changes of the NRMSE index for all values of the regularization parameter in the RELM, WRELM,

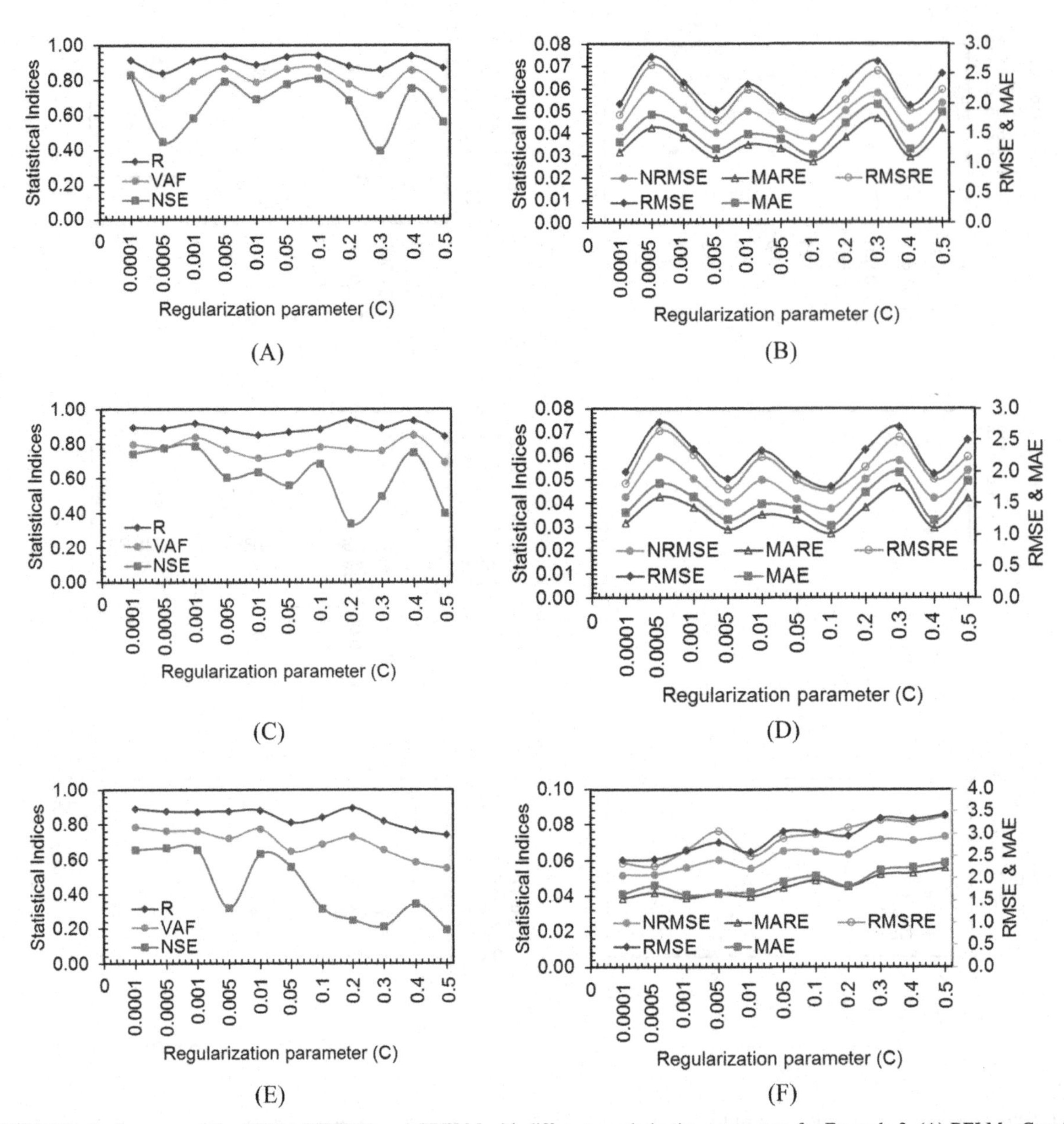

FIGURE 7.23 Performance of the RELM, WRELM, and ORELM with different regularization parameters for Example 3. (A) RELM—Correlation-based indices, (B) RELM—other indices, (C) WRELM—Correlation-based indices, (D) WRELM—other indices, (E) ORELM—Correlation-based indices, and (F) ORELM—other indices.

and ORELM is almost zero. In all methods, the trend of the relative error (i.e., RMSRE and MARE), as well as absolute indices (i.e., MAE and MARE), are relatively similar to each other. This figure shows that the most optimum value of the regularization parameter for the RELM, WERLM, and ORELM methods are 0.2, 0.01, and 0.5, respectively. It is observed that the obtained optimal values span an order of magnitude, however, still fall within the range of [0.0001, 0.5]. Therefore, it is strongly recommended to check the regularization parameters' values in a larger range if the desired accuracy is not achieved. Given that the accuracy of the ORELM model, in which the optimal value of C is equal to 0.5, is already acceptable (NSE = 0.95), increasing the value of C in this example is avoided.

7.4.1.5 Optimum value of the regularization parameters for Example 5

Fig. 7.25 shows the performance of the RELM, WRELM, and ORELM versus the regularization parameter for Example 5. Similar to the previous example, the trend of the NRMSE is relatively stable and does not show a clear tendency to favor any particular value of the regularization parameter. Similar to Examples 2—4, there is no clear trend in

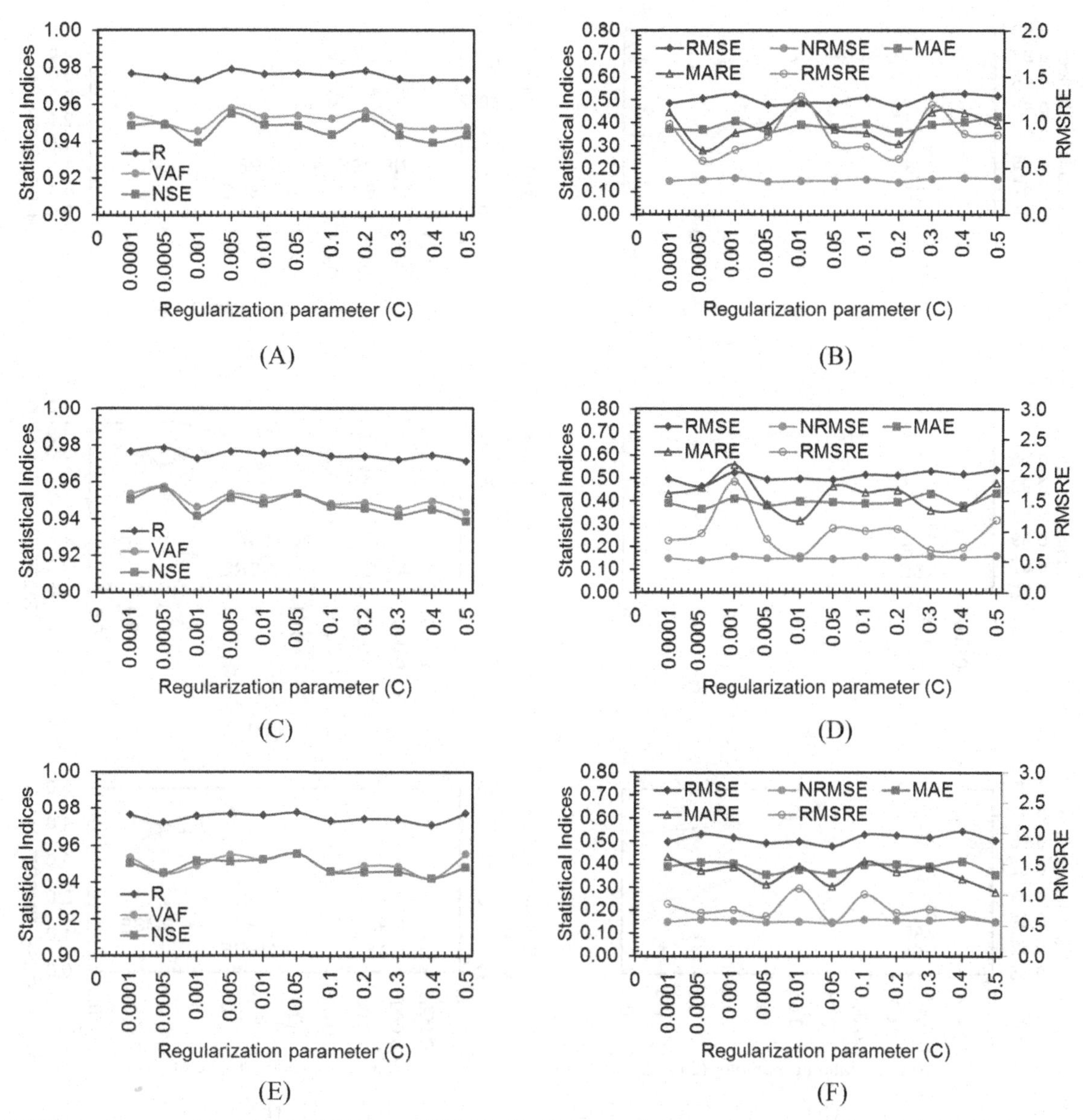

FIGURE 7.24 Performance of the RELM, WRELM, and ORELM with different regularization parameters for Example 4. (A) RELM—Correlation-based indices, (B) RELM—other indices, (C) WRELM—Correlation-based indices, (D) WRELM—other indices, (E) ORELM—Correlation-based indices, and (F) ORELM—other indices.

the model accuracy for different values of the regularization parameter. A comparison of the performance of different models using correlation-based indices shows that a decrease in accuracy is generally observed with increasing the value of C. However, this trend is not observed in other indices. For example, the minimum values of RMSRE in RELM, WRELM, and ORELM methods are 0.1, 0.001, and 0.0001, respectively.

7.4.1.6 General conclusion for the effect of regularization parameters at different examples

According to the above discussion regarding the selection of regularization parameters in five different example problems for the RELM, WRELM, and ORELM, it is concluded that there is no specific process in the quantification of the value of C. In order to arrive at an optimal value of this parameter, a trial and error investigation should be performed. It was observed in Example 3 that the model performance can be highly sensitive to the value of C in some cases, and incorrect selection of C can significantly reduce the model performance. For this reason, integration of evolutionary algorithms such as the genetic algorithm and particle swarm optimization may be necessary in order to achieve optimal values of this parameter. Practical experience has shown success in using these two algorithms for selecting the optimal C value.

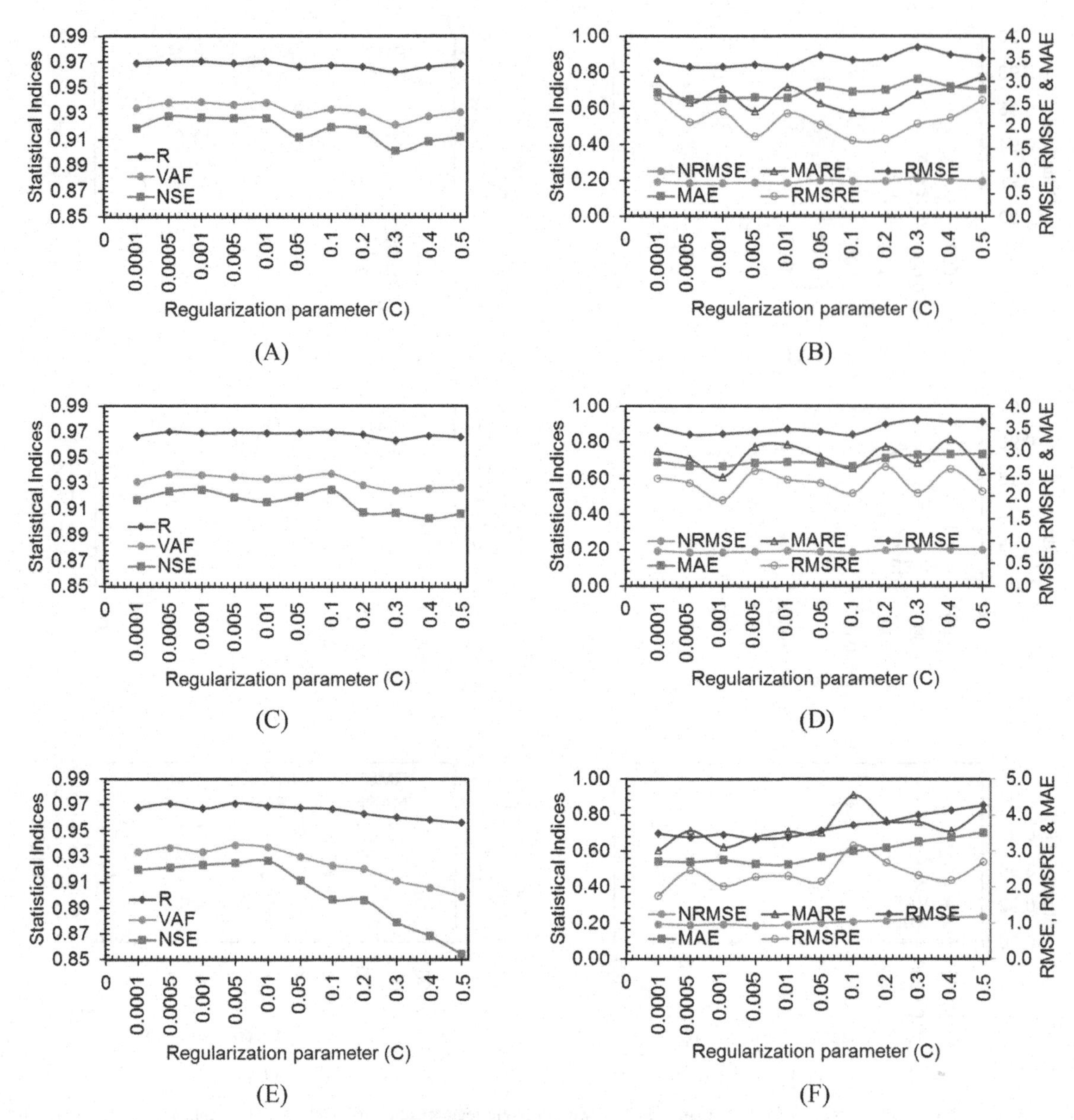

FIGURE 7.25 Performance of the RELM, WRELM, and ORELM with different regularization parameters for Example 5. (A) RELM—Correlation-based indices, (B) RELM—other indices, (C) WRELM—Correlation-based indices, (D) WRELM—other indices, (E) ORELM—Correlation-based indices, and (F) ORELM—other indices.

7.4.2 Weight function in the weighted regularized extreme learning machine

Fig. 7.26 evaluates the performance of the WRELM model when different weight functions are considered, as provided in Table 6.2. From Table 6.2 and Eq. (6.18), 10 weight functions will be considered in our analysis of the five examples. The weight functions to be evaluated are labeled as 1–10 in Fig. 7.26, which correspond to the use of the Bisquare, Huber, $\ell1$-norm, Andrews, Fair, Cauchy, Logistic, Talwar, and Welsch, respectively, in Eq. (6.18). From the statistical indices provided in the figure, it can be concluded that there is a significant impact of using different weight functions for each of the different examples. In Example 1, the best modeling performance is obtained when functions 6 and 7 are considered, while function number 7 is amongst the weakest results for Examples 2, 3, and 5. To compare the performance of the different functions, the best function for each of the examples is examined. For Example 1, the best-performing weight functions were numbers 6 and 7, while for Example 2 were functions 5 and 6. For Example 3, the best-performing weight functions were found to be numbers 6 and 9. It can be seen that function number 6 (Fair) has the best average performance in these three examples. In Example 4, it is difficult to select the best function

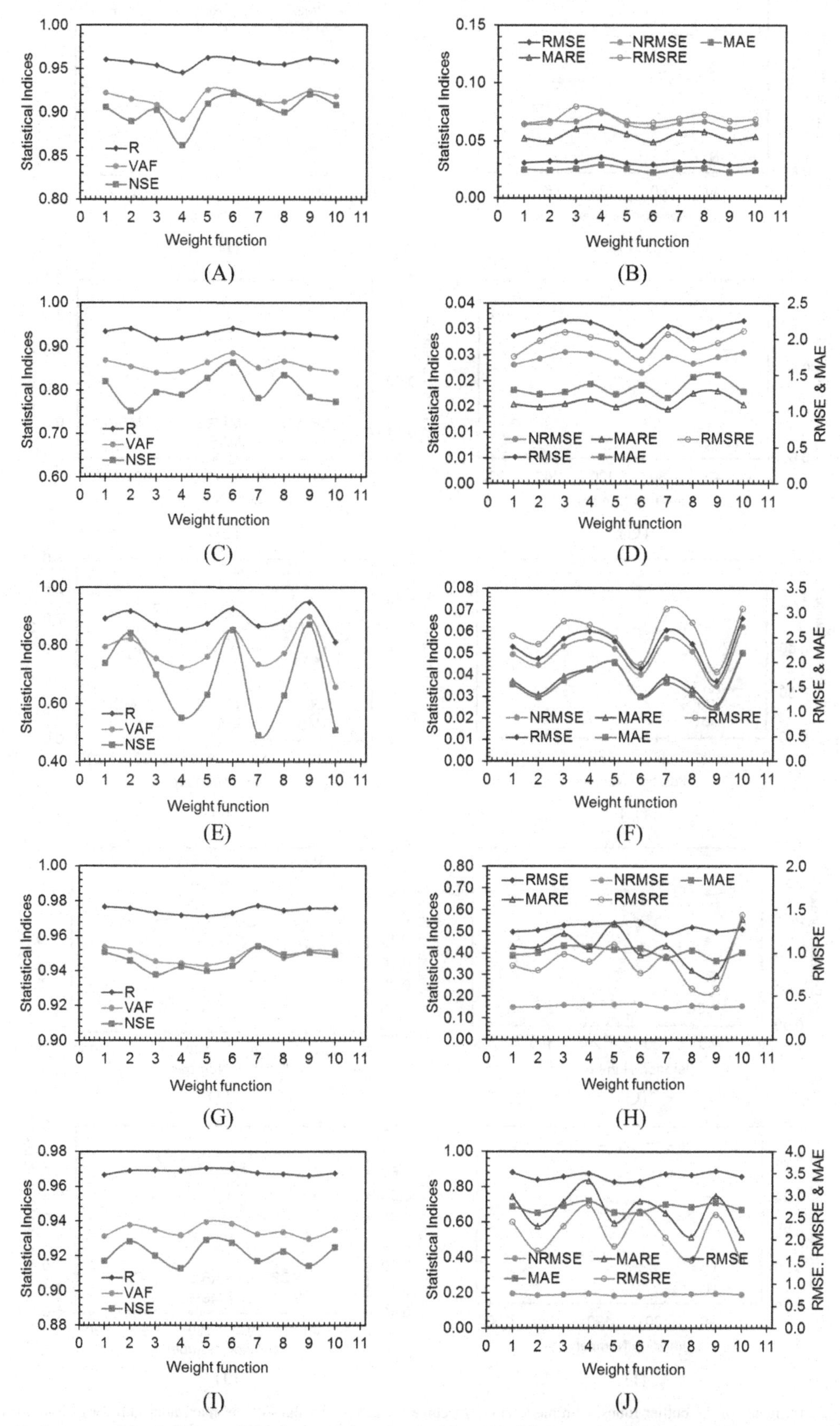

FIGURE 7.26 Performance of the weight regularized extreme learning machine with different weight functions. (A) Example 1—Correlation-based indices, (B) Example 1—Other indices, (C) Example 2—Correlation-based indices, (D) Example 2—Other indices, (E) Example 3—Correlation-based indices, (F) Example 3—Other indices, (G) Example 4—Correlation-based indices, (H) Example 4—Other indices, (I) Example 5—Correlation-based indices, and (J) Example 5—Other indices.

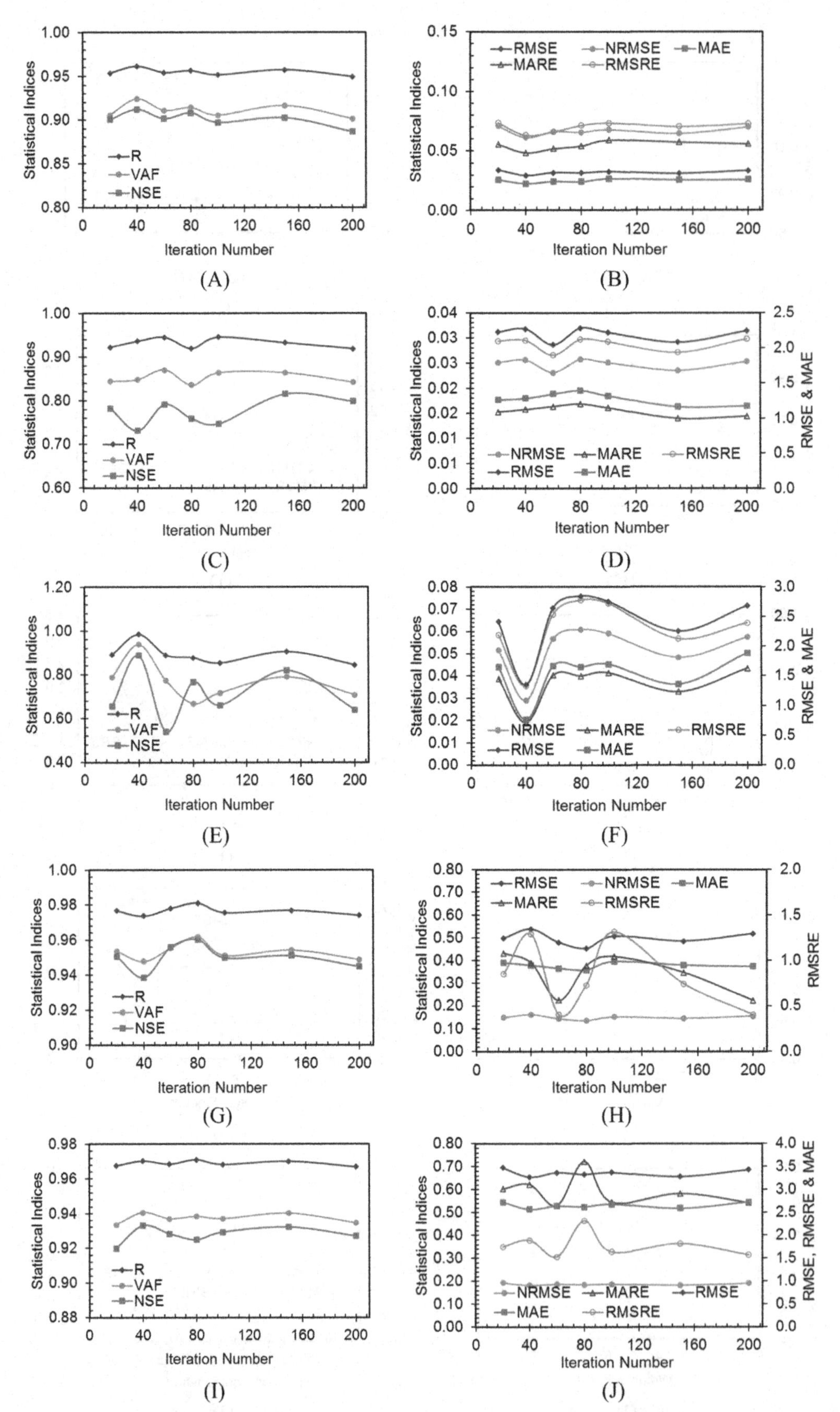

FIGURE 7.27 Performance of the outlier robust extreme learning machine for different values of the maximum number of iterations. (A) Example 1—Correlation-based indices, (B) Example 1—Other indices, (C) Example 2—Correlation-based indices, (D) Example 2—Other indices, (E) Example 3—Correlation-based indices, (F) Example 3—Other indices, (G) Example 4—Correlation-based indices, (H) Example 4—Other indices, (I) Example 5—Correlation-based indices, and (J) Example 5—Other indices.

because the results for different statistical indices yield conflicting results. This being said, functions 8, 9, and 3 can be selected as the general best-performing functions. Finally, in Example 5, functions 5 and 6 perform well, with function 5 being selected as the best. The result of comparing different weight functions in WRELM indicates that for different real-world problems, the performance of different functions must be examined to obtain the most optimal result. Also, in general, it can be said that function six, function, which had a fair performance for almost all examples, can be used as the starting point of a trial-and-error process for solving problems.

7.4.3 Maximum number of iterations in the outlier robust extreme learning machine

Fig. 7.27 demonstrates the performance of the ORELM when different values of the maximum iteration number are considered. This figure indicates that the change in the maximum number of iterations (MNIs) in Example 1 does not significantly affect the ORELM results, so that the amplitude of the NSE changes is (0.89, 0.91). Also, the mean absolute relative error (MARE) variation is less than 1%. In Example 2, the increase in MNI also does not show significant changes in the results of correlation-based indices (i.e., R, VAF, NSE), the relative indices (MARE, RMSRE), and other indices. Unlike Examples 1 and 2, the MNI effect is clearly seen in Example 3, so that the values of all indices have their best performance at MNI = 40. In Example 4, the best performance of correlation-based indices is MNI = 40, while in other indices, MNI = 60 and 200 provide the most accurate results. Similar to Example 4, the best performance of correlation-based indices and other indices is different in Example 5. It is not possible to accurately select a certain number of iterations as the optimal model. Therefore, it is suggested that the optimal MNI value in ORELM be selected using optimization algorithms and defining a multiindex objective function that considers several indices simultaneously.

7.5 Summary

As a result of the presence of outlying data points, limitations in the performance of the classical ELM model necessitate the development of novel approaches. Detailed MATLAB codings for three modified ELM-based models were developed to deal with outlying data points. The MATLAB codes are provided for all three of the developed ELM models, with each line of code being described in detail. One of the main advantages of the developed ELM-based codes in this chapter is introducing a calculator for the first time to apply the developed model to practical tasks. It should be noted that to employ the introduced calculator, there is no required knowledge of MATLAB or the mathematical background behind the ELM and its modified algorithms. Due to the fast training speed of the ELM, all of the developed ELM-based approaches can be run for a large number of iterations to find the optimum model. Moreover, the effect of the regularization parameters in the developed ELM-based model (i.e., RELM, WRELM, ORELM), the weight function in the WRELM, and the maximum number of iterations in the ORELM are also checked for five different examples with different numbers of samples (70–767) and input variables (two to six). After reading this chapter, readers can model using developed outlier-based ELM models in solving real-world problems. Also, given the experience gained through sensitivity analysis of the five real-world examples, the reader can easily use the ability of this method to model other issues for use in practical work.

Appendix 7.A Supporting information

Data on ELMs Model can be found in the online version at https://doi.org/10.1016/B978-0-443-15284-9.00005-7.

References

Adams, T. E., III, & Dymond, R. (2018). Evaluation and benchmarking of operational short-range ensemble mean and median streamflow forecasts for the Ohio River basin. *Journal of Hydrometeorology, 19*(10), 1689–1706. Available from https://doi.org/10.1175/JHM-D-18-0102.1.

Azimi, H., Bonakdari, H., & Ebtehaj, I. (2019). Design of radial basis function-based support vector regression in predicting the discharge coefficient of a side weir in a trapezoidal channel. *Applied Water Science, 9*(4), 78. Available from https://doi.org/10.1007/s13201-019-0961-5.

Azimi, H., Shabanlou, S., Ebtehaj, I., Bonakdari, H., & Kardar, S. (2017). Combination of computational fluid dynamics, adaptive neuro-fuzzy inference system, and genetic algorithm for predicting discharge coefficient of rectangular side orifices. *Journal of Irrigation and Drainage Engineering, 143*(7), 04017015. Available from https://doi.org/10.1061/(ASCE)IR.1943-4774.0001190.

Chen, H., He, L., Qian, W., & Wang, S. (2020). Multiple aerodynamic coefficient prediction of airfoils using a convolutional neural network. *Symmetry, 12*(4), 544. Available from https://doi.org/10.3390/sym12040544.

Diop, L., Samadianfard, S., Bodian, A., Yaseen, Z. M., Ghorbani, M. A., & Salimi, H. (2020). Annual rainfall forecasting using hybrid artificial intelligence model: integration of multilayer perceptron with whale optimization algorithm. *Water Resources Management*, *34*(2), 733–746. Available from https://doi.org/10.1007/s11269-019-02473-8.

Ebtehaj, I., & Bonakdari, H. (2016). Bed load sediment transport in sewers at limit of deposition. *Scientia Iranica*, *23*(3), 907–917. Available from https://doi.org/10.24200/sci.2016.2169.

Ebtehaj, I., & Bonakdari, H. (2017). Design of a fuzzy differential evolution algorithm to predict non-deposition sediment transport. *Applied Water Science*, *7*(8), 4287–4299. Available from https://doi.org/10.1007/s13201-017-0562-0.

Ebtehaj, I., Bonakdari, H., & Gharabaghi, B. (2019). A reliable linear method for modeling lake level fluctuations. *Journal of Hydrology*, *570*, 236–250. Available from https://doi.org/10.1016/j.jhydrol.2019.01.010.

Ebtehaj, I., Bonakdari, H., Moradi, F., Gharabaghi, B., & Khozani, Z. S. (2018). An integrated framework of extreme learning machines for predicting scour at pile groups in clear water condition. *Coastal Engineering*, *135*, 1–15. Available from https://doi.org/10.1016/j.coastaleng.2017.12.012.

Ebtehaj, I., Bonakdari, H., Safari, M. J. S., Gharabaghi, B., Zaji, A. H., Madavar, H. R., & Mehr, A. D. (2020). Combination of sensitivity and uncertainty analyses for sediment transport modeling in sewer pipes. *International Journal of Sediment Research*, *35*(2), 157–170. Available from https://doi.org/10.1016/j.ijsrc.2019.08.005.

Ebtehaj, I., Sammen, S. S., Sidek, L. M., Malik, A., Sihag, P., Al-Janabi, A. M. S., & Bonakdari, H. (2021a). Prediction of daily water level using new hybridized GS-GMDH and ANFIS-FCM models. *Engineering Applications of Computational Fluid Mechanics*, *15*(1), 1343–1361. Available from https://doi.org/10.1080/19942060.2021.1966837.

Ebtehaj, I., Soltani, K., Amiri, A., Faramarzi, M., Madramootoo, C. A., & Bonakdari, H. (2021b). Prognostication of shortwave radiation using an improved No-Tuned fast machine learning. *Sustainability*, *13*(14), 8009. Available from https://doi.org/10.3390/su13148009.

Elbadawi, M., Castro, B. M., Gavins, F. K., Ong, J. J., Gaisford, S., Pérez, G., & Goyanes, A. (2020). M3DISEEN: A novel machine learning approach for predicting the 3D printability of medicines. *International Journal of Pharmaceutics*, *590*, 119837. Available from https://doi.org/10.1016/j.ijpharm.2020.119837.

Erbir, M. A., & Ünver, H. M. (2021). The Do's and Don'ts for Increasing the Accuracy of Face Recognition on VGGFace2 Dataset. *Arabian Journal for Science and Engineering*, *46*(9), 8901–8911. Available from https://doi.org/10.1007/s13369-021-05693-6.

Essam, Y., Huang, Y. F., Ng, J. L., Birima, A. H., Ahmed, A. N., & El-Shafie, A. (2022). Predicting streamflow in Peninsular Malaysia using support vector machine and deep learning algorithms. *Scientific Reports*, *12*(1), 1–26. Available from https://doi.org/10.1038/s41598-022-07693-4.

Gholami, A., Bonakdari, H., Akhtari, A. A., & Ebtehaj, I. (2019). A combination of computational fluid dynamics, artificial neural network, and support vectors machines models to predict flow variables in curved channel. *Scientia Iranica*, *26*(2), 726–741. Available from https://doi.org/10.24200/sci.2017.4520.

Gholami, A., Bonakdari, H., Ebtehaj, I., & Akhtari, A. A. (2017). Design of an adaptive neuro-fuzzy computing technique for predicting flow variables in a 90 sharp bend. *Journal of Hydroinformatics*, *19*(4), 572–585. Available from https://doi.org/10.2166/hydro.2017.200.

Halpern-Wight, N., Konstantinou, M., Charalambides, A. G., & Reinders, A. (2020). *Training and testing of a single-layer LSTM network for near-future solar forecasting*, *10*(17), 5873. Available from https://doi.org/10.3390/app10175873.

Jahani, A., & Rayegani, B. (2020). Forest landscape visual quality evaluation using artificial intelligence techniques as a decision support system. *Stochastic Environmental Research and Risk Assessment*, *34*(10), 1473–1486. Available from https://doi.org/10.1007/s00477-020-01832-x.

Khozani, Z. S., Bonakdari, H., & Ebtehaj, I. (2017). An analysis of shear stress distribution in circular channels with sediment deposition based on Gene Expression Programming. *International Journal of Sediment Research*, *32*(4), 575–584. Available from https://doi.org/10.1016/j.ijsrc.2017.04.004.

Kumar, R., Singh, M. P., Roy, B., & Shahid, A. H. (2021). A comparative assessment of metaheuristic optimized extreme learning machine and deep neural network in multi-step-ahead long-term rainfall prediction for all-Indian regions. *Water Resources Management*, *35*(6), 1927–1960. Available from https://doi.org/10.1007/s11269-021-02822-6.

Mojtahedi, S. F. F., Ebtehaj, I., Hasanipanah, M., Bonakdari, H., & Amnieh, H. B. (2019). Proposing a novel hybrid intelligent model for the simulation of particle size distribution resulting from blasting. *Engineering with Computers*, *35*(1), 47–56. Available from https://doi.org/10.1007/s00366-018-0582-x.

Montes, C., Berardi, L., Kapelan, Z., & Saldarriaga, J. (2020). Predicting bedload sediment transport of non-cohesive material in sewer pipes using evolutionary polynomial regression–multi-objective genetic algorithm strategy. *Urban Water Journal*, *17*(2), 154–162. Available from https://doi.org/10.1080/1573062X.2020.1748210.

Nguyen, D. H., Le, X. H., Heo, J. Y., & Bae, D. H. (2021). Development of an extreme gradient boosting model integrated with evolutionary algorithms for hourly water level prediction. *IEEE Access*, *9*, 125853–125867. Available from https://doi.org/10.1109/ACCESS.2021.3111287.

Qasem, S. N., Ebtehaj, I., & Riahi Madavar, H. (2017a). Optimizing ANFIS for sediment transport in open channels using different evolutionary algorithms. *Journal of Applied Research in Water and Wastewater*, *4*(1), 290–298. Available from https://doi.org/10.22126/arww.2017.773.

Qasem, S. N., Ebtehaj, I., & Riahi Madavar, H. (2017b). Optimizing ANFIS for sediment transport in open channels using different evolutionary algorithms. *Journal of Applied Research in Water and Wastewater*, *4*(1), 290–298. Available from https://doi.org/10.22126/arww.2017.773.

Sihag, P., Esmaeilbeiki, F., Singh, B., Ebtehaj, I., & Bonakdari, H. (2019). Modeling unsaturated hydraulic conductivity by hybrid soft computing techniques. *Soft Computing*, *23*(23), 12897–12910. Available from https://doi.org/10.1007/s00500-019-03847-1.

Walton, R., Binns, A., Bonakdari, H., Ebtehaj, I., & Gharabaghi, B. (2019). Estimating 2-year flood flows using the generalized structure of the group method of data handling. *Journal of Hydrology*, *575*, 671–689. Available from https://doi.org/10.1016/j.jhydrol.2019.05.068.

Zeynoddin, M., Bonakdari, H., Ebtehaj, I., Azari, A., & Gharabaghi, B. (2020). A generalized linear stochastic model for lake level prediction. *Science of the Total Environment*, *138*, 015. Available from https://doi.org/10.1016/j.scitotenv.2020.138015.

Zhang, B., Li, W., Li, X. L., & Ng, S. K. (2018). Intelligent fault diagnosis under varying working conditions based on domain adaptive convolutional neural networks. *IEEE Access*, *6*, 66367–66384. Available from https://doi.org/10.1109/ACCESS.2018.2878491.

Zhang, X., Nguyen, H., Choi, Y., Bui, X. N., & Zhou, J. (2021). Novel extreme learning machine-multi-verse optimization model for predicting peak particle velocity induced by mine blasting. *Natural Resources Research*, *30*(6), 4735–4751. Available from https://doi.org/10.1007/s11053-021-09960-z.

Chapter 8

Online sequential non-tuned neural network—concept

8.1 Introduction

The extreme learning machine (ELM) is a unified framework for a single-layer feed-forward neural network (SLFFNN). In the current chapter, a new algorithm premised on the online sequential learning approach, known as the online sequential extreme learning machine (OSELM), is introduced for an SLFFNN. In general, the main learning framework of the OSELM is similar to the original ELM—the input weights and bias of hidden neurons are randomly allocated, and the output weights are analytically calculated. The difference between the algorithms of the OSELM and the original ELM is related to the calculation of the model output weight. To calculate this matrix of output weights, the ELM applied batch learning algorithms, while the OSELM considers an online learning approach. In this case, the training observations are sequentially presented to the learning algorithm, one by one or chunk-by-chunk (as a block of data). In the following section, a general background of the SLFFNN, especially the ELM, is reviewed. Following this, a detailed description of the batch learning approach utilized in the original ELM is presented, as well as the difference between batch learning and online learning. A brief review of the mathematical definition of the original ELM will be provided such that the main changes required to arrive at the OSELM are clear.

8.2 Main architectures of the single-layer feed-forward neural network

In recent decades, SLFFNNs have been used by many researchers across numerous scientific fields (Ebtehaj et al., 2016, 2019a; Azimi et al., 2017; Moeeni et al., 2017; Yaseen et al., 2018; Zeynoddin et al., 2018; Safari et al., 2019; Bonakdari et al., 2019a, 2019b) to solve real-world problems. There are two principal SLFFNN architectures commonly used in practice, which are distinguished by the type of hidden layer nodes: additive hidden nodes or those with radial basis functions (RBFs). The activation function for the RBF nodes can be any integrable piecewise continuous function, while for the additive nodes can be any bounded nonconstant piecewise continuous function (Huang et al., 2006a, 2019b).

Training algorithms are generally divided into two categories: batch-learning and online learning (Fig. 8.1). Most of the algorithms applied to train SLFFNN are of the batch-learning type. In this approach, model development is

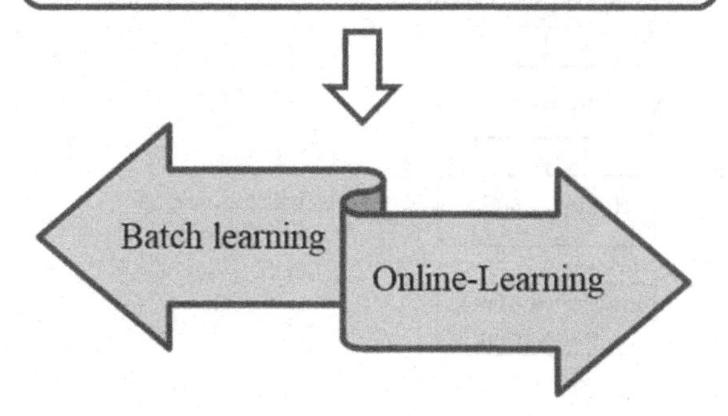

FIGURE 8.1 The structures of the single-layer feed-forward neural network training algorithms.

Machine Learning in Earth, Environmental and Planetary Sciences. DOI: https://doi.org/10.1016/B978-0-443-15284-9.00011-2

undertaken by considering the training data as a training batch, after which the testing data is used as a batch to evaluate the developed model performance. In contrast, in the online-learning approach model, training begins with an initial conjecture. One by one, the training samples are then used to calibrate the weights of the initial model.

8.2.1 The structure of batch and online learning algorithms

A flowchart displaying the structure of both batch and online learning algorithms is presented in Fig. 8.2. The differences between the two learning algorithms are clearly visible upon analysis of the schematic. In batch learning, the data is first divided into two categories: the training and testing samples. Model development begins by using the training samples, however, it should be noted that concurrent with model development, a percentage of training samples are used as validation samples against which the model performance is examined. After the development of the model, the performance is evaluated using the testing samples. In online learning, an initial model is first assumed, and the model is calibrated by entering the first case of the training samples. The performance of the model is then evaluated, and the model is re-calibrated using the next sample. This process continues until the last training sample has been evaluated.

8.2.2 The trade-off between applying online and batch learning algorithms

Some of the trade-offs between applying online and batch learning algorithms are as follows (Fig. 8.3):

(1) More space efficient and lower computational cost

In most online learning-based algorithms, a single pass on the entire data set is made during model development, while a multi-pass approach is employed for most batch learning-based algorithms. In most cases, algorithms premised on online learning are much faster than their equivalent algorithms developed based on batch learning. Furthermore, the required memory footprint for online learning-based approaches is much smaller than that of batch learning. In online learning, the data streams continuously into the learning algorithm and is used to update the developed model, while in batch learning, the model will re-consider all previous samples as well as the new data when updating the developed model.

(2) Easy to implement

Since the online learning-based model makes one pass over the data, the processing of data is one example at a time, sequentially.

(3) The production process is more difficult to maintain

In order to implement online learning-based algorithms, it is required that the algorithm is continuously passed data points for calibration and evaluation of the model.

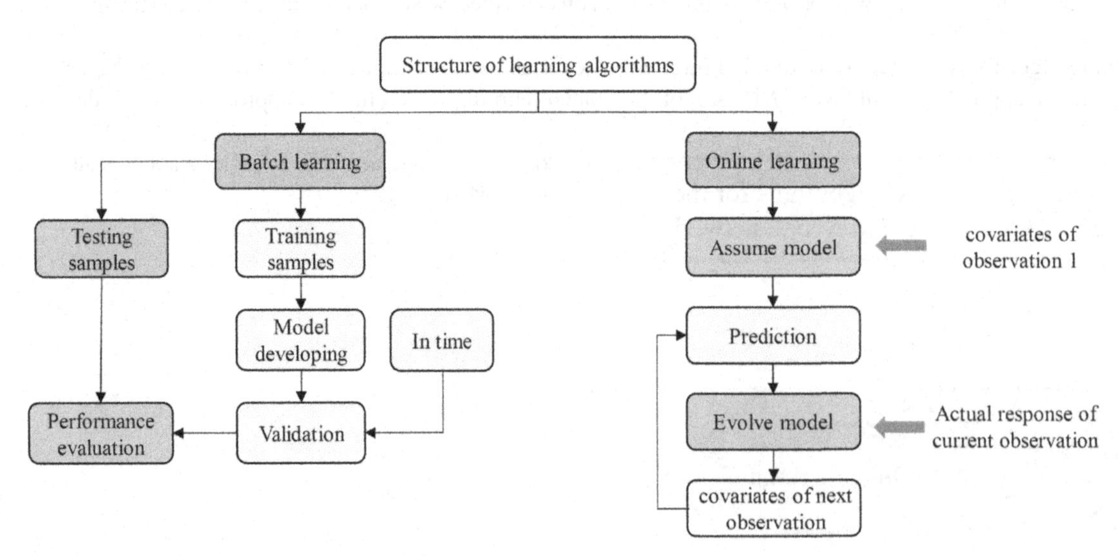

FIGURE 8.2 Structure of batch and online learning algorithms.

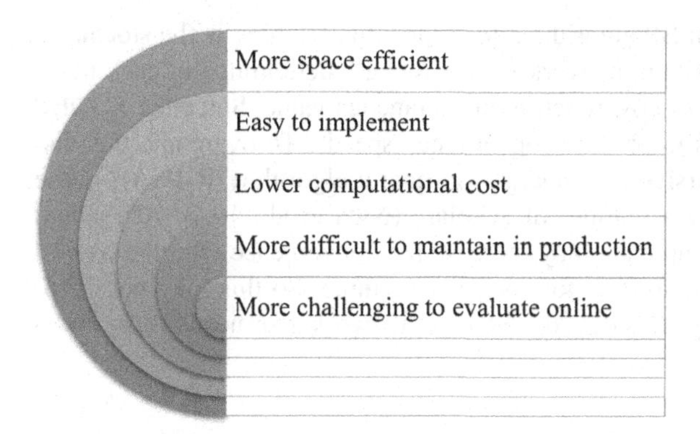

FIGURE 8.3 Trade-offs between applying online and batch learning algorithms.

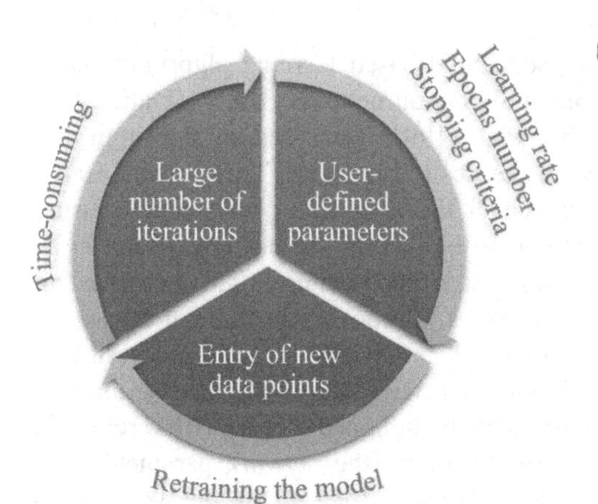

FIGURE 8.4 Main limitations of the batch learning algorithm.

(4) **More challenging to evaluate online**

In online learning-based algorithms, a subset of data cannot be separated and used as testing data because there are no distributional assumptions. Distributional assumptions mean that, if we were to separate any given subset of data from the total set of all samples, it would be assumed that this subset is representative of the entire population. It should be noted that testing data is generally applied to evaluate the performance of the calibrated model for unseen samples without recalibration of this model.

8.2.3 Principal limitations of batch learning

The main limitations of batch learning are summarized in Fig. 8.4. Given that modeling using batch learning-based algorithms may involve a large number of iterations, this method is relatively time-consuming. In fact, in most cases, the modeling time for batch learning-based algorithms can range from a few minutes to even several hours. In addition, the use of this learning method requires the user to determine the values of various parameters (i.e., stopping criteria, epochs number, learning rate, and other predefined parameters) to ensure convergence. Another consideration is that in batch learning, the entry of new data points requires that the new data be combined with the previous data to retrain the model, which itself requires a lot of time. In contrast, the developed model is recalibrated for online learning-based algorithms using only the new data point. Online learning-based algorithms have been widely used in many industrial applications where new data is generated at high rates and can be periodically added to the model as they are available. There is no need to retrain the model, rather the model needs only to be recalibrated to account for the new input data.

8.3 Development of the sequential-based learning algorithm

8.3.1 Stochastic gradient descent backpropagation

One of the most popular and widely used algorithms in SLFFNN training is the backpropagation (BP) algorithm (Ebtehaj et al., 2018, 2019b; Lillicrap & Santoro, 2019) and its variants with additive hidden nodes. BP is fundamentally a batch learning-based algorithm. One of the main BP algorithms for sequential learning is the stochastic gradient descent BP (SGDBP) (LeCun et al., 2012). In this algorithm, network parameters are determined in each iteration of the set of training iterations using first-order information of cost function instantaneous value. Since the SGDBP algorithm requires a large number of training samples, it suffers from slow convergence speeds. To overcome this disadvantage, the use of second-order information [such as the recursive Levenberg–Marquardt algorithm (RLMA)] in the learning process of the network parameters is recommended by a number of scholars (Ngia et al., 1998; Asirvadam et al., 2002; Shi et al., 2017). Although the second-order-based approach may increase the convergence rate, the overall modeling time will increase as the modeling process must be performed for each data sample. So this may pose problems in sequential learning in the case of data arriving quickly. Besides, the SGDBP network size needs to be fixed and predefined.

8.3.2 Resource allocation network

In order to train a feed-forward neural network (FFNN) with RBF nodes, sequential-based learning algorithms have been used extensively, one of the most popular of which include the resource allocation network (RAN) (Platt, 1991) and its developed methods (Suresh et al., 2010; Noor-A-Rahim et al., 2020). Platt (1991) generated an algorithm that assigns a new computational unit whenever unusual patterns are observed in the desired network. The network learns rapidly and easily by forming compact representations of unusual data. Whenever the algorithm observes an unusual data pattern, the allocator function identifies it and assigns a new unit to memorize the pattern. At any given point in its operation, the network possesses a current state which represents the cumulation of learned patterns. This network can be applied at any time during the learning process without repeating learning patterns. Moreover, the units of the generated network operate only on the local area of the input values space. As alluded to earlier, the learning processing in this network is done by regulating the parameters of the existing units and assigning new units in the case of unusual data patterns. In the event that the current patterns perform poorly, a new unit will be assigned to correct the response of the existing algorithm. In the case of acceptable performance of the current pattern, the network parameters are updated using the least mean square (LMS) algorithm. The authors compared the developed RAN with the well-known BP algorithm for forecasting the Mackey–Glass chaotic time series in terms of accuracy and training speed. They indicated that the RAN not only outperformed the BP algorithm in terms of accuracy but also found that the RAN had a faster speed than BP.

8.3.3 RAN training algorithm with an extended Kalman filter

Kadirkamanathan and Niranjan (1993) studied the problem of optimal sequential-based learning by considering the solution as a sequential estimation of an underlying function mapping to the optimal solution. First, the authors reach a suboptimum solution to the sequential approximation, which is then mapped by a growing Gaussian radial basis function (GRBF) network, which is similar to the RAN introduced by Platt (1991). The GRBF network adds hidden units for all samples. A growth criterion is developed to limit its growth using the function space approach. Second, Kadirkamanathan and Niranjan (1993) presented an enhanced version of the RAN algorithm. The authors replaced the LMS algorithm with an extended Kalman filter (EKF) algorithm, which is better suited to the function space approach used in determining the growth criterion. Comparative assessment of the RAN with the improved network demonstrated that the enhanced network, with fewer hidden units, outperformed RAN for time series forecasting and function approximation. The authors expressed that the improved network could be applied to the network with lower hidden neurons needed for a sequential-based learning problem.

8.3.4 The minimal resource allocation network

Yingwei et al. (1997) introduced a sequential-based learning algorithm for time series forecasting and function approximation using a minimal RBF neural network (RBFNN). The main idea of the developed algorithm was to combine the growth criterion of the RAN (Platt, 1991) with a pruning approach based on the relative contribution of each hidden unit to the overall network output. The comparison of the results of the RBFNN with those previous studies (Platt, 1991; Kadirkamanathan & Niranjan, 1993) using three different benchmarks show that RBFNN outperformed others.

Yingwei et al. (1998) compared the developed minimal RAN (M-RAN) with a multilayer FFNN trained with standard BP and dependence identification algorithm (Moody & Antsaklis, 1996). The comparison was made on several benchmark problems related to pattern classification and function approximation. The result of the comparison demonstrated better (or the same) classification/approximation precision with fewer hidden neurons and lower training time.

8.3.5 The growing and pruning radial basis function

Huang et al. (2004a) introduced a new algorithm for RBF networks referred to as the growing (Platt, 1991) and pruning (Yingwei et al., 1997), or (GAP)-RBF. The GAP-RBF was introduced considering a "significance" concept, where the significance of neuron links to the learning precision. The definition of this concept states that the significance of the desired neuron as the contribution of the neuron in the network output, on average, is greater than all the data received so far. An efficient and simple way to calculate the significance for uniformly distributed input data is done using a piecewise-linear estimation for the Gaussian function. The significance of the "nearest" neuron is considered for pruning and growing in the GAP-RBF algorithm. Comparison of the developed GAP-RBF with the sequential-based learning algorithms (Platt, 1991, Kadirkamanathan & Niranjan, 1993; Yingwei et al., 1997) for uniform and nonuniform input data indicates that the GAP-RBF significantly decreases computational cost, training time, and network size, simultaneously.

8.3.6 The generalized growing and pruning radial basis function

Huang et al. (2005) present the generalized GAP-RBF (GGAP-RBF) for RBF network learning using sequential learning algorithms. The authors employed the "significance" concept applied in the GAP-RBF (Huang et al., 2004a) to recognize parsimonious networks. Similar to GAP-RBF, the pruning and growing in this algorithm are defined based on connecting the required learning precision with the significance of the nearest neuron or by intentionally adding a new neuron. The significance of the neuron is a measure of the average information of the desired neuron. A comparison of the performance of the GGAP-RBF algorithm with that of the previously mentioned sequential-based learning algorithms demonstrates the superiority of the GGAP-RBF in terms of generalization performance, network size, and learning speed. The development of sequential-based learning algorithms is shown pictorially in Fig. 8.5. In this figure, the RAN is a resource allocation network, M-RAN is the minimal RAN, RANEKF is the RAN integrated with an EKF, GAP-RBF is the combination of the growing and pruning criterion as a new algorithm for the RBF network, and GGAP-RBF is the generalized GAP-RBF.

8.4 Main drawbacks of the classical sequential-based learning algorithms

Contrary to the SGDBP algorithm, the number of hidden nodes in the RAN RBF network and its developed algorithms is not predefined. In the RAN (Platt, 1991) and its variant based on the EKF (RANEKF) (Kadirkamanathan & Niranjan, 1993), the decision to add a new node is made according to the novelty of incoming data. The GGAP-RBF (Huang et al., 2005), GAP-RBF (Huang et al., 2004a), and M-RAN (Yingwei et al., 1998) consider additional nodes based on the novelty of the incoming data and prune insignificant nodes from the networks. The M-RAN, RANEKF, and RAN not only have slow learning speeds in the case of problems with large datasets but also require the input of several tunable control parameters (Huang et al., 2004a, 2005). The GAP-RBF (Huang et al., 2004a) and GGAP-RBF (Huang et al., 2005) use information regarding the distribution or range of the input samples to speed up the training

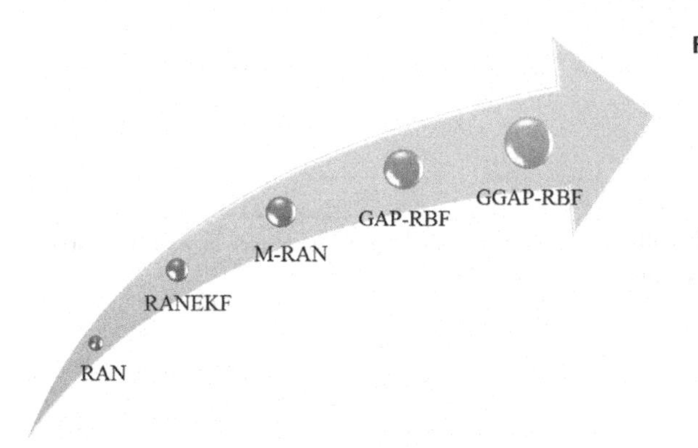

FIGURE 8.5 Development of sequential-based learning algorithms.

and simplify the algorithm. Therefore, the use of these two algorithms for large problems may be preferential as it is associated with reduced training speed (Liang et al., 2006).

The limitations of the existing sequential-learning-based algorithms are shown in Fig. 8.6. One of the main drawbacks of the existing sequential-learning-based algorithms is that they fundamentally consider data individually and not as a block of data, chunk-by-chunk. Furthermore, all sequential-based learning algorithms (BP-based and RAN-based) handle only a specific kind of hidden node (RBF or additive) and cannot handle both of them together.

8.5 Introduction to the online sequential extreme learning machine

The online sequential extreme learning machine (OSELM) (Liang et al., 2006) is a new algorithm that handles both RBF and additive hidden nodes in an integrated framework. The OSELM is capable of learning data one by one or chunk-by-chunk, with varying or fixed segment lengths. Another advantage of this algorithm is its ability to discard previously seen data through the training phase (sequential-based learning). The main features of the OSELM as a versatile sequential learning algorithm are as follows:

1. This algorithm has the capability to consider training samples one by one or chunk-by-chunk. The segment length of the chunks can be fixed or variable.
2. At any time, only the new observations (single or chunk) are learned.
3. The learned data (single or chunk) is discarded after completing the learning procedure.
4. No prior knowledge of the total number of training observations is available to the learning system.

The main feature of the OSELM algorithms is summarized in Fig. 8.7.

The OSELM emanates from the batch-based learning of the original ELM (Huang et al., 2004b, 2006a, 2006b, 2006c, 2006d; Huang & Siew, 2004), which was introduced for SLFFNNs with both RBF and additive hidden nodes. Comparison of the original ELM with BP as a classical learning algorithm indicates the higher performance of the original ELM in terms of the accuracy and training speed for both regression and classification-based problems (Huang & Siew, 2004; Huang et al., 2004b, 2006b, 2006c, 2006d). Two matrices are randomly assigned for both RBF and additive hidden nodes, while the output weight matrix is analytically determined through the training phase. In the OSELM with RBF hidden nodes, the centers and widths of nodes are randomly assigned, while in the OSELM with additive hidden neurons, the input weights and bias of hidden neurons are randomly assigned (Fig. 8.8). Unlike the sequential-based learning algorithms, only one parameter (i.e., number of hidden nodes) should be specified before modeling. In this way, the OSELM reduces the number of user-specified input variables that are required for input.

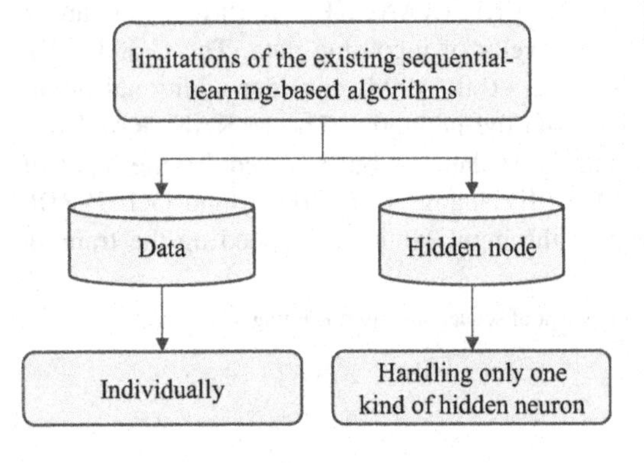

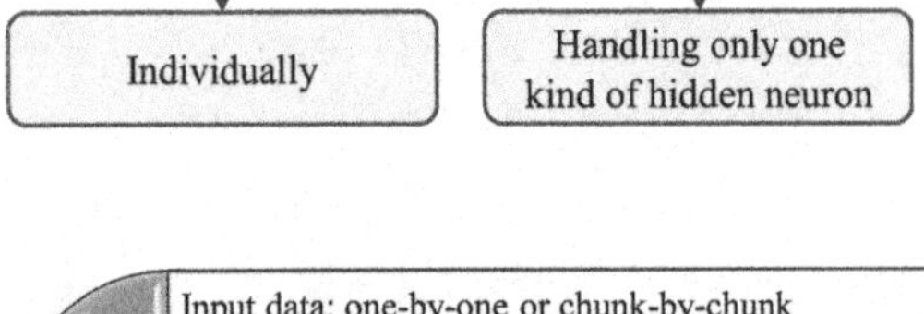

FIGURE 8.6 Limitations of the existing sequential-learning-based algorithms.

Input data: one-by-one or chunk-by-chunk
No prior knowledge regarding the number of observations
Only newly arrived observations are seen and learned
The observations are discarded after completing learning process

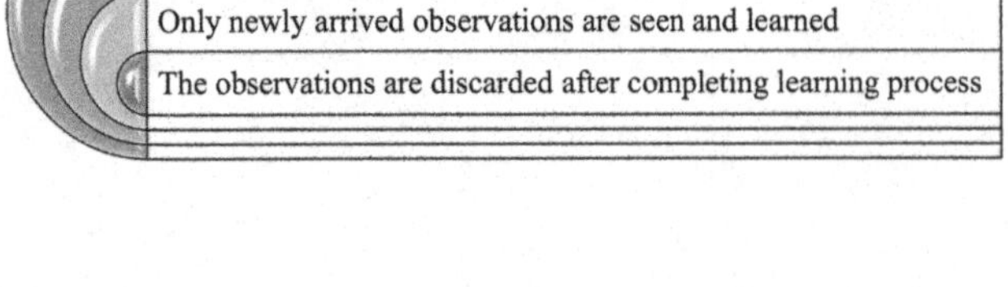

FIGURE 8.7 Main features of the online sequential extreme learning machine.

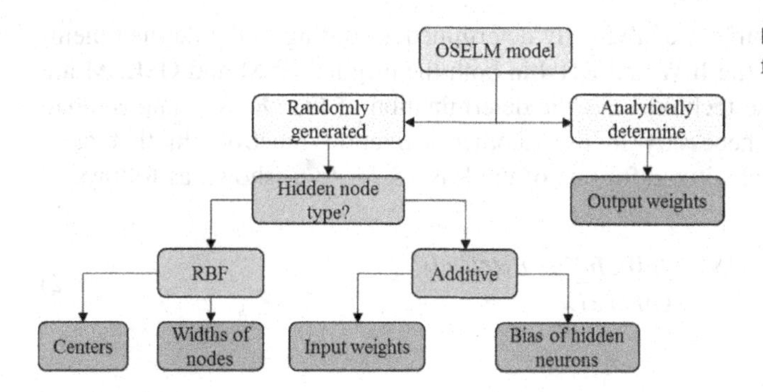

FIGURE 8.8 The online sequential extreme learning machine matrices for RBF and additive hidden nodes.

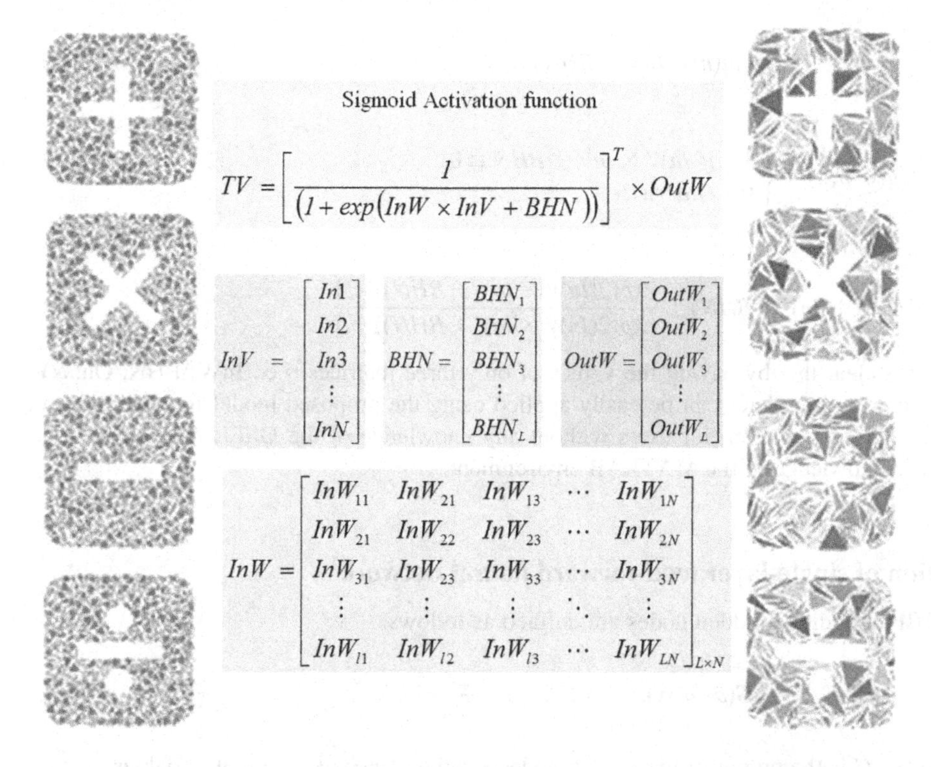

FIGURE 8.9 Schematic of the introduced calculator.

8.6 Mathematical description of the online sequential extreme learning machine

This section presents the general description of the batch-based ELM introduced by Huang et al. (Huang et al., 2004b, 2006a, 2006b, 2006c, 2006d; Huang & Siew, 2004) to provide the necessary foundation for the development of OSELM. First, a general mathematical explanation of SLFFNNs integrating both RBF and additive hidden nodes is presented. Following this, the mathematical description of the ELM model is provided. Finally, the developed OSELM algorithm is presented in detail in the final subsection.

This section is intended for both advanced and beginner users. Using the proposed calculator tool, beginner users can easily employ the developed code to solve nonlinear problems. With this calculator, the OSELM can be employed without any knowledge of the ELM and OSELM and with only a basic knowledge of MATLAB (Fig. 8.9).

Fig. 8.9 shows the schematic of the developed calculator for practical tasks. It should be noted that this figure considers the case where the Sigmoid function is applied as the activation function with the following form:

$$TV = \left[1/(1+exp(InW \times InV + BHN))\right]^T \times OutW \tag{8.1}$$

Here, TV is the target variable, InW, InV, BHN, and OutW are the input weight, input variable, bias of hidden neurons, and output weight, respectively. The InV matrix is known because it includes the training samples for all input variables, while the other matrices (i.e., InW, BHN, and OutW) are tuned through the training phase. The InW and

BHN matrices are randomly allocated, while the OutW matrix is analytically determined according to the defined methodology in different ELM-based techniques. For example, the InW and BHN in both the original ELM and OSELM are randomly generated, and the only difference between these techniques is the determination of the OutW. Suppose that, instead of using the Sigmoid activation function, it is necessary to use another activation function. In that case, Eq. (8.1) provided for the Sigmoid activation function is substituted for one of the Eqs. (8.2)–(8.6) shown as follows:

Triangular basis function:

$$TV = \begin{cases} 1 - |InW \cdot InV + BHN| & InW \cdot InV + BHN \geq 0 \\ 0 & Otherwise \end{cases} \tag{8.2}$$

Radial basis function:

$$TV = exp(-(InW \cdot InV + BHN)^2) \tag{8.3}$$

Sine function

$$TV = sin(nW \cdot InV + BHN) \tag{8.4}$$

Hard-limit function:

$$g(w, b, x) = \begin{cases} 1 & if\ InW \times InV + BHN \geq 0 \\ 0 & Otherwise \end{cases} \tag{8.5}$$

Tangent hyperbolic function:

$$TV = tanh(InW \times InV + BHN) = \frac{exp(2(InW \times InV + BHN)) - 1}{exp(2(InW \times InV + BHN)) + 1} \tag{8.6}$$

From the provided explanation, it is clear that by having the values of only three matrices (i.e., InW, BHN, OutW) plus the values of each input parameter, the calculator can be easily applied using the proposed model for unseen data without the need for any calibration. This is true even for users without any knowledge of the OSELM mathematical formulation and who have only a basic knowledge of the MATLAB environment.

8.6.1 Mathematical explanation of single-layer feed-forward neural network[1]

The output of an SLFFNN with L RBF or additive hidden nodes are defined as follows:

$$f_L(\mathbf{x}) = \sum_{i=1}^{L} \beta_i G(a_i, b_i, x), \quad x \in \mathbf{R}^n, \quad a_i \in \mathbf{R}^n \tag{8.7}$$

where L is the number of hidden nodes, G is the output of the hidden node, x is the vector of the input variables, a and b (which are known as learning parameters) are the input weights and the bias of the hidden nodes, respectively. Considering additive hidden nodes with $g(x)$ as an activation function (for example, Sigmoid), the output of the hidden node (i.e., G) is given as:

$$G(a_i, b_i, x) = g(a_i \cdot x + b_i), \quad b_i \in R \tag{8.8}$$

where a is the input weight, which links the input variables (x) to the hidden nodes, b is the bias of the hidden node and $a \cdot x$ is the inner product of the a and x in $\mathbf{R}^n$. For the RBF hidden node with $g(x)$ as an activation function (e.g., Sigmoid), the output of the hidden layer (i.e., G) is as follows:

$$G(a_i, b_i, x) = g(b_i \| x - a_i \|), \quad b_i \in R^+ \tag{8.9}$$

Here, b_i and a_i are the impact factor and center of the ith RBF node, respectively. The R^+ notation is used to denote the space of all positive real values. The inputs of all the RBF nodes include their impact factor and centroid, while their outputs are estimated using the radially symmetric function describing the distance between the node centroid and the input.

1. This section is provided for the advance users.

8.6.2 A quick overview of the mathematical definition of the extreme learning machine[2]

To train a model using supervised batch-based learning algorithms, a finite number of input-output samples are used as a training dataset. Consider N arbitrary discrete training samples as $(\mathbf{x}_i, \mathbf{t}_i) \in \mathbf{R}^n \times \mathbf{R}^m$. Here, the $\mathbf{x}_i$ is the input variable which is a $n \times 1$ vector (n is the number of input variables). $\mathbf{t}_i$ is considered to be the output variable, which is an $m \times 1$ vector (m is the number of output variables). In order to approximate the N training output variable(s), a SLFFNN with L hidden nodes should be considered. It is therefore implied that there are a_i, b_i, and β_i such that:

$$f_L(\mathbf{x}_j) = \sum_{i=1}^{L} \beta_i G(a_i, b_i, \mathbf{x}_j) = \mathbf{t}_j, \quad j = 1, 2, \ldots, N \tag{8.10}$$

where L is the number of hidden nodes, β_i is the weight that connects the ith hidden node to the output node, a_i is the weight that connects the inputs to the ith hidden node, $\mathrm{b_i}$ is the bias of the ith hidden nodes, t_j is the output variable(s), G is the output of the hidden node, and N is the number of training samples. The compact matrix-based form of the above-mentioned equation is as follows:

$$\mathbf{H}\beta = \mathbf{T} \tag{8.11}$$

where

$$\mathbf{H}(a_1, a_2, \ldots, a_L, b_1, b_2, \ldots, b_L, \mathbf{x}_1, \mathbf{x}_2, \ldots, \mathbf{x}_N) = \begin{bmatrix} G(a_1, b_1, \mathbf{x}_1) & \cdots & G(a_L, b_L, \mathbf{x}_1) \\ \vdots & \ddots & \vdots \\ G(a_1, b_1, \mathbf{x}_N) & \cdots & G(a_L, b_L, \mathbf{x}_N) \end{bmatrix}_{N \times L} \tag{8.12}$$

$$\beta = \begin{bmatrix} \beta_1^T \\ \vdots \\ \beta_L^T \end{bmatrix}_{L \times m} \tag{8.13}$$

$$\mathbf{T} = \begin{bmatrix} \mathbf{t}_1^T \\ \vdots \\ \mathbf{t}_N^T \end{bmatrix}_{N \times m} \tag{8.14}$$

It should be noted that m in Eqs. (8.13) and (8.14) is the number of the input variables. Therefore, if the problem is a single output, the value of this parameter is equal to one, and these two matrices [Eqs. (8.13)–(8.14)] are represented as single-column matrices. The $\mathbf{H}$ matrix provided in Eq. (8.12) is known as the output matrix of the hidden layer (Huang, 2003). The jth row of the hidden layer output matrix is equal to the output vector of the jth hidden layer with respect to the input x_j, while the ith column of the hidden layer output matrix is equal to the output vector of the ith hidden nodes with respect to inputs ($x_1, \ldots, x_N$).

Two fundamental principles are considered in the training and development of the ELM:

1. If the number of training samples exceeds the number of hidden layer nodes ($N > L$), the values of the hidden nodes (input weights and biases of hidden nodes for additive SLFFNN or centers and impact factor for RBF SLFFNN) are randomly assigned. By randomly determining the values of these two matrices, the value of the hidden layer output matrix can be easily calculated. Hence, the value of output weights is calculated using the pseudo-inverse of $\mathbf{H}$ to achieve a small positive error ($\varepsilon > 0$). As a result of the random allocation of the input weight and the bias of the hidden neurons, the only unknown in Eq. (8.11) is the output weight, β. In this case, a lengthy training process is avoided as only one unknown matrix must be isolated. Indeed, the nonlinear problem is solved through the resolution of a simple linear system.
2. If the number of hidden nodes and training samples are equal ($N = L$), then similar to the previous condition, the hidden node parameters (input weights and biases of hidden nodes for additive SLFFNN or centers and impact factor for RBF SLFFNN) are randomly assigned. With the allocation of the input weights and bias of the hidden nodes, the output weights values can be easily calculated by inverting the hidden node output matrix, $\mathbf{H}$. In this case, the output weight calculation is easily done in a one-step linear process, and there is no need for a lengthy training process where the network parameters (i.e., learning epochs, learning rates, etc.) must be determined prior to modeling.

2. This section is provided for the advance users.

Theorem I: Consider a SLFFNN with an activation function, g(x), which is infinitely differentiable over any interval of $\mathbf{R}$ with RBF or additive hidden nodes. For N input vectors $\{x_i | x_i \in \mathbf{R}^n, i = 1, 2..., N\}$, where n is the number of input variables and $\{(a_i, b_i)\}_{i=1}^{L}$ are the learning parameters (which are randomly generated with any continuous probability distribution), the output matrix of the hidden layer ($\mathbf{H}$) is invertible with a probability of one (Huang et al., 2006a).

This theorem indicates that an SLFFNN with L hidden nodes (RBF or additive) that are randomly produced has the ability to learn N distinct training samples with zero error.

Given that the number of hidden nodes (L) is generally less than the number of training samples (N) in a real application, the modeling error cannot be exactly zero. This fact is expressed in the following theory (Huang et al., 2006a).

Theorem II: For any small positive value ($\varepsilon > 0$) and an infinitely differentiable activation function ($g(x)$), there exists $N \geq L$ such that for N training input samples $\{x_i | x_i \in \mathbf{R}^n, i = 1, 2..., N\}$ and $\{(a_i, b_i)\}_{i=1}^{L}$ randomly produced according to a continuous probability distribution, $\|\mathbf{H}_{N\times L}\beta_{L\times m} - \mathbf{T}_{N\times m}\| < \varepsilon$ with probability one. As a result of Theorem I, if there exists $L \geq N$, then the training error is exactly zero such that $\|\mathbf{H}_{N\times L}\beta_{L\times m} - \mathbf{T}_{N\times m}\| = 0$. According to Theorem II, when the number of training samples, N, is greater than the number of hidden nodes, L, then $\|\mathbf{H}_{N\times L}\beta_{L\times m} - \mathbf{T}_{N\times m}\| < \varepsilon$. It should be noted that m is the number of output variables, and therefore, for the commonly encountered multiinput-single-output problem, m is equal to one.

In accordance with the above Theorems, the RBF (impact factors and centers) and additive (input weights and bias of hidden neurons) hidden nodes of the SLFFNN are randomly allocated through the training phase. Thus, Eq. (8.11) becomes a linear equation with the output weight (β) being estimated through a linear system as follows:

$$\beta = \mathbf{H}^{+}\mathbf{T} \tag{8.15}$$

Here, the $\mathbf{H}^{+}$ is the Moore–Penrose generalized inverse (MPGI) of $\mathbf{H}$ (Rao & Mitra, 1971). To estimate the MPGI of a matrix, the iterative method, orthogonalization method, orthogonal projection method (OPM), and singular value decomposition (SVD) could be applied (Rao & Mitra, 1971). Due to the use of searching and iterations, both orthogonalization and iterative methods have limitations. When $\mathbf{H}^{+} = (\mathbf{H}^T\mathbf{H})^{-1}\mathbf{H}^T$ and $\mathbf{H}^T\mathbf{H}$ are nonsingular, the OPM can be applied. However, in many practical situations, $\mathbf{H}^T\mathbf{H}$ tends to become singular, and as a result, the OPM may not perform well in all practical applications. Since the SVD method can be applied for all possible situations to estimate the MPGI of a matrix (i.e., $\mathbf{H}$ in the current study), it is considered in the implementation of Eq. (8.15) for ELM.

According to the proposed theorems, in order to calculate output weights using Eq. (8.15), training samples must be available, as well as the value of the output of the hidden nodes ($\mathbf{H}$). The output of the hidden nodes is dependent on the matrix of input weights and the bias of the hidden neurons, which are randomly generated for both additive and RBF hidden nodes. Therefore, the calculation of output weights using training samples and the output of the hidden nodes ($\mathbf{H}$) is easily done in a linear process. According to Eq. (8.15), which is provided to calculate output weights in the original ELM, all training samples could be considered to estimate output weights. Therefore, the ELM is defined as a batch-based learning approach. Huang et al. (2006b) applied an incremental method to analyze the universal approximation capability of the original ELM. In this study, the authors applied batch-based learning and incrementally (one by one) increased the number of hidden nodes within the network. The results of this last study implied that SLFFNNs using randomly generated RBF or additive hidden nodes with piecewise continuous activation functions can estimate any continuous function. In the original ELM implementation, the activation functions for RBF and additive hidden nodes can be any integrable and bounded nonconstant (respectively) piecewise continuous functions.

8.6.3 Mathematical definition of the online sequential extreme learning machine[3]

The original ELM, described in detail in the previous section, uses all N training samples to calibrate the model in one step. Indeed, the training methodology employs the batch-based learning procedure described at the beginning of this chapter. In many real-world practical applications (i.e., time series-based data), the training samples may be obtained one by one or chunk by chunk. Furthermore, as more data from the process we are attempting to model becomes available, it is desirable to update the model to best reflect the current information we have at our disposal. Therefore, the developed model may need to be recalibrated throughout the modeling process in order to consider new samples as

3. This section is provided for the advance users.

they become available to us. To account for this re-calibration, the online sequential-based learning process of the ELM (i.e., OSELM) is developed in the current section.

The least-squares solution of Eq. (8.11) was provided through consideration of Eq. (8.15), where the output weight matrix (β) is determined. In the OSELM, the rank of the output of the hidden nodes is considered to be the number of hidden nodes (i.e., Rank(**H**) = L). The rank of a matrix $b \times d$ is a nonnegative integer and cannot be greater than b or d. In other words, $rank(A) \leq \min(b, d)$. Using the condition (Rank(**H**) = L), the MPGI ($\mathbf{H}^+$) of Eq. (8.15) is given as follows:

$$\mathbf{H}^+ = (\mathbf{H}^T\mathbf{H})^{-1}\mathbf{H}^T \tag{8.16}$$

$\mathbf{H}^+$ in the above equation is termed the "left pseudo-inverse of H" – this is based on the fact that $\mathbf{H}^+\mathbf{H} = \mathbf{I_L}$, where $\mathbf{I_L}$ is a unit matrix with the dimension of L.

If $\mathbf{H}^T\mathbf{H}$ oves towards singularity, it can be made nonsingular by decreasing the network size (increasing the number of hidden layer nodes, L) or increasing the number of training samples (i.e., N). By substituting Eq. (8.16) into Eq. (8.15), the network output weight is calculated as follows:

$$\beta = (\mathbf{H}^T\mathbf{H})^{-1}\mathbf{H}^T\mathbf{T} \tag{8.17}$$

Eq. (8.17) represents the least-square solution (LSS) of Eq. (8.11). If the LSS given by Eq. (8.17) is implemented sequentially, then the OSELM model is produced.

Consider a block of training samples as the initial training input, $N_0 = \{(\mathbf{x}_i, \mathbf{t}_i)\}_{i=1}^{N_0}$ with $N_0 > L$. To implement modeling using the batch-based original ELM, the training process must simply minimize the following equation in order to calculate the output weight:

$$E_{OSELM} = \min \|\mathbf{H_0}\beta - \mathbf{T_0}\| \tag{8.18}$$

where

$$\mathbf{T}_0 = \begin{bmatrix} \mathbf{t}_1^T \\ \vdots \\ \mathbf{t}_{N_0}^T \end{bmatrix}_{N_0 \times m} \tag{8.19}$$

$$\mathbf{H}_0 = \begin{bmatrix} G(a_1, x_1, b_1) & \cdots & G(a_L, x_1, b_L) \\ \vdots & \ddots & \vdots \\ G(a_1, x_{N_0}, b_1) & \cdots & G(a_L, x_{N_0}, b_L) \end{bmatrix}_{N_0 \times L} \tag{8.20}$$

It should be noted that m is the number of output variables, so that for a problem with only one output, its value is equal to one.

According to Theorem II, the solution of Eq. (8.18) is as follows:

$$\beta^{(0)} = \mathbf{K}_0{}^{-1}\mathbf{H}_0{}^{-T}\mathbf{T}_0 \tag{8.21}$$

where

$$\mathbf{K}_0 = \mathbf{H}_0{}^T\mathbf{H}_0 \tag{8.22}$$

Consider another block of data with N_1 samples given by $N_1 = \{(\mathbf{x}_i, \mathbf{t}_i)\}_{i=N_0+1}^{N_0+N_1}$. Thus, the minimization problem defined by Eq. (8.18) becomes the following:

$$E_{OSELM} = \min \left\| \begin{bmatrix} \mathbf{H_0} \\ \mathbf{H_1} \end{bmatrix} \beta - \begin{bmatrix} \mathbf{T_0} \\ \mathbf{T_1} \end{bmatrix} \right\| \tag{8.23}$$

where

$$\mathbf{T}_0 = \begin{bmatrix} \mathbf{t}_{N_0+1}^T \\ \vdots \\ \mathbf{t}_{N_0+N_1}^T \end{bmatrix}_{N_0 \times m} \tag{8.24}$$

$$\mathbf{H}_0 = \begin{bmatrix} G(a_1, x_{N_0+1}, b_1) & \cdots & G(a_L, x_{N_0+1}, b_L) \\ \vdots & \ddots & \vdots \\ G(a_1, x_{N_0+N_1}, b_1) & \cdots & G(a_L, x_{N_0+N_1}, b_L) \end{bmatrix}_{N_0 \times L} \tag{8.25}$$

Considering the N_0 and N_1 as two blocks of the dataset, the matrix of the output weights is calculated as follows:

$$\beta^{(1)} = \mathbf{K}_0^{-1}\begin{bmatrix}\mathbf{H}_0\\ \mathbf{H}_1\end{bmatrix}^{-T}\begin{bmatrix}\mathbf{T}_0\\ \mathbf{T}_1\end{bmatrix} \tag{8.26}$$

where

$$\mathbf{K}_1 = \begin{bmatrix}\mathbf{H}_0\\ \mathbf{H}_1\end{bmatrix}^{T}\begin{bmatrix}\mathbf{H}_0\\ \mathbf{H}_1\end{bmatrix} \tag{8.27}$$

To introduce an ELM model based on the sequential learning, $\beta^{(1)}$ must be given as a function of $\mathbf{T_1}$, $\mathbf{H_1}$, and $\mathbf{K_1}$. For this purpose, $\mathbf{K_1}$ is rewritten as follows:

$$\mathbf{K}_1 = \begin{bmatrix}\mathbf{H}_0^T & \mathbf{H}_1^T\end{bmatrix}\begin{bmatrix}\mathbf{H}_0\\ \mathbf{H}_1\end{bmatrix} = \mathbf{K}_0 + \mathbf{H}_1^T\mathbf{H}_1 \tag{8.28}$$

and

$$\begin{aligned}\begin{bmatrix}\mathbf{H}_0\\ \mathbf{H}_1\end{bmatrix}^{T}\begin{bmatrix}\mathbf{T}_0\\ \mathbf{T}_1\end{bmatrix} &= \mathbf{H}_0^T\mathbf{H}_0 + \mathbf{H}_1^T\mathbf{T}_1\\ &= \mathbf{K}_0\mathbf{K}_0^{-1}\mathbf{H}_0^T\mathbf{H}_0 + \mathbf{H}_1^T\mathbf{T}_1\\ &= \mathbf{K}_0\beta^{(0)} + \mathbf{H}_1^T\mathbf{T}_1\end{aligned} \tag{8.29}$$

By substitution of Eqs. (8.29) and (8.28), one can arrive at Eq. (8.30) as shown below:

$$\begin{aligned}\begin{bmatrix}\mathbf{H}_0\\ \mathbf{H}_1\end{bmatrix}^{T}\begin{bmatrix}\mathbf{T}_0\\ \mathbf{T}_1\end{bmatrix} &= \mathbf{K}_0\beta^{(0)} + \mathbf{H}_1^T\mathbf{T}_1\\ &= \left(\mathbf{K}_1 - \mathbf{H}_1^T\mathbf{T}_1\right)\beta^{(0)} + \mathbf{H}_1^T\mathbf{T}_1\\ &= \mathbf{K}_1\beta^{(0)} - \mathbf{H}_1^T\mathbf{T}_1\beta^{(0)} + \mathbf{H}_1^T\mathbf{T}_1\end{aligned} \tag{8.30}$$

By combining Eqs. (8.30) and (8.26), $\beta^{(1)}$ is obtained as a function of $\mathbf{T_1}$, $\mathbf{H_1}$, $\mathbf{K_1}$ and $\beta^{(0)}$. It should be noted that $\mathbf{K_1}$ [Eq. (8.28)] is a function of $\mathbf{H_1}$ and $\mathbf{K_0}$.

$$\begin{aligned}\beta^{(1)} &= \mathbf{K}_1^{-1}\begin{bmatrix}\mathbf{H}_0\\ \mathbf{H}_1\end{bmatrix}^{T}\begin{bmatrix}\mathbf{T}_0\\ \mathbf{T}_1\end{bmatrix}\\ &= \mathbf{K}_1^{-1}\left(\mathbf{K}_1\beta^{(0)} - \mathbf{H}_1^T\mathbf{H}_1\beta^{(0)} + \mathbf{H}_1^T\mathbf{T}_1\right)\\ &= \beta^{(0)} + \mathbf{K}_1^{-1}\mathbf{H}_1^T\left(\mathbf{T}_1 - \mathbf{H}_1\beta^{(0)}\right)\end{aligned} \tag{8.31}$$

The above equations are presented for the case where training samples are considered as two chunks. To generalize these equations, a recursive algorithm, which is very similar to the recursive least-squares algorithm (Chong & Zak, 2001), is proposed below to update the LSS when new data arrives.

For a positive k and N_{k+1} number of samples at the $(k + 1)$th chunk, the $(k + 1)$th chunk of the dataset is considered as $N_{k+1} = \left\{(\mathbf{x}_i, \mathbf{t}_i)\right\}_{i=\left(\sum_{j=0}^{k+1} N_j\right)+1}^{\sum_{j=0}^{k+1} N_j}$. The recalibrated output weight for $(k + 1)$th chunk is calculated as follows:

$$\beta^{(k+1)} = \beta^{(k)} + \mathbf{K}_{k+1}^{-1}\mathbf{H}_{k+1}^T\left(\mathbf{T}_{k+1} - \mathbf{H}_{k+1}\beta^{(k)}\right) \tag{8.32}$$

where

$$\mathbf{K}_{k+1} = \mathbf{K}_k + \mathbf{H}_{k+1}^T\mathbf{H}_{k+1} \tag{8.33}$$

It is clear that to calculate the output weight after consideration of the $(k + 1)$th chunk, only the values $\mathbf{T}$, $\mathbf{H}$, and β are needed at the kth chunk. The $\mathbf{T}$ and $\mathbf{H}$ at the $(k + 1)$th chunk are calculated as follows:

$$\mathbf{T}_{k+1} = \begin{bmatrix}\mathbf{t}^T_{\left(\sum_{j=0}^{k} N_j\right)+1}\\ \vdots\\ \mathbf{t}^T_{\left(\sum_{j=0}^{k+1} N_j\right)}\end{bmatrix}_{N_{k+1}\times m} \tag{8.34}$$

$$\mathbf{H}_{k+1} = \begin{bmatrix} G\left(a_1, x_{\left(\sum_{j=0}^{k} N_j\right)+1}, b_1\right) & \cdots & G\left(a_L, x_{\left(\sum_{j=0}^{k} N_j\right)+1}, b_L\right) \\ \vdots & \ddots & \vdots \\ G\left(a_1, x_{\left(\sum_{j=0}^{k+1} N_j\right)}, b_1\right) & \cdots & G\left(a_L, x_{\left(\sum_{j=0}^{k+1} N_j\right)}, b_L\right) \end{bmatrix}_{N_{k+1} \times L} \tag{8.35}$$

In order to calculate β using Eq. (8.32), the parameter $\mathbf{K}_{k+1}^{-1}$ is needed, which may be obtained by considering Eq. (8.33). Therefore, to calculate $\mathbf{K}_{k+1}^{-1}$, Eq. (8.33) is rewritten as follows (Golub & Loan, 1996):

$$\mathbf{K}_{k+1}^{-1} = \left(\mathbf{K}_k + \mathbf{H}_{k+1}{}^T \mathbf{H}_{k+1}\right)^{-1} = \mathbf{K}_k{}^{-1} - \mathbf{K}_k{}^{-1}\mathbf{H}_{k+1}\left(\mathbf{I} + \mathbf{H}_{k+1}\mathbf{K}_k{}^{-1}\mathbf{H}_{k+1}{}^T\right)^{-1}\mathbf{H}_{k+1}{}^T\mathbf{K}_k{}^{-1} \tag{8.36}$$

Considering $\mathbf{P}_{k+1} = \mathbf{K}_{k+1}{}^{-1}$, the above-mentioned equation is rewritten as follows:

$$\mathbf{P}_{k+1} = \mathbf{P}_k - \mathbf{P}_k\mathbf{H}_{k+1}\left(\mathbf{I} + \mathbf{H}_{k+1}\mathbf{P}_k\mathbf{H}_{k+1}{}^T\right)^{-1}\mathbf{H}_{k+1}{}^T\mathbf{P}_k \tag{8.37}$$

Eq. (8.33) can now be rewritten as follows:

$$\beta^{(k+1)} = \beta^{(k)} + \mathbf{P}_{k+1}\mathbf{H}_{k+1}{}^T\left(\mathbf{T}_{k+1} - \mathbf{H}_{k+1}\beta^{(k)}\right) \tag{8.38}$$

Eq. (8.38) is the generalized form of a recursive algorithm for calculating the output weights.

If the rank($\mathbf{H_0}$) = L, the OSELM can attain similar training performance (generalization accuracy and training error) as the original ELM. To make rank($\mathbf{K_0}$) = L and rank($\mathbf{H_0}$) = L, the number of initialization data (N_O) must be more than the hidden layer nodes (L). The OSELM algorithm is summarized in the next section.

8.6.4 Proposed online sequential extreme learning machine algorithm

The first step in the proposed OSELM algorithm is to select the type of hidden nodes (additive or RBF). Following this, the activation function (g) type and the number of hidden layer nodes (L) should be specified. The problem data $N = \{(\mathbf{x}_i, \mathbf{t}_i) | \mathbf{x}_i \in \mathbf{R}^n, \mathbf{t}_i \in \mathbf{R}^m\}$ will then be considered sequentially. In general, the modeling process for the OSELM consists of two main phases: initialization and sequential learning.

In the initialization phase, the output of the hidden nodes matrix ($\mathbf{H_0}$), the structure of which is determined using initialization samples, is calculated randomly. This matrix is used to calculate output weights through the sequential-based learning phase. The number of samples considered as initial samples should not be less than the number of hidden neurons (L). According to Theorem I, the rank of a number of distinct training samples is given by rank ($\mathbf{H_0}$) = L. In fact, if the number of hidden nodes is equal to L, L training samples are sufficient for analysis. However, in the case that the training samples are not distinct, more training samples may be needed.

In most cases, the number of training samples in the initialization phase is equal to or close to the number of hidden nodes (L). After determining the output of the hidden node during the initialization phase, the sequential learning phase begins. In this phase, the data is considered chunk-by-chunk or one-by-one and following its use for model calibration of the output weights is discarded.

Step I: Initialization phase:

In the initialization phase of the OSELM, a chunk of data, $N_0 = \{(\mathbf{x}_i, \mathbf{t}_i)\}_{i=1}^{N_0}$, smaller than the number of hidden nodes $L \geq N_O$ is considered for analysis. The initialization phase of the OSELM is implemented as follows:

1. The hidden nodes, including the impact factor (b_i) and centers ($\mathbf{a_i}$) for RBF nodes, and the bias of hidden neurons (b_i) and input weights ($\mathbf{a_i}$) for additive hidden nodes, are randomly assigned. It should be noted that **I** is the unit matrix with a dimension of L and is, therefore, in the range of [1, L].
2. After the random allocation of the hidden nodes, the output matrix of the initial hidden layer ($\mathbf{H_0}$) is calculated based on the selected activation function type. The final form of this matrix is as follows:

$$\mathbf{H}_0 = \begin{bmatrix} G(a_1, x_1, b_1) & \cdots & G(a_L, x_1, b_L) \\ \vdots & \ddots & \vdots \\ G(a_1, x_{N_0}, b_1) & \cdots & G(a_L, x_{N_0}, b_L) \end{bmatrix}_{N_0 \times L} \tag{8.39}$$

3. Using the calculated initial output matrix of the hidden nodes and the corresponding target values, ($\mathbf{T}_0 = [\mathbf{t}_1, \ldots, \mathbf{t}_{N_0}]^T$), the initial output weights ($\beta^{(0)}$) are calculated as follows by considering Eqs. (8.21) and (8.37):

$$\beta^{(0)} = \mathbf{P}_0\mathbf{H}_0^T\mathbf{T}_0 = (\mathbf{H}_0^T\mathbf{H}_0)^{-1} \times \mathbf{H}_0^T\mathbf{T}_0 \tag{8.40}$$

4. set $k = 1$. Recall that k is the counter used to identify each chunk of data during the sequential learning phase

Step II: Sequential learning phase:

In this step, the new chunk of data (i.e., $(k+1)$th chunk) is presented as $N_{k+1} = \{(\mathbf{x}_i, \mathbf{t}_i)\}_{i=\left(\sum_{j=0}^{k+1} N_j\right)+1}^{\sum_{j=0}^{k+1} N_j}$ where N_{k+1} is the number of training samples for the $(k+1)$th chunk of data, ($\mathbf{H}_{k+1}$).

1. After receiving the new chunk of data, the output matrix of the hidden nodes for the $(k+1)$th chunk should be evaluated using Eq. (8.36).
2. Set $\mathbf{T}_{k+1} = [\mathbf{t}_{\left(\sum_{j=0}^{k} N_j\right)+1}, \ldots, \mathbf{t}_{\sum_{j=0}^{k+1} N_j}]^T$
3. After calculating $\mathbf{H}_{k+1}$ for the $(k+1)$th chunk of data, the output weight $\beta^{(k+1)}$ is calculated using Eqs. (8.37) and (8.38).
4. set $k = k + 1$
5. Repeat all steps associated with *Step II: Sequential Learning Phase*
6. The size of the chunk of data considered at each sequential step is not required to be identical such that the number of training samples in the $(k+1)$th chunk (i.e., N_{k+1}) could differ from the number of training samples in the kth chunk (i.e., N_k). For the case where the training samples are considered on a one-by-one basis, $N_{k+1} = 1$. In this case, Eqs. (8.37) and (8.38) can be rewritten as follows (Golub & Loan, 1996):

$$\mathbf{P}_{k+1} = \mathbf{P}_k - \frac{\mathbf{P}_k\mathbf{h}_{k+1}\mathbf{h}_{k+1}^T\mathbf{P}_k}{1 + \mathbf{h}_{k+1}^T\mathbf{P}_k\mathbf{h}_{k+1}} \tag{8.41}$$

$$\beta^{(k+1)} = \beta^{(k)} + \mathbf{P}_{k+1}\mathbf{h}_{k+1}(\mathbf{t}_{k+1}^T - \mathbf{h}_{k+1}^T\beta^{(k)}) \tag{8.42}$$

where

$$\mathbf{h}_{k+1} = \left[G(a_1, x_{k+1}, b_1) \quad \cdots \quad G(a_L, x_{k+1}, b_L)\right]_{N_0 \times L} \tag{8.43}$$

It should be noted that $\mathbf{P}_k$ and $\beta^{(k)}$ are computed using SVD in order to make the OSELM more robust and able to handle the case where $\mathbf{I} + \mathbf{H}_{k+1}\mathbf{P}_k\mathbf{H}_{k+1}^T$ and/or $\mathbf{H}_0$ are near singular or singular.

According to the steps outlined above, the OSELM algorithm can be conceptually understood using the flowchart presented in Fig. 8.10. In order to begin the modeling process, the data must first be read and loaded into the MATLAB environment. The data can be loaded from the MATLAB environment or outside sources such as excel or text files. Once the data has been read, it should be split into two categories; training samples and testing samples. When splitting the data into these two classes, a certain minimum percentage of the data should be considered for model training and development to provide sufficient model experience with different data ranges. The remaining data should be considered as testing samples. A study performed by Ebtehaj et al. (2020) examined the different ratios for training samples (50%–90%) in detail. From the study results, it was observed that for lower percentages of training samples, the model performance considering the training data was low, but it was more likely to approach similar performance levels with both the training and testing sets. Conversely, increasing the percentage of training samples increases the performance of the model during training and model development. However, there is a greater likelihood that the model may overfit, which significantly reduces the generalizability of the proposed model and makes it virtually impossible to use on unseen data. The selection of training and testing samples is typically done randomly so that the model experiences a range of different conditions throughout the training phase. In the event that the data set is a time series, the order of the training and testing data must be consecutive so that if 70% of the data is considered as training data, the first 70% of the data is considered for model training while the last 30% of the data is considered for model testing.

The second step is the parameter definition. The parameters presented in this step are divided into two general categories; "General parameters," including the activation function type and the number of hidden neurons, and "OSELM parameters," consisting of the number of initial training data samples and the number of blocks that the user wishes to split the data into. There are a large number of activation functions which may be selected for use in the model development according to the type of data. Six of the most common activation functions are provided in Eqs. (8.1)–(8.6) and

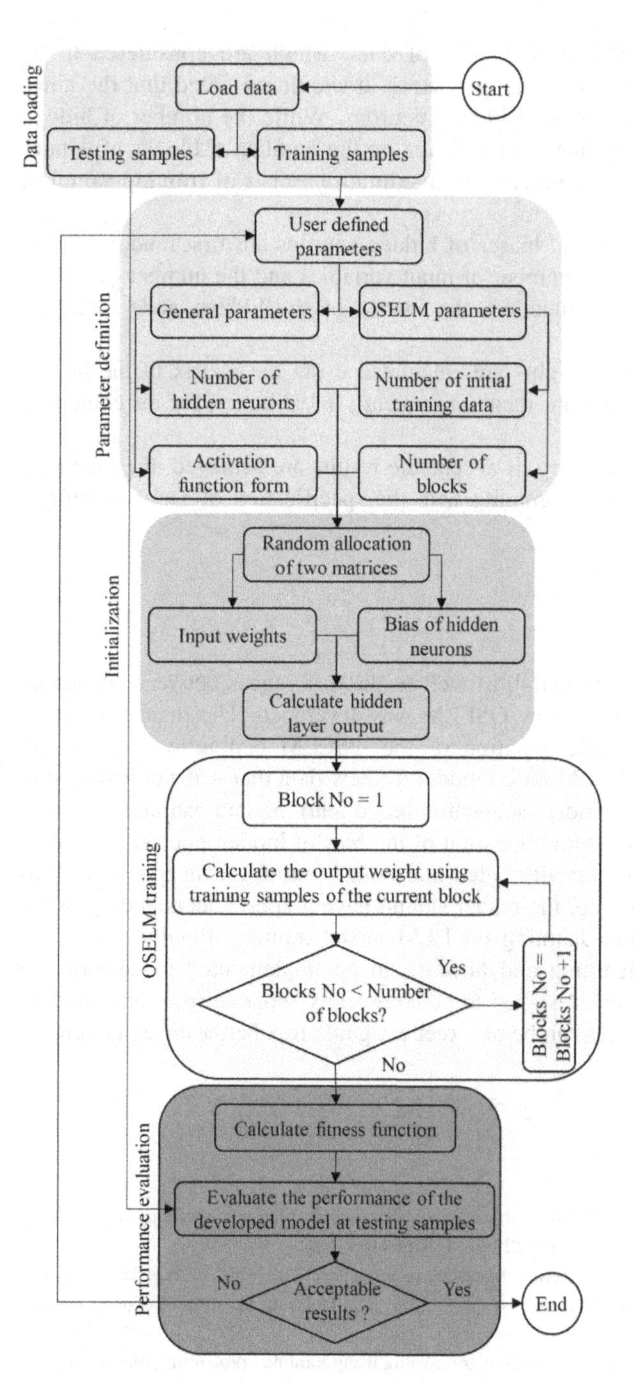

FIGURE 8.10 Flowchart of the online sequential extreme learning machine algorithm.

include sigmoid, triangular basis function, RBF, sine function, hard-limit function, and hyperbolic tangent function. The number of hidden neurons should be selected through a trial-and-error approach to find a model that can estimate the target variable with acceptable performance for unseen data (i.e., testing samples). It should be noted that the minimum number of hidden neurons is one but can be increased to a maximum defined by recent studies (Zeynoddin et al., 2020; Ebtehaj et al., 2021; Ebtehaj & Bonakdari, 2022) using Eq. (8.44).

$$NHN < \frac{NTrS}{(InV + 2)} \tag{8.44}$$

where InV is the number of input variables, NTrS is the number of training samples and NHN is the number of hidden neurons.

The number of initial training data samples and the number of blocks of data, which are considered to be OSELM parameters, should also be optimized through a process of trial and error. It should be noted that the number of initial training data samples should be more than the number of hidden neurons. While the number of hidden neurons restricts the number of initial training data, there is, in fact, no limitation on the number of blocks of data to consider. The smallest and largest values for this parameter are one and the maximum number of training samples, respectively.

The third step is initialization. In this step, the input weights and biases of hidden neurons are first randomly allocated. The architecture of these matrices is generated based on the number of input variables and the number of hidden neurons. Using these two matrices and the selected activation function, the matrix of the hidden node output is calculated.

The fourth step is OSELM training. In this step, the output weights are calculated using the matrix of the hidden node output and the initial training samples. The output weights are then subsequently recalibrated for all chunks of data considered during model development.

The performance of the trained model is evaluated in the fifth step. If acceptable results are achieved, the modeling is finished. Otherwise, the modeling should be performed again, beginning with the specification of the user-defined parameters.

8.7 Summary

In the current chapter, sequential-based learning was discussed in detail, as well as the differences between sequential and batch learning. Using the sequential-based learning concept, a new OSELM was developed. The main concept of this algorithm as well as the detail regarding MATLAB implementation of the OSELM coding were provided. Moreover, a calculator is introduced to apply the developed OSELM-based model for new data that was not involved in the model calibration period. The developed OSELM model considers sequential-based learning and can also easily be employed with a large number of iterations to overcome the random allocation of the bias of hidden neurons and input weights. After reading this chapter, the reader should be familiar with batch and online learning concepts as well as their differences. From the discussion of batch and online learning, the reader should have a good understanding of the limitations of batch learning that have driven interest in online learning for ELM model training. Finally, the reader was introduced to the mathematical concept of sequential learning and how it can be implemented to develop the OSELM method. Although the chapters presented in this text are designed to be completely separate from one another, sufficient knowledge of the material contained within this chapter can be an excellent guide to a better understanding of the next chapter, where OSELM coding is detailed line by line.

References

Asirvadam, V.S., McLoone, S.F., & Irwin, G.W. (2002, September). Parallel and separable recursive Levenberg-Marquardt training algorithm. In *Proceedings of the 12th IEEE workshop neural network signal process* (no. 4–6, pp. 129–138) September 2002.

Azimi, H., Bonakdari, H., & Ebtehaj, I. (2017). Sensitivity analysis of the factors affecting the discharge capacity of side weirs in trapezoidal channels using extreme learning machines. *Flow Measurement and Instrumentation, 54*, 216–223. Available from https://doi.org/10.1016/j.flowmeasinst.2017.02.005.

Bonakdari, H., Ebtehaj, I., Samui, P., & Gharabaghi, B. (2019a). Lake water-level fluctuations forecasting using minimax probability machine regression, relevance vector machine, gaussian process regression, and extreme learning machine. *Water Resources Management, 33*(11), 3965–3984. Available from https://doi.org/10.1007/s11269-019-02346-0.

Bonakdari, H., Qasem, S. N., Ebtehaj, I., Zaji, A. H., Gharabaghi, B., & Moazamnia, M. (2019b). An expert system for predicting the velocity field in narrow open channel flows using self-adaptive extreme learning machines. *Measurement*. Available from https://doi.org/10.1016/j.measurement.2019.107202.

Chong, E. K. P., & Zak, S. H. (2001). *An introduction to optimization*. New York: Wiley.

Ebtehaj, I., & Bonakdari, H. (2022). A reliable hybrid outlier robust non-tuned rapid machine learning model for multi-step ahead flood forecasting in Quebec, Canada. *Journal of Hydrology*, 128592. Available from https://doi.org/10.1016/j.jhydrol.2022.128592.

Ebtehaj, I., Bonakdari, H., Zaji, A. H., & Sharafi, H. (2019a). Sensitivity analysis of parameters affecting scour depth around bridge piers based on the non-tuned, rapid extreme learning machine method. *Neural Computing and Applications, 31*(12), 9145–9156. Available from https://doi.org/10.1007/s00521-018-3696-6.

Ebtehaj, I., Bonakdari, H., & Gharabaghi, B. (2019b). A reliable linear method for modeling lake level fluctuations. *Journal of Hydrology, 570*, 236–250. Available from https://doi.org/10.1016/j.jhydrol.2019.01.010.

Ebtehaj, I., Bonakdari, H., Safari, M. J. S., Gharabaghi, B., Zaji, A. H., Madavar, H. R., & Mehr, A. D. (2020). Combination of sensitivity and uncertainty analyses for sediment transport modeling in sewer pipes. *International Journal of Sediment Research, 35*(2), 157–170. Available from https://doi.org/10.1016/j.ijsrc.2019.08.005.

Ebtehaj, I., Bonakdari, H., & Zaji, A. H. (2016). An expert system with radial basis function neural network based on decision trees for predicting sediment transport in sewers. *Water Science and Technology, 74*(1), 176–183. Available from https://doi.org/10.2166/wst.2016.174.

Ebtehaj, I., Bonakdari, H., & Zaji, A. H. (2018). A new hybrid decision tree method based on two artificial neural networks for predicting sediment transport in clean pipes. *Alexandria Engineering Journal, 57*(3), 1783–1795. Available from https://doi.org/10.1016/j.aej.2017.05.021.

Ebtehaj, I., Soltani, K., Amiri, A., Faramarzi, M., Madramootoo, C. A., & Bonakdari, H. (2021). Prognostication of shortwave radiation using an improved no-tuned fast machine learning. *Sustainability, 13*(14), 8009. Available from https://doi.org/10.3390/su13148009.

Golub, G. H., & Loan, C. F. V. (1996). *Matrix computations* (3rd ed.). Baltimore, MD: The Johns Hopkins Univ. Press.

Huang, G. B. (2003). Learning capability and storage capacity of two-hidden-layer feedforward networks. *IEEE Transactions on Neural Networks, 14* (2), 274–281. Available from https://doi.org/10.1109/TNN.2003.809401.

Huang, G. B., Saratchandran, P., & Sundararajan, N. (2004a). An efficient sequential learning algorithm for growing and pruning RBF (GAP-RBF) networks. *IEEE Transactions on Systems, Man, and Cybernetics, part B (Cybernetics), 34*(6), 2284–2292. Available from https://doi.org/10.1109/tsmcb.2004.834428.

Huang, G.B., Zhu, Q.Y., & Siew, C.K. (2004b, July). Extreme learning machine: a new learning scheme of feedforward neural networks. In *2004 IEEE international joint conference on neural networks (IEEE Cat. No. 04CH37541)* (Vol. 2, pp. 985–990). IEEE. https://doi.org/10.1109/IJCNN.2004.1380068.

Huang, G. B., Saratchandran, P., & Sundararajan, N. (2005). A generalized growing and pruning RBF (GGAP-RBF) neural network for function approximation. *IEEE Transactions on Neural Networks, 16*(1), 57–67. Available from https://doi.org/10.1109/TNN.2004.836241.

Huang, G.B., & Siew, C.K. (2004, December). Extreme learning machine: RBF network case. In *ICARCV 2004 8th control, automation, robotics and vision conference, 2004.* (Vol. 2, pp. 1029–1036). IEEE. https://doi.org/10.1109/ICARCV.2004.1468985.

Huang, G. B., Zhu, Q. Y., & Siew, C. K. (2006a). Extreme learning machine: theory and applications. *Neurocomputing, 70*(1–3), 489–501. Available from https://doi.org/10.1016/j.neucom.2005.12.126.

Huang, G. B., Zhu, Q. Y., & Siew, C. K. (2006b). Real-time learning capability of neural networks. *IEEE Trans. Neural Networks, 17*(4), 863–878. Available from https://doi.org/10.1109/tnn.2006.875974.

Huang, G. B., Zhu, Q. Y., Mao, K. Z., Siew, C. K., Saratchandran, P., & Sundararajan, N. (2006c). Can threshold networks be trained directly? *IEEE Transactions on Circuits and Systems II: Express Briefs, 53*(3), 187–191. Available from https://doi.org/10.1109/TCSII.2005.857540.

Huang, G. B., Chen, L., & Siew, C. K. (2006d). Universal approximation using incremental constructive feedforward networks with random hidden nodes. *IEEE Transactions of the Neural Networks, 17*(4), 879–892. Available from https://doi.org/10.1109/TNN.2006.875977.

Kadirkamanathan, V., & Niranjan, M. (1993). A function estimation approach to sequential learning with neural networks. *Neural Computation, 5*(6), 954–975. Available from https://doi.org/10.1162/neco.1993.5.6.954.

LeCun, Y. A., Bottou, L., Orr, G. B., & Müller, K. R. (2012). Efficient backprop. *Lecture Notes Comput. Sci, 1524*, 9–50. Available from https://doi.org/10.1007/978-3-642-35289-8_3, 1998.

Liang, N. Y., Huang, G. B., Saratchandran, P., & Sundararajan, N. (2006). A fast and accurate online sequential learning algorithm for feedforward networks. *IEEE Transactions on Neural Networks, 17*(6), 1411–1423. Available from https://doi.org/10.1109/TNN.2006.880583.

Lillicrap, T. P., & Santoro, A. (2019). Backpropagation through time and the brain. *Current Opinion in Neurobiology, 55*, 82–89. Available from https://doi.org/10.1016/j.conb.2019.01.011.

Moeeni, H., Bonakdari, H., & Ebtehaj, I. (2017). Integrated SARIMA with neuro-fuzzy systems and neural networks for monthly inflow prediction. *Water Resource Management, 31*(7), 2141–2156. Available from https://doi.org/10.1007/s11269-017-1632-7.

Moody, J. O., & Antsaklis, P. J. (1996). The dependence identification neural network construction algorithm. *IEEE Transactions on Neural Networks, 7*(1), 3–15. Available from https://doi.org/10.1109/72.478388.

Ngia, L.S., Sjoberg, J., & Viberg, M. (1998, November). Adaptive neural nets filter using a recursive levenberg-marquardt search direction. In *Proceedings of the asilomar conference of the signals, systems, and computations* (vol. 1–4, pp. 697–701). Nov. 1998. https://doi.org/10.1109/ACSSC.1998.75d0952.

Noor-A-Rahim, M., Liu, Z., Lee, H., Ali, G. M. N., Pesch, D., & Xiao, P. (2020). A survey on resource allocation in vehicular networks. *IEEE Transactions on Intelligent Transportation Systems.* Available from https://doi.org/10.1109/TITS.2020.3019322.

Platt, J. (1991). A resource-allocating network for function interpolation. *Neural Computation, 3*(2), 213–225. Available from https://doi.org/10.1162/neco.1991.3.2.213.

Rao, C. R., & Mitra, S. K. (1971). *Generalized inverse of matrices and its applications.* New York: John Wiley & Sons. Inc.

Safari, M. J. S., Ebtehaj, I., Bonakdari, H., & Es-haghi, M. S. (2019). Sediment transport modeling in rigid boundary open channels using generalize structure of group method of data handling. *Journal of Hydrology, 577*, 123951. Available from https://doi.org/10.1016/j.jhydrol.2019.123951.

Shi, X., Feng, Y., Zeng, J., & Chen, K. (2017). Chaos time-series prediction based on an improved recursive Levenberg–Marquardt algorithm. *Chaos, Solitons & Fractals, 100*, 57–61. Available from https://doi.org/10.1016/j.chaos.2017.04.032.

Suresh, S., Dong, K., & Kim, H. J. (2010). A sequential learning algorithm for self-adaptive resource allocation network classifier. *Neurocomputing, 73*(16–18), 3012–3019. Available from https://doi.org/10.1016/j.neucom.2010.07.003.

Yaseen, Z. M., Deo, R. C., Ebtehaj, I., & Bonakdari, H. (2018). Hybrid data intelligent models and applications for water level prediction. *Handbook of research on predictive modeling and optimization methods in science and engineering*, 121–139. Available from https://doi.org/10.4018/978-1-5225-4766-2.ch006, IGI Global.

Yingwei, L., Sundararajan, N., & Saratchandran, P. (1997). A sequential learning scheme for function approximation using minimal radial basis function neural networks. *Neural Computation*, *9*(2), 461–478. Available from https://doi.org/10.1162/neco.1997.9.2.461.

Yingwei, L., Sundararajan, N., & Saratchandran, P. (1998). Performance evaluation of a sequential minimal radial basis function (RBF) neural network learning algorithm. *IEEE Transactions on neural networks*, *9*(2), 308–318. Available from https://doi.org/10.1109/72.661125.

Zeynoddin, M., Bonakdari, H., Azari, A., Ebtehaj, I., Gharabaghi, B., & Madavar, H. R. (2018). Novel hybrid linear stochastic with non-linear extreme learning machine methods for forecasting monthly rainfall a tropical climate. *Journal of Environmental Management*, *222*, 190–206. Available from https://doi.org/10.1016/j.jenvman.2018.05.072.

Zeynoddin, M., Bonakdari, H., Ebtehaj, I., Azari, A., & Gharabaghi, B. (2020). A generalized linear stochastic model for lake level prediction. *Science of the Total Environment*, *138*, 015. Available from https://doi.org/10.1016/j.scitotenv.2020.138015.

Chapter 9

Online sequential nontuned neural network—coding and implementation

9.1 Summary of the online sequential extreme learning machine-based modeling algorithm

In this section, all of the steps involved in online sequential extreme learning machine (OSELM)-based modeling of a real-world problem in the MATLAB environment are discussed in detail. Fig. 9.1 presents a flow chart which summarize the required steps for OSELM implementation in the MATLAB environment. Like almost all machine learning-based approaches, the modeling process begins by loading all the data samples and input variables as well as their corresponding target variable(s). Once the data set has been loaded into the MATLAB environment, the samples are split into training and testing subsets. Data division was presented in detail in the previous chapter, but briefly, it is performed by using different percentages for the training and testing stages. The training percentages could be selected to be between 50% and 90% of the total data set (i.e., 50%–10% for the testing stage) (Arabameri et al., 2020; Ebtehaj et al., 2020; Lanera et al., 2019). It is generally recommended to split training and testing data according to the ratio of 70%–30% (Ebtehaj et al., 2016; Ebtehaj, Sammen, Sidek, et al., 2021; Ebtehaj, Soltani, Amiri et al., 2021; Grégoire et al., 2022; Saha et al., 2021; Thottakkara et al., 2016) or 80%–20% (Bonakdari, Ebtehaj, et al., 2020; Bonakdari, Gholami, et al., 2020; Herrera et al., 2019; Mojtahedi et al., 2019; Ratzinger et al., 2018). Using this approach, either 30%–20% of samples are randomly selected for model testing and verification while the remaining samples are considered for model calibration (i.e., training samples). Four different matrices are defined according to the selected split percentage for allocating data to the training or testing sets: training inputs, training output, testing inputs, and testing outputs.

The second step is the general definition of the OSELM parameters, including OSELM type, iteration number, number of hidden neurons, number of blocks, number of initial training samples, and activation function type. Following the definition of the OSELM parameters, the next step is the initialization of the OSELM. The initialization and training stages form the core of the OSELM model development. The first step in the initialization is the random allocation of the matrices of input weights and the bias of hidden neurons. Once these have been generated, the orthogonal of these matrices (i.e., input weights and bias of hidden neurons) can be calculated based on the user input in the parameter definition stage (i.e., whether to consider orthogonal matrices or not). Recall that whenever the transpose of a square matrix with real numbers or elements equals its inverse matrix, it is said to be orthogonal (Fig. 9.2). Using the matrices of input weights and bias of the hidden neurons, as well as the defined activation function type from the second step (i.e., general definition), the output matrix of the hidden layer can be computed. The output weights are then calculated using the output matrix of the hidden layer. It should be made clear that, during the initialization stage, all calculations are performed by considering the initial training sample block and not the entirety of the training set. Once the initial output weights have been determined, they are then recalibrated using other training samples through a defined block size in the training step. The training step is continued until all training samples have been considered for input weight recalibration. The initialization and training steps are repeated N times, where N is the number of defined iteration numbers by the user. After the modeling is complete, the results of all iterations are saved, and the quantitative and qualitative performance of the developed model is evaluated for training and testing stages.

9.2 Implementation of the online sequential extreme learning machine in the MATLAB environment

In order to employ the developed OSELM algorithms in solving nonlinear real-world problems, a MATLAB-based package is provided. It should be noted that this package is provided for both advanced and beginner users. The

Machine Learning in Earth, Environmental and Planetary Sciences. DOI: https://doi.org/10.1016/B978-0-443-15284-9.00007-0

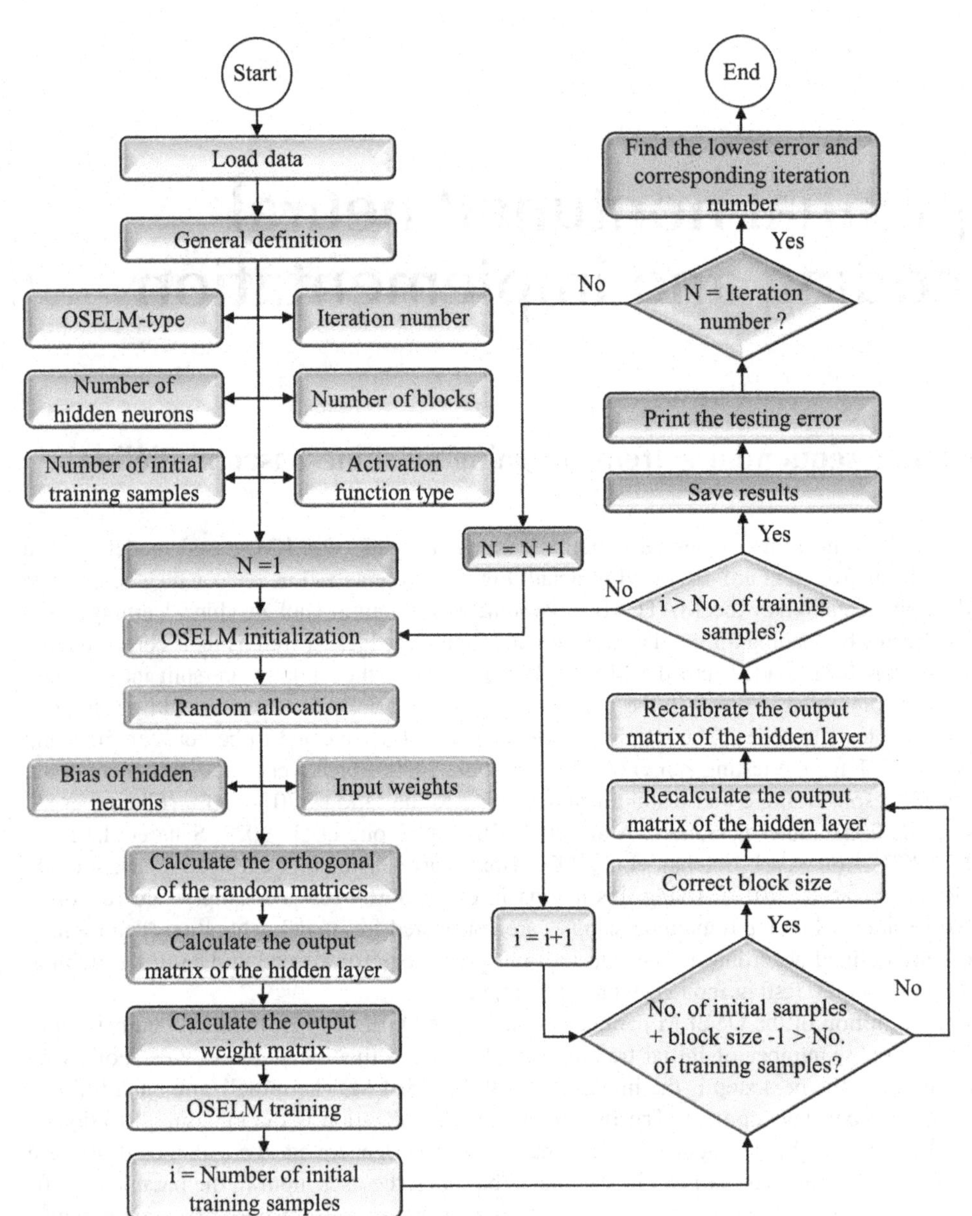

FIGURE 9.1 A summary of the steps required for online sequential extreme learning machine-based modeling.

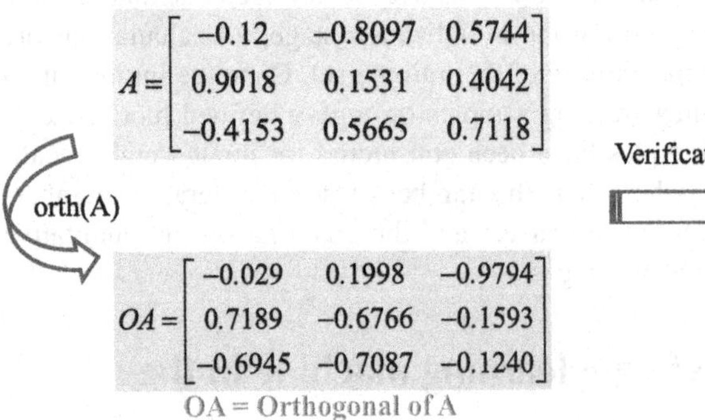

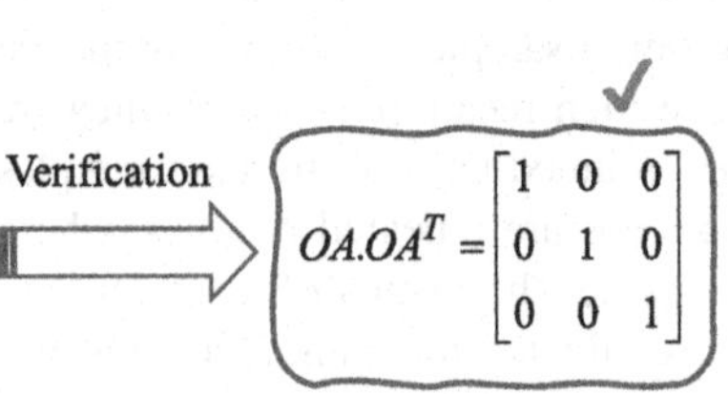

FIGURE 9.2 An example of an orthogonal matrix.

beginner users can easily run the proposed main OSELM m-file, where they are only required to adjust the user-defined value of different parameters to model the desired problem. In addition, a calculator is provided that can be used to apply the developed model for unseen data. In addition to the provided explanations for beginners, advanced users can run the model line-by-line to verify the modeling process as well as to develop it to better suit their specific applications.

In this section, the line-by-line implementation of OSELM in MATLAB is provided in detail. Beginner users can follow this section without following the detailed MATLAB coding explanations provided in the shaded gray boxes. The provided package consists of four MATLAB functions which are detailed in Table 9.1. The "PlotResults" and "Statistical_Indices" functions are applied during the visual and numerical evaluation of the developed model. "Calculator" is provided to apply the OSELM-based model for unseen data. The "OSELM" function is the core of the developed approach for modeling real-world problems.

9.2.1 Data loading from Microsoft Excel

The implementation of the OSELM in the MATLAB environment is done using the "OSELM" m-file and the functions provided in Table 9.1. The first step is loading the dataset, including the inputs and targets for the training and testing stages. This is detailed in Code 9.1 below, and its schematic is provided in Fig. 9.3. All data sets are read from an excel file that contains four sheets so that the train inputs, train targets, test inputs, and test targets are provided in sheet1 to sheet4, respectively (Fig. 9.4). MATLAB's built-in `xlsread` function is employed to read the data, as seen in Lines 2–5 of Code 9.1. The inputs at the training and testing stages are of the size $n \times N$, where n and N denote the number of input variables and the number of samples, respectively.

TABLE 9.1 Required online sequential extreme learning machine functions in the provided package.

Function name	Definition
Calculator	Used to apply the developed model to previously unseen data
OSELM	OSELM-based modeling, training, and testing
PlotResults	Provide the visual results
Statistical_Indices	Provides a numerical evaluation of model performance by computing different statistical indices

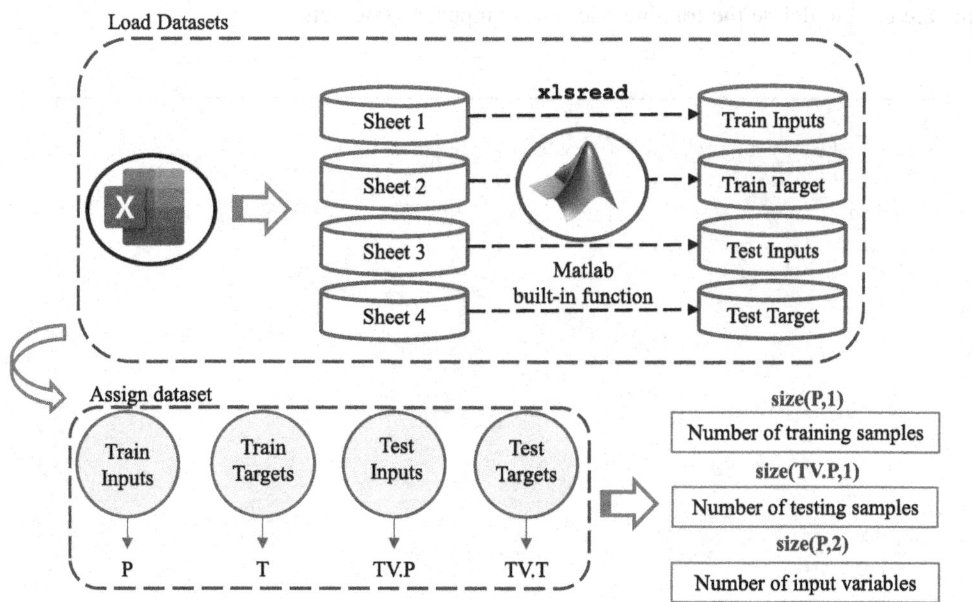

FIGURE 9.3 Schematic representation of Code 9.1.

Train Inputs

	A	B	C	D
1	In1	In2	In3	In4
2	5	30	20000	35
3	5	30	20000	30
4	5	30	16500	35
5	5.5	30	20000	30
6	5	30	20000	25
7	5	30	16500	30
8	6	30	20000	30
9	5.5	30	20000	25
10	6	30	16500	35

Train Targets

	A	B	C	D
1	Out			
2	28.62033			
3	31.09058			
4	32.70322			
5	33.22552			
6	33.28823			
7	34.72283			
8	35.13146			
9	35.18745			
10	36.41666			

Test Inputs

	A	B	C	D
1	In1	In2	In3	In4
2	5.5	30	20000	35
3	6	30	20000	35
4	5.5	30	16500	35
5	5	30	20000	20
6	5.5	30	16500	30
7	5.5	30	20000	20
8	5.5	30	16500	25
9	5	30	9000	35
10	6	30	16500	25

Test Targets

	A	B	C	D
1	Out			
2	31.02008			
3	33.16257			
4	34.66519			
5	35.24325			
6	36.46812			
7	36.93268			
8	38.07182			
9	39.11637			
10	39.45756			

FIGURE 9.4 Format of Excel workbook for data loading in the online sequential extreme learning machinemodel.

Similarly, the output(s) at the training and testing stages are of the size $m \times N$, where m and N denote the number of target variables and the number of samples, respectively. It should be noted that the examples presented in this text consider only a singular output variable. It is highly recommended to define problems with one output to prevent the effect of the random allocation of the input weights and bias of hidden neurons at each iteration. Because these matrices are randomly assigned in a given iteration, the performance of the proposed model may be acceptable for one output while performing poorly for another. To simplify the calling of the dataset throughout the program, in lines 8–11, `TrainInputs` is assigned to `P`, `TrainTargets` to `T`, `TestInputs` to `TV.P`, and `TestTargets` to `TV.T`. The number of training samples, testing samples, and input variables is computed in lines 13–15, respectively. Fig. 9.4 demonstrates an example of the four excel sheets that are used to define the training and testing inputs and outputs.

Code 9.1

```
1   %% Load Datasets
2   TrainInputs  = xlsread('Example3','sheet1');
3   TrainTargets = xlsread('Example3','sheet2');
4   TestInputs   = xlsread('Example3','sheet3');
5   TestTargets  = xlsread('Example3','sheet4');
6
7   % Assign dataset
8   P=TrainInputs;
9   T=TrainTargets;
10  TV.P=TestInputs;
11  TV.T=TestTargets;
12
13  nTrainingData=size(P,1);
14  nTestingData=size(TV.P,1);
15  nInputNeurons=size(P,2);
```

9.2.2 Definition of online sequential extreme learning machine type

The next step is the selection of the OSELM-type, Regression or Classification. To select the problem type under consideration, the questdlg command is employed. This MATLAB command can be used to store up to three input options, with one of the options being defined as the default. The default option is defined as the final input argument of the `questdlg` command, such that for two or three options, the third and fourth input arguments are defined as the default option. Code 9.2 shows the segment of the OSELM code used to determine the OSELM type. The schematic representation of Code 9.2 is provided in Fig. 9.5.

Code 9.2

```
1    %% OSELM Type
2    Option{1}='Regression';
3    Option{2}='Classification';
4    ANSWER=questdlg('Regression or Classification ?',...
5                    'OSELM Type',...
6                    Option{1},Option{2},Option{1});
7    pause(0.1);
8    switch ANSWER
9        case Option{1}
10           OSELM_Type = 0;
11       case Option{2}
12           OSELM_Type = 1;
13   end
```

The structure of the questdlg command is presented in Code 9.3. The input arguments of the command are as follows: the question string (qstring), the title (title), one to three strings representing the options (str1, str2, str3), and the default string (default). Fig. 9.6 shows the dialog box used to select the OSELM type that the user will be presented with when the m file is executed.

Code 9.3

```
1    button = questdlg(qstring,title,str1,str2,str3,default)
```

Considering the input arguments of the questdlg command outlined above, the two possibilities are listed in lines 2–3 as "Regression" or "Classification." Line 4–6 details the questdlg input – the qstring is specified as "Regression or Classification?," the title is specified as "OSELM Type," and "Option{1}" is given as the default value.

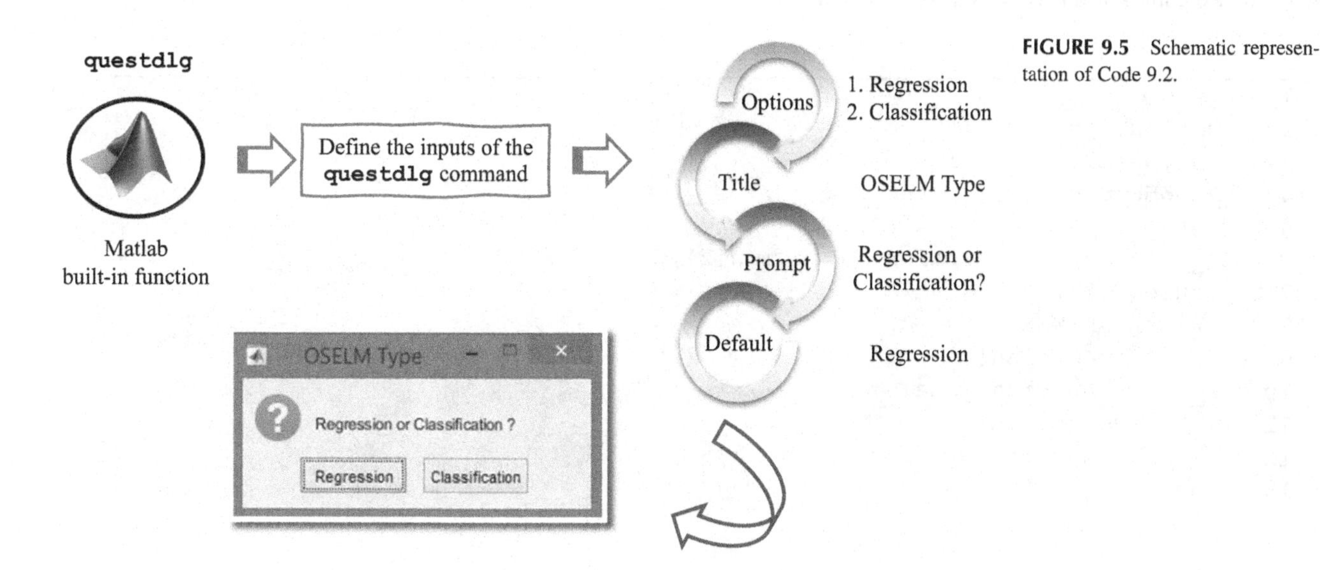

FIGURE 9.5 Schematic representation of Code 9.2.

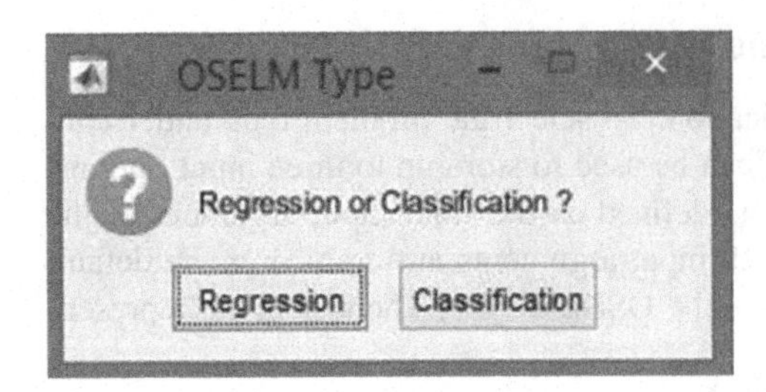

FIGURE 9.6 Dialog box for online sequential extreme learning machine type selection.

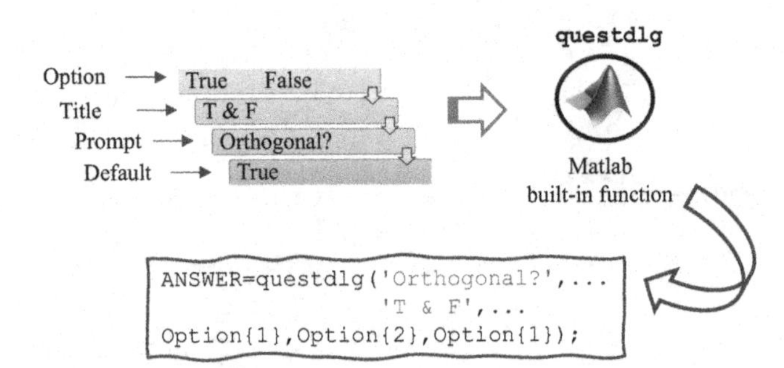

FIGURE 9.7 Schematic representation of Code 9.4.

9.2.3 Applying the orthogonal function

After the definition of the problem type being considered (i.e., regression or classification), the user should determine whether or not to consider orthogonality, a well-known technique in machine learning applications (Wu et al., 2022; Yarrakula & Dabbakuti, 2022). An example of an orthogonal matrix was previously shown in Fig. 9.2. As was seen for the selection of the problem type, the questdlg command (detailed in Code 9.4) is used to prompt the user whether to apply orthogonal matrices. In this section of the code, the qstring is set to "Orthogonal?," the dialog box title is "T & F," and the default value is set to true (Option{1}). The selection of "True" results in the algorithm calculating the orthogonal of the randomly allocated hidden nodes matrices (i.e., bias of hidden neurons and input weights). If the user selects "False," the randomly allocated hidden node matrices are applied in subsequent steps without orthogonalization. The application of orthogonality to the initial randomly assigned matrices was not considered in the original version of the OSELM, however, the model developed in this text considers it as a selective user-defined parameter. The schematic representation of Code 9.4 is provided in Fig. 9.7. Furthermore, the resulting question dialog box that is output from the execution of this code is also presented in Fig. 9.8.

Code 9.4

```
1   %% Orthogonal
2   Option{1}='True';
3   Option{2}='False';
4   ANSWER=questdlg('Orthogonal?',...
5                        'T & F',...
6                        Option{1},Option{2},Option{1});
7   pause(0.1);
8   switch ANSWER
9       case Option{1}
10          nn.orthogonal = 'true';
11      case Option{2}
12          nn.orthogonal = 'false';
13  end
```

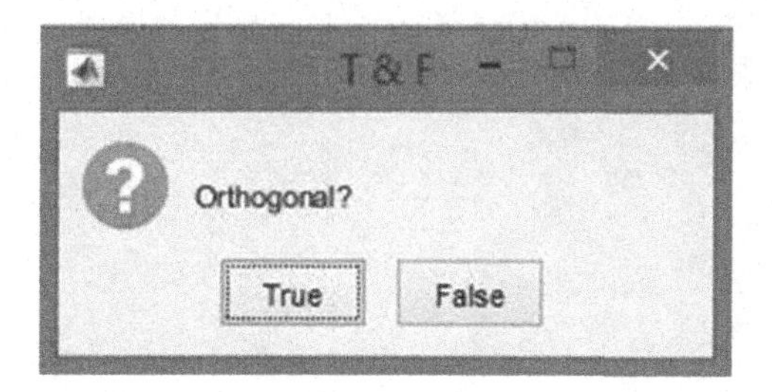

FIGURE 9.8 Question dialog box generated using the "questdlg" command regarding orthogonalization of randomly assigned matrices.

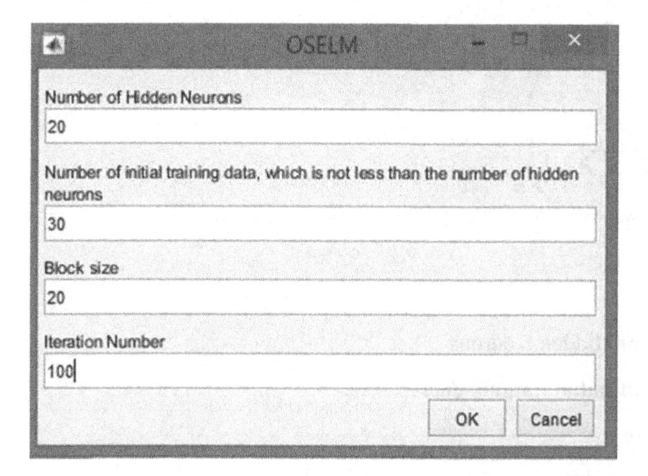

FIGURE 9.9 Dialog box generated using the `inputdlg` command to assign the values for four OSELM parameters.

9.2.4 Online sequential extreme learning machine parameters

In the next step, the user should define four general and OSELM-specific parameters, including the number of hidden neurons (NHNs), the number of initial training data, the block size, and the iteration number. The NHN should be selected in such a way that the proposed model not only performs well in estimating the value of the target parameter related to training samples but also has good generalizability to be able to estimate testing samples that have no role in the model training. The selection should also be done in such a way that the model performance during both the training and testing phases is balanced and similar to one another. Therefore, considering the minimum value (i.e., one) for this parameter will result in low modeling accuracy, while a large NHN will result in overfitting, which has been observed in various studies in the ML field (Bonakdari & Ebtehaj, 2021; Ebtehaj, Bonakdari, & Kisi, et al., 2021; Ebtehaj & Bonakdari, 2022a). It should be noted that the maximum allowable value of NHN is calculated using Eq. (8.44) (Ebtehaj, Soltani, Amiri, et al., 2021; Ebtehaj & Bonakdari, 2022b; Zeynoddin et al., 2020). In order to select the NHN that satisfies both the generalizability and predictive accuracy constraints, a trial-and-error approach should be used to determine the range of permissible NHN values. Another parameter that must be defined is the number of initial training data. The value of this parameter is used for the initial modeling of OSELM, with it being generally suggested that its value be greater than the NHN. The remaining training samples (after subtraction of the number of initial training data) are then converted into multiple chunks using the defined block size. The model output weights are recalibrated for each chunk of data processed. The final parameter is the iteration number. A sufficient number of iterations should be considered to overcome the impact of the random allocation of the hidden node matrices. Indeed, applying this parameter leads to many modeling iterations so that the effect of the random allocation of hidden layer nodes is eliminated as much as possible. Due to the fact that the training speed in OSELM is very high and the model is generally executed in several seconds each time, it is suggested that the value of this parameter be considered at least equal to 1000, however, it can be up to 10,000–20,000.

In order to assign these user-defined input parameters to the model, the inputdlg command is applied as shown in Code 9.5. The dialog box that this command produces upon execution of the program is shown in Fig. 9.9. The inputdlg command is applied because the number of user inputs exceeds the three permissible by the questdlg command. The Prompt, title, and default value are specified in lines 2–6, and called upon in line 7 by the inputdlg command. The values returned by the user are stored in the variables P1-P4. Sequentially, in lines 14–17, the character strings returned are converted to their respective numerical values by the str2num function. The schematic representation of Code 9.5 is provided in Fig. 9.10.

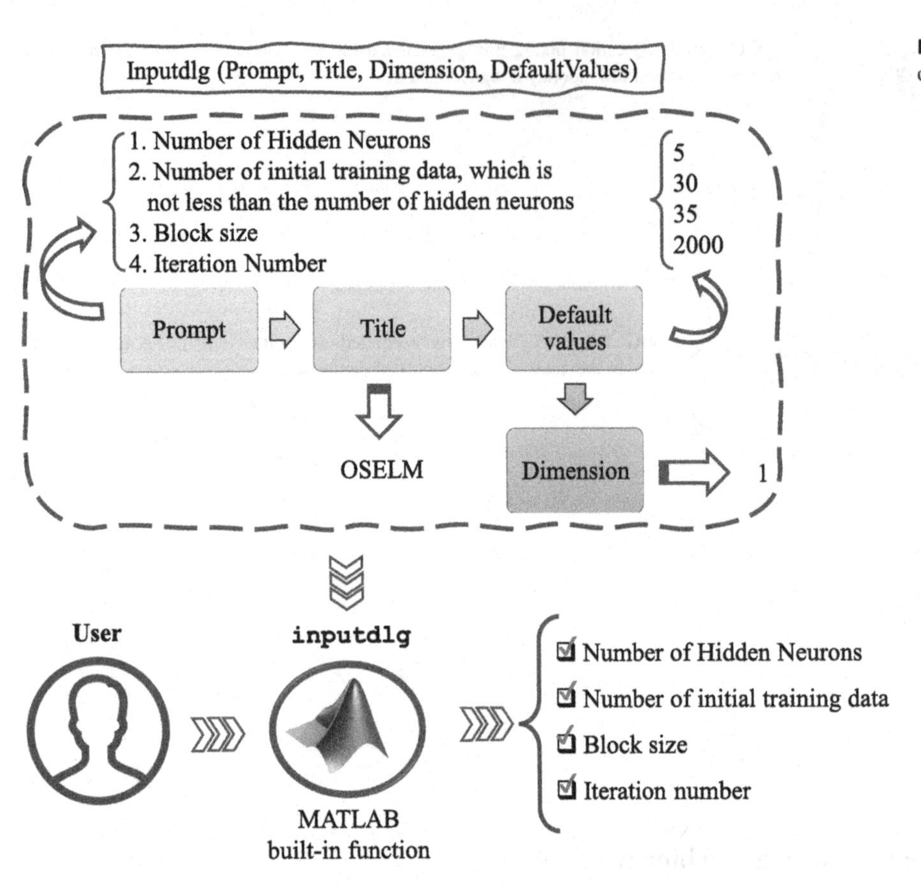

FIGURE 9.10 Schematic representation of Code 9.5.

Code 9.5

```
1   %% User-defined parameters
2   Prompt={'Number of Hidden Neurons',...
3       'Number of initial training data, which is not less than the number
    of hidden neurons',...
4       'Block size',...
5       'Iteration Number'};
6   Title='OSELM';
7   DefaultValues={'20','30','20','100'};
8   PARAMS=inputdlg(Prompt,Title,1,DefaultValues);
9   P1=PARAMS{1};
10  P2=PARAMS{2};
11  P3=PARAMS{3};
12  P4=PARAMS{4};
13
14  nHiddenNeurons=[str2num(P1)];
15  N0 = [str2num(P2)];
16  Block = [str2num(P3)];
17  N = str2num(P4);
```

The last step during model initialization prior to the training stage is to select the activation function. The activation functions that may be applied in this code include the hyperbolic tangent (tanh) (Ebtehaj et al., 2018; Maimaitiyiming et al., 2019; Ratnawati et al., 2020), triangular basis (Tribas) (Azimi et al., 2021; Sattar et al., 2019), radial basis (Radbas) (Samal et al., 2022; Tripathi et al., 2020), hard-limit (Hardlim) (Suchithra & Pai, 2020; Zeynoddin et al.,

2018), sigmoid (Sig) (Calabrò et al., 2021; Ebtehaj et al., 2017), and sine (Sin) (Owolabi & Abd Rahman, 2021) functions. Since the questdlg command can only contain three entries, the spaceList command is employed to list the five most common activation functions. The MATLAB code for the selection of the activation function is shown in Code 9.6. The schematic representation of Code 9.6 is provided in Fig. 9.11. The window that the user is prompted with during program execution is demonstrated in Fig. 9.12.

Code 9.6

```
1   spaceList = {'Hyperbolic tangent','Triangular basis','Radial
    basis','Hard-limit','Sigmoid','Sine'};
2         [idx, tf] = listdlg('ListString', spaceList,...
              'SelectionMode', 'Single', 'PromptString', 'Select item',
3   'Initialvalue', 1,'Name', 'Activation Function Type','ListSize',[300
    100]);
4         if idx == 1
5             ActivationFunction='tanh';
6         elseif idx == 2
7             ActivationFunction='tribas';
8         elseif idx == 3
9             ActivationFunction='radbas';
10        elseif idx == 4
11            ActivationFunction='hardlim'
12        elseif idx == 5
13            ActivationFunction='sig';
14        elseif idx == 6
15            ActivationFunction='sin' ;
16        end
```

9.2.5 Modeling time in online sequential extreme learning machine

After defining the ELM type, whether to consider orthogonality, the number of hidden neurons, the number of initial training samples, the block size, and the iteration number, modeling with the OSELM is commenced. The true

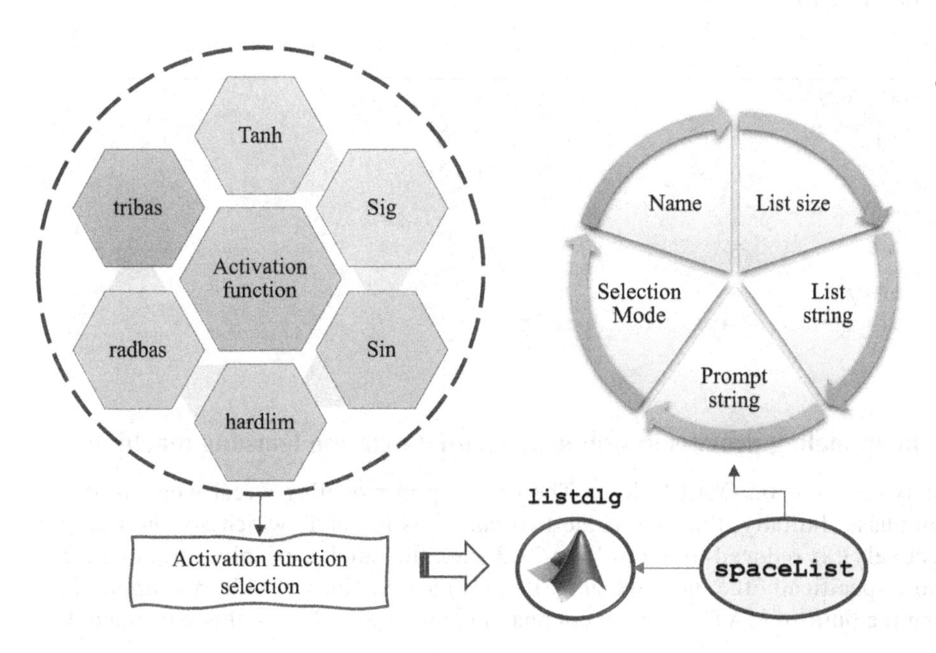

FIGURE 9.11 Schematic representation of Code 9.6.

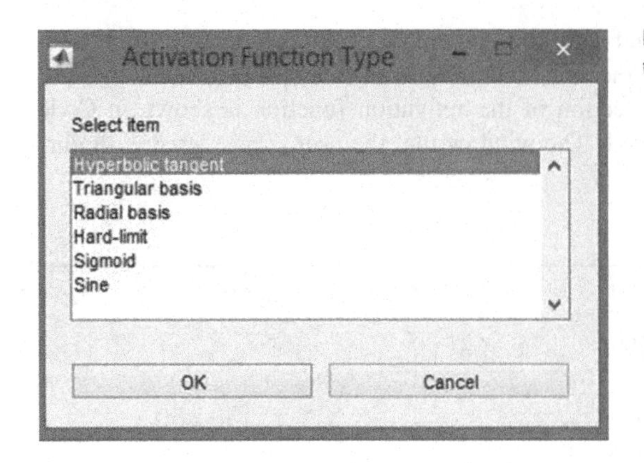

FIGURE 9.12 List selection dialog box generated using the `listdlg` command to select the activation function type.

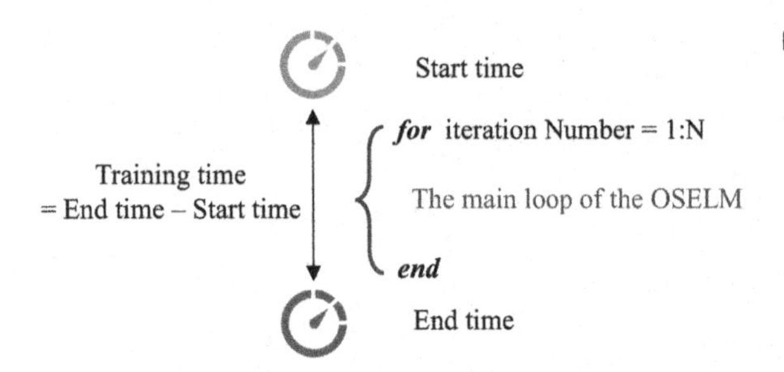

FIGURE 9.13 Schematic representation of Code 9.7.

core of the modeling step is related to the embedded loop, which iterates the OSELM modeling for the user-defined number of iterations. The detailed line-by-line coding of the main loop of OSELM modeling is presented in the coming discussion. It should be noted that in order to calculate the modeling time for all the considered iterations, the built-in cputime command is used, as shown in Code 9.7. Before the start of the main loop as well as after its end, the cputime is calculated, and the difference between the two values is considered to be the iteration modeling time. In Code 9.7, line 1 determines the initial cputime at the commencement of the modeling, while line 4 represents the main body of the code. Here, the OSELM code would be inserted. In line 7, the cputime is computed again, following the program execution, and in line 8, the difference is computed. The schematic representation of Code 9.7 is presented in Fig. 9.13.

Code 9.7

```
1   start_time_train=cputime;
2
3   for i=1: N
4       %%% Main loop
5   end
6
7   end_time_train=cputime;
8   TrainingTime=end_time_train-start_time_train;
```

9.2.6 Initialization phase of the modeling process in online sequential extreme learning machine

Code 9.8 presents the coding syntax for the main OSELM loop. The first step in modeling a real-world problem using OSELM is the initialization phase. Initially, the size of the two variables P and T, which are the training inputs and training targets (respectively), is reduced to N0 in lines 2–3 (i.e., the number of initial training samples). After this, two matrices, more specifically the input weights (i.e., IW) and the bias of hidden neurons (i.e., Bias), are randomly assigned using the built-in MATLAB rand command (Lines 5–6). To use this command, the

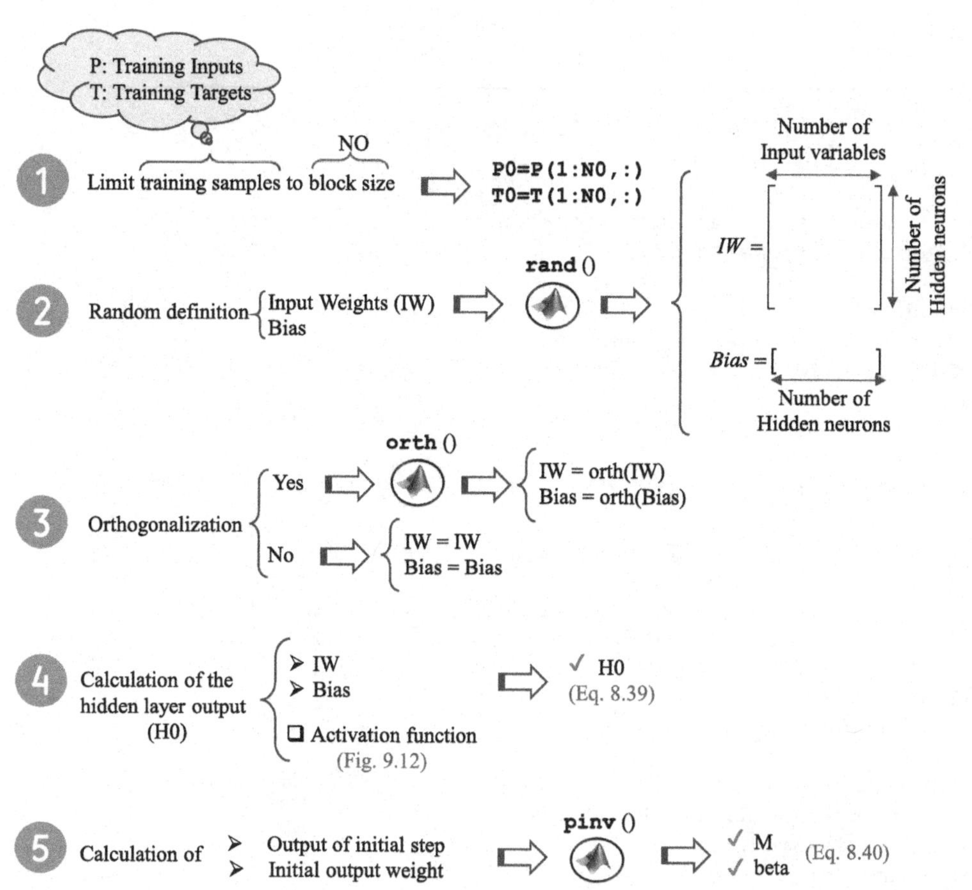

FIGURE 9.14 Schematic representation of Code 9.8.

dimensions of the desired matrix must also be specified. The number of rows and columns in the IW matrix are equal to the NHN and the number of input variables (respectively), while the Bias matrix has one column and a number of rows equaling the NHN. After randomly assigning the values of the two matrices (IW and Bias), the orthogonal of these two matrices will be taken based on whether the user selected "true" or "false" as the model input (See Fig. 9.8). For this purpose, the if-then conditional commands are used in lines 8–14. If the user has specified the orthogonalization of the two matrices, the orthogonal of these two matrices is calculated using the built-in orth command. Otherwise, the orthogonal of the matrix will not be considered. Next, according to the type of activation function selected in Fig. 9.12, the output matrix of the hidden layer is calculated (H0). The output M and Beta (lines 55–56) are considered to be the output of the initialization phase in MATLAB, where M is the output matrix (H0), and Beta is the matrix of initial output weights. It should be noted that both M and Beta are presented in Eq. (8.40). The M is a function of the output matrix of the hidden layer (H0), while the Beta is a function of H0 and initial training targets (T0). The schematic representation of Code 9.8 is presented in Fig. 9.14.

Code 9.8

```
1   %%%%%%%%%%% step 1 Initialization Phase
2   P0=P(1:N0,:);
3   T0=T(1:N0,:);
4
5   IW = rand(nHiddenNeurons,nInputNeurons)*2-1;
6   Bias = rand(1,nHiddenNeurons)*2-1;
7
8       if nn.orthogonal
9           if nHiddenNeurons > nInputNeurons
10              IW = orth(IW);
11          else
12              IW = orth(IW')';
13          end
14          Bias=orth(Bias);
15      end
16
17  switch lower(ActivationFunction)
18      case{'sig'}
19          V=P0*IW'; ind=ones(1,size(P0,1));
20          BiasMatrix=Bias(ind,:);
21          V=V+BiasMatrix;
22          H0 = 1./(1+exp(-V));
23
24      case{'tanh'}
25          V=P0*IW'; ind=ones(1,size(P0,1));
26          BiasMatrix=Bias(ind,:);
27          V=V+BiasMatrix;
28          H0 = tanh(V);
29
30      case{'sin'}
31          V=P0*IW'; ind=ones(1,size(P0,1));
32          BiasMatrix=Bias(ind,:);
33          V=V+BiasMatrix;
34          H0 = sin(V);
35
36      case{'tribas'}
37          V=P0*IW'; ind=ones(1,size(P0,1));
38          BiasMatrix=Bias(ind,:);
39          V=V+BiasMatrix;
40          H0 = tribas(V);
41
42      case{'radbas'}
43          V=P0*IW'; ind=ones(1,size(P0,1));
44          BiasMatrix=Bias(ind,:);
45          V=V+BiasMatrix;
46          H0 = radbas(V);
47
48      case{'hardlim'}
49          V=P0*IW'; ind=ones(1,size(P0,1));
50          BiasMatrix=Bias(ind,:);
51          V=V+BiasMatrix;
52          H0 = hardlim(V);
53          H0 = double(H0);
54  end
55  M = pinv(H0' * H0);
56  beta = pinv(H0) * T0;
57
```

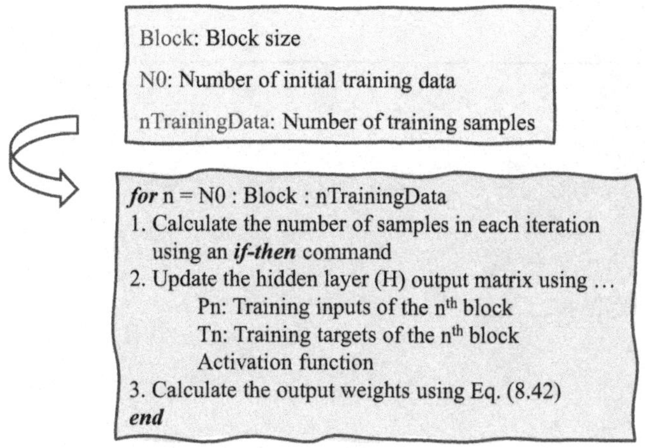

FIGURE 9.15 Schematic representation of Code 9.9.

9.2.7 Sequential learning phase of the online sequential extreme learning machine modeling process

In the initialization phase, the output weight matrix was calculated using the initial training data. The modeling process in this last phase is almost the same as the methodology provided for the original ELM, except that only part of the total data specified during data loading is used as initial training samples. After the initialization phase, the algorithm enters the sequential learning phase, which is presented in Code 9.9. In this phase, sequential learning is performed in the main loop based on the block size that the user determines at the beginning of modeling.

The number of iterations for sequential learning is defined based on the block size (block) and the range between the N0 parameter (i.e., initial training samples) and the maximum number of training samples (i.e., nTrainingData). After defining this domain, the sum of block size and the value of *n*, the minimum value of which is equal to the number of initial samples and the greater of which is equal to the maximum number of training samples, is examined. If the value of *n* + Block-1 is greater than the maximum number of training samples, the value of the input and output parameters of the problem (P and T) is in the range n and the maximum number of training samples. Otherwise, the value of these two parameters (P and T) is in the range *n* and *n* + Block-1. After defining the problem inputs (P) and the target parameter (T), with the matrices of the bias of hidden neurons (i.e., Bias) and the input weights (i.e., IW) having been defined in the previous phase, the value of the V matrix is calculated in line 13. Recall that the V is the output of the hidden layers for an iteration prior to the activation function having been applied. After calculating this matrix and using the activation function selected by the user, the output matrix value of the hidden neurons (i.e., H) is calculated in lines 18–30. With the value of H, which is recalculated in the second phase, the value of the output weight (i.e., Beta) is calculated using Eqs. (8.37) and (6.38) in lines 33–34. This process continues until the defined cycle is completed so that the output weight value is recalibrated using the new data received at each step. The schematic representation of Code 9.9 is presented in Fig. 9.15.

Code 9.9

```
1    %%%%%%%%%%%% step 2 Sequential Learning Phase
2    for n = N0 : Block : nTrainingData
3        if (n+Block-1) > nTrainingData
4            Pn = P(n:nTrainingData,:);
5            Tn = T(n:nTrainingData,:);
6            Block = size(Pn,1);            %%%% correct the block size
7            clear V;                       %%%% correct the first dimension of V
8        else
9            Pn = P(n:(n+Block-1),:);
10           Tn = T(n:(n+Block-1),:);
11       end
12
13       V=Pn*IW';
14       ind=ones(1,size(Pn,1));
15       BiasMatrix=Bias(ind,:);
16       V=V+BiasMatrix;
17
18       switch lower(ActivationFunction)
19           case{'sig'}
20               H = 1./(1+exp(-V));
21           case{'tanh'}
22               H = tanh(V);
23           case{'sin'}
24               H = sin(V);
25           case{'tribas'}
26               H = tribas(V);
27           case{'radbas'}
28               H = radbas(V);
29           case{'hardlim'}
30               H = hardlim(V);
31
32       end
33       M = M - M * H' * (eye(Block) + H * M * H')^(-1) * H * M;
34       beta = beta + M * H' * (Tn - H * beta);
35   end
```

9.2.8 Finding the optimum model from the online sequential extreme learning machine training process

Using the estimated output weights, the developed model is employed to calculate the target value for both training and testing samples. The modeling process that has thus far been described, including the initialization, the sequential learning phase, as well as the calculation of the value of the target parameter for training and testing samples, is performed iteratively according to the number of iterations defined by the user at the commencement of modeling. The values of train outputs and test outputs calculated by the developed OSELM are stored for each iteration. In addition, the three basic OSELM matrices are stored, including the input weights, the bias of hidden neurons, and the output weights. Recall that the first two of these matrices were randomly determined while the matrix of outputs weights was computed during the model training operation. The model error value for testing samples is also shown in the command window in each iteration. After the modeling is completed, the best-performing model using the testing error during the specified number of iterations is located using the find command shown in Code 9.10. The schematic representation of this code is provided in Fig. 9.16.

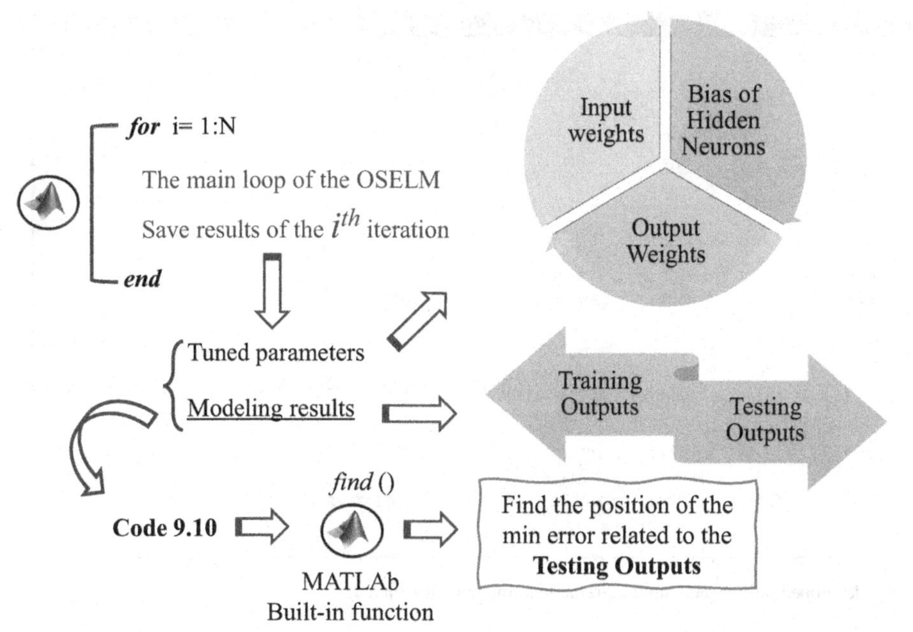

FIGURE 9.16 Schematic representation of code 9.10.

Code 9.10

```
1    BestPosition = find(Error==min(Error));
```

After finding the best model, the modeling time, the minimum error obtained, and the position of the best model are displayed in the command window using the built-in disp command shown in Code 9.11.

Code 9.11

```
1    display(['      ']);
2    disp(['Training Time = ' num2str(TrainingTime)])
3    display(['      ']);
4    disp(['Min Error is ' num2str(min(Error)) ' at position '
     num2str(BestPosition)]);
5    disp(['Training Time = ' num2str(TrainingTime)])
```

9.2.9 Quantitative and qualitative exhibition of the online sequential extreme learning machine results

In Fig. 9.17 below, it can be seen that by using the "Statistical_Indices" function, the results of eleven statistical indices are calculated for the training and testing stages and are displayed in the command window. In this figure, the best cost for each iteration is shown in the form of "Iteration A: Best Cost = B" where A and B denote the iteration number and the cost of iteration A, respectively. The training time required to consider the user-defined number of iterations is presented in the command window, as well as the best position and its corresponding cost value. The statistical indices of the model for this best position are also provided.

Fig. 9.18 presents the results of the developed OSELM model for the training and testing stages both in terms of qualitative and quantitative assessment. In this figure, we can see that a new window in MATLAB displays three subfigures to the user. The first subfigure is positioned across the top of the window and plots the observed and predicted output (estimated by the developed OSELM model) values for all samples (training or testing). From the figure, we can see that the number of samples is provided on the abscissa while the observed and predicted output values are plotted on the ordinate. The subfigure in the bottom left of the window displays the computed error for

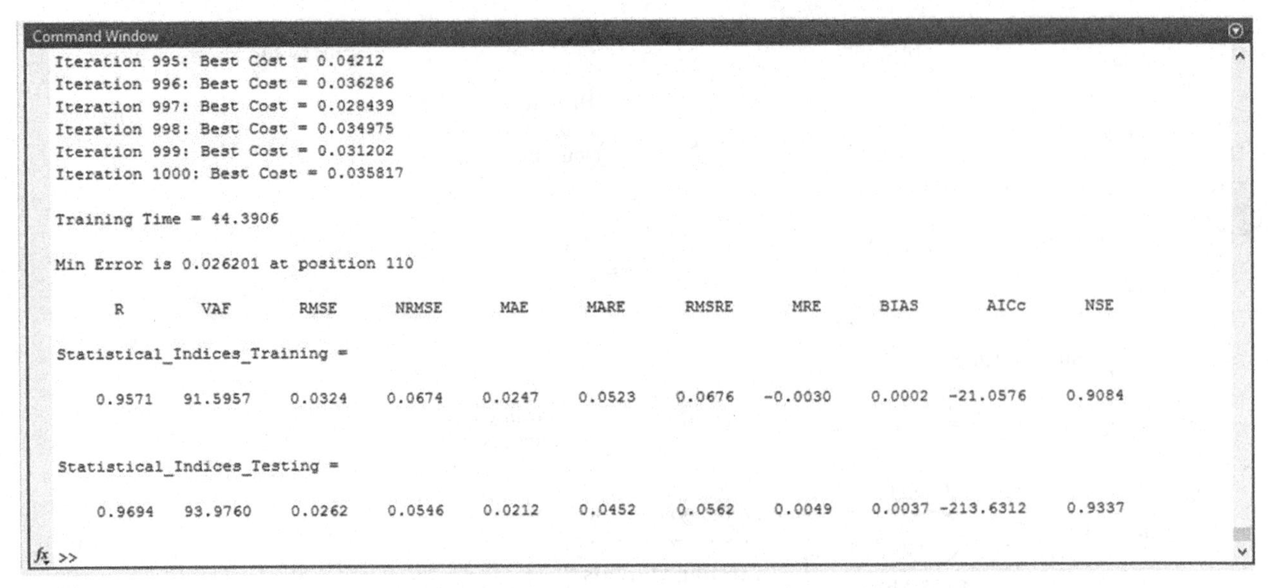

FIGURE 9.17 Command window output of the developed online sequential extreme learning machine model.

each sample considered in the developed OSELM model. In this case, the number of data samples is plotted on the abscissa while the observed error is plotted on the ordinate axis. In addition, the mean square error (MSE) and root mean square error (RMSE) of the developed model are also provided at the top of this subfigure. The subfigure shown on the bottom right-hand side of the window displays the histogram of the errors and the related Normal distribution. Moreover, the mean and standard deviation (StD) of the error is also provided at the top of this subfigure.

Fig. 9.19 presents the scatter plots of the developed OSELM-based model for the training and testing stages. The abscissa displays the observed values (i.e., Target), while the ordinate demonstrates the estimated values by the developed OSELM model (i.e., Output). The dashed Y = T line demonstrates the perfect correlation and is useful in providing a rapid assessment of the model performance in predicting the target variable. If all of the circles in this figure are positioned on the Y = T line, the developed OSELM model contains zero error. The blue fit line indicates the best linear relationship between the target and output values. The equation describing this linear relationship is displayed on the ordinate axis for the convenience of the reader. In addition, the Pearson correlation coefficient of the developed model for the training and testing stages is provided at the top of each subfigure.

Fig. 9.20 shows the error of the developed OSELM model for all iterations (i.e., 1000 iterations for the current examples). From this figure, we can see that the highest error is almost two times greater than the lowest one (for the example being considered). From this, it can be concluded that it is necessary to consider a large number of iterations in order to overcome the initial random allocation of the input weights and bias of the hidden neurons.

9.3 Online sequential extreme learning machine calculator

One of the main advantages of developed OSELM is the provision of a calculator that can be used to easily compute the value of the target parameter for future inputs that may not be available during modeling.

This calculator is again developed using the MATLAB environment. The first window that the user is prompted with is shown in Fig. 9.21 below and asks the user to define the number of input variables. After defining the number of model input variables, a new window opens, which prompts the user to enter the value of each of the input variables (Fig. 9.22). Following this, the calculator outputs the predicted result. Indeed, the use of this calculator does not require any knowledge of OSELM or MATLAB and only requires the ability to run an m-file.

The coding process of the calculator presented in this section is shown in Code 9.12. First, the modeling results stored under the name Results are called in line 6. Using the inputdlg command, the user is prompted to specify the number of input variables for the problem being considered in line 11 (Fig. 9.21). The use of the inputdlg command has

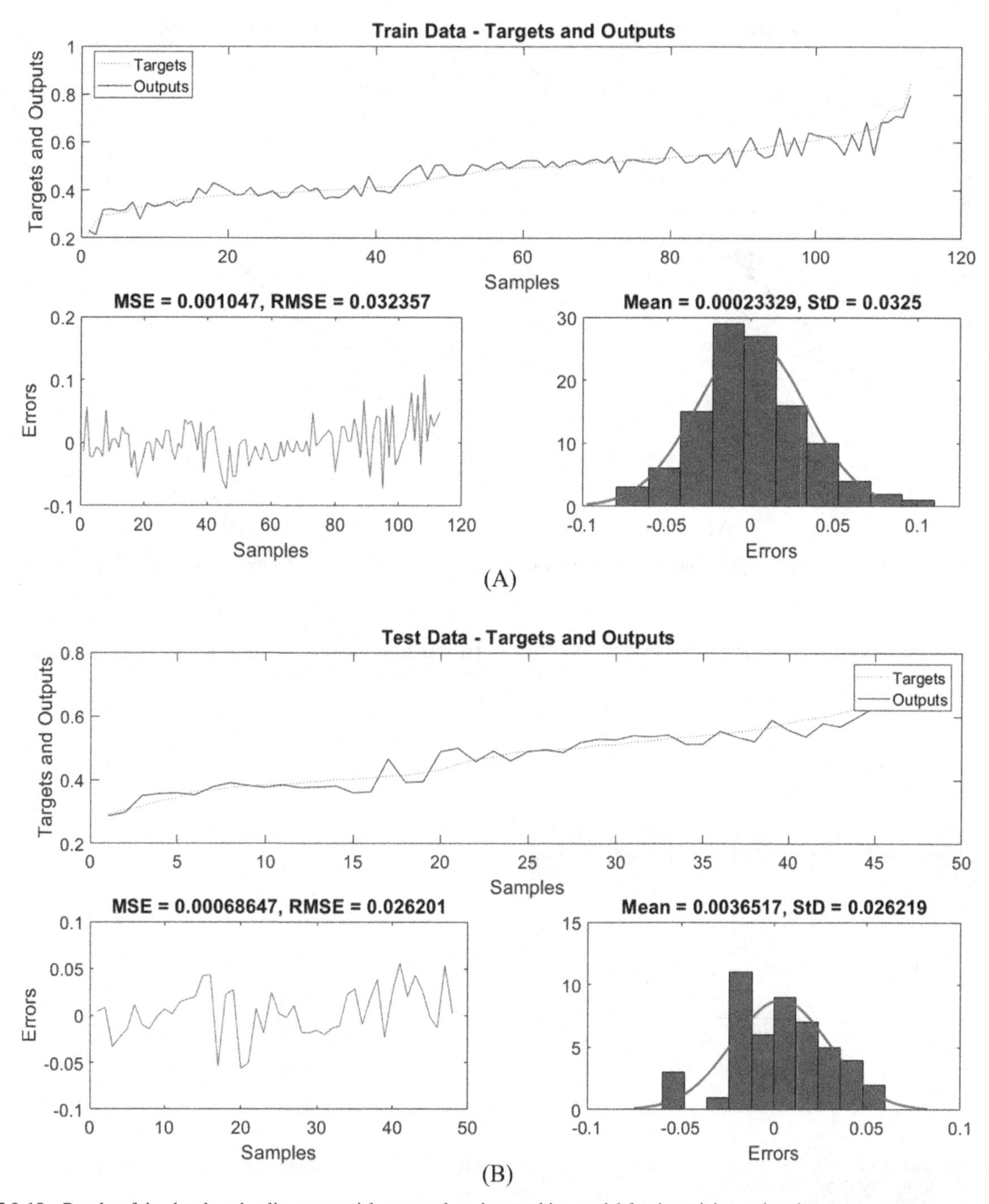

FIGURE 9.18 Results of the developed online sequential extreme learning machine model for the training and testing stages.

previously been described in detail. Once the number of input variables has been defined, the value of each input parameter is entered by the user, again using the Inputdlg command. This is coded in lines 15–139 (Code 9.12b to Code 9.12k) for up to 10 input variables. Next, according to the specified activation function type, which is also stored in the Results file, the output value of the problem is calculated (Code 9.12 l). The schematic representation of Code 9.12 is presented in Fig. 9.23.

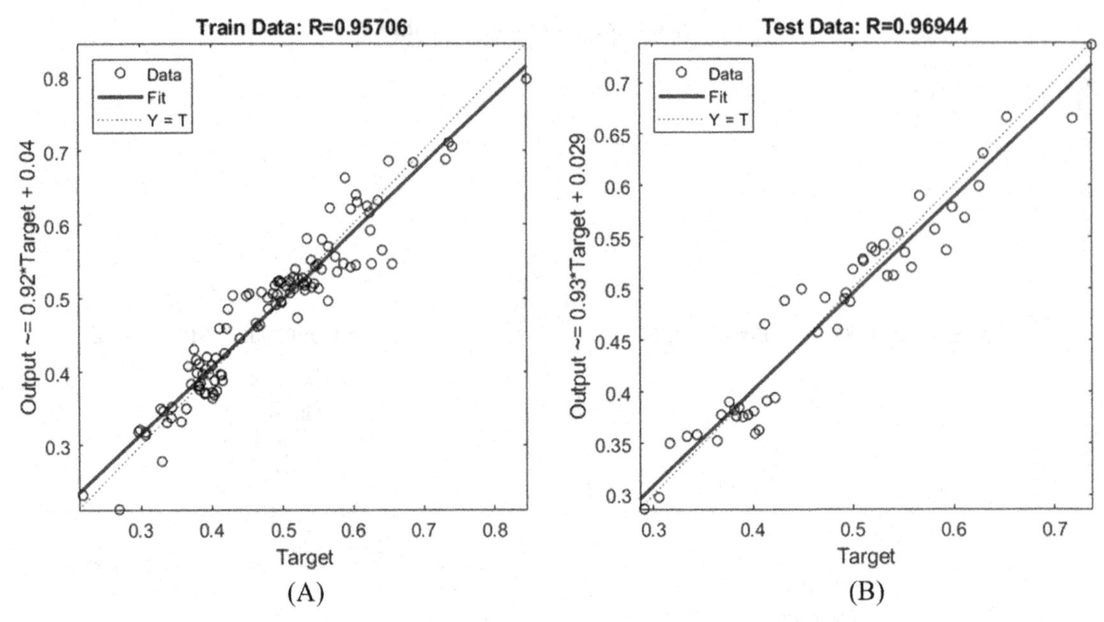

FIGURE 9.19 Scatter plot of the developed online sequential extreme learning machine-based model for the training and testing stages.

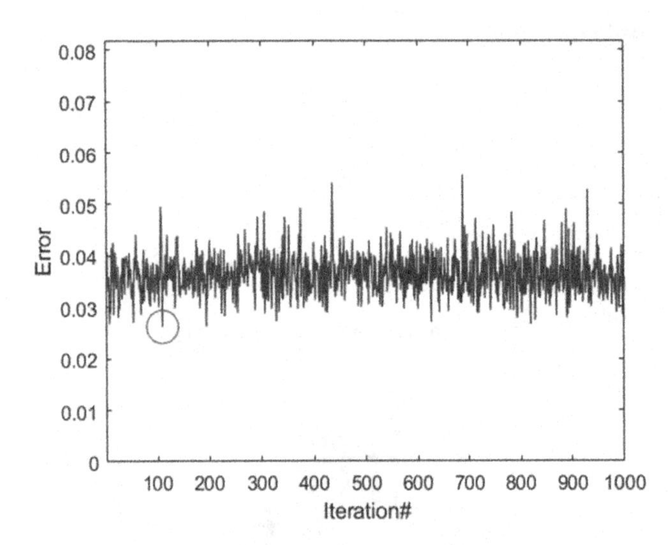

FIGURE 9.20 Error of the developed online sequential extreme learning machine for each iteration.

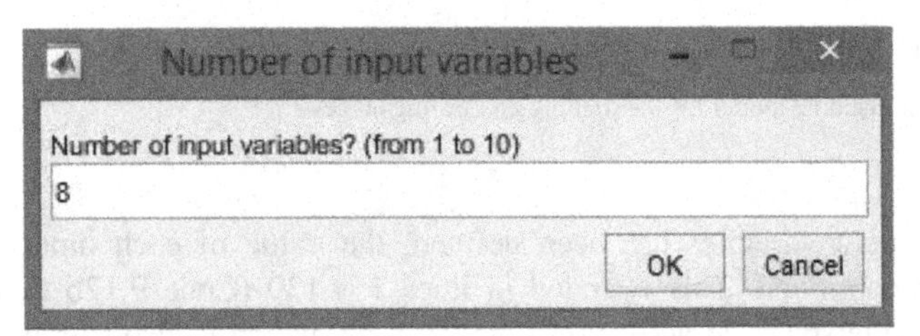

FIGURE 9.21 Calculator dialog box used to assign the number of input variables.

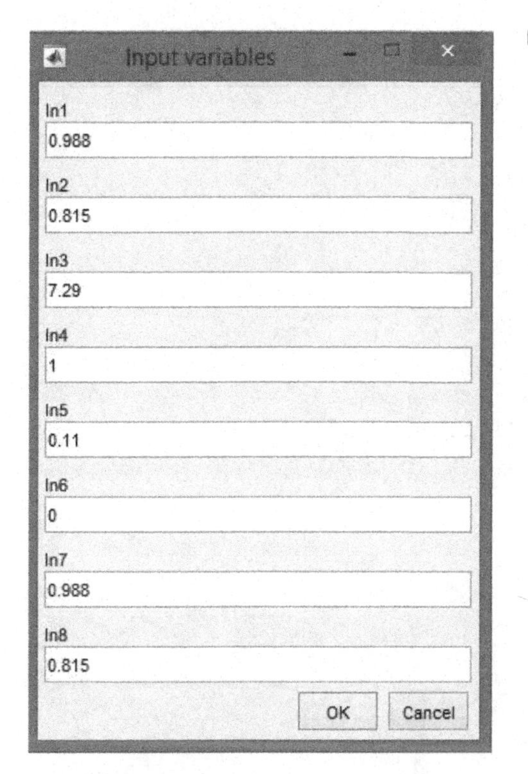

FIGURE 9.22 Calculator dialog box used to assign the value of the input variables.

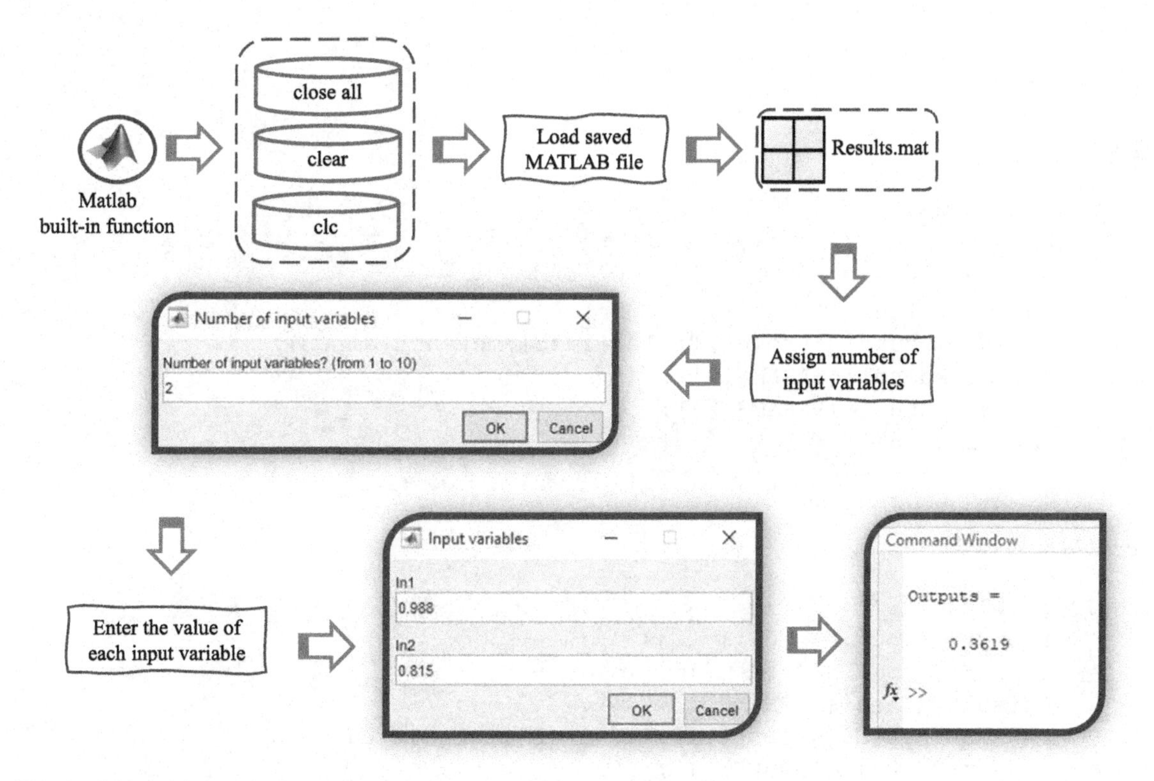

FIGURE 9.23 Schematic representation of Code 9.12.

Code 9.12a

```
1   close all
2   clear
3   clc
4
5   %% Load data
6   load('Results')
7
8   Prompt={'Number of input variables? (from 1 to 10)'};
9   Title='Number of input variables';
10  DefaultValues={'6'};
11  PARAMS=inputdlg(Prompt,Title,[1 56],DefaultValues);
12  NIV=str2double(PARAMS{1});
```

Code 9.12b

```
13
14  %% Output Calculation
15  if NIV==1
16      Prompt={'In1'};
17      Title='Input variables';
18      DefaultValues={'0.988'};
19      PARAMS=inputdlg(Prompt,Title,[1 45],DefaultValues);
20      In1=str2double(PARAMS{1});
21      InV=[In1];
22
```

Code 9.12c

```
23  elseif NIV==2
24      Prompt={'In1','In2'};
25      Title='Input variables';
26      DefaultValues={'0.988','0.815'};
27      PARAMS=inputdlg(Prompt,Title,[1 45],DefaultValues);
28      In1=str2double(PARAMS{1});
29      In2=str2double(PARAMS{2});
30      InV=[In1 In2];
31
```

Code 9.12d

```
32  elseif NIV==3
33      Prompt={'In1','In2','In3'};
34      Title='Input variables';
35      DefaultValues={'0.988','0.815','7.29'};
36      PARAMS=inputdlg(Prompt,Title,[1 45],DefaultValues);
37      In1=str2double(PARAMS{1});
38      In2=str2double(PARAMS{2});
39      In3=str2double(PARAMS{3});
40      InV=[In1 In2 In3];
41
```

Code 9.12e

```
42   elseif NIV==4
43       Prompt={'In1','In2','In3','In4'};
44       Title='Input variables';
45       DefaultValues={'0.988','0.815','7.29','1'};
46       PARAMS=inputdlg(Prompt,Title,[1 45],DefaultValues);
47       In1=str2double(PARAMS{1});
48       In2=str2double(PARAMS{2});
49       In3=str2double(PARAMS{3});
50       In4=str2double(PARAMS{4});
51       InV=[In1 In2 In3 In4];
52
```

Code 9.12f

```
53   elseif NIV==5
54       Prompt={'In1','In2','In3','In4','In5'};
55       Title='Input variables';
56       DefaultValues={'0.988','0.815','7.29','1','0.11'};
57       PARAMS=inputdlg(Prompt,Title,[1 45],DefaultValues);
58       In1=str2double(PARAMS{1});
59       In2=str2double(PARAMS{2});
60       In3=str2double(PARAMS{3});
61       In4=str2double(PARAMS{4});
62       In5=str2double(PARAMS{5});
63       InV=[In1 In2 In3 In4 In5];
64
```

Code 9.12g

```
65   elseif NIV==6
66       Prompt={'In1','In2','In3','In4','In5','In6'};
67       Title='Input variables';
68       DefaultValues={'0.988','0.815','7.29','1','0.11','0'};
69       PARAMS=inputdlg(Prompt,Title,[1 45],DefaultValues);
70       In1=str2double(PARAMS{1});
71       In2=str2double(PARAMS{2});
72       In3=str2double(PARAMS{3});
73       In4=str2double(PARAMS{4});
74       In5=str2double(PARAMS{5});
75       In6=str2double(PARAMS{6});
76       InV=[In1 In2 In3 In4 In5 In6];
77
```

Code 9.12h

```
78   elseif NIV==7
79        Prompt={'In1','In2','In3','In4','In5','In6','In7'};
80        Title='Input variables';
81        DefaultValues={'0.988','0.815','7.29','1','0.11','0','0.988'};
82        PARAMS=inputdlg(Prompt,Title,[1 45],DefaultValues);
83
84        In1=str2double(PARAMS{1});
85        In2=str2double(PARAMS{2});
86        In3=str2double(PARAMS{3});
87        In4=str2double(PARAMS{4});
88        In5=str2double(PARAMS{5});
89        In6=str2double(PARAMS{6});
90        In7=str2double(PARAMS{7});
91        InV=[In1 In2 In3 In4 In5 In6 In7];
92
```

Code 9.12i

```
93   elseif NIV==8
94        Prompt={'In1','In2','In3','In4','In5','In6','In7','In8'};
95        Title='Input variables';
96
     DefaultValues={'0.988','0.815','7.29','1','0.11','0','0.988','0.815'};
97        PARAMS=inputdlg(Prompt,Title,[1 45],DefaultValues);
98        In1=str2double(PARAMS{1});
99        In2=str2double(PARAMS{2});
100       In3=str2double(PARAMS{3});
101       In4=str2double(PARAMS{4});
102       In5=str2double(PARAMS{5});
103       In6=str2double(PARAMS{6});
104       In7=str2double(PARAMS{7});
105       In8=str2double(PARAMS{8});
106       InV=[In1 In2 In3 In4 In5 In6 In7 In8];
107
```

Code 9.12j

```
108  elseif NIV==9
109       Prompt={'In1','In2','In3','In4','In5','In6','In7','In8','In9'};
110       Title='Input variables';

111  DefaultValues={'0.988','0.815','7.29','1','0.11','0','0.988','0.815','7.2
     9'};
112       PARAMS=inputdlg(Prompt,Title,[1 45],DefaultValues);
113       In1=str2double(PARAMS{1});
114       In2=str2double(PARAMS{2});
115       In3=str2double(PARAMS{3});
116       In4=str2double(PARAMS{4});
117       In5=str2double(PARAMS{5});
118       In6=str2double(PARAMS{6});
119       In7=str2double(PARAMS{7});
120       In8=str2double(PARAMS{8});
121       In9=str2double(PARAMS{9});
122       InV=[In1 In2 In3 In4 In5 In6 In7 In8 In9];
123
```

Code 9.12k

```
124 elseif NIV==10
125 Prompt={'In1','In2','In3','In4','In5','In6','In7','In8','In9','In10'};
126     Title='Input variables';
127 DefaultValues={'0.988','0.815','7.29','1','0.11','0','0.988','0.815','7.2
    9','1'};
128     PARAMS=inputdlg(Prompt,Title,[1 45],DefaultValues);
129     In1=str2double(PARAMS{1});
130     In2=str2double(PARAMS{2});
131     In3=str2double(PARAMS{3});
132     In4=str2double(PARAMS{4});
133     In5=str2double(PARAMS{5});
134     In6=str2double(PARAMS{6});
135     In7=str2double(PARAMS{7});
136     In8=str2double(PARAMS{8});
137     In9=str2double(PARAMS{9});
138     In10=str2double(PARAMS{10});
139     InV=[In1 In2 In3 In4 In5 In6 In7 In8 In9 In10];
140 else
141     disp("The number of input variables is not in the range of [1 10]");
142     return
143 end
144
```

Code 9.12l

```
145 tempH = (InW*InV')+BHI;
146
147 switch lower(ActivationFunction)
148     case{'s','sig','sigmoid'}     % Sigmoid
149         H = 1 ./ (1 + exp(-tempH));
150         Outputs = H'*OutW;
151     case{'t','tanh'}              % Hyperbolic Tangent
152         H = tanh(tempH);
153         Outputs = H'*OutW;
154     case {'sin','sine'}           % Sine
155         H = sin(tempH);
156         Outputs = H'*OutW;
157     case {'hardlim'}              % Hard Limit
158         H = double(hardlim(tempH));
159         Outputs = H'*OutW;
160     case {'tribas'}               % Triangular basis function
161         H = tribas(tempH);
162         Outputs = H'*OutW;
163     case {'radbas'}               % Radial basis function
164         H = radbas(tempH);
165         Outputs = H'*OutW;
166 end
```

9.4 Evaluating the effect of the online sequential extreme learning machine parameters on model performance

This section surveyed the effect of the user-defined parameters including the number of initial training data and the block size/number of blocks. The other parameters, including iteration number, NHN, activation function type, and consideration of orthogonality, have been surveyed in the previous chapters. Since the examples in the current chapter use the same data set as the previous chapters, the optimal values of each input parameter are summarized in Fig. 9.24.

9.4.1 Number of initial training samples

This subsection surveyed the effect of the number of initial training samples (NITS) on the OSELM performance. It should be noted that during the sensitivity analysis, all of the user-defined parameters are fixed except NITS. The values of the fixed parameters are taken from the optimal values presented in Fig. 9.24 for all examples. For the purpose of this exercise, the block size is considered as 20.

9.4.1.1 Optimum number of initial training samples for Example 1

Fig. 9.25 shows the effect of the number of initial training samples on the results of OSELM for Example 1 see in Appendix 9.A. According to this figure, the values of correlation-based indices (R, VAF, and NSE) do not have a defined trend, so that the highest value of VAF and R is recorded at NITS = 45, while the maximum value of NSE is calculated at NITS = 30 and 75. The performance of other indices, including RMSE, NRMSE, RMSRE, MAE, and MARE, is considered optimal at NITS = 75. Given that the difference between the maximum and minimum values of each index for different NITS is very small, it can be considered that the best overall performance occurs at NITS = 75. Therefore, it is concluded that changes in NITS value (positive or negative) are not directly correlated with changes in the model performance. In this case, the best OSELM performance is obtained at NITS = 75, which is not the minimum or maximum value of the NITS. For this data set, it can be said that the NITS value has only a marginal impact on model performance, and regardless of the value chosen, a similar result, as determined from the statistical indices, is obtained.

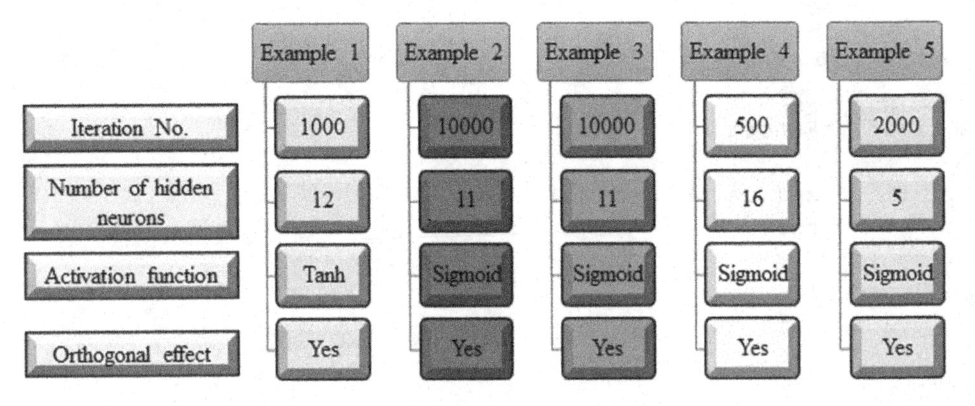

FIGURE 9.24 Optimum values of online sequential extreme learning machine model parameters for all examples.

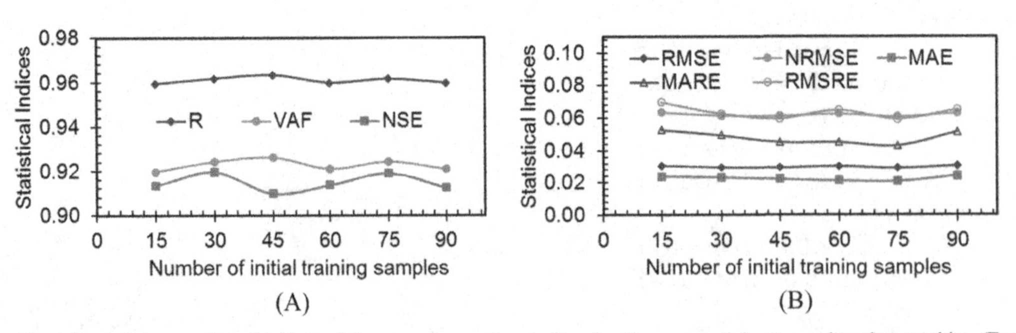

FIGURE 9.25 The effect of the number of initial training samples on the results of online sequential extreme learning machine (Example 1 see in Appendix 9.A).

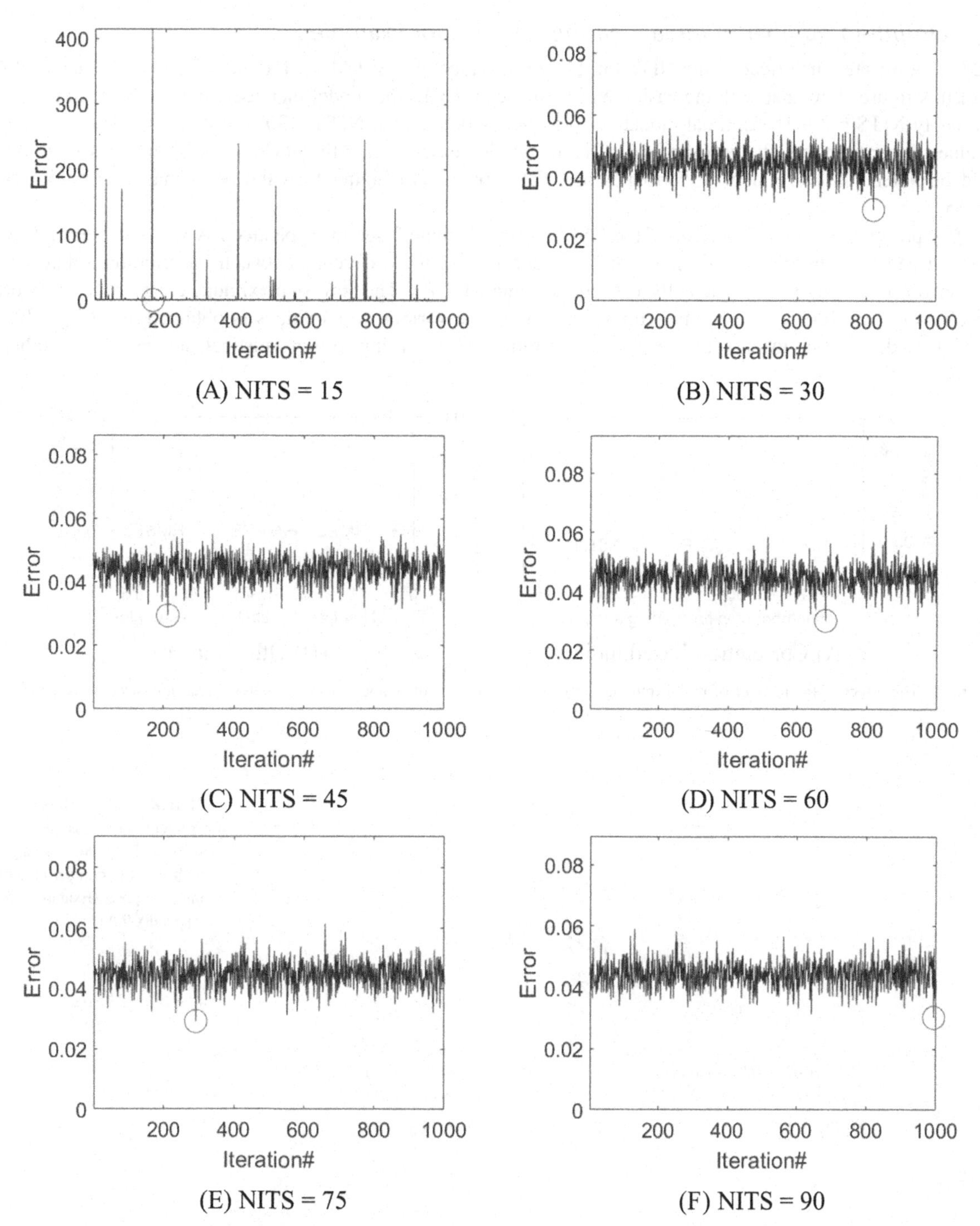

FIGURE 9.26 Modeling error in different iterations for the online sequential extreme learning machine with variable number number of initial training samples (Example 1 see in Appendix 9.A).

Fig. 9.26 indicates the error of the OSELM models when considering different numbers of initial training samples for Example 1 see in Appendix 9.A. The ratio of minimum and maximum errors recorded in all iterations for 30 $<$NITS $<$90 varies around 2, while for NITS = 15, this ratio is more than 1300. In fact, the results of this figure show that the small value of NITS in Example 1 see in Appendix 9.A affects the amplitude of the modeling error fluctuations at different iterations. Indeed, this is due to the significant effect of a small number of NITS on randomly generating input weights and the bias of hidden neurons. Due to the highly variable magnitude of modeling error observed for different iterations with NITS $<$ 30, it is advisable for the user to select higher NITS values for improved stability.

9.4.1.2 *Optimum number of initial training samples for Example 2*

Fig. 9.27 demonstrates the effect of the NITS on the performance of OSELM for Example 2 see in Appendix 9.A. The results of this figure show that with increasing NITS, the accuracy of the model increases in all indices up to a value of approximately NITS = 30. The optimal model performance is obtained at NITS = 30, and it can be clearly seen that for NITS values larger than this, the performance of the model decreases. While the model performance is seen to decrease, it should be noted that the model performance at NITS = 60 is still higher than the performance at a low value of NITS = 15.

Fig. 9.28 presents the modeling error of the OSELM for Example 2 see in Appendix 9.A with different NITS. From the figure, it can be seen that for all values of NITS considered, the difference between the minimum and maximum recorded errors is so large that it actually exceeds a value of 10^{41}. The lowest maximum error value is recorded at NITS = 60, while the highest maximum error value among the various NITS values is obtained at NITS = 30, which was previously determined to have the best model performance according to the statistical indices. On the other hand,

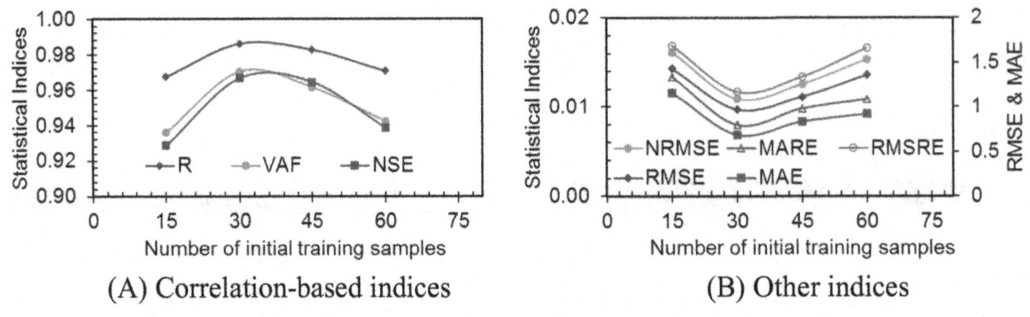

FIGURE 9.27 The effect of the number of initial training samples on the results of online sequential extreme learning machine (Example 2 see in Appendix 9.A).

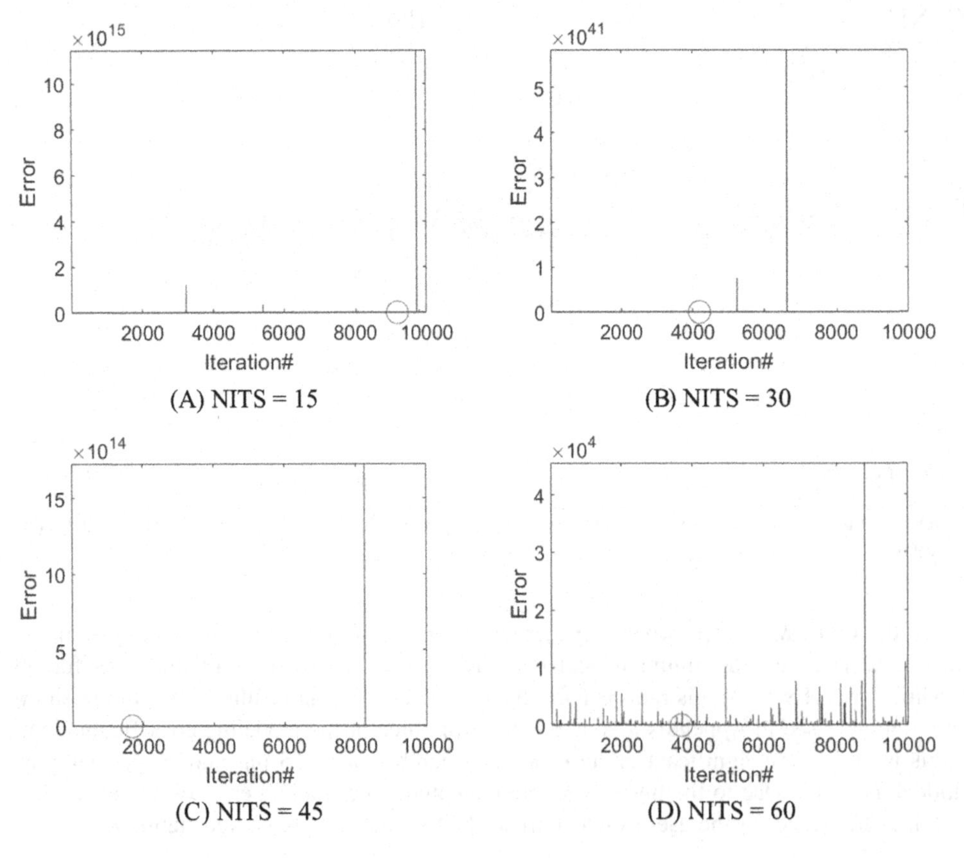

FIGURE 9.28 Modeling error in different iterations for the online sequential extreme learning machine with variable number of initial training samples (Example 2 see in Appendix 9.A).

the maximum errors for NITS = 15 and 45 are similar to each other, although their performance (in terms of statistical indices) remains very different. Therefore, it can be concluded from these results that there is no direct relationship between the amplitude of error changes and the accuracy of the superior model. In fact, using multiple iterations in the modeling process considered in this study for OSELM modeling has minimized the effects of randomly determining the matrices of input weights and bias of hidden neurons.

9.4.1.3 Optimum number of initial training samples for Example 3

Fig. 9.29 surveyed the effect of different NITS on the performance of OSELM for Example 3 see in Appendix 9.A. The changes in NITS for this example have an almost damped harmonic impact on the model performance, such that changing this parameter from 15 to 60 leads to an increase in model performance, while the weakest model performance is obtained at NITS = 75. In this case, the overall best model performance can be considered to occur at NITS = 60. Similar to the case of Example 1 see in Appendix 9.A, changes in the NITS do not have a significant relationship with the accuracy of the model. For this reason, prior to beginning the modeling process of the problem under consideration, it is not possible to define a specific range of NITS values to consider. Therefore, the optimal value of this parameter should be determined through a trial-and-error process. Similar to Example 2 see in Appendix 9.A, the observed error changes associated with each iteration are so large that the maximum value recorded is over 10^{136} (Fig. 9.30).

9.4.1.4 Optimum value of initial training samples for Example 4

Fig. 9.31 studies the effect of different NITS on the performance of the OSELM for Example 4 see in Appendix 9.A. Increasing the NITS from 15 to 30 was found to reduce the modeling accuracy, while a further increase from 30 to 45 was found to slightly improve the model accuracy. While the accuracy increased, the best performance among these three values remained NITS = 15. In general, changing the value of NITS from 45 to 60 to 90 had an upward trend in model performance, with the exception of the RMSRE at NITS = 75. In general, the best performance is obtained at NITS = 90, which has the greatest level of correlation indices (R, VAF, NSE) and the lowest level of statistical error indices. According to Fig. 9.32, in this example (as was the case for Examples 2 and 3 see in Appendix 9.A), the difference in errors between different iterations is significantly large such that the maximum error value is related to NITS = 75 (error = 10^{68}) and the minimum value is related to NITS = 90 (error = 10^{8}). In fact, the iteration process has been able to solve the problem of randomly determining matrices in the OSELM.

9.4.1.5 Optimum value of initial training samples for Example 5

Fig. 9.33 demonstrates the effect of different NITS on the results of OSELM for Example 5 see in Appendix 9.A. Increasing the NITS leads to decreasing correlation-based indices (R, VAF, NSE) in the range of 15 <NITS <60, but these indices trend upwards beyond a NITS value of 60. The trend presented in these three indices is also observed in NRMSE, RMSE, and MAE indices. The two indices MARE and RMSRE, which are relative indices, experience the greatest change in the range of 15 < NITS <30. According to consideration of all the statistical indices simultaneously, the best performance of the OSELM in modeling Example 5 see in Appendix 9.A is obtained at NITS = 30.

The error changes at all iterations are provided in Fig. 9.34. The minimum and maximum variations of the values recorded in this example are much smaller than the previous examples so that the maximum-to-minimum

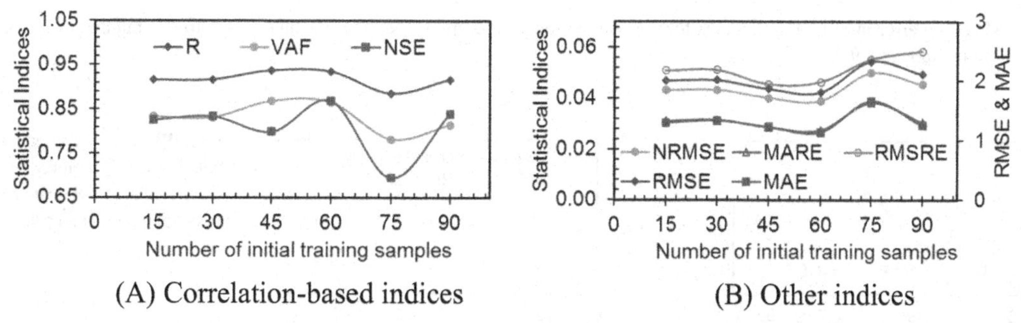

FIGURE 9.29 The effect of the number of initial training samples on the results of online sequential extreme learning machine (Example 3 see in Appendix 9.A).

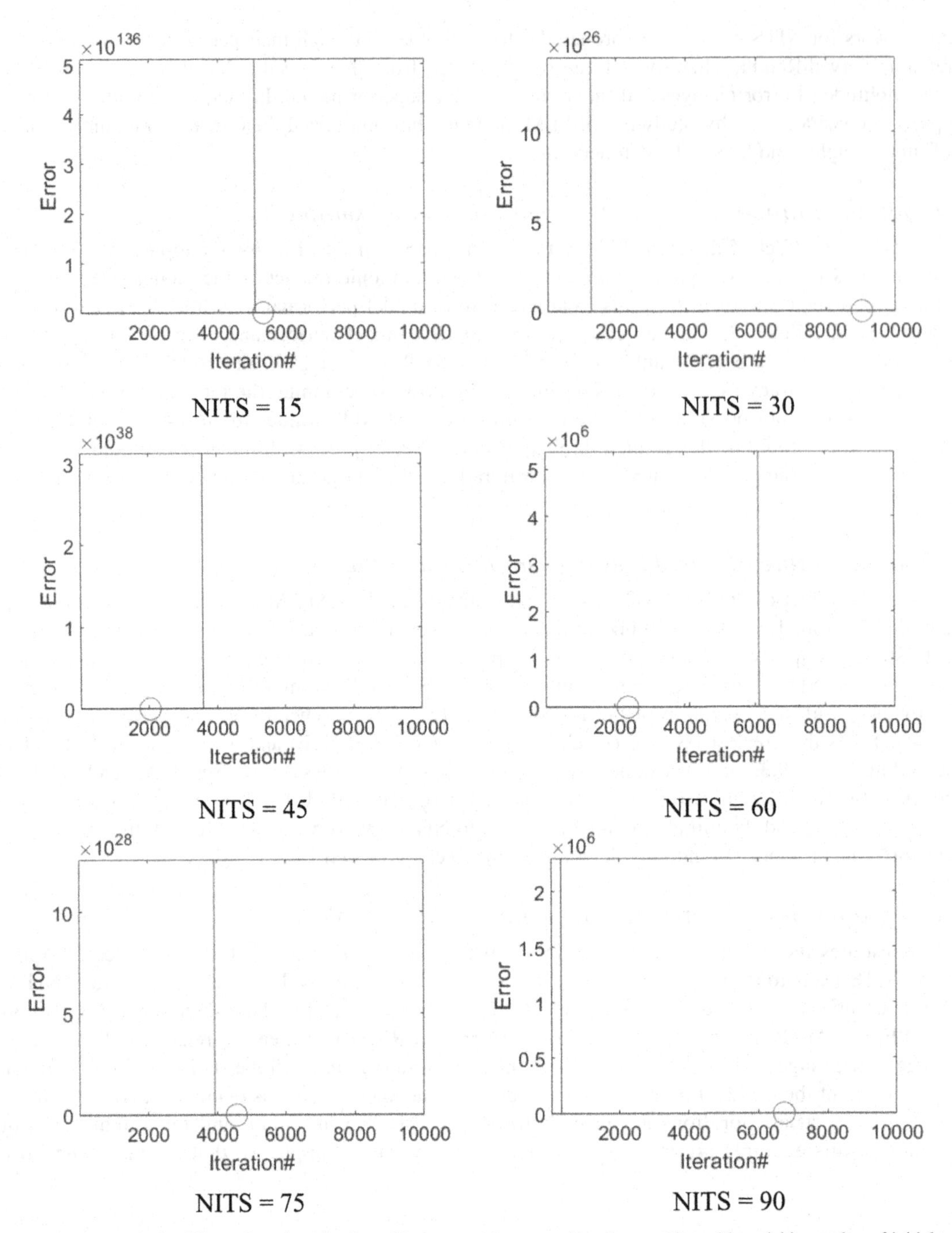

FIGURE 9.30 Modeling error in different iterations for the online sequential extreme learning machine with variable number of initial training samples (Example 3 see in Appendix 9.A).

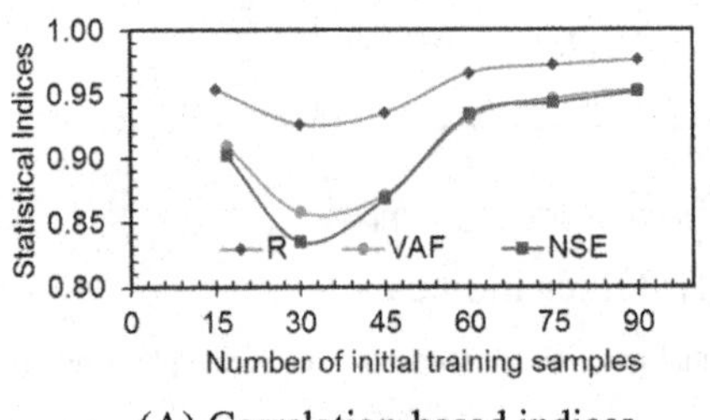

(A) Correlation-based indices

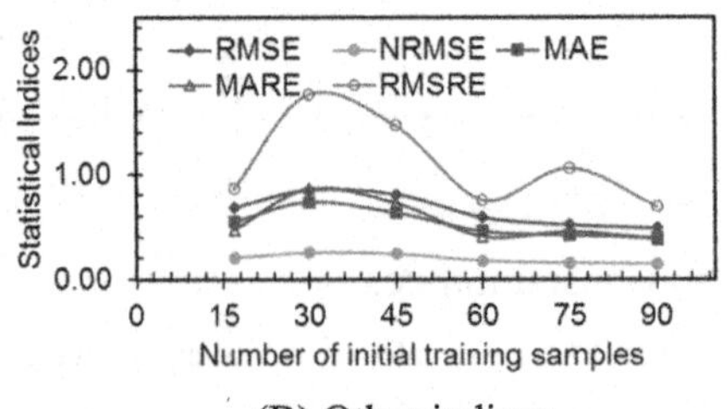

(B) Other indices

FIGURE 9.31 The effect of number of initial training samples on the results of online sequential extreme learning machine (Example 4 see in Appendix 9.A).

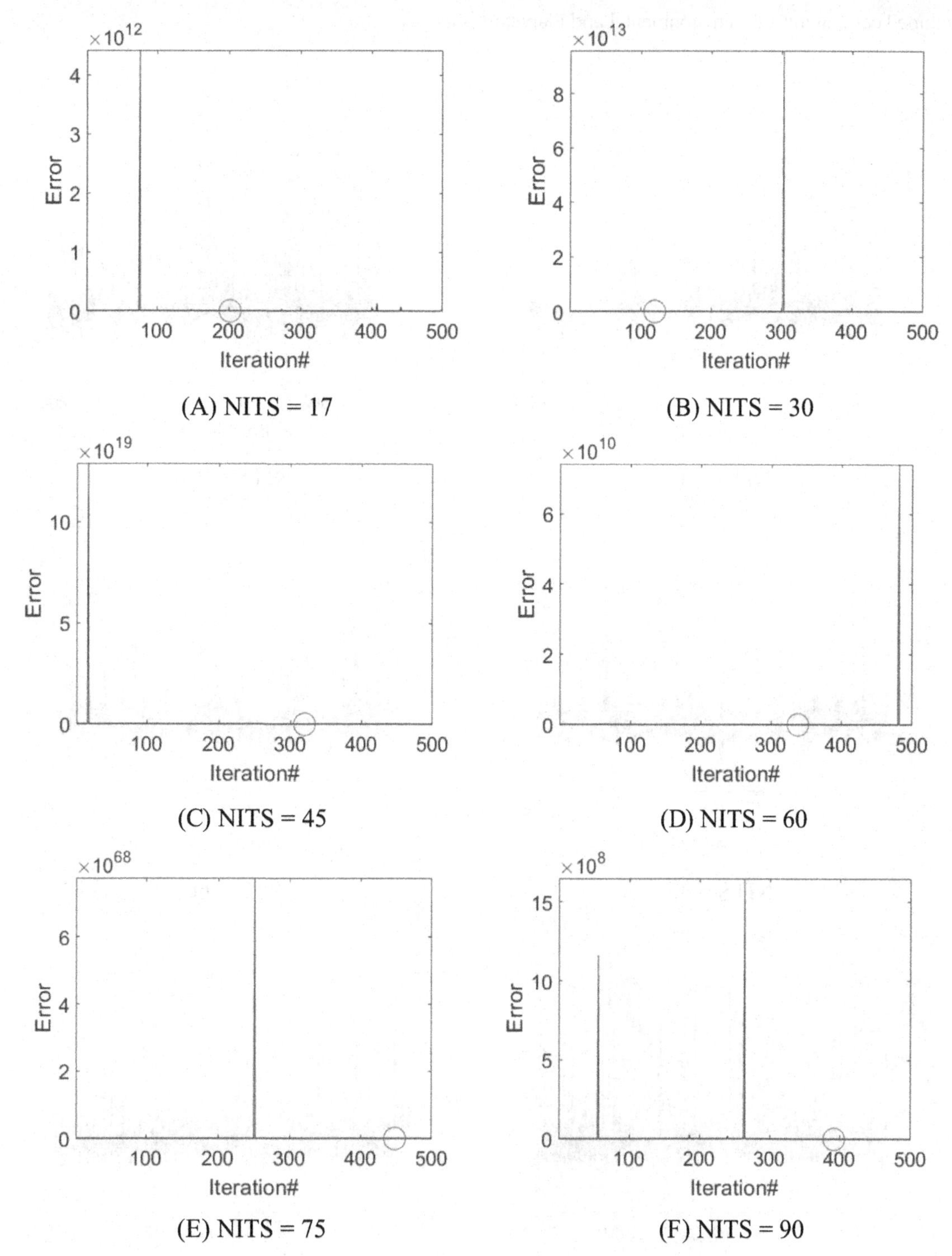

FIGURE 9.32 Modeling error in different iterations for the online sequential extreme learning machine with variable number of initial training samples (Example 4 see in Appendix 9.A).

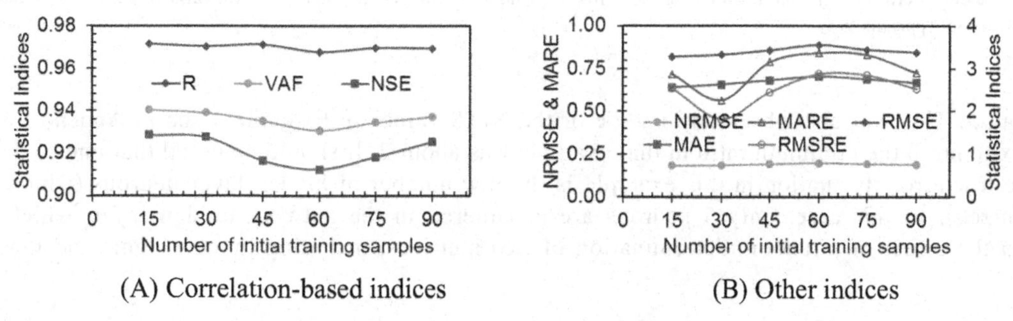

FIGURE 9.33 The effect of the number of initial training samples on the results of online sequential extreme learning machine (Example 5 see in Appendix 9.A).

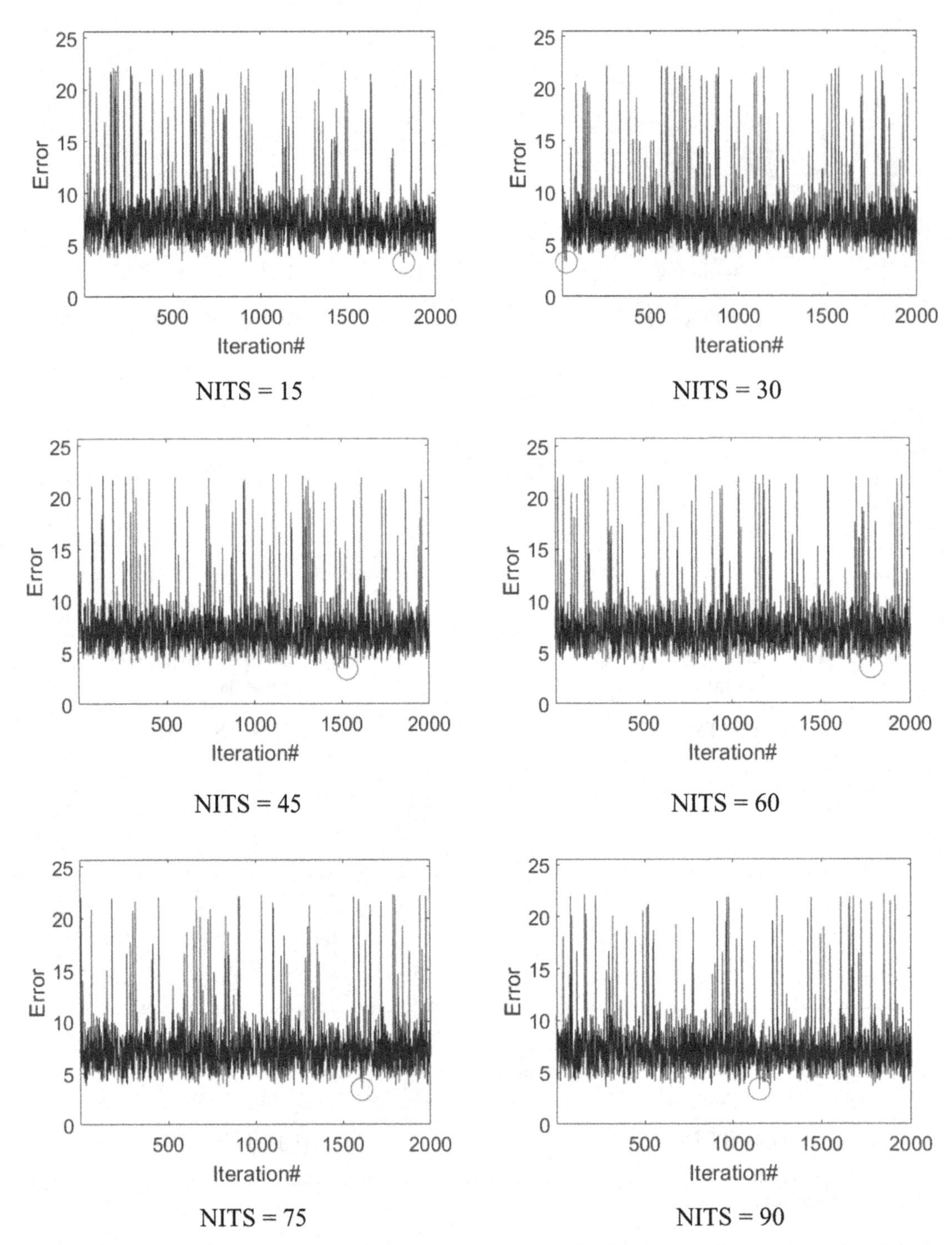

FIGURE 9.34 Modeling error in different iterations for the online sequential extreme learning machine with variable number of initial training samples (Example 5 see in Appendix 9.A).

ratio is about 5. This was also observed in some of the NITS values in Example 1 see in Appendix 9.A, except that the maximum to the minimum ratio in that example was about 2. It should be noted that one of the main reasons for the low error fluctuation in this example is the low number of hidden layer neurons (which was previously optimized). In this case, only 5 neurons are considered in the network hidden layer, which leads to a reduction in the effect of a random determination of two matrices, including input weights and bias of hidden neurons.

9.4.2 Block size

In this subsection, the effect of block size on the performance of the OSELM model is examined using five different examples. It should be noted that all the OSELM parameters are considered according to Fig. 9.23. Moreover, the NITS value is decided according to the results obtained in the previous section.

9.4.2.1 *Optimum block size for Example 1*

Fig. 9.35 demonstrates the effect of block size on the performance of the OSELM for Example 1 see in Appendix 9.A. The values of the statistical indices presented in this figure show that in order to achieve optimal model performance, it is necessary to examine different values of block size. The optimal model should have the highest value of correlation-based indicators (R, VAF, NSE) as well as having the lowest possible value of the other statistical indices (RMSE, NRMSE, MAE, MARE, RMSRE). In fact, the results of this model show that increasing or decreasing the block size has no significant relationship with the performance of the model, so that the best performance of the model is obtained by considering block size = 30, although the model results at block size = 20 are also very close to block size = 30.

9.4.2.2 *Optimum block size for Example 2*

Fig. 9.36 demonstrates the effect of block size on the performance of the OSELM for Example 2 see in Appendix 9.A. According to this figure, it is observed that the change in block size will bring the periodic oscillatory performance of indices so that in all indices, the highest value of the correlation-based index and other indices (absolute and relative) is obtained in block size = 20. The difference between the maximum and minimum relative error values (MARE) is about 0.63%, which is a small value. In fact, in this example, although block size changes have had significant effects on correlation-based indicators, the relative error changes are considered to be very small.

9.4.2.3 *Optimum block size for Example 3*

Fig. 9.37 surveyed the effect of block size on the performance of the OSELM for Example 3 see in Appendix 9.A. It can be seen that the effect of increasing the block size does not have a constant trend when considering the statistical indices, however, in general, they increase up to a value of block size = 30. The difference between the highest correlation coefficient (block size = 30) and the lowest (block size = 10) is about 7%. For the relative-based indices, the

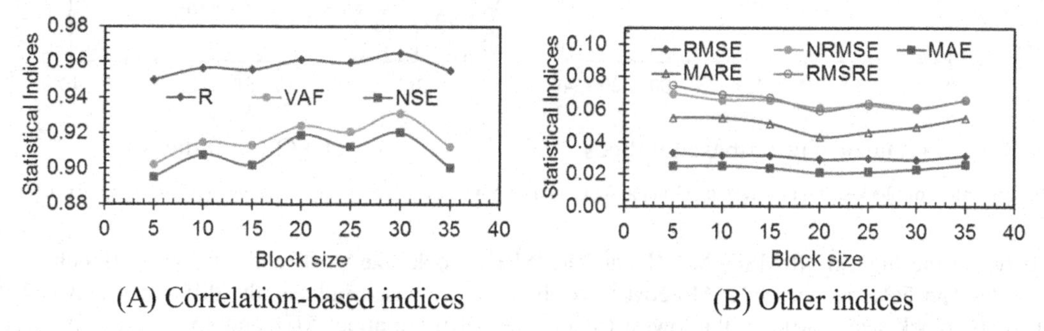

FIGURE 9.35 The effect of block size on the results of online sequential extreme learning machine (Example 1 see in Appendix 9.A).

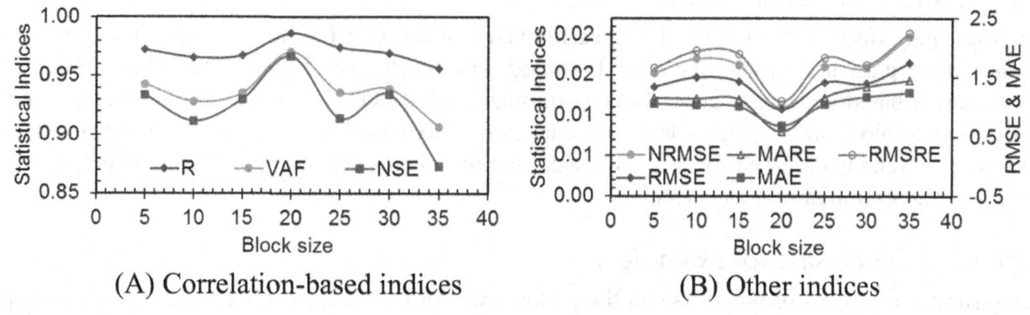

FIGURE 9.36 The effect of block size on the results of online sequential extreme learning machine (Example 2 see in Appendix 9.A).

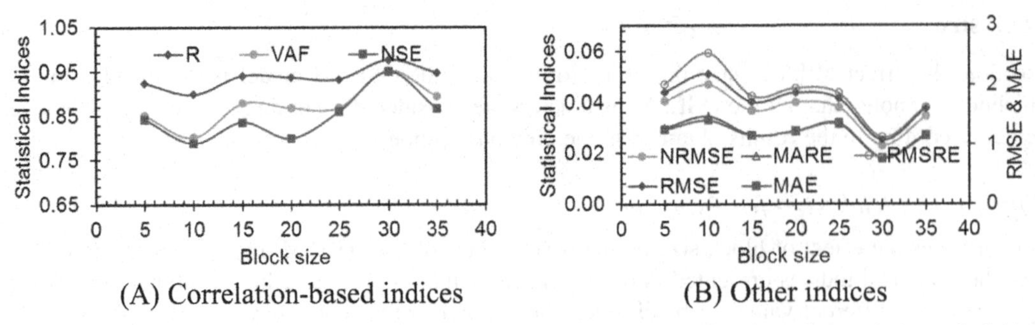

FIGURE 9.37 The effect of block size on the results of online sequential extreme learning machine (Example 3 see in Appendix 9.A).

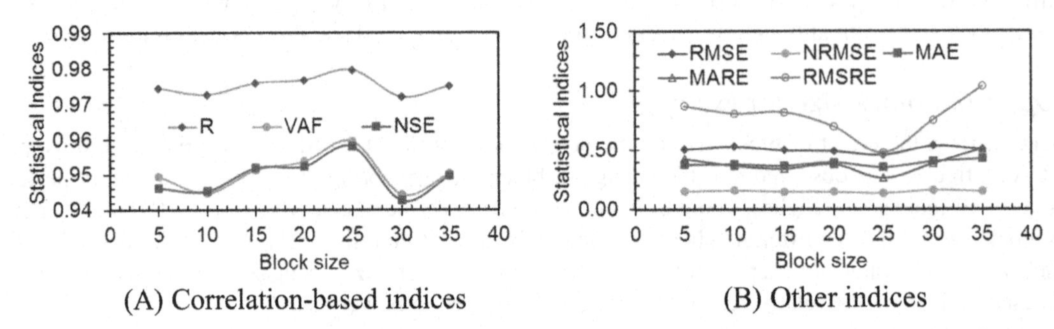

FIGURE 9.38 The effect of block size on the results of online sequential extreme learning machine (Example 4 see in Appendix 9.A).

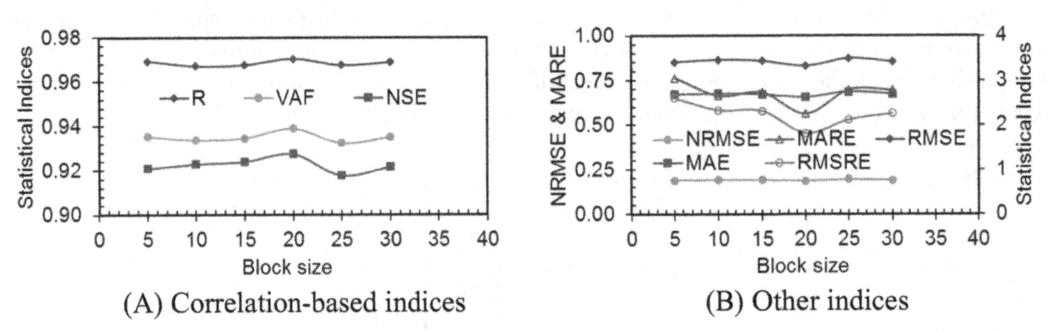

FIGURE 9.39 The effect of block size on the results of online sequential extreme learning machine (Example 5 see in Appendix 9.A).

difference between the highest NRMSE, MARE, and RMSRE (block size = 30) and the lowest (block size = 10) are about 51%, 48%, and 56%, respectively. Moreover, for the absolute-based indices, the difference between the highest RMSE and MAE (block size = 30) and the lowest (block size = 10) are about 51% and 45%, respectively. Indeed, the relative difference between the maximum value to the minimum value is significant.

9.4.2.4 *Optimum block size for Example 4*

Fig. 9.38 surveyed the effect of block size on the performance of the OSELM for Example 4 see in Appendix 9.A. Changes in the indices presented in this figure are U-shaped with block size changes. This form is more clear for the RMSE and correlation-based indices. In general, the performance of the OSELM model in this example increases with increasing block size to block size = 25 and decreases after as the block size increases to 30. In fact, similar to previous examples, no specific trend is observed for this parameter, and in order to achieve the optimal model, it is necessary to examine different values through trial and error.

9.4.2.5 *Optimum block size for Example 5*

Fig. 9.39 demonstrates the effect of block size on the performance of the OSELM for Example 5 see in Appendix 9.A. The OSELM function in this example is quite similar to example 4. In both examples, a U-shaped pattern is observed.

The difference, in this case, is that the model performance increases to a value of block size = 20, while the worst model performance is observed at block size = 25.

9.5 Summary

In this chapter, the reader was first introduced to the differences between batch and sequential learning, after which a detailed examination of sequential-based learning was provided. This chapter details the development of an OSELM based on the sequential-based learning concept. Implementation of the OSELM coding in the MATLAB environment was provided in step-by-step detail. A calculator tool was introduced to allow the reader to use the model with data not included in the calibration period of the developed OSELM-based. The introduced OSELM in the current chapter integrated sequential-based learning with a large number of iterations to overcome the random assignment of the matrices of input weights and bias of hidden neurons. The number of initial training samples and block size must be defined to apply the developed OSELM model (in addition to activation function type, the number of hidden neurons, and iteration numbers). The effects of these parameters were studied for five examples with two to six input variables and a wide-ranging number of samples. After completing this chapter, the reader has the ability to model different real-world problems using the OSELM. Moreover, the reader can use the calculator tool presented in this chapter to apply it to a wide variety of practical problems with ease.

Appendix 9.A Supporting information

Data on OSELM can be found in the online version at https://doi.org/10.1016/B978-0-443-15284-9.00005-7.

References

Arabameri, A., Asadi Nalivan, O., Chandra Pal, S., Chakrabortty, R., Saha, A., Lee, S., Pradhan, B., & Tien Bui, D. (2020). Novel machine learning approaches for modelling the gully erosion susceptibility. *Remote Sensing*, *12*(17), 2833. Available from https://doi.org/10.3390/rs12172833.

Azimi, H., Shiri, H., & Malta, E. R. (2021). A non-tuned machine learning method to simulate ice-seabed interaction process in clay. *Journal of Pipeline Science and Engineering*, *1*(4), 379–394. Available from https://doi.org/10.1016/j.jpse.2021.08.005.

Bonakdari, H., & Ebtehaj, I. (2021). Discussion of "time-series prediction of streamflows of malaysian rivers using data-driven techniques" by Siraj Muhammed Pandhiani, Parveen Sihag, Ani Bin Shabri, Balraj Singh, and Quoc Bao Pham. *Journal of Irrigation and Drainage Engineering*, *147* (9), 07021014. Available from https://doi.org/10.1061/(ASCE)IR.1943-4774.0001602.

Bonakdari, H., Ebtehaj, I., Gharabaghi, B., Sharifi, A., & Mosavi, A. (2020). Prediction of discharge capacity of labyrinth weir with gene expression programming. In *Proceedings of SAI intelligent systems conference* (pp. 202–217). Springer, Cham. https://doi.org/10.1007/978-3-030-55180-3_1.

Bonakdari, H., Gholami, A., Mosavi, A., Kazemian-Kale-Kale, A., Ebtehaj, I., & Azimi, A. H. (2020b). A novel comprehensive evaluation method for estimating the bank profile shape and dimensions of stable channels using the maximum entropy principle. *Entropy*, *22*(11), 1218. Available from https://doi.org/10.3390/e22111218.

Calabrò, F., Fabiani, G., & Siettos, C. (2021). Extreme learning machine collocation for the numerical solution of elliptic PDEs with sharp gradients. *Computer Methods in Applied Mechanics and Engineering*, *387*, 114188. Available from https://doi.org/10.1016/j.cma.2021.114188.

Ebtehaj, I., & Bonakdari, H. (2022a). Discussion of "multivariate drought forecasting in short-and long-term horizons using mspi and data-driven approaches" by Pouya Aghelpour, Ozgur Kisi, and Vahid Varshavian. *Journal of Hydrologic Engineering*, *27*(11), 07022007. Available from https://doi.org/10.1061/(ASCE)HE.1943-5584.0002216.

Ebtehaj, I., & Bonakdari, H. (2022b). A reliable hybrid outlier robust non-tuned rapid machine learning model for multi-step ahead flood forecasting in Quebec, Canada. *Journal of Hydrology*, 128592. Available from https://doi.org/10.1016/j.jhydrol.2022.128592.

Ebtehaj, I., Bonakdari, H., & Kisi, O. (2021). Discussion of "ANFIS modeling with ICA, BBO, TLBO, and IWO optimization algorithms and sensitivity analysis for predicting daily reference evapotranspiration" by Maryam Zeinolabedini Rezaabad, Sadegh Ghazanfari, and Maryam Salajegheh. *Journal of Hydrologic Engineering*, *26*(12), 07021006. Available from https://doi.org/10.1061/(ASCE)HE.1943-5584.0002141.

Ebtehaj, I., Bonakdari, H., Moradi, F., Gharabaghi, B., & Khozani, Z. S. (2018). An integrated framework of extreme learning machines for predicting scour at pile groups in clear water condition. *Coastal Engineering*, *135*, 1–15. Available from https://doi.org/10.1016/j.coastaleng.2017.12.012.

Ebtehaj, I., Bonakdari, H., Safari, M. J. S., Gharabaghi, B., Zaji, A. H., Madavar, H. R., & Mehr, A. D. (2020). Combination of sensitivity and uncertainty analyses for sediment transport modeling in sewer pipes. *International Journal of Sediment Research*, *35*(2), 157–170. Available from https://doi.org/10.1016/j.ijsrc.2019.08.005.

Ebtehaj, I., Bonakdari, H., Zaji, A. H., Bong, C. H., & Ab Ghani, A. (2016). Design of a new hybrid artificial neural network method based on decision trees for calculating the Froude number in rigid rectangular channels. *Journal of Hydrology and Hydromechanics*, *64*(3), 252. Available from https://doi.org/10.1515/johh-2016-0031.

Ebtehaj, I., Sammen, S. S., Sidek, L. M., Malik, A., Sihag, P., Al-Janabi, A. M. S., & Bonakdari, H. (2021). Prediction of daily water level using new hybridized GS-GMDH and ANFIS-FCM models. *Engineering Applications of Computational Fluid Mechanics*, *15*(1), 1343–1361. Available from https://doi.org/10.1080/19942060.2021.1966837.

Ebtehaj, I., Sattar, A. M., Bonakdari, H., & Zaji, A. H. (2017). Prediction of scour depth around bridge piers using self-adaptive extreme learning machine. *Journal of Hydroinformatics*, *19*(2), 207–224. Available from https://doi.org/10.2166/hydro.2016.025.

Ebtehaj, I., Soltani, K., Amiri, A., Faramarzi, M., Madramootoo, C. A., & Bonakdari, H. (2021). Prognostication of shortwave radiation using an improved no-tuned fast machine learning. *Sustainability*, *13*(14), 8009. Available from https://doi.org/10.3390/su13148009.

Grégoire, G., Fortin, J., Ebtehaj, I., & Bonakdari, H. (2022). Novel hybrid statistical learning framework coupled with random forest and grasshopper optimization algorithm to forecast pesticide use on golf courses. *Agriculture*, *12*(7), 933. Available from https://doi.org/10.3390/agriculture12070933.

Herrera, G. P., Constantino, M., Tabak, B. M., Pistori, H., Su, J. J., & Naranpanawa, A. (2019). Data on forecasting energy prices using machine learning. *Data in Brief*, *25*, 104122. Available from https://doi.org/10.1016/j.dib.2019.104122.

Lanera, C., Berchialla, P., Sharma, A., Minto, C., Gregori, D., & Baldi, I. (2019). Screening PubMed abstracts: is class imbalance always a challenge to machine learning? *Systematic Reviews*, *8*(1), 317. Available from https://doi.org/10.1186/s13643-019-1245-8.

Maimaitiyiming, M., Sagan, V., Sidike, P., & Kwasniewski, M. T. (2019). Dual activation function-based extreme learning machine (ELM) for estimating grapevine berry yield and quality. *Remote Sensing*, *11*(7), 740. Available from https://doi.org/10.3390/rs11070740.

Mojtahedi, S. F. F., Ebtehaj, I., Hasanipanah, M., Bonakdari, H., & Amnieh, H. B. (2019). Proposing a novel hybrid intelligent model for the simulation of particle size distribution resulting from blasting. *Engineering with Computers*, *35*(1), 47–56. Available from https://doi.org/10.1007/s00366-018-0582-x.

Owolabi, T. O., & Abd Rahman, M. A. (2021). Prediction of band gap energy of doped graphitic carbon nitride using genetic algorithm-based support vector regression and extreme learning machine. *Symmetry*, *13*(3), 411. Available from https://doi.org/10.3390/sym13030411.

Ratnawati, D.E., Marjono, Widodo, & Anam, S. (2020). Comparison of activation function on extreme learning machine (ELM) performance for classifying the active compound. In *AIP conference proceedings* (Vol. 2264, No. 1, p. 140001). AIP Publishing LLC. https://doi.org/10.1063/5.0023872.

Ratzinger, F., Haslacher, H., Perkmann, T., Pinzan, M., Anner, P., Makristathis, A., & Dorffner, G. (2018). Machine learning for fast identification of bacteraemia in SIRS patients treated on standard care wards: a cohort study. *Scientific Reports*, *8*(1), 12233. Available from https://doi.org/10.1038/s41598-018-30236-9.

Saha, S., Roy, J., Pradhan, B., & Hembram, T. K. (2021). Hybrid ensemble machine learning approaches for landslide susceptibility mapping using different sampling ratios at East Sikkim Himalayan, India. *Advances in Space Research*, *68*(7), 2819–2840. Available from https://doi.org/10.1016/j.asr.2021.05.018.

Samal, D., Dash, P. K., & Bisoi, R. (2022). Modified added activation function based exponential robust random vector functional link network with expanded version for nonlinear system identification. *Applied Intelligence*, *52*(5), 5657–5683. Available from https://doi.org/10.1007/s10489-021-02664-0.

Sattar, A., Ertuğrul, Ö. F., Gharabaghi, B., McBean, E. A., & Cao, J. (2019). Extreme learning machine model for water network management. *Neural Computing and Applications*, *31*(1), 157–169. Available from https://doi.org/10.1007/s00521-017-2987-7.

Suchithra, M. S., & Pai, M. L. (2020). Improving the prediction accuracy of soil nutrient classification by optimizing extreme learning machine parameters. *Information Processing in Agriculture*, *7*(1), 72–82. Available from https://doi.org/10.1016/j.inpa.2019.05.003.

Thottakkara, P., Ozrazgat-Baslanti, T., Hupf, B. B., Rashidi, P., Pardalos, P., Momcilovic, P., & Bihorac, A. (2016). Application of machine learning techniques to high-dimensional clinical data to forecast postoperative complications. *PLoS One*, *11*(5), e0155705. Available from https://doi.org/10.1371/journal.pone.0155705.

Tripathi, D., Edla, D. R., Kuppili, V., & Bablani, A. (2020). Evolutionary extreme learning machine with novel activation function for credit scoring. *Engineering Applications of Artificial Intelligence*, *96*, 103980. Available from https://doi.org/10.1016/j.engappai.2020.103980.

Wu, Y., Chen, Y., & Tian, Y. (2022). Incorporating empirical orthogonal function analysis into machine learning models for streamflow prediction. *Sustainability*, *14*(11), 6612. Available from https://doi.org/10.3390/su14116612.

Yarrakula, M., & Dabbakuti, J. R. K. (2022). Modeling and prediction of TEC based on multivariate analysis and kernel-based extreme learning machine. *Astrophysics and Space Science*, *367*(3), 34–41. Available from https://doi.org/10.1007/s10509-022-04062-5.

Zeynoddin, M., Bonakdari, H., Azari, A., Ebtehaj, I., Gharabaghi, B., & Madavar, H. R. (2018). Novel hybrid linear stochastic with non-linear extreme learning machine methods for forecasting monthly rainfall a tropical climate. *Journal of Environmental Management*, *222*, 190–206. Available from https://doi.org/10.1016/j.jenvman.2018.05.072.

Zeynoddin, M., Bonakdari, H., Ebtehaj, I., Azari, A., & Gharabaghi, B. (2020). A generalized linear stochastic model for lake level prediction. *Science of the Total Environment*, *138*, 015. Available from https://doi.org/10.1016/j.scitotenv.2020.138015.

Chapter 10

Self-adaptive evolutionary of non-tuned neural network—concept

10.1 Development process of single-layer feedforward neural network

10.1.1 Classical neural network

Over the past decade, a large number of machine learning (ML) techniques have been developed to solve real-world problems, especially those related to earth, environmental and planetary sciences (Azimi et al., 2017a, 2018; Ebtehaj, Bonakdari, et al., 2017; Gholami et al., 2017, 2019b; Khozani et al., 2017; Sihag et al., 2019). The neural network is one of the most fundamental and widely used ML techniques. During their investigation of the critical variables impacting the estimation accuracy of sediment transport at clean pipe conditions (i.e., without deposition) using neural networks, Ebtehaj and Bonakdari (2016a) developed a multilayer perceptron neural network and evaluated three different training algorithms. The three training algorithms considered in the study include Levenberg–Marquardt, resilient back-propagation, and variable learning rate. A comparison of the performance of the different back-propagation (in terms of the correlation coefficient) suggested that the Levenberg–Marquardt algorithm outperformed the resilient back-propagation and variable learning rate algorithms. In addition, it was demonstrated by the authors that the developed ML technique was more robust than existing classical methods, which had until that point been identified as the most accurate approach for modeling sediment transport in clean pipes. Scour depth forecasting using ML is another field of Civil/Environmental engineering which attracts the attention of many researchers. Sharafi et al. (2016) considered a large number of experimental and field datasets when studying the effect of six different kernel functions on the performance of a support vector machine in estimating scour depth for different pier shapes. Moradi et al. (2019) compared three different adaptive neuro-fuzzy inference systems (ANFIS) for the estimation of scour depth, including sub-clustering (SC), grid-partitioning, and fuzzy c-means clustering. Although almost all models gave good results, the best performance was related to the ANFIS-SC.

10.1.2 Non-tuned fast neural network

Despite the introduction of many new ML methods, the use of neural networks is still considered an efficient and powerful approach to solving a wide array of problems. For this reason, many researchers have tried to combine neural networks with other methods to form a hybrid model (Ebtehaj et al., 2016, 2018; Moeeni et al., 2017). Alternatively, some authors seek to increase the modeling performance of neural networks using optimization algorithms (Bonakdari & Ebtehaj, 2014; Ebtehaj & Bonakdari, 2016b).

A new algorithm for a single-layer feed-forward neural network (SLFFNN) was introduced by Huang et al. (2004, 2006a), known as the extreme learning machine (ELM). The ELM has caught the attention of many researchers because it not only has a fast training speed but also has noteworthy generalizability. The reason behind the fast training speed of the ELM is the random generation of the bias of hidden neurons (*BHNs*) and the input weights (*InWs*) (recall the *InW* matrix connects the input feature to the hidden neurons while the bias of the hidden neurons is simply a matrix containing the biases associated with the hidden neurons). Indeed, by randomly generating these matrices, a previously nonlinear problem becomes linear. In this new problem, the only component left to determine is the output weight (*OutW*) [a matrix that connects the hidden layer neurons to the output variable(s)], which is done using the least square algorithm (Bonakdari et al., 2019a; Ebtehaj et al., 2017b). By converting the nonlinear problem into a linear one, the ELM overcomes existing limitations of the gradient-based algorithms (i.e., back-propagation), including slow convergence performance (Bonakdari & Ebtehaj, 2016; Melo & Watada, 2016), stagnation in local minima, and adjustment

Machine Learning in Earth, Environmental and Planetary Sciences. DOI: https://doi.org/10.1016/B978-0-443-15284-9.00010-0

(Abba et al., 2020; Ebtehaj et al., 2018) of different user-defined parameters. The main advantages associated with the ELM are summarized in Fig. 10.1.

10.1.3 Limitation of the first generation non-tuned fast neural network

In the ELM, the number of hidden neurons is the only user-defined parameter that must be defined prior to the commencement of modeling. According to the user-defined number of hidden neurons and the randomly generated *InWs* and *BHNs* (these matrices remain unchanged throughout the modeling process), the output of this layer is also determined using random variables (recall that the *InWs* and *BHNs* are two main matrices in ELM that are randomly generated). It has been shown that a large number of nonoptimal nodes can be identified that have a small contribution in minimizing the cost function. Huang et al. (2006a), based on this finding, emphasized that consideration of the ELM algorithm leads to the use of a greater number of hidden nodes (in most cases) during the training of feed-forward neural networks than classical approaches such as the Levenberg–Marquardt algorithm (Hagan & Menhaj, 1994; Levenberg, 1944).

A conceptual model of the ELM modeling process is presented in Fig. 10.2. As can be seen in the figure, the matrices of the *InW* and *BHNs* are randomly allocated by defining the number of hidden layer neurons. The "random generation" section is colored in red to indicate that no optimization is undertaken for these matrices in the classical ELM. Following this, the output of the hidden layer is determined by applying the activation function to the result of each hidden neuron. It is clear from the figure that the *BHNs* and *InW* matrices were determined without any prior experience

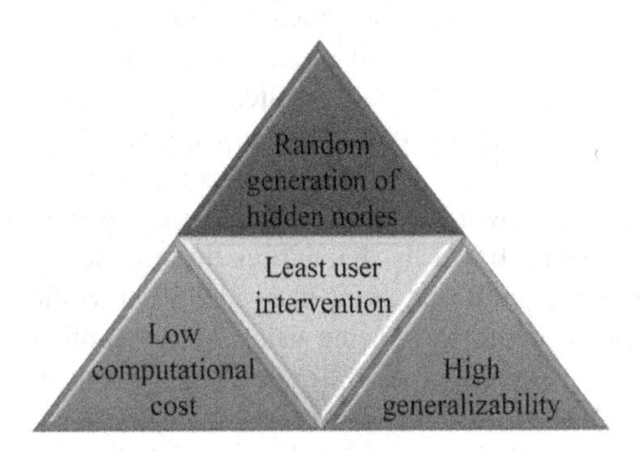

FIGURE 10.1 Primary advantages of the extreme learning machine algorithm.

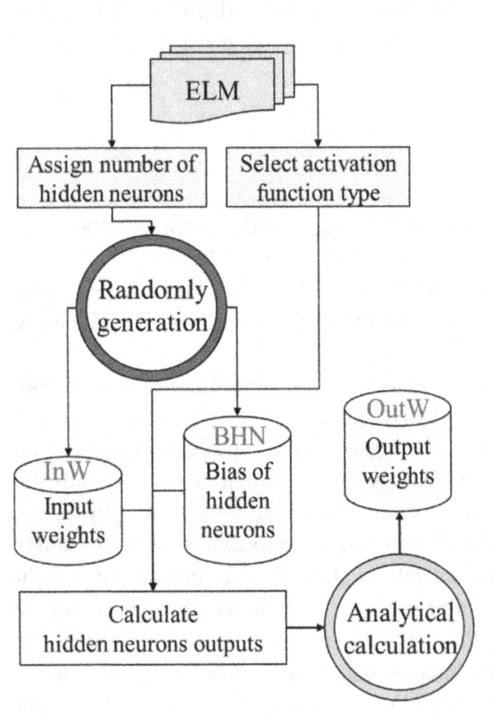

FIGURE 10.2 Conceptual model of the extreme learning machine process.

with the data of the problem under consideration. Using the computed output of the hidden layer matrix, the output weight matrix is calculated using the least square algorithm. The dimensions of the different matrices involved depend on the number of input variables and the number of hidden neurons defined. The dimensions are given by *InV* × *NHN*, *NHN*, and *NHN* for the *InW*, the *BHN*, and the *OutWs*, respectively. Consequently, the output of the ELM modeling process is given by a combination of parameters that are both randomly defined as well as determined analytically. If we define k to be an output of the model, it is composed of analytically and randomly defined parameters with dimensionalities given below:

$$k = InV \times NHN + NHN + NHN \tag{10.1}$$

Two first terms in Eq. (10.1) are randomly determined (*InV* × *NHN* for the *InW* and *NHN* for the *BHN*). Eq. (10.1) could be rewritten as follows:

$$k = k_R + k_A \tag{10.2}$$

where k_R and k_A are the randomly and analytically determined values, respectively, and defined as follows:

$$k_R = InV \times NHN + NHN = NHN(InV + 1) \tag{10.3}$$

$$k_A = NHN \tag{10.4}$$

Consequently, the ratio of randomly allocated parameters to the total parameters is calculated as follows:

$$\begin{aligned} R &= \frac{k_R}{k_R + k_A} = \frac{(InV \times NHN + NHN)}{(InV \times NHN + NHN) + NHN} = \frac{NHN(InV+1)}{NHN(InV+2)} \\ &= \frac{NHN(InV+2) - NHN}{NHN(InV+2)} = 1 - \frac{NHN}{NHN(InV+2)} = 1 - \frac{1}{InV+2} \end{aligned} \tag{10.5}$$

According to Eq. (10.5), it is observed that if the problem has only one input ($InV = 1$), more than 66% of the values that are optimized during the training process, are randomly assigned (Ebtehaj et al., 2021a). As the number of input variables increases, this percentage also increases so that if $InV = 5$, more than 85% of the values obtained are obtained randomly, which is a significant percentage.

Although the ELM has several advantages, including a lower level of user intervention, a reduced number of user-defined parameters, a low computational cost, and high generalizability, random allocation of the *InW* and the *BHN* matrices can, in some cases, reduce the generalizability and require a greater number of hidden neurons (Huang et al., 2006a). Therefore, to increase the performance of the ELM, it is necessary to remedy the problems caused by the random allocation of *InW* and *BHN* matrices through optimization.

10.1.4 Optimization of the classical non-tuned neural network

10.1.4.1 Differential evolution algorithm

In the past decade, evolutionary algorithms (EAs) have been widely applied in the optimization of feed-forward neural networks, including particle swarm optimization (Bonakdari et al., 2019b; Qasem et al., 2017), differential evolution (Kumar & Singh, 2018), genetic algorithm (Ebtehaj & Bonakdari, 2014), firefly algorithm (Wang et al., 2017), and imperial competitive algorithm (Ebtehaj et al., 2021b). A widely used EA is the differential evolution (DE) algorithm (Storn & Price, 1997) which is a simple and efficient population-based stochastic algorithm for global optimization. These characteristics lend themself to frequent application in the optimization of different ML techniques. The main advantages of the DE algorithm are summarized in Fig. 10.3.

The structure of the DE algorithm is simple as it involves only three operators, which will be discussed in detail in the next subsections. Moreover, the implementation of this algorithm is simplistic and it has been reported that the quality of the DE solutions is high (Ebtehaj & Bonakdari, 2017). The DE algorithm designed for a stochastic direct search is also able to handle numerous cost functions, including multimodal, nonlinear, and nondifferentiable. The direct search-based methods could be easily applied to experimental minimization. Moreover, the use of the DE algorithm by the user is easy because there are few control variables when implemented for minimization. Consecutive independent tests of DE considering experimental data covering a range of conditions indicate consistent convergence to the global minimum (Storn & Price, 1997). Furthermore, Singh and Kumar (2016) indicate that DE algorithms are not susceptible to becoming entrapped in local minima.

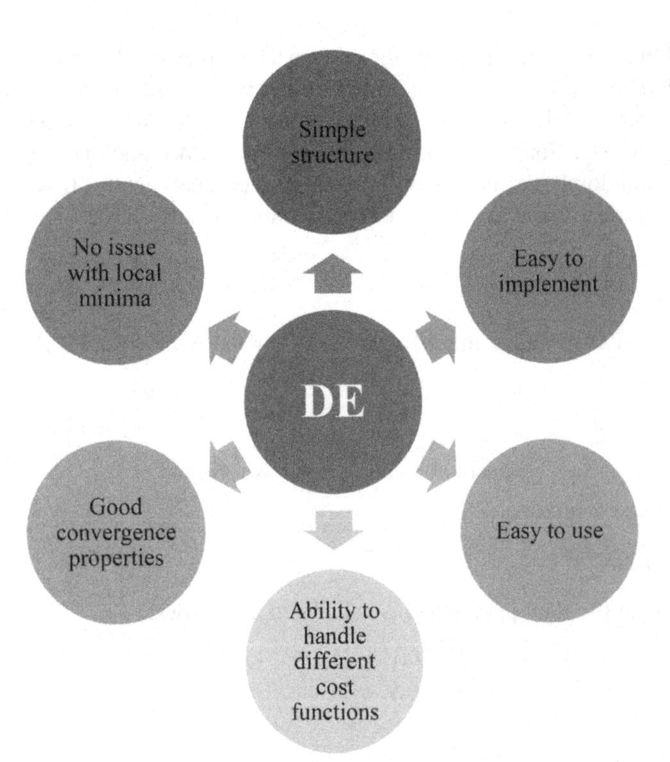

FIGURE 10.3 Principal advantages of the differential evolution algorithm.

10.1.4.2 Different versions of the optimized non-tuned neural network

The ELM algorithm randomly generates two matrices, namely the *InW* and the *BHN*. Using these matrices and the activation function, the output of the hidden neuron is calculated. Indeed, the *InW* and the *BHN* matrices are randomly populated without any prior experience with the data. Eq. (10.5) shows that at least 66% of all model parameters are randomly generated when considering the smallest-sized model (a single input neuron). As the ELM model becomes more complicated and considers greater numbers of hidden neurons, even higher percentages of model parameters are randomly allocated. To overcome this limitation, Zhu et al. (2005) optimized the ELM by considering DE as a powerful well-known optimization algorithm, which resulted in the evolutionary ELM (E-ELM). The authors applied DE to optimize the matrix of *InWs* while the *OutWs* were estimated analytically using the Moore–Penrose generalized inverse, as is typical in classical ELM techniques. The authors reported that the E-ELM not only has higher generalizability when compared to the conventional ELM, but also provides a more compact model with a smaller number of hidden nodes.

Although the E-ELM method has improved the performance of the classical ELM, the use of this method has limitations that, if not properly addressed, may adversely affect the performance of the model. The main drawback of the E-ELM is the manual selection of the DE control parameters (i.e., scaling factor and crossover rate) and trial vector generation strategies. In the E-ELM, empirical evidence is used to manually select the control parameters, while a simple random generation approach is employed to generate a trial vector.

10.1.4.3 Self-adaptive optimization of the non-tuned neural network

Many scholars (Azimi et al., 2017b; Gholami et al., 2018, 2019a) have shown that the performance of the DE algorithm is highly dependent on the choice of its control parameters and strategy of trial vector generation, the improper selection of which may lead to stagnation or premature convergence. In order to achieve acceptable accuracy in solving a variety of problems, the optimal values of the control parameters and strategy of the trial vector generation must be found. When these values are held constant across varying conditions, the obtained results may not be acceptable in some instances. Consequently, a self-adaptive evolutionary extreme learning machine (SaE-ELM) (Cao et al., 2012) was developed to overcome the limitation of the existing E-ELM regarding the selection of DE parameters. In the SaE-ELM, the hidden nodes (*InW* and *BHN*) are optimized using the self-adaptive DE (Qin et al., 2009), while the *OutWs* are calculated analytically using the Moore–Penrose generalized inverse (MPGI).

It should be noted that in the self-adaptive DE algorithm, the control parameters and the strategies of the trial vector generation are adapted in a strategy pool by learning from their previous experiences in generating promising solutions.

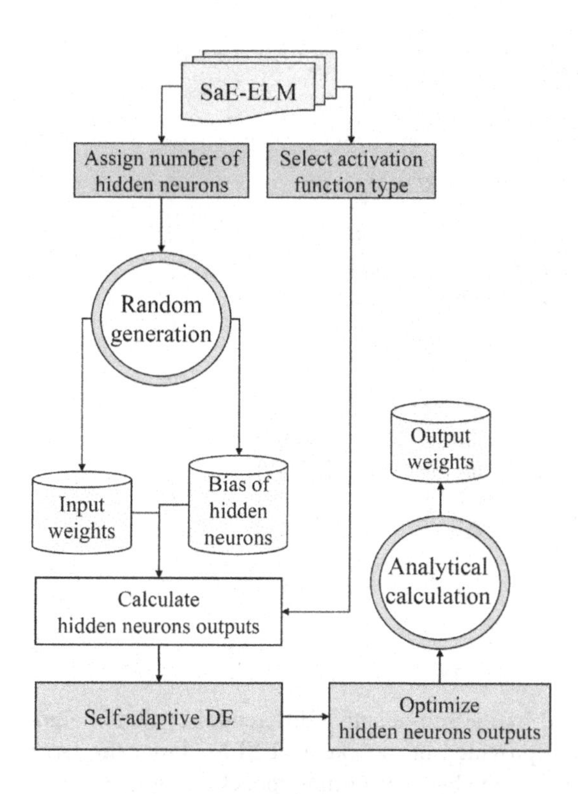

FIGURE 10.4 Conceptual diagram of the self-adaptive evolutionary extreme learning machine modeling process.

Fig. 10.4 presents the conceptual model of the SaE-ELM modeling process. It can be seen that the "randomly generated" portion of the code, which was previously colored in red for the classic ELM method (Fig. 10.2), is shown in green in this figure. This indicates that the randomly generated values are optimized using self-adaptive DE. Additionally, similar to the ELM, the user must specify only the number of hidden neurons and the activation function type, while all other parameters are determined automatically by the model in a self-adapted manner.

10.2 Mathematical explanation of self-adaptive evolutionary extreme learning machine

Given that the previous chapters provide a comprehensive formula-based derivation for the ELM along with its coding, this chapter will only provide a general overview of ELM and the basic elements that are needed to develop the SaE-ELM. The reader is referred to Chapter 4 for a complete mathematical derivation of the classical ELM model. In this section, the reader will be introduced to the DE algorithm as well as the use of this algorithm in the optimization of the SLFFNN. Finally, the developed SaE-ELM method, which uses the self-adaptive DE algorithm to optimize the hidden neurons of the individual ELM, is fully described.

The explanations presented in this chapter are divided into two general categories according to the level of the reader: beginner and advanced. Content provided for the advanced user includes details on the mathematical formulation of the individual ELM and the DE methods, the development of the SaE-ELM method, as well as the coding of these algorithms in the MATLAB environment. For the beginner user, a calculator is presented, allowing for simple implementation of the developed SaE-ELM to various real-world problems with different numbers of inputs. In order to use this calculator, beginner users do not need any knowledge of any of the individual ELM, DE algorithm, or SaE-ELM methods and need only to be able to run MATLAB software with an understanding of the user-defined inputs.

10.2.1 Conceptual model of the developed self-adaptive optimized neural network

A conceptual model of the developed calculator is presented in Fig. 10.5. The reader should note that the activation function considered in this figure is the hyperbolic tangent function, although alternative activation functions may be considered. A full list of the activation functions that may be employed in the developed calculator along with their mathematical formulas is provided in Fig. 10.6. The required values to calculate the target variable (i.e., *TV* in Fig. 10.5) include the *InV*, *InW*, *BHN*, and *OutW*, which denote input variables, *InWs*, the *BHNs*, and *OutWs*,

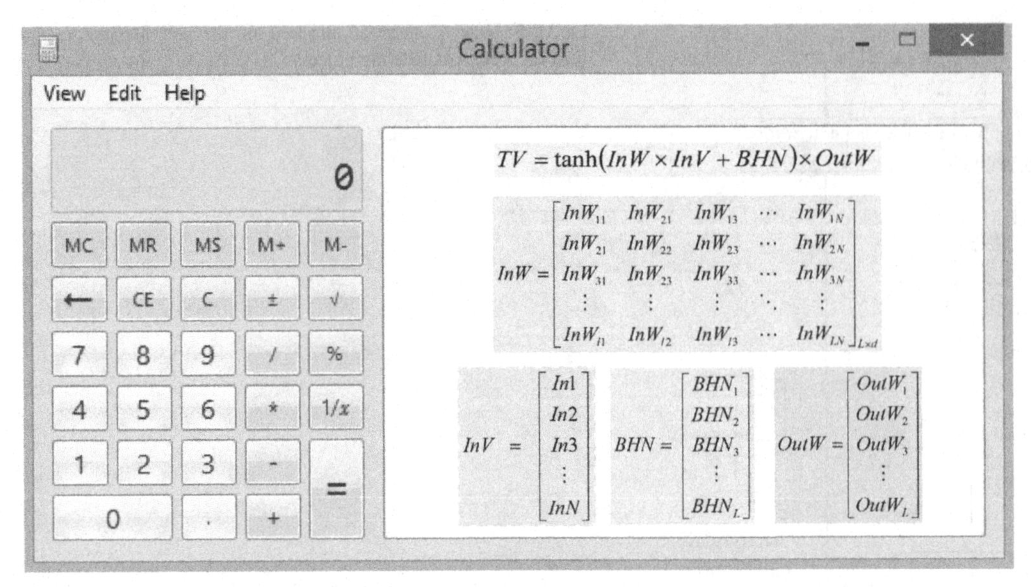

FIGURE 10.5 Conceptual diagram of the developed calculator.

respectively. The *InW* and *BHN* are randomly allocated and optimized by the self-adaptive differential evolution algorithm, while the *OutW* is calculated using the least square algorithm provided in the classical ELM. From the large number of iterations considered during the development of the SaE-ELM, the best-performing model during the testing stage is selected as the optimum model. Indeed, the *InW*, *BHN*, and *OutW* of the optimum model are saved and employed in the developed calculator. Using these matrices and the assignment of the input variables(s) (i.e., *InV*), the target value is estimated using the developed model.

10.2.2 Non-tuned rapid machine learning model[1]

Let's consider N training samples in the form of $\{(\boldsymbol{x}_i, \boldsymbol{t}_i)\}_{i=1}^{N}$ where $\boldsymbol{t}_i \in \mathbf{R}^m$, $\boldsymbol{x}_i \in \mathbf{R}^d$, and m and d denote the number of output and input variable(s), respectively. For an SLFFNN with L hidden neurons and activation function, $g()$, the output is calculated as follows:

$$\boldsymbol{O}_i = \sum_{j=1}^{L} \beta_j g(\boldsymbol{a}_j, b_j, \boldsymbol{x}_i), \; i = 1, 2..., N \tag{10.6}$$

where $b_j \in \mathbf{R}$ and $\boldsymbol{a}_j \in \mathbf{R}^d$ ($j = 1, 2,..., L$) are the learning parameters related to the jth hidden node, $\beta_j \in \mathbf{R}^m$ denotes a matrix that connects the jth hidden node to the output node, and $g(\boldsymbol{a}_j, b_j, \boldsymbol{x}_i)$ is the output of the jth hidden node with respect to the $\boldsymbol{x}_i$.

To train the SLFFNN using a minimizing cost function [Eq. (10.7)], the network parameter W is iteratively updated using classic gradient descent (GD) based training algorithm [Eq. (10.8)]:

$$E_{ELM} = \sum_{i=1}^{N} \left\| \boldsymbol{O}_i - \boldsymbol{t}_i \right\| < \varepsilon \tag{10.7}$$

$$W_{k+1} = W_k - \eta \frac{\partial E(W_k)}{\partial W_k} \tag{10.8}$$

where the η is the learning rate, and W is a set of learning parameters $(\boldsymbol{a}_j, b_j, \beta_i)$. Although the application of this algorithm in training SLFFNN may result in acceptable results for a number of applications, it also has its limitations, including slow convergence rate, overfitting, local minima, etc.

1. This section is provided for the advance users.

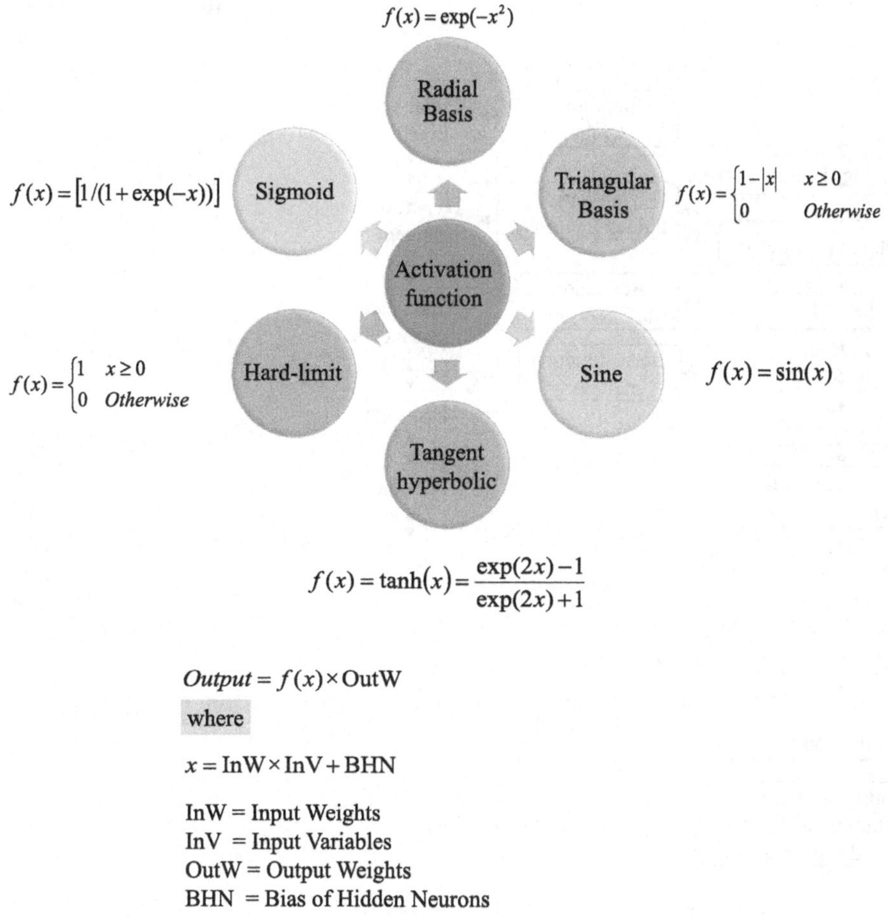

$f(x) = \exp(-x^2)$

Radial Basis

$f(x) = [1/(1+\exp(-x))]$ Sigmoid

Triangular Basis $f(x) = \begin{cases} 1-|x| & x \geq 0 \\ 0 & Otherwise \end{cases}$

Activation function

$f(x) = \begin{cases} 1 & x \geq 0 \\ 0 & Otherwise \end{cases}$ Hard-limit

Sine $f(x) = \sin(x)$

Tangent hyperbolic

$f(x) = \tanh(x) = \frac{\exp(2x)-1}{\exp(2x)+1}$

$Output = f(x) \times \mathrm{OutW}$

where

$x = \mathrm{InW} \times \mathrm{InV} + \mathrm{BHN}$

InW = Input Weights
InV = Input Variables
OutW = Output Weights
BHN = Bias of Hidden Neurons

FIGURE 10.6 Mathematical formulation of different activation functions.

Contrary to the results presented in training the SLFFNN using the GD algorithm, the training of an SLFFNN with L hidden neurons and the activation function $g()$ by the ELM can approximate all N training samples with zero error:

$$E_{ELM} = \sum_{i=1}^{N} \|\mathbf{O}_i - \mathbf{t}_i\| = 0 \tag{10.9}$$

Indeed, there exist $\mathbf{a}_j$, b_j and β_i such that:

$$\mathbf{H}\beta = \mathbf{T} \tag{10.10}$$

where $\mathbf{T} = (\mathbf{t}_1, \mathbf{t}_2, \ldots, \mathbf{t}_N)$ is the target output, $\beta = (\beta_1, \beta_2, \ldots, \beta_L)$ denotes the matrix of *OutWs*, and $\mathbf{H}$ is the matrix of the hidden layer output, which is defined as follows:

$$\mathbf{H} = \begin{bmatrix} g(a_1, b_1, \mathbf{x}_1) & \cdots & g(a_L, b_L, \mathbf{x}_1) \\ \vdots & \ddots & \vdots \\ g(a_1, b_1, \mathbf{x}_N) & \cdots & g(a_L, b_L, \mathbf{x}_N) \end{bmatrix}_{N \times L} \tag{10.11}$$

The SLFFNN works as a universal approximator by using the random assignment of hidden node parameters (*InWs* and *BHNs*) and adjusting the *OutWs*. According to Huang et al. (2004, 2006a), the hidden node parameters are randomly allocated. Therefore, the hidden layer output (**H**), as well as the target output (T), are known, and the only unknown parameter in Eq. (10.10) is the matrix of *OutWs* (β). Consequently, this system is reduced to a series of linear equations where the *OutWs* are the unknown parameter and must be estimated through model training. The least-square solution of this linear system [Eq. (10.10)] is analytically determined as follows:

$$\beta = \mathbf{H}^{+}\mathbf{T} \tag{10.12}$$

where $\mathbf{H}^{+}$ is the Moore–Penrose generalized inverse of the hidden layer output (**H**) matrix.

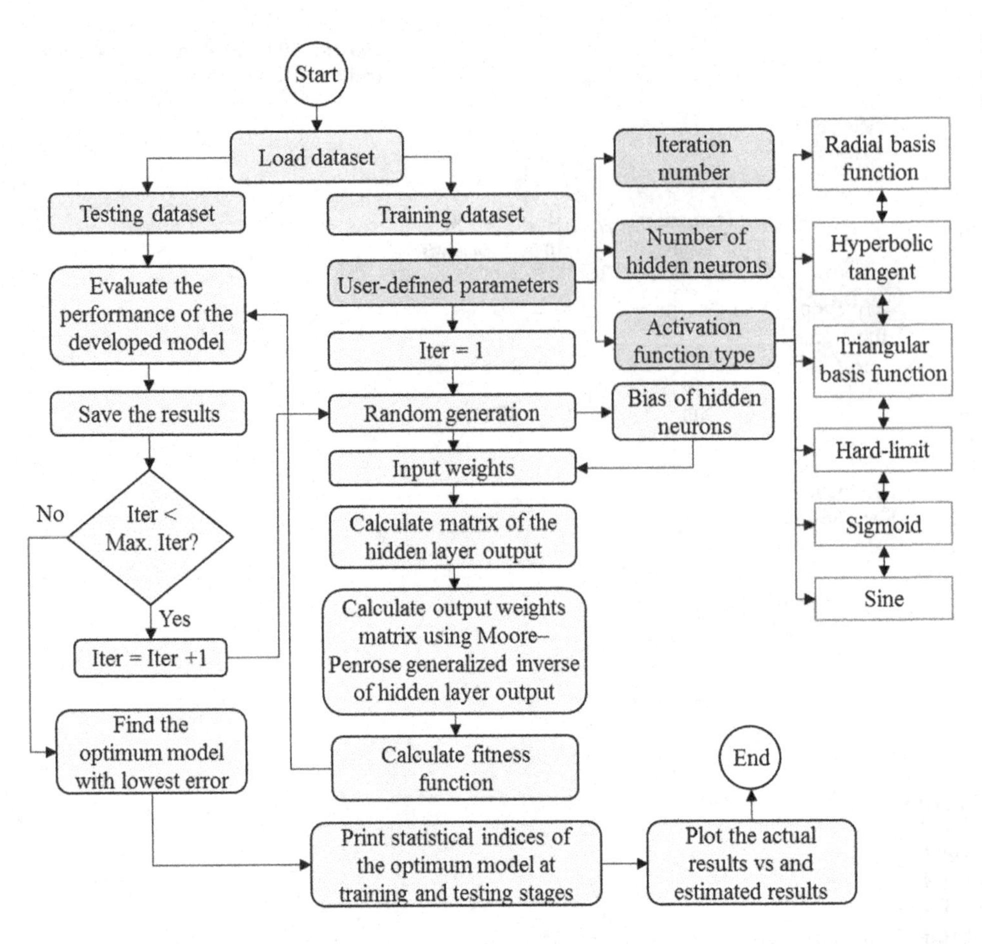

FIGURE 10.7 Flowchart of the classical extreme learning machine.

According to Eq. (10.5), in the most optimistic case (i.e., only one input), random generation of the *InWs* and *BHNs* represent 66% of the total parameters and may significantly impact model results. Although the random allocation of the input matrices leads to fast modeling (only taking several seconds in most cases), there may be instances of reduced modeling performance associated with this allocation. Therefore, the parameters obtained randomly before the ELM training process are optimized using the DE algorithm. The flowchart of the classical ELM is presented in Fig. 10.7.

10.2.3 Differential evolution[2]

The DE algorithm minimizes the cost function [Eq. (10.13)] with regards to the target vector ($\boldsymbol{\theta} \in \mathbf{R}^D$ where D is the dimension of the desired problem) using *NP* (i.e., number of population) evolving individual target vectors to reach the global optimum (Storn & Price, 1997).

$$E_{DE} = \min_{\boldsymbol{\theta} \in \mathbf{R}^D} f(\boldsymbol{\theta}) \tag{10.13}$$

The ith candidature parameter at the Gth generation is defined as follows:

$$\boldsymbol{\theta}_{i,G} = \left[\theta_{i,G}^1, \ldots, \theta_{i,G}^D\right] \; i = 1, \ldots, NP \tag{10.14}$$

where $\boldsymbol{\theta}$ is the target vector, G is the generation number, and NP is the population number. The optimization process in the DE algorithm includes initialization, mutation, crossover, and selection which are defined in the following subsections.

2. This section is provided for advanced users.

10.2.3.1 Step 1: Initialization

The first step is the generation of the initial population. When generating this population, it is important that the entire search space be covered. Indeed, it must be constrained by a predefined minimum ($\boldsymbol{\theta}_{\min}$) and maximum ($\boldsymbol{\theta}_{\max}$) of the parameter bounds. A set of *NP* individual target vectors $\boldsymbol{\theta}_{i,G}$ for the DE algorithm is defined as follows:

$$\boldsymbol{\theta}_{i,G} = \boldsymbol{\theta}_{\min} + rand(0,1)\cdot(\boldsymbol{\theta}_{\max} - \boldsymbol{\theta}_{\min}) \tag{10.15}$$

where $\boldsymbol{\theta}_{\min} = \left[\theta^1_{\min},\ldots,\theta^D_{\min}\right]$ and $\boldsymbol{\theta}_{\max} = \left[\theta^1_{\max},\ldots,\theta^D_{\max}\right]$ are the bounds of the considered parameters and *rand*(0,1) represents a uniformly distributed random value in the range [0, 1].

10.2.3.2 Step 2: Mutation

Subsequent to the problem initialization, the existing target vectors are perturbed to generate new mutant vectors. A mutant vector is produced with respect to each individual target vector, $\boldsymbol{\theta}_{i,G}$, at the current generation via different mutation strategies. According to Storn and Price (1997), to produce the mutant vector $\boldsymbol{\nu}_{i,G}$, four strategies are utilized:

Strategy 1: DE/rand/1

$$\boldsymbol{\nu}_{i,G} = \boldsymbol{\theta}_{r^i_1,G} + F.\left(\boldsymbol{\theta}_{r^i_2,G} - \boldsymbol{\theta}_{r^i_3,G}\right) \tag{10.16}$$

Strategy 2: DE/rand-to-best/2

$$\boldsymbol{\nu}_{i,G} = \boldsymbol{\theta}_{r^i_1,G} + F.\left(\boldsymbol{\theta}_{best,G} - \boldsymbol{\theta}_{r^i_1,G}\right) + F.\left(\boldsymbol{\theta}_{r^i_2,G} - \boldsymbol{\theta}_{r^i_3,G}\right) + F.\left(\boldsymbol{\theta}_{r^i_4,G} - \boldsymbol{\theta}_{r^i_5,G}\right) \tag{10.17}$$

Strategy 3: DE/rand/2

$$\boldsymbol{\nu}_{i,G} = \boldsymbol{\theta}_{r^i_1,G} + F.\left(\boldsymbol{\theta}_{r^i_2,G} - \boldsymbol{\theta}_{r^i_3,G}\right) + F.\left(\boldsymbol{\theta}_{r^i_4,G} - \boldsymbol{\theta}_{r^i_5,G}\right) \tag{10.18}$$

Strategy 4: DE/current-to-rand/1

$$\boldsymbol{\nu}_{i,G} = \boldsymbol{\theta}_{i,G} + K.\left(\boldsymbol{\theta}_{r^i_1,G} - \boldsymbol{\theta}_{i,G}\right) + F.\left(\boldsymbol{\theta}_{r^i_2,G} - \boldsymbol{\theta}_{r^i_3,G}\right) \tag{10.19}$$

where $r^i_1, r^i_2, r^i_3, r^i_4, r^i_5$ are mutually exclusive integer numbers that are randomly produced in the range of 1 and *NP* so that they are different from the index *i*, *K* is a control parameter that is randomly generated in the range of [0, 1], *F* is the scaling factor and generally selected in the range of [0, 2], and $\boldsymbol{\theta}_{\text{best},\,G}$ is the individual vector with the best fitness value in the population at *G*th generation. A survey of pertinent available literature (Qin et al., 2009; Storn & Price, 1997) indicates that different vector generation strategies perform differently at solving different real-world optimization problems. The main limitation of strategy 1 (i.e., DE/rand/1) is its slow convergence speed, although due to its powerful exploration capacity, it has been shown to be suited for solving multimodal problems (Qin et al., 2009; Storn & Price, 1997).

A "global" optimal solution is the best feasible solution to an optimization problem. Multimodal optimization is a branch of optimization concerned with finding multiple locally optimal solutions when optimizing multimodal functions. Recall that multimodal functions are those with multiple local optima. Sometimes these local optimal solutions are also of interest to the decision-maker as, in the case of complex problems, finding the global optimal solution is quite challenging. Strategy 2 (i.e., DE/rand-to-best/2) converges quickly using the best solution found so far and performs well on unimodal problems. However, this strategy generally gets stuck in the local optimum and results in premature convergence when applied to solving multimodal problems. One-difference-vector-based strategies (i.e., strategy 1, DE/rand/1) could result in a further away from optimal perturbation than the two-difference-vectors-based strategies considered in strategy 2 (i.e., DE/rand-to-best/2) and strategy 3 (i.e., DE/rand/2). It should be noted that the computational cost of the one-difference-vector-based strategies is lower than two-difference-based strategies. Strategy 4 (i.e., DE/current-to-rand/1), a rotation-invariant strategy, is proficient in solving multiobjective optimization problems (Qin et al., 2009).

10.2.3.3 Step 3: Crossover

The next step after generating a mutant vector using one of several different mutation strategies is crossover. The need for applying crossover on the mutant vectors is to increase the population diversity of the perturbed target vectors.

Considering the mutant vector at the Gth generation $v_{i,G} = \left[v_{i,G}^1, \ldots, v_{i,G}^D\right]$, the trial vector $u_{i,G} = \left[u_{i,G}^1, \ldots, u_{i,G}^D\right]$ is generated using the crossover equation as follows:

$$u_{i,G}^j = \begin{cases} v_{i,G}^j, & \text{if}(\text{rand}_j \leq CR) \text{ or } (j = j_{\text{rand}}) \\ \theta_{i,G}^j, & \text{Otherwise} \end{cases} \tag{10.20}$$

where CR $(0 \leq CR \leq 1)$ is the crossover rate, $rand_j$ is the jth assessment of a uniform random number generation in the range of [0, 1], and j_{rand} is a random integer value from $[1, D]$ where D is the dimension of the parameter vector. The CR is applied to control the fraction of the parameter values copied from the mutant vector, while j_{rand} is employed to guarantee that at least one parameter exists in the trial vector $\boldsymbol{u}_{i,G}$ differing from the target vector $\boldsymbol{\theta}_{i,G}$.

10.2.3.4 Step 4: Selection

After applying the mutation and crossover operators, the performance of each trial vector against the corresponding target vector in the current population is evaluated using the fitness function. If the trial vector has a higher fitness level than the corresponding target vector, the trial vector will remain in the population for the next generation. Otherwise, the target vector will replace the trial vector and enter the population of the next generation. The selection operation is presented as follows:

$$\theta_{i,G+1} = \begin{cases} u_{i,G}, & \text{if } f(u_{i,G}) \leq f(\theta_{i,G}) \\ \theta_{i,G}, & \text{Otherwise} \end{cases} \tag{10.21}$$

The application of mutation and crossover operators, as well as fitness checks, is continued until the maximum number of iterations is reached. The flowchart of the DE algorithm is presented in Fig. 10.8.

10.2.4 Proposed self-adaptive evolutionary extreme learning machine technique[3]

Application of the DE algorithm with the E-ELM requires that the mutant vector generation strategy and its associated parameters, including crossover rate (*CR*), scaling factor (*F*), and population number (*NP*) are assigned before modeling. Generally, the easiest way to determine the mutation strategy and its associated parameters is to use the trial-and-error process. One can quickly comprehend that this can lead to a large number of different combinations of control strategies and parameters, leading to a high model computational cost. Moreover, different mutant vector generation strategies with particular control parameters may lead to better model performance, however, it may be too computationally expensive to verify all combinations during a trial-and-error process.

A large number of studies have employed the DE algorithm to optimize the SLFFNN (Baioletti et al., 2020; Bonakdari et al., 2020). Zhu et al. (2005) integrated DE with ELM to introduce an evolutionary extreme learning machine (E-ELM). In the E-ELM, the hidden node parameters are selected using DE, while the *OutWs* are analytically calculated similarly to the classical ELM approach. Subudhi and Jena (2008) combined the DE and Levenberg–Marquardt algorithms to introduce DE-LM for training a feedforward network. In the DE-LM algorithm, the network input and *OutW* vectors are optimized by DE through an iteration process. Following the completion of the specified number of iterations, the mutant vector with the lowest cost function value is selected and considered as the initial value of the LM algorithm. Comparison of the E-ELM, DE-LM, and individual ELM indicates that E-ELM has higher generalization performance with a lower number of hidden neurons than individual ELM and lower computation cost than DE-ELM (Zhu et al., 2005).

10.2.4.1 Drawbacks of the classical optimization algorithm

While these algorithms (i.e. E-ELM, DE-LM) work well in system identifications, classifications, and regressions, their main limitation is the need for manual selection of the control parameters (for DE) and mutation strategies. The control parameters include the scaling factor (*F*) and the crossover rate (*CR*), which must both be manually specified in the E-ELM and DE-LM algorithms. Furthermore, the mutation vector generation strategy for both the E-ELM and DE-LM algorithms is "DE/rand/1," which has a slow convergence speed and a worse perturbation compared to the other proposed strategies (i.e., DE/rand-to-best/2 and DE/rand/2). As discussed by many researchers (Gämperle et al., 2002; Price et al., 2006; Zaharie, 2003; Zelinka & Lampinen, 2000), unsuitable selection of the control parameters and mutant

3. This section is provided for advanced users.

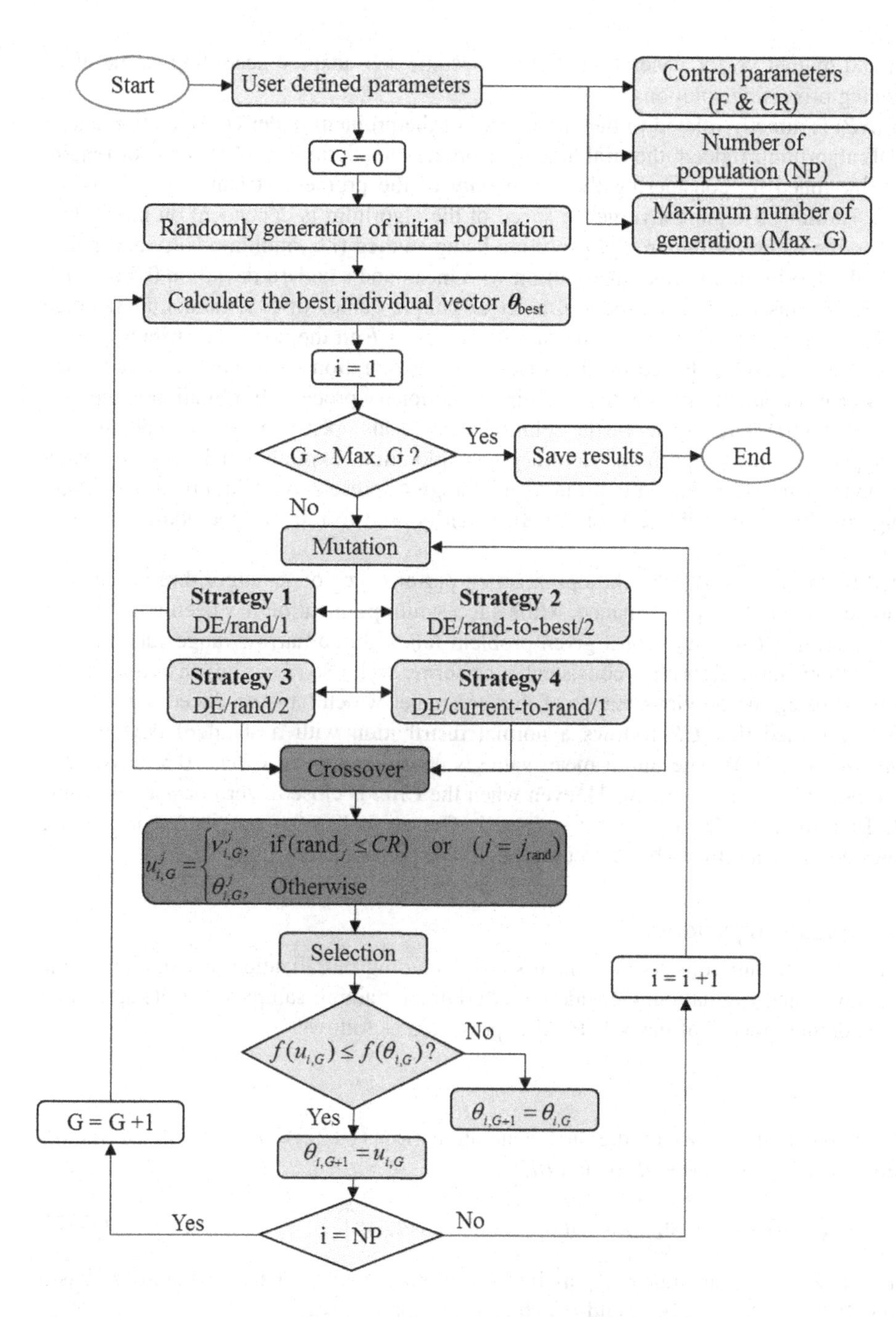

FIGURE 10.8 Flowchart of the differential evolution algorithm.

vector generation strategy may result in premature convergence or stagnation. Indeed, different control parameters (i.e., scaling factor and crossover rate) and vector generation strategies may perform drastically differently when faced with the varying conditions associated with different optimization problems. For this reason, the manual selection of control parameters and the designation of a fixed strategy for mutant vector generation in the DE-LM and E-ELM algorithms may not be suitable approaches for different applications.

10.2.4.2 Proposed solution

To overcome the above-mentioned drawbacks of the existing DE-based optimization approaches (i.e., DE-LM and E-ELM), Cao et al. (2012) integrated the self-adaptive differential evolutionary (SADE) algorithm (Qin et al., 2009) with the ELM (i.e., SaE-ELM) for SLFFNN optimization. In this approach, SADE algorithms are applied to optimize the hidden layer outputs (i.e., *InWs* and *BHNs*), while the individual ELM is employed to derive the network *OutWs*.

The control parameters (*CR* and *F*) and mutant vector generation strategy become self-adapted according to the algorithm's previous experience in producing promising solutions.

The required population number (*NP*) is directly related to the complexity of the problem under consideration and is therefore left as an input to the SADE algorithm. Indeed, the trial-and-error process for optimizing *NP* does not require significant effort and can be easily fine-tuned by considering the complexity of the problem at hand as well as the values cited in the relevant literature. Additionally, the convergence speed of the algorithm is dependent on the scaling factor(*F*) used, while *CR* is sensitive to the nature of the problem solution being studied (i.e., multimodality and unimodality). To approximate *F* in the SADE algorithm, a normal distribution with mean and standard deviation 0.5 and 0.3, respectively, is applied (*N*(0.5,0.3)). Using this normal distribution, a set of unique values of *F* is randomly sampled and used in the current population. With a probability of 0.997, the sample values of *F* in the normal distribution must fall into the range of [− 0.4, 1.4] (Qin et al., 2009). Based on this range, both exploration (with large *F* values) and exploitation (with small *F* values) power is sustained throughout the entire evolutionary process. Exploration is the process of searching for the global optimal through large spans of the solution space. This operation serves to prevent the search algorithm from becoming trapped in a local optimum. Contrary to exploration, exploitation is a local search which discovers local spaces of solutions. It should be noted that the *K* in strategy 4 [presented in Eq. (10.19)] is a randomly generated number in the range of [0, 1] (Subudhi & Jena, 2008) which is used to remove one additional user-defined parameter.

According to the results presented by Price et al. (2006), the optimization performance of the algorithm is sensitive to the selection of *CR*—poor selection can reduce the performance, while successful optimization has been noted with a good selection. In general, the best-performing *CR* values for a given problem fall within a narrow range such that for all *CR* values within this range, the optimization algorithm consistently performs well. So, for a given problem, the range of *CR* values is gradually adjusted using the previous values of this parameter which have produced mutant vectors efficaciously. Similar to *F*, it is assumed that *CR* follows a normal distribution with a standard deviation of $Std = 0.1$ and a mean value of *CRm* (*N*(*CRm*, *Std*)). The initial mean value is considered to be $CRm = 0.5$. Given that the values produced by *N*(*CRm*, *Std*) should be in the range [0, 1], even when the *CRm* is close to zero or one, the value of *Std* should be considered small. Therefore, the *Std* is considered as 0.1. Qin et al. (2009) expressed that a minor change in the *Std* at *N*(*CRm*, *Std*) does not have a remarkable effect on the SADE performance.

10.2.4.3 Main steps of the optimization process

The modeling process for the SaE-ELM algorithm consists of 4 main steps, including initialization, calculation of the *OutWs* and *RMSE*, mutation and crossover, and evaluation. Considering *N* arbitrary training samples, the *g*() activation function, and *L* hidden neurons, the modeling process of the SaE-ELM is presented as follows:

10.2.4.3.1 Step 1: Initialization

The first step is initializing the *NP* population vectors of the first generation (Eq. (10.22)). The initialized vectors encompass all of the network hidden node parameters (i.e., *InW* and *BHNs*).

$$\theta_{k,G} = \left[\boldsymbol{a}^T_{1,(k,G)}, \boldsymbol{a}^T_{2,(k,G)}, \ldots, \boldsymbol{a}^T_{L,(k,G)}, b_{1,(k,G)}, b_{2,(k,G)}, \ldots, b_{L,(k,G)}\right] \tag{10.22}$$

where $\boldsymbol{a}_j$ ($j = 1, 2, \ldots, L$) and b_j ($j = 1, 2, \ldots, L$) are randomly assigned, L is the number of hidden neurons, k is a counter for the number of population (*NP*) ($k = 1, 2, \ldots, NP$), and G is the generation number.

10.2.4.3.2 Step 2: Calculation of the output weights and root mean square error

Following initialization of the *InWs* and *BHNs*, the *OutW* matrix and root mean square error (*RMSE*) for all vectors in the defined population are calculated for each generation as shown below:

$$\beta_{k,G} = H^{+}_{k,G} T \tag{10.23}$$

$$RMSE_{k,G} = \sqrt{\frac{\sum_{i=1}^{N} \left| \sum_{j=1}^{L} \beta_j g\left(\boldsymbol{a}_{j,(k,G)}, b_{j,(k,G)}, \boldsymbol{x}_i\right) - \boldsymbol{t}_i \right|}{m \times N}} \tag{10.24}$$

where $H_{k,G}^{+}$ is the MPGI of $H_{k,G}$ and can be written as

$$H_{k,G} = \begin{bmatrix} g(a_{1,(k,G)}, b_{1,(k,G)}, x_1) & \cdots & g(a_{L,(k,G)}, b_{L,(k,G)}, x_1) \\ \vdots & \ddots & \vdots \\ g(a_{1,(k,G)}, b_{1,(k,G)}, x_N) & \cdots & g(a_{L,(k,G)}, b_{L,(k,G)}, x_N) \end{bmatrix} \quad (10.25)$$

The best population vector in the first generation (i.e., $\boldsymbol{\theta}_{best,1}$) is stored and is selected according to the lowest *RMSE* value at the current generation ($RMSE_{\boldsymbol{\theta}_{best,1}}$). To evaluate the mutant vector as the subsequent generations, the following equation is defined:

$$\theta_{k,G+1} = \begin{cases} u_{k,G+1} \ if \ RMSE_{\boldsymbol{\theta}_{k,G}} - RMSE_{\boldsymbol{\theta}_{k,G+1}} > \varepsilon.RMSE_{\boldsymbol{\theta}_{k,G}} \\ u_{k,G+1} \ if \ \left|RMSE_{\boldsymbol{\theta}_{k,G}} - RMSE_{\boldsymbol{\theta}_{k,G+1}}\right| < \varepsilon.RMSE_{\boldsymbol{\theta}_{k,G}} \ and \ \left|\beta_{\boldsymbol{u}_{k,G+1}}\right| < \left|\beta_{\boldsymbol{\theta}_{k,G}}\right| \\ \theta_{k,G} \ Otherwise \end{cases} \quad (10.26)$$

10.2.4.3.3 Step 3: Mutation and crossover

A candidate pool is made using four strategies [Eqs. (10.16)–(10.19)]. The strategy of the trial vector generation is selected from the pool for each target vector in the current generation with respect to the probability of each strategy ($p_{l,\,G}$ where $l = 1, 2, 3, 4$, and G is the generation number). Considering the learning period (*LP*) as a fixed number of generations, the $p_{l,G}$ is updated as follows:

$$p_{l,G} = \begin{cases} \dfrac{1}{4} & G \le LP \\ \dfrac{S_{l,G}}{\sum_{l=1}^{4} S_{l,G}} & G > LP \end{cases} \quad (10.27)$$

and

$$S_{l,G} = \frac{\sum_{g=G-LP}^{G-1} ns_{l,g}}{\sum_{g=G-LP}^{G-1} ns_{l,g} + \sum_{g=G-LP}^{G-1} nf_{l,g}} + \varepsilon \ (l = 1, 2, 3, 4) \quad (10.28)$$

where $nf_{l,g}$ is the number of trial vectors generated at the gth generation that are discarded in the next generation, while $ns_{l,g}$ is the number of trial vectors generated at the gth generation that are successfully entered in the next generation. The failure and success of *LP* generations are saved. The ε, which is a small positive fixed value, is defined to avoid the possibility of a null success rate. When the number of iterations exceeds the *LP* generations, the initial records are deleted and replaced by new numbers in the current generations.

After the definition of criteria to update the generation strategy of the trial vector, the crossover rate (*CR*) and scaling parameter (*F*) are randomly produced. In order to do so, the $N(0.5, 0.1)$ and $N(0.5, 0.3)$ normal distributions are applied for *CR* and *F*, respectively. The appropriate values of *CR* are typically located in a small range, and the mean value of this parameter is gradually tuned with respect to the previous values of the *CR* that were effective arriving in the next generation.

10.2.4.3.4 Step 4: Evaluation

Eq. (10.26) is applied to assess all produced trial vectors at $(G+1)$th generation ($\boldsymbol{u}_{k,G+1}$). It should be noted that the provided ε in this equation is a small positive tolerance rate. The results of Bartlett (1998) indicate that considering smaller weights in neural networks results in better generalization performance. Therefore, the norm of the *OutW* (i.e., $\|\boldsymbol{\beta}\|$) is considered as another criterion for trial vector selection. Step 3 (i.e., mutation and crossover) and step 4 (i.e., evaluation) are repeated until the stopping criteria have been reached, which is defined by a performance goal or the maximum number of iterations.

Because the population number (*NP*) is highly dependent on the complexity of the real-world problem being considered, it was left as a user-defined parameter in addition to the number of generations. Moreover, a validation dataset that has no overlap with the training samples is applied in step 1 (i.e., initialization) and adopted in the evolutionary process to avoid the overfitting problem. The flowchart of the SaE-ELM model is presented in Fig. 10.9.

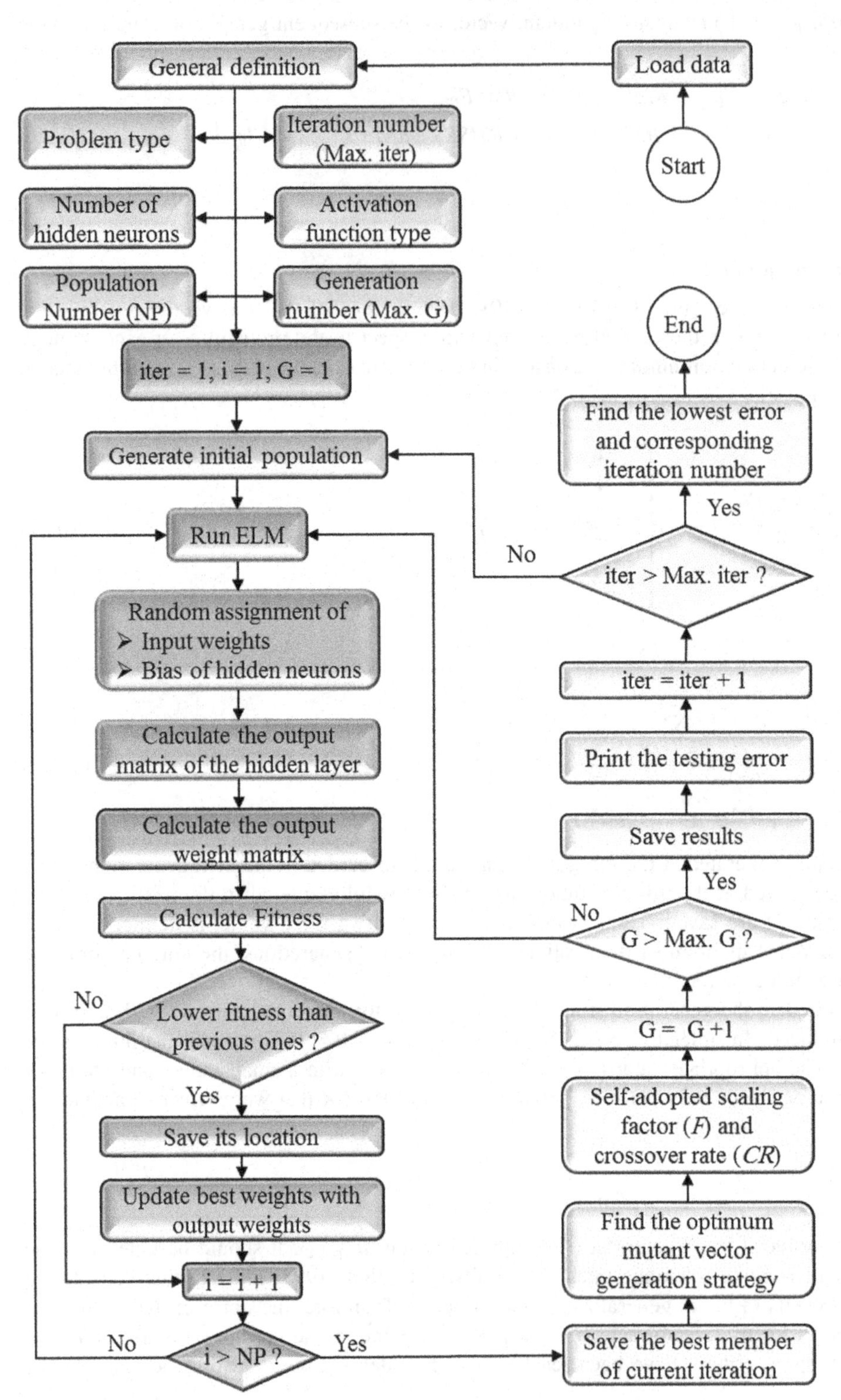

FIGURE 10.9 Flowchart of the self-adaptive evolutionary extreme learning machine algorithm.

10.3 Summary

This chapter provided a new self-adaptive optimized version of the ELM, known as SaE-ELM. After reviewing the principals of the ELM and its main limitations, a proposed solution to overcome these limitations was provided by integrating the self-adaptive differential evolution and original ELM. Regarding the proposed solution, detailed explanations of the mathematical formulation of the SaE-ELM were provided. Furthermore, a calculator is introduced to apply the developed SaE-ELM-based model for future applications that were not involved in the model development process. The developed SaE-ELM not only overcame the main limitation of the original ELM, but also was able to overcome the random generation of the *BHNs* and *InWs* by applying a large number of iterations. After reading this chapter, the reader is familiar with the development process of the single-layer neural network, including classical neural networks and non-tuned fast neural networks. Furthermore, they understand the main limitations of the first generation of the non-tuned fast neural network and can find a solution to solve the existing problems using optimization tools. Finally, the reader has learned an optimal method that can not only perform well in solving complex problems but also has a high modeling speed.

References

Abba, S. I., Pham, Q. B., Usman, A. G., Linh, N. T. T., Aliyu, D. S., Nguyen, Q., & Bach, Q. V. (2020). Emerging evolutionary algorithm integrated with kernel principal component analysis for modeling the performance of a water treatment plant. *Journal of Water Process Engineering, 33*, 101081. Available from https://doi.org/10.1016/j.jwpe.2019.101081.

Azimi, H., Bonakdari, H., & Ebtehaj, I. (2017a). Sensitivity analysis of the factors affecting the discharge capacity of side weirs in trapezoidal channels using extreme learning machines. *Flow Measurement and Instrumentation, 54*, 216–223. Available from https://doi.org/10.1016/j.flowmeasinst.2017.02.005.

Azimi, H., Bonakdari, H., Ebtehaj, I., Talesh, S. H. A., Michelson, D. G., & Jamali, A. (2017b). Evolutionary Pareto optimization of an ANFIS network for modeling scour at pile groups in clear water condition. *Fuzzy Sets and Systems, 319*, 50–69. Available from https://doi.org/10.1016/j.fss.2016.10.010.

Azimi, H., Bonakdari, H., Ebtehaj, I., Gharabaghi, B., & Khoshbin, F. (2018). Evolutionary design of generalized group method of data handling-type neural network for estimating the hydraulic jump roller length. *Acta Mechanica, 229*(3), 1197–1214. Available from https://doi.org/10.1007/s00707-017-2043-9.

Baioletti, M., Di Bari, G., Milani, A., & Poggioni, V. (2020). Differential evolution for neural networks optimization. *Mathematics, 8*(1), 69. Available from https://doi.org/10.3390/math8010069.

Bartlett, P. L. (1998). The sample complexity of pattern classification with neural networks: the size of the weights is more important than the size of the network. *IEEE Transactions on Information Theory, 44*(2), 525–536. Available from https://doi.org/10.1109/18.661502.

Bonakdari, H., & Ebtehaj, I. (2014). Study of sediment transport using soft computing technique. In *7th international conference on fluvial hydraulics, river flow* (pp. 933–940).

Bonakdari, H., & Ebtehaj, I. (2016). A comparative study of extreme learning machines and support vector machines in prediction of sediment transport in open channels. *International Journal of Engineering, 29*(11), 1499–1506.

Bonakdari, H., Ebtehaj, I., Samui, P., & Gharabaghi, B. (2019a). Lake water-level fluctuations forecasting using minimax probability machine regression, relevance vector machine, Gaussian process regression, and extreme learning machine. *Water Resources Management, 33*(11), 3965–3984. Available from https://doi.org/10.1007/s11269-019-02346-0.

Bonakdari, H., Moeeni, H., Ebtehaj, I., Zeynoddin, M., Mahoammadian, A., & Gharabaghi, B. (2019b). New insights into soil temperature time series modeling: linear or nonlinear. *Theoretical and Applied Climatology, 135*(3–4), 1157–1177. Available from https://doi.org/10.1007/s00704-018-2436-2.

Bonakdari, H., Qasem, S. N., Ebtehaj, I., Zaji, A. H., Gharabaghi, B., & Moazamnia, M. (2020). An expert system for predicting the velocity field in narrow open channel flows using self-adaptive extreme learning machines. *Measurement, 151*, 107202. Available from https://doi.org/10.1016/j.measurement.2019.107202.

Cao, J., Lin, Z., & Huang, G. B. (2012). Self-adaptive evolutionary extreme learning machine. *Neural Processing Letters, 36*(3), 285–305. Available from https://doi.org/10.1007/s11063-012-9236-y.

Ebtehaj, I., & Bonakdari, H. (2014). Comparison of genetic algorithm and imperialist competitive algorithms in predicting bed load transport in clean pipe. *Water Science and Technology, 70*(10), 1695–1701. Available from https://doi.org/10.2166/wst.2014.434.

Ebtehaj, I., & Bonakdari, H. (2016a). Bed load sediment transport estimation in a clean pipe using multilayer perceptron with different training algorithms. *KSCE Journal of Civil Engineering, 20*(2), 581–589. Available from https://doi.org/10.1007/s12205-015-0630-7.

Ebtehaj, I., & Bonakdari, H. (2016b). Assessment of evolutionary algorithms in predicting non-deposition sediment transport. *Urban Water Journal, 13*(5), 499–510. Available from https://doi.org/10.1080/1573062X.2014.994003.

Ebtehaj, I., & Bonakdari, H. (2017). Design of a fuzzy differential evolution algorithm to predict non-deposition sediment transport. *Applied Water Science, 7*(8), 4287–4299. Available from https://doi.org/10.1007/s13201-017-0562-0.

Ebtehaj, I., Bonakdari, H., & Zaji, A. H. (2018a). A new hybrid decision tree method based on two artificial neural networks for predicting sediment transport in clean pipes. *Alexandria Engineering Journal*, *57*(3), 1783–1795. Available from https://doi.org/10.1016/j.aej.2017.05.021.

Ebtehaj, I., Bonakdari, H., Moradi, F., Gharabaghi, B., & Khozani, Z. S. (2018b). An integrated framework of extreme learning machines for predicting scour at pile groups in clear water condition. *Coastal Engineering*, *135*, 1–15. Available from https://doi.org/10.1016/j.coastaleng.2017.12.012.

Ebtehaj, I., Bonakdari, H., Shamshirband, S., Ismail, Z., & Hashim, R. (2017a). New approach to estimate velocity at limit of deposition in storm sewers using vector machine coupled with firefly algorithm. *Journal of Pipeline Systems Engineering and Practice*, *8*(2), 04016018. Available from https://doi.org/10.1061/(ASCE)PS.1949-1204.0000252.

Ebtehaj, I., Sattar, A. M., Bonakdari, H., & Zaji, A. H. (2017b). Prediction of scour depth around bridge piers using self-adaptive extreme learning machine. *Journal of Hydroinformatics*, *19*(2), 207–224. Available from https://doi.org/10.2166/hydro.2016.025.

Ebtehaj, I., Bonakdari, H., Zaji, A. H., Bong, C. H. J., & Ab Ghani, A. (2016). Design of a new hybrid artificial neural network method based on decision trees for calculating the Froude number in rigid rectangular channels. *Journal of Hydrology and Hydromechanics, 64(3), 252–260.* Available from https://doi.org/10.1515/johh-2016-0031.

Ebtehaj, I., Soltani, K., Amiri, A., Faramarzi, M., Madramootoo, C. A., & Bonakdari, H. (2021a). Prognostication of shortwave radiation using an improved No-Tuned fast machine learning. *Sustainability*, *13*(14), 8009. Available from https://doi.org/10.3390/su13148009.

Ebtehaj, I., Bonakdari, H., Zaji, A. H., & Gharabaghi, B. (2021b). Evolutionary optimization of neural network to predict sediment transport without sedimentation. *Complex & Intelligent Systems*, *7*, 401–416. Available from https://doi.org/10.1007/s40747-020-00213-9.

Gämperle, R., Müller, S. D., & Koumoutsakos, P. (2002). A parameter study for differential evolution. *Advances in Intelligent Systems, Fuzzy Systems, Evolutionary Computation*, *10*(10), 293–298.

Gholami, A., Bonakdari, H., Ebtehaj, I., & Akhtari, A. A. (2017). Design of an adaptive neuro-fuzzy computing technique for predicting flow variables in a 90 sharp bend. *Journal of Hydroinformatics*, *19*(4), 572–585. Available from https://doi.org/10.2166/hydro.2017.200.

Gholami, A., Bonakdari, H., Ebtehaj, I., Gharabaghi, B., Khodashenas, S. R., Talesh, S. H. A., & Jamali, A. (2018). A methodological approach of predicting threshold channel bank profile by multi-objective evolutionary optimization of ANFIS. *Engineering Geology*, *239*, 298–309. Available from https://doi.org/10.1016/j.enggeo.2018.03.030.

Gholami, A., Bonakdari, H., Ebtehaj, I., Talesh, S. H. A., Khodashenas, S. R., & Jamali, A. (2019a). Analyzing bank profile shape of alluvial stable channels using robust optimization and evolutionary ANFIS methods. *Applied Water Science*, *9*(3), 40. Available from https://doi.org/10.1007/s13201-019-0928-6.

Gholami, A., Bonakdari, H., Zeynoddin, M., Ebtehaj, I., Gharabaghi, B., & Khodashenas, S. R. (2019b). Reliable method of determining stable threshold channel shape using experimental and gene expression programming techniques. *Neural Computing and Applications*, *31*(10), 5799–5817. Available from https://doi.org/10.1007/s00521-018-3411-7.

Hagan, M. T., & Menhaj, M. B. (1994). Training feedforward networks with the Marquardt algorithm. *IEEE transactions on Neural Networks*, *5*(6), 989–993. Available from https://doi.org/10.1109/72.329697.

Huang, G. B., Zhu, Q. Y., & Siew, C. K. (2004). Extreme learning machine: a new learning scheme of feedforward neural networks. In *2004 IEEE international joint conference on neural networks* (*IEEE Cat. No. 04CH37541*) (Vol. 2, pp. 985–990). IEEE. https://doi.org/10.1109/IJCNN.2004.1380068.

Huang, G. B., Zhu, Q. Y., & Siew, C. K. (2006a). Extreme learning machine: theory and applications. *Neurocomputing*, *70*(1–3), 489–501. Available from https://doi.org/10.1016/j.neucom.2005.12.126.

Khozani, Z. S., Bonakdari, H., & Ebtehaj, I. (2017). An analysis of shear stress distribution in circular channels with sediment deposition based on gene expression programming. *International Journal of Sediment Research*, *32*(4), 575–584. Available from https://doi.org/10.1016/j.ijsrc.2017.04.004.

Kumar, J., & Singh, A. K. (2018). Workload prediction in cloud using artificial neural network and adaptive differential evolution. *Future Generation Computer Systems*, *81*, 41–52. Available from https://doi.org/10.1016/j.future.2017.10.047.

Levenberg, K. (1944). A method for the solution of certain non-linear problems in least squares. *Quarterly of applied mathematics*, 2(2), 164–168. Available from http://www.jstor.org/stable/43633451.

Melo, H., & Watada, J. (2016). Gaussian-PSO with fuzzy reasoning based on structural learning for training a neural network. *Neurocomputing*, *172*, 405–412. Available from https://doi.org/10.1016/j.neucom.2015.03.104.

Moeeni, H., Bonakdari, H., & Ebtehaj, I. (2017). Integrated SARIMA with neuro-fuzzy systems and neural networks for monthly inflow prediction. *Water Resources Management*, *31*(7), 2141–2156. Available from https://doi.org/10.1007/s11269-017-1632-7.

Moradi, F., Bonakdari, H., Kisi, O., Ebtehaj, I., Shiri, J., & Gharabaghi, B. (2019). Abutment scour depth modeling using neuro-fuzzy-embedded techniques. *Marine Georesources & Geotechnology*, *37*(2), 190–200. Available from https://doi.org/10.1080/1064119X.2017.1420113.

Price, K., Storn, R. M., & Lampinen, J. A. (2006). *Differential evolution: a practical approach to global optimization*. Berlin, Germany: Springer-Verlag.

Qasem, S. N., Ebtehaj, I., & Bonakdari, H. (2017). Potential of radial basis function network with particle swarm optimization for prediction of sediment transport at the limit of deposition in a clean pipe. *Sustainable Water Resources Management*, *3*(4), 391–401. Available from https://doi.org/10.1007/s40899-017-0104-9.

Qin, A. K., Huang, V. L., & Suganthan, P. N. (2009). Differential evolution algorithm with strategy adaptation for global numerical optimization. *IEEE transactions on Evolutionary Computation*, *13*(2), 398–417. Available from https://doi.org/10.1109/TEVC.2008.927706.

Sharafi, H., Ebtehaj, I., Bonakdari, H., & Zaji, A. H. (2016). Design of a support vector machine with different kernel functions to predict scour depth around bridge piers. *Natural Hazards*, *84*(3), 2145–2162. Available from https://doi.org/10.1007/s11069-016-2540-5.

Sihag, P., Esmaeilbeiki, F., Singh, B., Ebtehaj, I., & Bonakdari, H. (2019). Modeling unsaturated hydraulic conductivity by hybrid soft computing techniques. *Soft Computing*, *23*(23), 12897–12910. Available from https://doi.org/10.1007/s00500-019-03847-1.

Singh, A., & Kumar, S. (2016). Differential evolution: An overview. In *Proceedings of fifth international conference on soft computing for problem solving* (pp. 209–217). Springer, Singapore. Available from https://doi.org/10.1007/978-981-10-0448-3_17.

Storn, R., & Price, K. (1997). Differential evolution–a simple and efficient heuristic for global optimization over continuous spaces. *Journal of global optimization*, *11*(4), 341–359. Available from https://doi.org/10.1023/A:1008202821328.

Subudhi, B., & Jena, D. (2008). Differential evolution and levenberg marquardt trained neural network scheme for nonlinear system identification. *Neural Processing Letters*, *27*(3), 285–296. Available from https://doi.org/10.1007/s11063-008-9077-x.

Wang, D., Luo, H., Grunder, O., Lin, Y., & Guo, H. (2017). Multi-step ahead electricity price forecasting using a hybrid model based on two-layer decomposition technique and BP neural network optimized by firefly algorithm. *Applied Energy*, *190*, 390–407. Available from https://doi.org/10.1016/j.apenergy.2016.12.134.

Zaharie, D. (2003). Control of population diversity and adaptation in differential evolution algorithms. In Matousek, R., & Osmera, P. (Eds.), *Proceeding of mendel 9th international conference of soft computing* (Vol. 9, pp. 41–46). Brno, Czech Republic, June 2003.

Zelinka, I., & Lampinen, J. (2000). On stagnation of the differential evolution algorithm. In Ošmera, P. (Ed.), *Proceedings of mendel, 6th international mendel conference on soft computing* (pp. 76–83). 2002.

Zhu, Q. Y., Qin, A. K., Suganthan, P. N., & Huang, G. B. (2005). Evolutionary extreme learning machine. *Pattern Recognition*, *38*(10), 1759–1763. Available from https://doi.org/10.1016/j.patcog.2005.03.028.

Chapter 11

Self-adaptive evolutionary of non-tuned neural network—coding and implementation

11.1 Implementation of the self-adaptive evolutionary extreme learning machine in detail

In this subsection, all of the necessary steps for modeling a real-world problem using the developed SaE-ELM are presented, including a detailed line-by-line presentation of the coding techniques in the MATLAB® environment. The MATLAB package for the SaE-ELM model contains five different functions, which are organized into three groups, as shown in Table 11.1. The first of these three groups are the ELM-based codes, which include the "SaE-ELM" and "ELM_X" functions. The "SaE-ELM" function is the main function used to model a problem based on the SaE-ELM technique. The second function in the ELM-based codes category is the "ELM_X" function, which is called by the "SaE-ELM" function to optimize the output weights matrix through evolutionary-based optimization. The inputs of this function include the ELM parameters (i.e., number of hidden neurons and activation function type) as well the initialized matrices (i.e., input weights and bias of hidden neurons). The second group of functions is related to the visual and numerical output of the results, which are performed using the "PlotResults" and "Statistical_Indices" functions, respectively. The final group is the calculator, which contains the novel "Calculator" function used to apply the modeling results for unseen data which was not used during model development.

For the beginner user, an understanding of the user-defined input parameters, including ELM (i.e., activation function type and the number of hidden neurons) and SaE-ELM parameters (i.e., generation number and number of population), is all that is required to apply the model to real-world applications using the calculator tool. Therefore, there is no requirement for the beginner user to follow the detailed explanations of the SaE-ELM coding in MATLAB, which are provided in the gray boxes. Data on SaE-ELM can be found at Appendix 11A.

11.1.1 Loading of training and testing data

Before commencing the modeling process, it is necessary to call the data used to model the desired problem. It should be noted that data recall can be done in two ways: (1) calling all the data and handling them in two groups of training and testing samples using an appropriate method (there are several methods for dividing data into training and testing sets as mentioned in Chapter 2, of which the simplest and most commonly used is random selection without replacing) and (2) classification of training and test data before calling by the model so that each training and testing samples are called separately. In this study, the second method is used because data classification may have been done before modeling was started. For example, if the goal is to compare two models, all training and testing samples must be exactly the same. Furthermore, suppose we intend to compare the performance of a machine learning (ML) approach (i.e., SaE-ELM) for a given problem by altering the input parameters. In that case, all training and testing samples must be the same.

The percentage of samples used for model training can be selected between 50% and 90% of the total dataset (i.e., 50%–10% of the total data would be used for the testing stage) (Arabameri et al., 2020; Ebtehaj et al., 2020; Lanera et al., 2019). It is generally recommended to split training and testing data according to the ratio of 70%–30% (Ebtehaj et al., 2016, Ebtehaj, Sammen et al., 2021; Ebtehaj, Soltani et al., 2021; Grégoire et al., 2022; Saha et al., 2021; Thottakkara et al., 2016) or 80%–20% (Bonakdari, Ebtehaj et al., 2020; Bonakdari, Gholami, et al., 2020; Herrera et al., 2019; Mojtahedi et al., 2019; Ratzinger et al., 2018).

Machine Learning in Earth, Environmental and Planetary Sciences. DOI: https://doi.org/10.1016/B978-0-443-15284-9.00004-5

TABLE 11.1 Classification and definitions of functions employed in the developed self-adaptive evolutionary extreme learning machine (SaE-ELM) model.

Groups	Function name	Definitions
ELM-based approaches	ELM_X	Provide the individual ELM
Results	SaE-ELM	The main code of the developed SaE-ELM
	PlotResults	Provide the visual results
Calculator	Statistical_Indices	Provide the numerical results
	Calculator	A calculator tool used for the application of the developed SaE-ELM model to unseen datasets in real-world problems

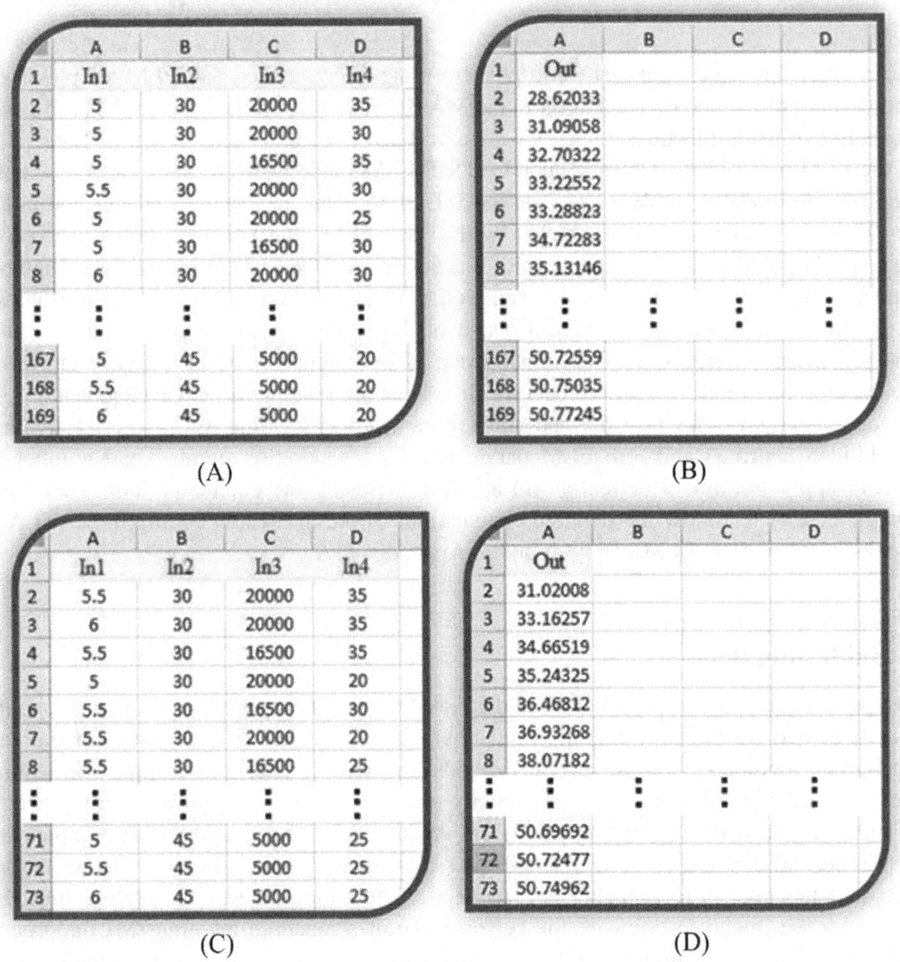

	A	B	C	D
1	In1	In2	In3	In4
2	5	30	20000	35
3	5	30	20000	30
4	5	30	16500	35
5	5.5	30	20000	30
6	5	30	20000	25
7	5	30	16500	30
8	6	30	20000	30
⋮	⋮	⋮	⋮	⋮
167	5	45	5000	20
168	5.5	45	5000	20
169	6	45	5000	20

(A)

	A	B	C	D
1	Out			
2	28.62033			
3	31.09058			
4	32.70322			
5	33.22552			
6	33.28823			
7	34.72283			
8	35.13146			
⋮	⋮	⋮	⋮	⋮
167	50.72559			
168	50.75035			
169	50.77245			

(B)

	A	B	C	D
1	In1	In2	In3	In4
2	5.5	30	20000	35
3	6	30	20000	35
4	5.5	30	16500	35
5	5	30	20000	20
6	5.5	30	16500	30
7	5.5	30	20000	20
8	5.5	30	16500	25
⋮	⋮	⋮	⋮	⋮
71	5	45	5000	25
72	5.5	45	5000	25
73	6	45	5000	25

(C)

	A	B	C	D
1	Out			
2	31.02008			
3	33.16257			
4	34.66519			
5	35.24325			
6	36.46812			
7	36.93268			
8	38.07182			
⋮	⋮	⋮	⋮	⋮
71	50.69692			
72	50.72477			
73	50.74962			

(D)

FIGURE 11.1 Preparation of the inputs and output for the training and testing phases using an Excel file. ((A) Training Inputs, (B) Training Target, (C) Testing Inputs, (D) Testing Taget).

To load data in the developed SaE-ELM technique, the input samples and corresponding output for both the training and testing phases are read from an Excel file using the `xlsread` command. The Excel file is divided into four sheets, where sheets 1– 4 are related to training inputs, training targets, testing inputs, and testing targets, respectively. Fig. 11.1 demonstrates the format of the Excel workbook, which in our example, contains 169 training points and 73 testing points. The structure of the `xlsread` command is shown in Code 11.1. Indeed, the `File Name` and `Sheet name` should be replaced with the specific location associated with each of the four sample subsets for data loading.

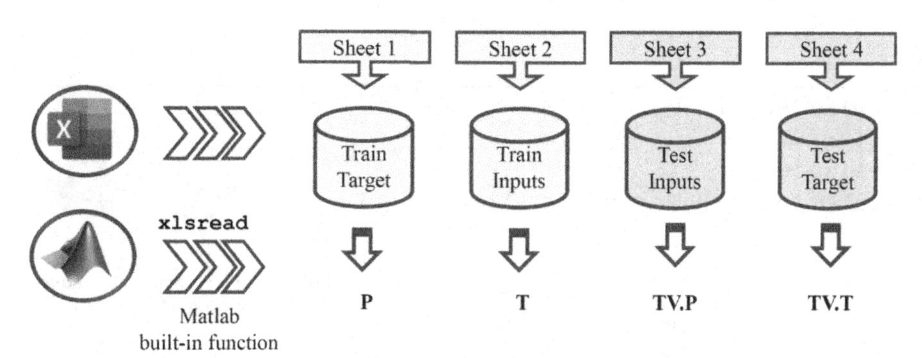

FIGURE 11.2 Schematic representation of Code 11.2.

Code 11.1

```
1   xlsread('File Name','Sheet name')
```

It can be seen in Code 11.2 that the training input and target values are saved as P and T (respectively), while the testing phase maintains the same notation but with TV affixed at the beginning. According to Fig. 11.1, for both the training and testing phases, the number of input variables and target variables are $d = 4$ and $m = 1$, respectively. Recall that the loading of data from the reference Excel file is column-based. For example, the output (for a problem with only one target variable) is a matrix with one column and N training (or testing) samples ($N \times d$). The coding employed in the development of the SaE-ELM requires that the data is in row format instead of column format, and therefore the dimensions of the values called from the Excel file must be changed to $d \times N$. This is done using the transpose operation ($'$), which can be seen before the semicolon of each line of Code 11.2. The schematic representation of Code 11.2 is provided in Fig. 11.2.

Code 11.2

```
1   %% Load data
2   P    = xlsread('Example1','sheet1')';    % TrainInputs
3   T    = xlsread('Example1','sheet2')';    % TrainTargets
4   TV.P = xlsread('Example1','sheet3')';    % TestInputs
5   TV.T = xlsread('Example1','sheet4')';    % TestTargets
```

After calling the training and testing samples, the number of samples included for each phase is calculated using the size command. The general format of this command is shown in Code 11.3.

Code 11.3

```
1   size(P,D)
```

In this command, P is the parameter for which we desire the size of, and D indicates the desired dimension. For example, if we have a two-dimensional matrix, a value of $D = 1$ indicates that the size of the row should be calculated, while $D = 2$ indicates that the size of the column should be calculated. In Code 11.4, we can see that to determine the number of training and testing samples D (in the size command) should be set to 2, while to determine the number of input neurons, it is set to a value of 1. As discussed earlier, when loading the data from Microsoft Excel, the transpose operation is performed. For this reason, the rows and columns of these matrices (i.e., `P` and `TV.P`) indicate the number of input variables and the number of samples, respectively. It is clear that to determine the number of training and testing data `T` can be used instead of `P` and `TV.T` instead of `TV.P` because the number of input and output samples of each phase are equal. When determining the number of input neurons `P` could be replaced with `TV.P`, the reason being that the number of problem inputs in both the training and testing phases must be equal. The schematic representation of Code 11.4 is provided in Fig. 11.3.

Code 11.4

```
1    NumberofTrainingData=size(P,2);
2    NumberofTestingData=size(TV.P,2);
3    NumberofInputNeurons=size(P,1);
4    NumberofValidationData = round(NumberofTestingData/5);
```

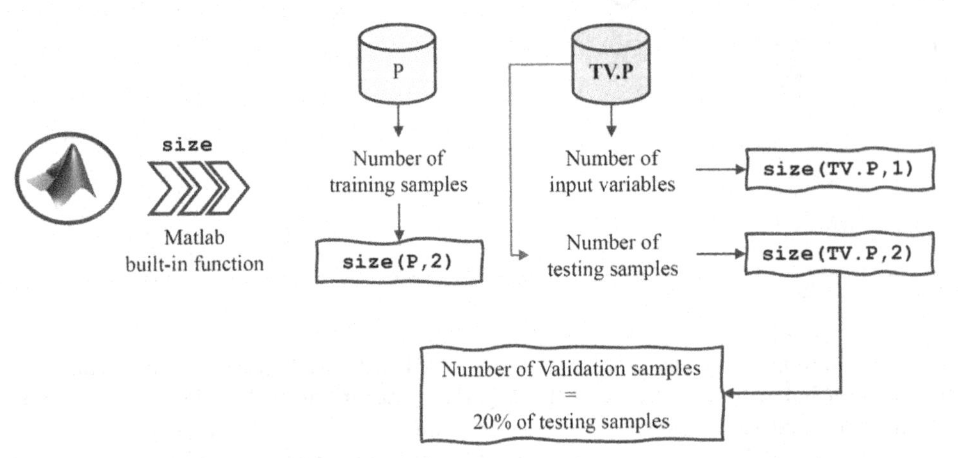

FIGURE 11.3 Schematic representation of Code 11.4.

11.1.2 Choosing the validation dataset

In the proposed data splitting method, the data called to the MATLAB environment from the reference Excel workbook have already been split into training and testing data subsets. However, to evaluate the performance of the model during the evolutionary optimization process, it is necessary to use validation samples. In this study, 20% of the test data was used for validation. Regarding the validation data, two important points must be mentioned.

The first point to discuss is the selection of 20% as the fraction of test data to be used for validation. There is no obligation to use this percentage: the reason for its selection is based on the authors' experience in evaluating the developed method while solving various real-world problems. It is recommended that if the test phase does not achieve acceptable accuracy in comparison with the training phase, the fraction of testing data considered for validation be increased incrementally up to 50% in a trial-and-error process. Increasing the percentage of validation data beyond a maximum of 50% will significantly limit the number of testing data cases. Therefore, if satisfactory modeling results cannot be achieved while considering 50% of the testing data as validation data, then two approaches may be suggested. The first approach is to increase the number of training samples, which will require the collection of more field or laboratory data. This is not always possible, however, due to time and financial constraints on data collection. The second solution would be to adjust the ratio of training and testing data to be closer to each other. In this way, a greater percentage of the testing samples could be considered as validation samples without severely limiting the number of testing data cases. For example, the percentage of training samples can be taken as 50% of the total data, while the remaining 50% can be split in half for testing samples and validation samples.

The second point to note is regarding the selection of validation samples from among the testing samples. Since validation samples are used only to evaluate the performance of the trained model during the evolutionary optimization process, these data have virtually no role in model training and only monitor the performance of the model during the optimization process. Therefore, the only samples involved in model training are training samples, and validation samples, like testing samples, are applied to examine the generalizability of the model.

11.1.3 Definition of main parameters in the self-adaptive optimization process

After calling the data to the MATLAB environment from our reference Excel workbook (i.e., inputs and targets for both the training and testing stages), the modeling parameters for the SaE-ELM method should be defined. The items

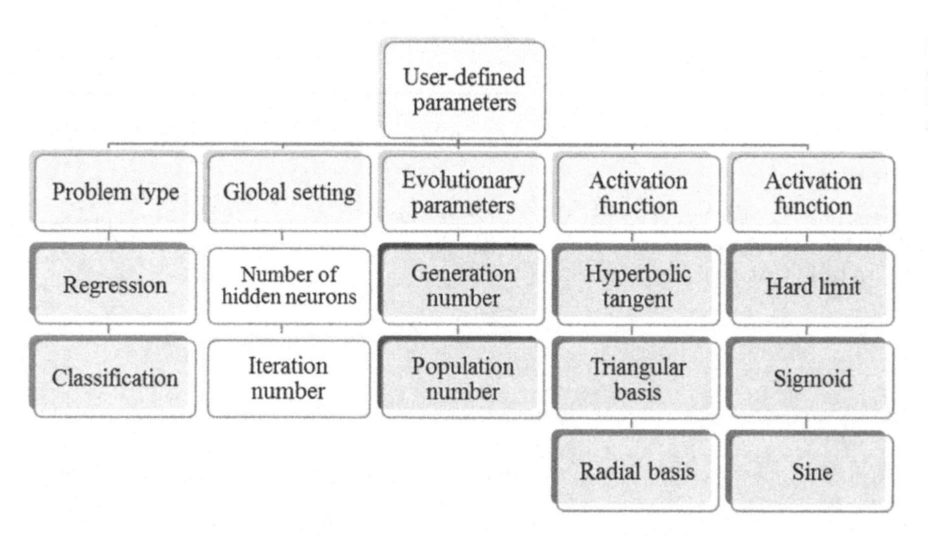

FIGURE 11.4 Definition of the user-defined parameters in the developed SaE-ELM. *SaE-ELM*, Self-adaptive evolutionary extreme learning machine.

that must be specified prior to modeling can be divided into four general categories, including problem type (i.e., regression or classification), global settings (i.e., number of hidden neurons and iteration number), evolutionary parameters (i.e., Generation number and population number), and activation function type (Fig. 11.4). Regarding the problem type selection, the type of problem to be solved through implementing the developed code should be specified, which can be either regression or classification.

11.1.3.1 Problem type selection

The questdlg command is used to select the problem type. This MATLAB command can be used to store up to three input options, with one of the options being defined as the default. The general format for this command is as follows:

Code 11.5

```
1    A = questdlg(qstring,title,str1,str2,str3,default)
```

The input arguments of the questdlg command are as follows: the question string (qstring), the title (title), one to three strings representing the possible options (str1-str3), and the default string (default). The first input argument for the questdlg command is termed qstring and allows the user to display the desired text message using a character string. The remaining input arguments will be different depending on the specific application of this command, with either two (**str1**, **str2**) or three (**str1**, **str2**, **str3**) selected modes possible. If the purpose of the question is to select an option from more than three options, other commands such as spaceList should be used. Note that the last input of questdlg command is always considered to be the default answer. Therefore, for cases where the user inputs two or three possible selections, the total number of entries within the input argument is equal to three or four, respectively, to account for the default answer.

As shown in Code 11.6, the first step in using the questdlg command is to define the two options, with Option {1} being regression and Option {2} being classification (lines 2–3). Next, we enter the qstring input, which is our text message posed to the reader, with the question "Regression or Classification?." Based on the questdlg structure, **Option {1}** and **Option {2}** are considered as str1 and str2 (respectively), and the default response will be considered as Option {1} (see line 6). According to the user response to the question dialog box, the Elm_Type is set, and the problem type (i.e., regression or classification) is specified. The schematic representation of Code 11.6 is provided in Fig. 11.5. Fig. 11.6 presents the resulting dialog box generated from this algorithm. Note that the default option is highlighted in blue in the question dialog box (i.e., Option {1}—"Regression").

Code 11.6

```
1   %% Problem Type
2   Option{1}='Regression';
3   Option{2}='Classification';
4   ANSWER=questdlg('Regression or Classification ?',...
5                   'Problem Type',...
6                   Option{1},Option{2},Option{1});
7   pause(0.1);
8   switch ANSWER
9       case Option{1}
10          Elm_Type = 0;
11          REGRESSION = 0;
12      case Option{2}
13          Elm_Type = 1;
14          CLASSIFIER = 1;
15  end
```

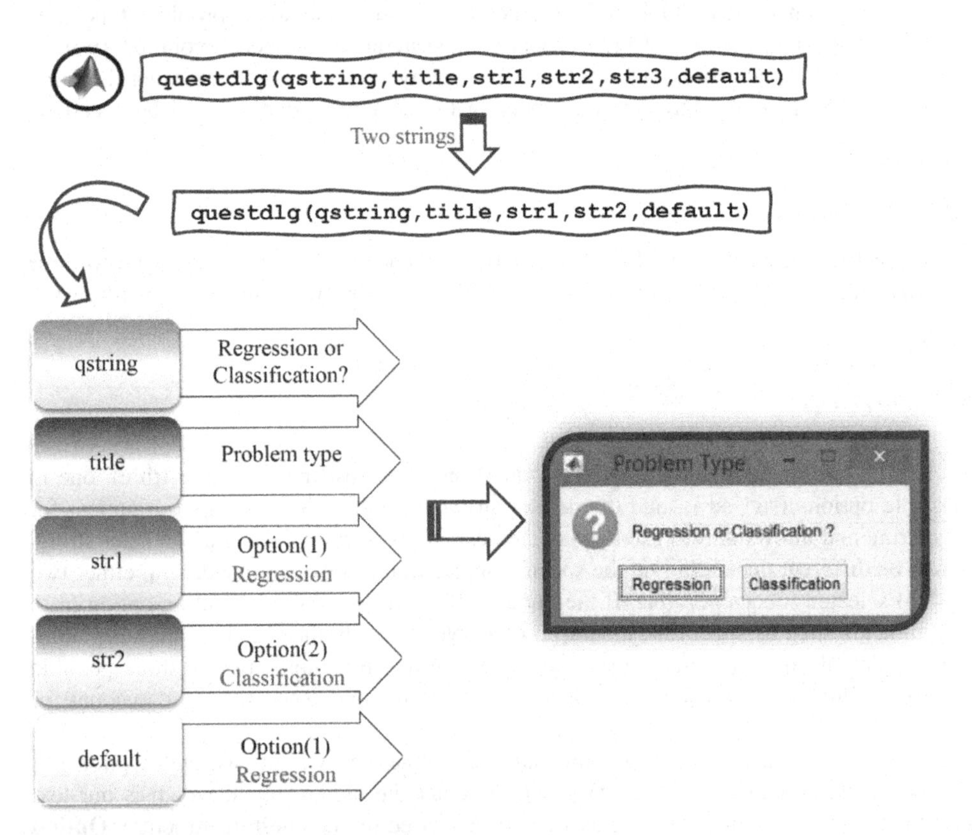

FIGURE 11.5 Schematic representation of Code 11.6.

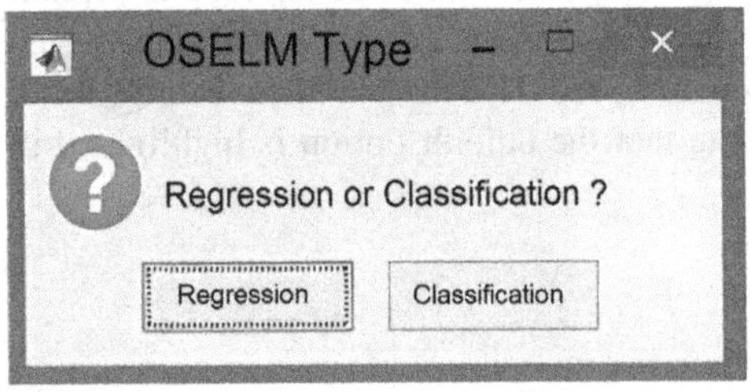

FIGURE 11.6 Resulting question dialog box generated using the questdlg command for problem-type selection.

11.1.3.2 Global parameters of the non-tuned machine learning model

The next step is determining the global settings, which include the number of hidden neurons and the iteration number. The iteration number is a parameter used to overcome the random generation of input weights and bias of hidden neurons. Recall that the random generation of these two matrices accounts for at least 66% of all tuned parameters during the SaE-ELM training process [Eq. (10.5)]. The simplest solution to overcome the random generation of the input weights and bias of hidden neurons is to consider a sufficient number of iterations. In this way, the random generation of the two matrices does not have a negative impact on the generalizability or accuracy of the model during the training and testing phases, which should perform similarly to one another. Given that the modeling time in the SaE-ELM method is very low and that this method is known for its fast training speed, considering a large number of iterations is not problematic for this model. It is suggested that the value of this parameter be considered as large as possible at the beginning of modeling so that the effect of changes in other parameters on modeling can be well investigated.

The second global settings parameter to be defined is the number of hidden neurons. According to Eq. (10.1), the number of tuned parameters through the training phase is as follows:

$$k = NHN(InV + 2) \tag{11.1}$$

where NHN is the number of hidden neurons and InV is the number of input variables.

Prior to initiating the NHN before modeling, the upper and lower limits of this parameter should be defined. The minimum NHN is one, while the maximum NHN must be specified in such a way that the number of tuned parameters through the training phase (k) is less than the number of training samples. In this way, the overfitting problem, which has been observed in various studies in the ML field (Bonakdari & Ebtehaj, 2021; Ebtehaj et al., 2021c; Ebtehaj & Bonakdari, 2022a) is avoided (Azimi et al., 2018). The maximum NHN is calculated as follows:

$$k = NHN(InV + 2) < TrS \tag{11.2}$$

where TrS is the number of training samples and all other parameters have been previously defined.

Then,

$$NHN < \frac{TrS}{InV + 2} \tag{11.3}$$

Therefore, when determining the value of NHN, the abovedefined relationship [Eq. (11.3)] should be considered.

In situations where an acceptable answer is not obtained using the number of existing training samples and the number of input variables specified, there are two ways to increase accuracy: increase the number of training samples and/or decrease the number of input variables.

To initialize the two global settings parameters (number of hidden neurons and iteration number), the inptdlg command is used. The general format of this command is as follows:

Code 11.7

```
15    PARAMS=inputdlg(prompt,dlg_title,num_lines,defAns);
```

where prompt is the text edit field that allows the modeler to request the desired parameter(s) to the user, dlg_title is the title of the dialog box, num_lines is defined as [row column], and defAns is the default value for each question. The input argument num_lines is used to define the edit field height and width when it is input as an array.

Code 11.8 demonstrates the coding used to request the global parameters, including the number of hidden neurons and iteration number using the inputdlg command. It should be noted that the number of default values and prompts must contain the same number of entries. For example, the prompt in line 2 is "{'Number of Hidden Neurons,' 'Iteration Number'}" which contains two entries. Consequently, the default values variable must also be composed of two entries, which in this case are specified as "{'20,' '50'}" for the NHN and iteration number, respectively. Since there are two input values stored into the variable PARAMS, it should be recalled using the syntax PARAMS{1} and PARAMS{2} for the first (i.e., Number of Hidden Neurons) and second (i.e., Iteration Number) prompt, respectively. The schematic representation of Code 11.8 is provided in Fig. 11.7. Additionally, the resulting dialog window is presented in Fig. 11.8. Note that the default values for the NHN and iteration number are populated within the text edit field when the window is prompted to the user.

Code 11.8

```
1   %% Global parameters
2   Prompt={'Number of Hidden Neurons','Iteration Number'};
3   Title='Global Setting';
4   DefaultValues={'20', '50'};
5   PARAMS=inputdlg(Prompt,Title,[1 40],DefaultValues);
6   pause(0.1);
7   NumberofHiddenNeurons=str2double(PARAMS{1});
8   N=str2double(PARAMS{2});
```

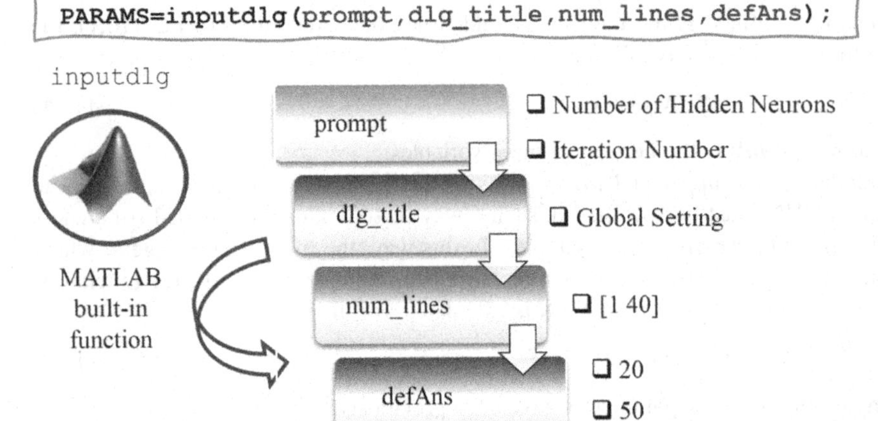

FIGURE 11.7 Schematic representation of Code 11.8.

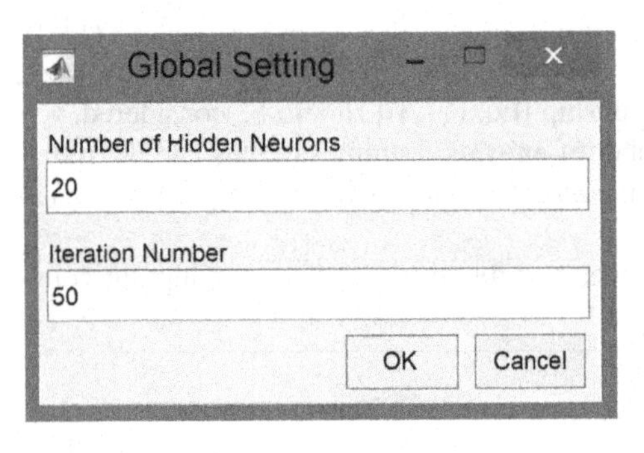

FIGURE 11.8 The dialog box resulting from the inputdlg command for defining the global setting parameters.

11.1.3.3 Activation function types

Following the definition of the problem type and the global settings, the activation function type (which is central to all of the ELM-based techniques) must be defined by the user. A simple way to allow the user to input the activation function type is by using the questdlg command, which was previously used during the selection of the problem type. The activation functions that may be applied in this code include the sine (Sin) (Owolabi & Abd Rahman, 2021), sigmoid (Sig) (Calabrò et al., 2021; Ebtehaj et al., 2017), hard-limit (Hardlim) (Suchithra & Pai, 2020; Zeynoddin et al., 2018), triangular basis (Tribas) (Azimi et al., 2021; Sattar et al., 2019), radial basis (Radbas) (Tripathi et al., 2020; Samal et al., 2022), and hyperbolic tangent (tanh) (Ebtehaj et al., 2018; Maimaitiyiming et al., 2019; Ratnawati et al., 2020) functions. As can be seen, six activation functions are considered for our purposes, which complicates the user definition process as the questdlg command is limited to three options. For this reason, the listdlg command must be applied. The listdlg command creates a list selection dialog box allowing the user to select from the six activation functions. The general format of this command is as follows:

Code 11.9

```
1   [Selection,ok] = listdlg(Name,Value)
```

Using the above general format of the listdlg command and considering the six different activation function types available, selecting the desired function is accomplished using Code 11.10. The spacelist function is used below to store the list of character strings associated with the six activation functions. spacelist was previously used in Chapter 9 for the OSELM initiation. The ListString input argument is used to display the list of activation functions contained within our spacelist variable. The second input, SelectionMode, is used to control how many activation functions may be selected by the user at one time. The possible values for this entry are Single or Multiple, with Single being the default value applied in this program.

The next argument to the listdlg command is PromptString, which is used to display a dialog above the selection box. In our case, it has been set to display "Select item," which can be seen in Fig. 11.9. The Initialvalue argument is a vector of indices which sets the default activation function to be used by the SaE-ELM. According to the acceptable performance of the Sigmoid activation function in the environmental field, the default value of Initialvalue is considered 5, which is the index number associated with the sigmoid function. The Name argument is used to define the dialog box title. The default value of it is an empty character vector. In our case, the dialog box has been entitled "Activation Function Type." The final input argument is ListSize, which is used to define the list box size in pixels. It is specified as a two-element vector [width height]. The default list box size is [160 300], however, it should be adjusted according to the length and nature of the text presented within the box (so that the text fits within the list box). In Code 11.10, the size of the list box is specified as [300 100]. The output of Code 11.10 is presented in Fig. 11.9, while the schematic representation of Code 11.10 is provided in Fig. 11.10.

Code 11.10

```
1   %% Activation function selection
2   spaceList = {'Hyperbolic tangent','Triangular basis','Radial
    basis','Hard-limit','Sigmoid','Sine'};
3       [idx, ~] = listdlg('ListString', spaceList,...
          'SelectionMode', 'single', 'PromptString', 'Select item',
4   'Initialvalue', 5,'Name', 'Activation Function Type','ListSize',[300
    100]);
5       if idx == 1
6           ActivationFunction='tanh';
7       elseif idx == 2
8           ActivationFunction='tribas';
9       elseif idx == 3
10          ActivationFunction='radbas';
11      elseif idx == 4
12          ActivationFunction='hardlim';
13      elseif idx == 5
14          ActivationFunction='sig';
15      elseif idx == 6
16          ActivationFunction='sin' ;
17      end
```

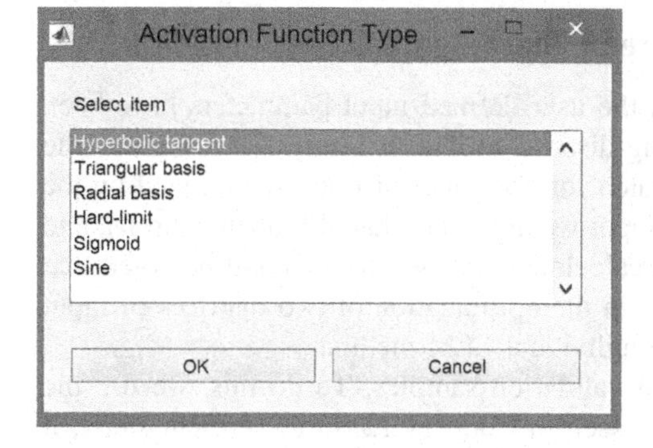

FIGURE 11.9 The selection box resulting from the listdlg command for activation function selection.

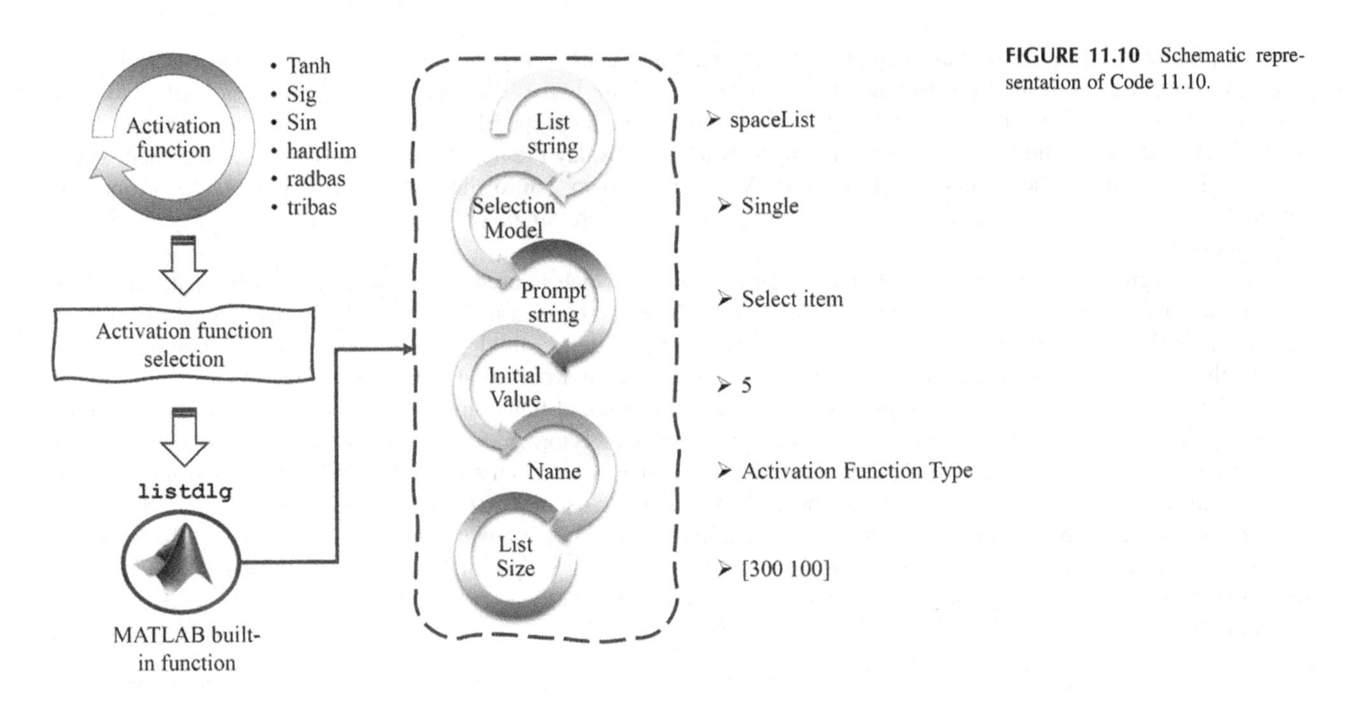

FIGURE 11.10 Schematic representation of Code 11.10.

11.1.3.4 Evolutionary parameters

The determination of the SaE-ELM parameters is the final step in the assignment of user-defined parameters prior to the commencement of modeling. Since the scaling factor (*F*) and crossover rate (*CR*) are assigned using the self-adaptive algorithm, the remaining parameters to be defined by the user include the population number and the generation number. Entry of these parameters by the user is performed using the inputdlg command. The MATLAB-based code for the definition of the SaE-ELM parameters is provided in Code 11.11. It should be noted that the commands used to read these parameters are exactly the same as was used for the global setting determination (NHN and Iteration Number). The schematic representation and output of Code 11.11 are provided in Figs. 11.11 and 11.12, respectively.

Code 11.11

```
1   %% Evolutionary Parameters
2   Prompt={'Generation Number'...
3       'Population Number'};
4           Title='SaE-ELM Parameters';
5           DefaultValues={'20','20'};
6           PARAMS=inputdlg(Prompt,Title,[1 50],DefaultValues);
7           pause(0.1);
8           Max_Gen=str2double(PARAMS{1});
9           NP=str2double(PARAMS{2});
```

11.1.4 Main coding loop of the self-adaptive evolutionary extreme learning machine

Up to this point, the data has been read into the MATLAB environment, the user-defined input parameters have been assigned, and the activation function type has been selected. The upcoming discussion will therefore be focused on the main SaE-ELM loop used during the modeling process. This loop is repeated for a number of times determined by the user-defined iteration number. In each iteration, three different matrices (input weights, the bias of hidden neurons, and output weights) are optimized, with the accuracy of the model also being calculated and stored. It should be noted that the main difference between the SaE-ELM method and individual ELM is in the optimization of two matrices of input weights and bias of hidden neurons, which were randomly allocated in the individual ELM method.

The first step is to determine the input and output data that form the validation samples. To do this, we use the NumberofValidationData variable, which was previously described, and is seen in Code 11.12. In the present study, it

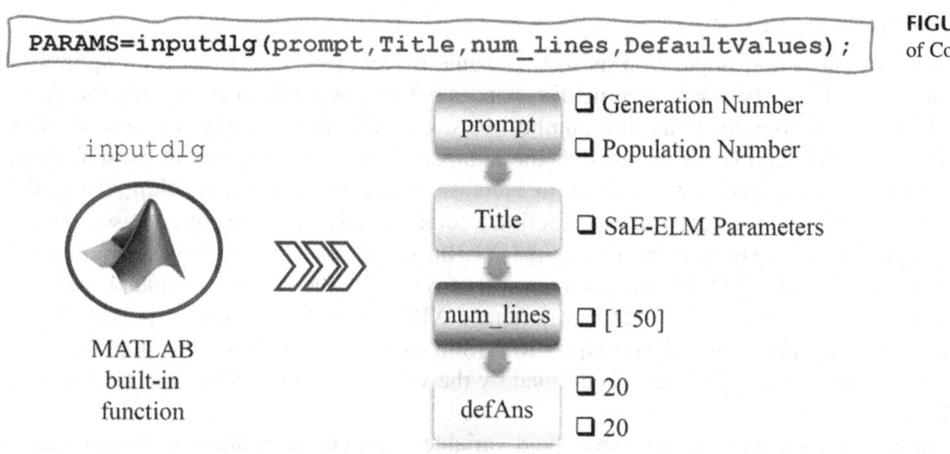

FIGURE 11.11 Schematic representation of Code 11.11.

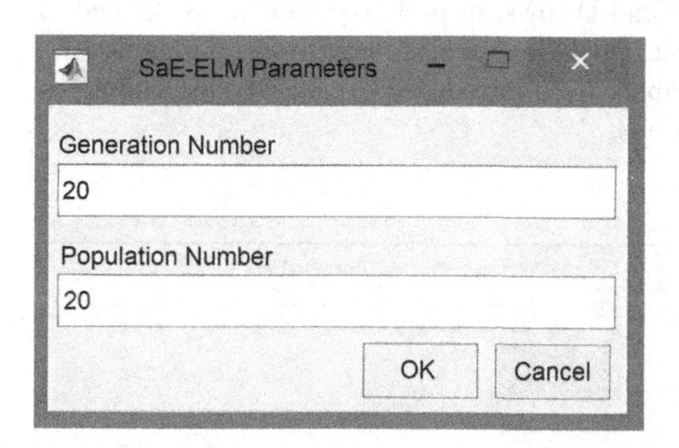

FIGURE 11.12 The dialog box resulting from the inputdlg command for determining evolutionary parameters.

was suggested to select 20% of the total test samples as NumberofValidationData. Knowing the value of this parameter, the selection of the validation samples is made according to Code 11.12, with the input and output variables stored in the **VV** variable. In Code 11.12, the ":"is used to make the algorithm assess all rows of data stored in TV.P and TV.T, while 1:NumberofValidationData tells it to select the data contained in rows 1 to NumberofValidationData.

Code 11.12

```
1    VV.P = TV.P(:,1:NumberofValidationData);
2    VV.T = TV.T(:,1:NumberofValidationData);
```

Next, the dimension of the optimization problem, which is given by the number of parameters that must be optimized during the training process, must be determined. The number of parameters to be optimized is defined according to the character of the input weights and bias of hidden neurons. The number of columns of the input weights matrix is equal to the problem inputs and the number of rows is equal to the number of hidden neurons. Besides, the bias of hidden neurons is a columnar matrix whose number of rows is equal to the number of hidden neurons. Therefore, the sum of this parameter is similar to Eq. (10.3), which is stored in the variable *D*, as shown in Code 11.13.

Another parameter that must be initialized by the algorithm is the "tolerance." The tolerance is a constant number that represents a minimum value of acceptable error, and is applied throughout the optimization process to find the global optimum solution. When the tolerance is met, the algorithm will stop, and the optimal solution will have been met. Cao et al. (2012) suggested that a constant value of 0.002 should be used throughout the modeling process.

Code 11.13

```
1    D=NumberofHiddenNeurons*(NumberofInputNeurons+1);
2    tolerance = 0.0002;
```

11.1.4.1 Initialization of the main problem

Once the dimension (D) has been set, the initial population as well as some placeholder arrays that are required in the optimization process, are initialized. First, the population matrix, pop, is defined as a zero-matrix using the zeros command. The zeros command allows the user to define the number of rows and columns using the format zeros (row, column). The population number (NP) and D are assigned as the number of rows and columns in the definition of the pop variable. The reason for defining this zero matrix is to increase the speed of the algorithm. Using the Xrmin and Xrmax (bounds of the optimization space which are defined automatically based on the dataset) as the lower and upper bounds of the optimization spaces, respectively, the XRRmin and XRRmax variables are defined using the repmat command in lines 2–3 of Code 11.14 with a dimension of $NP \times D$. The repmat command will create copies of Xrmin and Xrmax and store them in an $NP \times D$ matrix named XRRmin and XRRmax, respectively. All of the entries in XRRmin (and XRRmax) are identical and equal to Xrmin (and Xrmax). Next, the **pop** variable, which was initialized as an $NP \times D$ matrix with zero values, is defined by the values XRRmin, XRRmax, as well as a random term obtained using NP and D.

To store the population during the optimization process, the popold variable is generated in line 6 as a zero-matrix with dimensions equal to the size (pop). It should be noted that pop is a matrix with a dimension of $NP \times D$. The val and DE_gbest variables are defined as a column-based matrix with NP and D rows, respectively. The val is defined to create and reset the cost array, while the DE_gbest is defined as the best population member ever. Using the numst, the variables aaaa, pfit, and ccm are also randomly assigned. These variables were introduced to simplify the coding in Code 11.14 and will be described in further detail in the upcoming sections.

Code 11.14

```
1   pop = zeros(NP,D);
2   XRRmin=repmat(Xrmin,NP,D);
3   XRRmax=repmat(Xrmax,NP,D);
4   pop=XRRmin+(XRRmax-XRRmin).*rand(NP,D);
5
6   popold     = zeros(size(pop));
7   val        = zeros(1,NP);
8   DE_gbest   = zeros(1,D);
9
10  numst=4;
11  aaaa=cell(1,numst);
12  pfit=ones(1,numst);
13  ccm = 0.5*ones(1,numst);
```

11.1.4.2 Non-tuned fast machine learning coding

Following initialization of the population as well as some placeholder arrays, the best member after initialization is evaluated using Code 11.15. At first, the best iteration (ibest) is defined as 1 (i.e., the first population member). The evaluation of the initialized population is begun with the implementation of the individual ELM, which is defined as the ELM_X function. The objective of this function is to calculate two variables: (1) val(1), and (2) OutputWeight. The val(1) variable is the fitness of the implemented individual ELM, while the OutputWeight variable is the output weight matrix that is analytically calculated during the training phase of the individual ELM. The calculated val(1) is saved as DE_gbestval. This parameter is used to save the best objective function value so far. To check the remaining population members, a loop is constructed with an index that spans from 2 to the population number (NP). In this loop, the individual ELM is run for NP-1 times. For each population member (2 to NP), the fitness is evaluated and compared with the best objective function value so far (i.e., DE_gbestval). If the fitness of the current member (i.e., val(i)) is better than the best objective function value so far (i.e., DE_gbestval), the DE_gbestval and bestweight are updated. The location of this member is also saved as ibest = i. After all of the population members have been checked, the best member is stored in the DE_gbest variable using the location of its member (i.e., ibest) as pop(ibest,:). The schematic representation of Code 11.15 is provided in Fig. 11.13.

Code 11.15

```
1    ibest   = 1;
     [val(1),OutputWeight]  =
2    ELM_X(Elm_Type,pop(ibest,:),P,T,VV,NumberofHiddenNeurons,ActivationFuncti
     on);
3
4    DE_gbestval = val(1);
5    bestweight = OutputWeight;
6
7    for i=2:NP
       [val(i),OutputWeight] =
8    ELM_X(Elm_Type,pop(i,:),P,T,VV,NumberofHiddenNeurons,ActivationFunction);
9
10     if (val(i) < DE_gbestval)
11        ibest   = i;
12        DE_gbestval = val(i);
13        bestweight = OutputWeight;
14     end
15   end
```

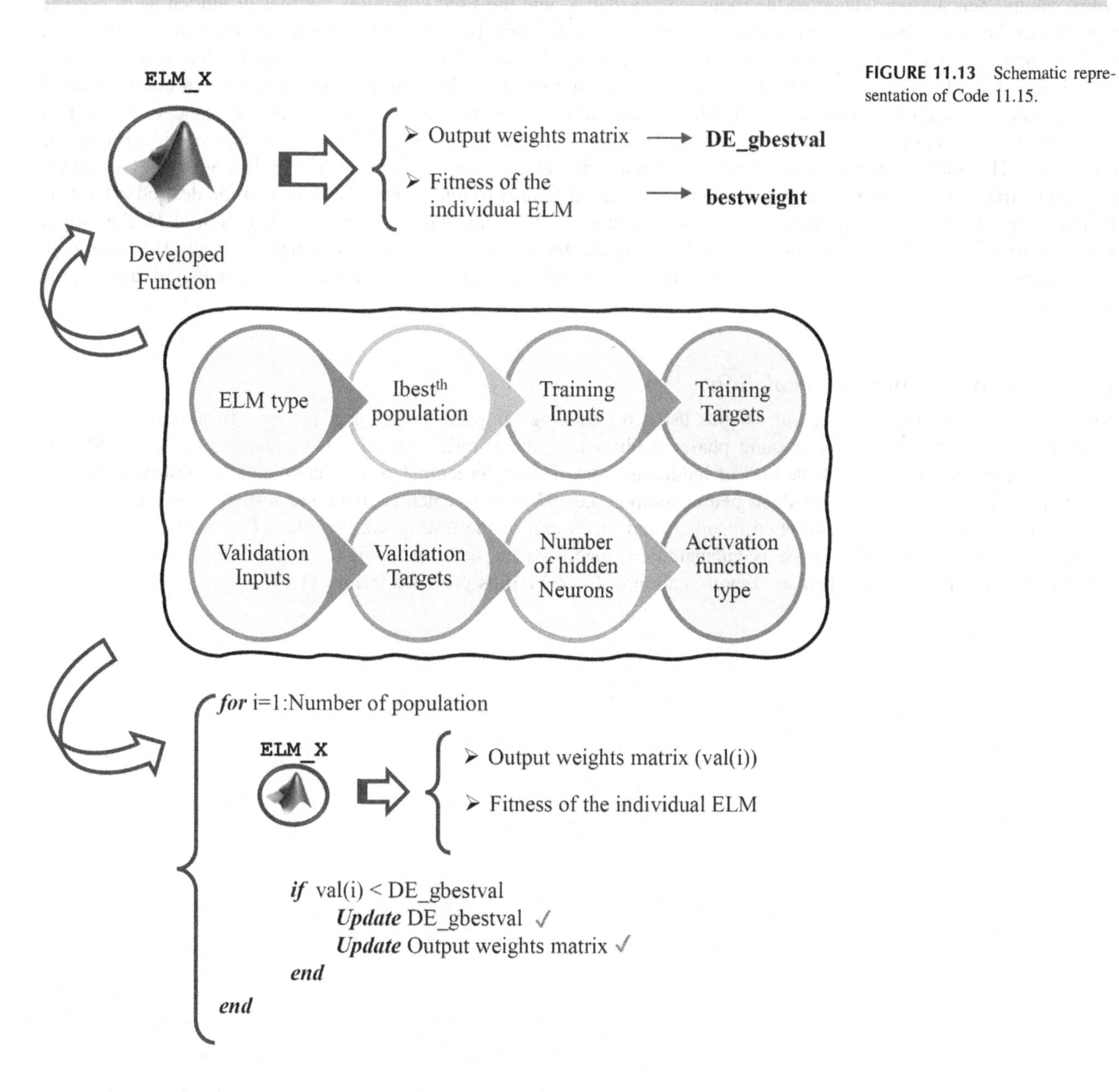

FIGURE 11.13 Schematic representation of Code 11.15.

In Code 11.15, the ELM_X function, which contains the individual ELM, was called. In the discussion next, the inputs and outputs of this function, as well as the required coding, are discussed. The inputs of the ELM_X function are seen in line 2 to be Elm_Type, weight_bias, P, T, TV, NumberHiddenNeurons, and ActivationFunction, while Fitness and OutputWeight are the outputs of this function. The Elm_Type argument is the ELM type (i.e., Classification of Regression) that was specified by the user (see Fig. 11.6). The weight_bias argument is a randomly generated value which is obtained by the combination of input weights and bias of hidden neurons matrices. P and T are the training inputs and targets (respectively), while TV is the validation sample inputs and targets. The NumberHiddenNeurons argument is the number of hidden neurons, ActivationFunction is the activation function type, Fitness is the fitness of the developed ELM model [based on the root mean square error (RMSE) index], and OutputWeight is the output weight matrix.

In this function, the number of input neurons, training samples, and testing samples are first calculated. The weight_bias, which has one row and NHN $\times$ (InV + 1) (where NHN is the number of hidden neurons and INV is the number of input variables) columns, is reshaped to temp_weight_bias with NHN rows and (InV + 1) columns using the reshape command. Using the newly defined matrix (i.e., temp_weight_bias), the input weights and bias of hidden neurons are redefined as InputWeight and BiasofHiddenNeurons, respectively. Indeed, the first six columns in the temp_weight_bias are considered as the input weight matrix, and the final column is the bias of hidden neurons. The tempH variable is calculated by multiplying the input weight matrix (i.e., InputWeight) by the training samples (i.e., P). The matrix of bias of hidden neurons is required to extend to match the dimension of the hidden neurons output matrix (i.e., H), which has a number of rows equal to the number of hidden neurons and a number of columns equal to the number of training samples. To do that, the ind variable is generated using the ones command with one row and a number of columns equal to the number of training samples. The extended form of the bias of hidden neurons (i.e., BiasofHiddenNeurons) generated using the ind command is stored as the variable BiasMatrix. The newly defined matrix (i.e., BiasMatrix) is summed with the current tempH to update tempH. Based on the desired activation function type, the matrix of the hidden neurons' output (i.e., H) is calculated. After the calculation of H, the output weight matrix is analytically calculated by multiplying the Moore–Penrose generalized inverse of the H matrix with the training target (i.e., T). The Moore–Penrose generalized inverse of the H matrix is calculated using pinv command.

11.1.4.3 Fitness function calculation

After calculating the matrix of output weights using the training samples, tempH_test, ind, and BiasMatrix are calculated in the same manner as the training phase. Additionally, the tempH_test variable is updated using BiasMatrix, which is the extended version of the bias of hidden neurons. Using the tempH_test variable and the desired activation function type, the matrix of the hidden neuron output (i.e., H_test) is calculated for the testing phase. Finally, the H_test and OutputWeight are multiplied together, and the output of the testing samples (i.e., TY) is calculated. Using RMSE, the fitness of the testing phase is calculated and together with the output weight matrix, is considered the output of the ELM_X function. The schematic representation of Code 11.16 is provided in Fig. 11.14.

Code 11.16

```
1   function [Fitness,OutputWeight] = ELM_X (Elm_Type, weight_bias, P, T, TV,
    NumberHiddenNeurons,ActivationFunction)
2   NumberInputNeurons=size(P, 1);
3   NumberofTrainingData=size(P, 2);
4   NumberofTestingData=size(TV.P, 2);
5   
6   temp_weight_bias=reshape(weight_bias, NumberHiddenNeurons,
    NumberInputNeurons+1);
7   InputWeight=temp_weight_bias(:, 1:NumberInputNeurons);
8   BiasofHiddenNeurons=temp_weight_bias(:,NumberInputNeurons+1);
9   tempH=InputWeight*P;
10  ind=ones(1,NumberofTrainingData);
11  BiasMatrix=BiasofHiddenNeurons(:,ind);
12  tempH=tempH+BiasMatrix;
13  clear BiasMatrix
14  
15  switch lower(ActivationFunction)
16      case{'s','sig','sigmoid'}
17          H = 1 ./ (1 + exp(-tempH));
18      case{'t','tanh'}
19          H = tanh(tempH);
20      case {'sin','sine'}
21          %%%%%%%% Sine
22          H = sin(tempH);
23      case {'hardlim'}
24          %%%%%%%% Hard Limit
25          H = double(hardlim(tempH));
26      case {'tribas'}
27          %%%%%%%% Triangular basis function
28          H = tribas(tempH);
29      case {'radbas'}
30          %%%%%%%% Radial basis function
31          H = radbas(tempH);
32          %%%%%%%% More activation functions can be added here
33  end
34  
35  clear tempH;
36  OutputWeight=pinv(H') * T';
37  
38  tempH_test=InputWeight*TV.P;
39  ind=ones(1,NumberofTestingData);
40  BiasMatrix=BiasofHiddenNeurons(:,ind);      %   Extend the bias matrix
    BiasofHiddenNeurons to match the dimension of H
41  tempH_test=tempH_test + BiasMatrix;
42  
43  switch lower(ActivationFunction)
44      case{'s','sig','sigmoid'}
45          H_test = 1 ./ (1 + exp(-tempH_test));
46      case{'t','tanh'}
47          H_test = tanh(tempH_test);
48      case {'sin','sine'}
49          %%%%%%%% Sine
```

```
50            H_test = sin(tempH_test);
51        case {'hardlim'}
52            %%%%%%%% Hard Limit
53            H_test = double(hardlim(tempH_test));
54        case {'tribas'}
55            %%%%%%%% Triangular basis function
56            H_test = tribas(tempH_test);
57        case {'radbas'}
58    %%%%%%%% Radial basis function
59            H_test = radbas(tempH_test);
60            %%%%%%%% More activation functions can be added here
61    end
62
63    TY=(H_test' * OutputWeight)';
64
65    Fitness=sqrt(mse(TV.T - TY));
```

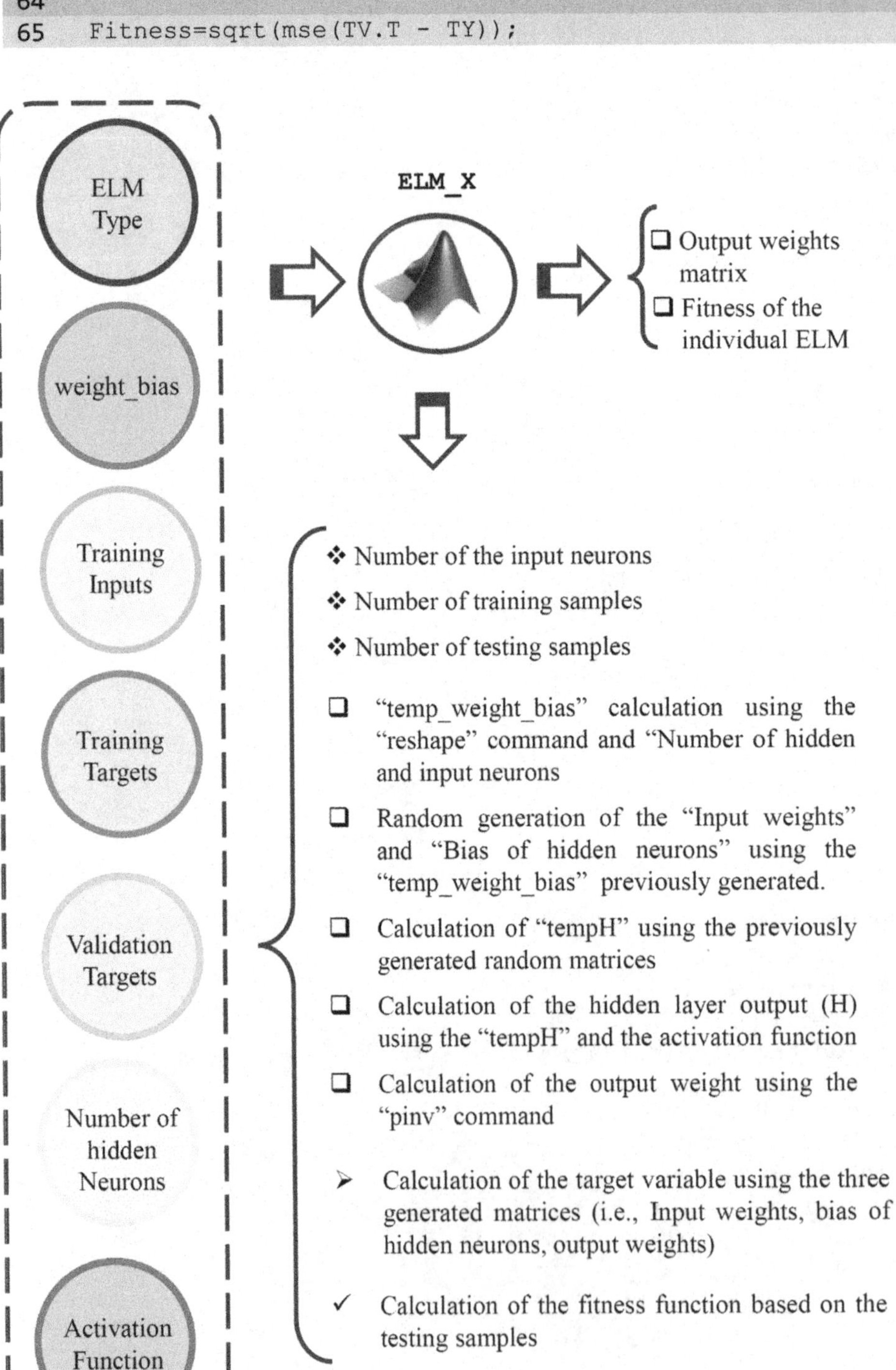

FIGURE 11.14 Schematic representation of Code 11.16.

11.1.4.4 Self-adaptive differential evolution

Following the initialization step and evaluation of the best member, the self-adaptive differential evolution (SADE) is applied to minimize the fitness of this optimization algorithm. In general, popold is the population that will be competing for fitness against the newly generated populations through the optimization process. The pop variable is used to store these newly emerging populations.

To begin the SADE phase, it is first required to initialize some arrays that will be called throughout the optimization process. The first arrays are entitled pm1 to pm5. These are initialized population matrices 1—5 and are defined using a zero-matrix of dimension NP × D (where NP and D are the population number and dimension, respectively). The bm array is initialized to save the global best of the optimization algorithm, while the ui array is initialized to store the intermediate values of the perturbed vectors. Like the population matrices, the bm and ui arrays are initialized using a zero matrix of dimension $NP \times D$. The mui and mpo arrays are also initialized as zero-matrices with identical dimensions to all previous arrays. The rot and rotd are rotated in the index array with sizes *NP* and *D*, respectively. rt and rtd are zero matrices with *NP* and *D* dimensions, respectively. a1 to a5 are also zero matrix index arrays with the dimension of $NP \times NP$.

Code 11.17

```
1   pm1  = zeros(NP,D);
2   pm2  = zeros(NP,D);
3   pm3  = zeros(NP,D);
4   pm4  = zeros(NP,D);
5   pm5  = zeros(NP,D);
6   bm   = zeros(NP,D);
7   ui   = zeros(NP,D);
8   mui  = zeros(NP,D);
9   mpo  = zeros(NP,D);
10  rot  =  (0:1:NP-1);
11  rotd=  (0:1:D-1);
12  rt   = zeros(NP);
13  rtd  = zeros(D);
14  a1   = zeros(NP);
15  a2   = zeros(NP);
16  a3   = zeros(NP);
17  a4   = zeros(NP);
18  a5   = zeros(NP);
```

In Code 11.18, the optimization process for generation number one to the maximum number of generations is performed using a while loop. The MATLAB coding outlined in the upcoming pages is provided for a single generation. Upon entry to the while loop, the iteration number is considered to be one (i.e., iter = 1). A series of commands are executed during the while loop until the iteration number exceeds the maximum number of generations (i.e., iter < Max_Gen).

Code 11.18

```
1    iter = 1;
2    while iter < Max_Gen
3
4          COMMANDS
5
6    end
```

The discussion next will surround the MATLAB-based coding that forms the <COMMANDS> in line 4 above, which are executed during the while loop.

At first, the old population which was stored as pop is assigned to the variable popold. Following this, the index pointer array is defined as ind using the built-in randperm command. The randperm(n) command returns a row vector containing a random permutation of the integers from 1 to *n* inclusive. To shuffle the location of vectors, the variable a1 is defined using randperm(NP), where NP is the population number. Using the rem command, the indices are rotated by ind(1) (or ind(2), ind(3), and ind(4)) positions. Following this, the a1 (or a2, a3, a4) is rotated and saved as a2 (or a3, a4, a5). The rem(a,b) command returns the remainder after dividing **a** by **b**, where **a** is the dividend and **b** is the divisor. Populations 1 to 5 are shuffled and saved as pm1 to pm5, respectively. In line 19 of Code 11.19, DE_gbest is copied as the best population member ever using the repmat command. The DE_gbest is copied NP (i.e., number of population) times in NP rows.

Code 11.19

```
1    popold = pop;
2         ind = randperm(4);
3         a1   = randperm(NP);
4         rt = rem(rot+ind(1),NP);
5         a2   = a1(rt+1);
6         rt = rem(rot+ind(2),NP);
7         a3   = a2(rt+1);
8         rt = rem(rot+ind(3),NP);
9         a4   = a3(rt+1);
10        rt = rem(rot+ind(4),NP);
11        a5   = a4(rt+1);
12
13        pm1 = popold(a1,:);
14        pm2 = popold(a2,:);
15        pm3 = popold(a3,:);
16        pm4 = popold(a4,:);
17        pm5 = popold(a5,:);
18
19        bm = repmat(DE_gbest,NP,1);
```

In Code 11.20, cc for the maximum number of generations (numst) is calculated using a for loop in line 3. First, the cc_tmp variable is initiated in line 2 without any input. Using a nested **for** loop iterated from one to NP (i.e., number of population), the tt variable is calculated using the normrnd command. The generic format of this command is given by normrnd (mu, sigma), and generates random numbers from the normal distribution with mean parameter mu and standard deviation parameter sigma. As seen next, the mu and sigma for generating tt in our example are equal to ccm of the *i*th iteration and 0.1, respectively. As mentioned previously, the ccm is calculated using the numst variable and the ones command. After implementation of the MATLAB code below for all population members (NP), the tt variable is saved in cc_tmp. Therefore, the cc_tmp variable is a matrix with NP rows and one column. For all numst, the cc_tmp are saved in the variable CC, which is a matrix with NP rows and numst columns. The schematic representation of Code 11.20 is provided in Fig. 11.15.

Code 11.20

```
1    for i=1:numst
2            cc_tmp=[];
3            for k=1:NP
4                tt=normrnd(ccm(i),0.1);
5                while tt>1 | tt<0
6                    tt=normrnd(ccm(i),0.1);
7                end
8                cc_tmp=[cc_tmp;tt];
9            end
10           cc(:,i)=cc_tmp;
11       end
```

By randomly defining rr and calculating spacing as the inverse of NP (number of population), the randnums parameter may be defined using the mod command. The general format of the mod command is $b = \text{mod}(a,m)$. The function returns the remainder after dividing a by m, where a is the dividend and m is the divisor. This function is often called the modulo operation, which can be expressed as $b = a - m.*\text{floor}(a./m)$. In the event of the divisor being zero, the mod function follows the convention that mod(a,0) returns a. The normfit variable is calculated as pfit divided by sum(pfit). It should be noted that pfit was defined initially as a matrix of ones with a single row and numst columns as pfit = ones (1, numst). Other parameters, including partsum, count, and stpool, are also defined to apply in the next lines of the SaE-ELM MATLAB-based code.

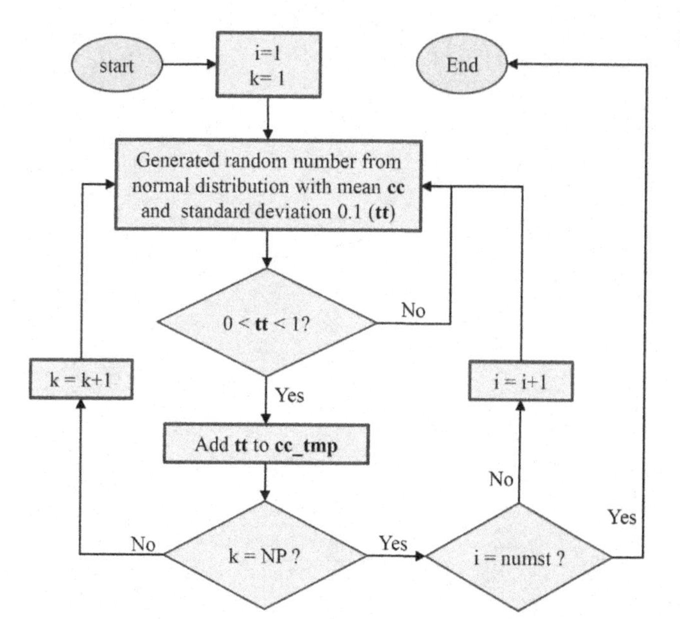

FIGURE 11.15 Schematic representation of Code 11.20.

Code 11.21

```
rr=rand;
spacing=1/NP;
randnums=sort(mod(rr:spacing:1+rr-0.5*spacing,1));

normfit=pfit/sum(pfit);
partsum=0;
count(1)=0;
stpool=[];
```

Using a for loop, as shown in Code 11.22, an iterative process is repeated for one to the length of pfit times to calculate the stpool. To calculate the stpool variable, the partsum, count, and select variables must be computed at each iteration. The partsum is updated at each iteration using normfit(i).

Both pfit and normfit are matrices with one column and an identical number of rows. Unlike the other parameters in this loop, the count is commenced at $i + 1$, because it was previously initiated as zero in the last code segment (i.e., count(1) = 0). The select parameter is the difference in count in the current and previous iterations. Finally, using select, stpool is updated in each iteration. The schematic representation of Code 11.22 is provided in Fig. 11.16.

Code 11.22

```
for i=1:length(pfit)
    partsum=partsum+normfit(i);
    count(i+1)=length(find(randnums<partsum));
    select(i,1)=count(i+1)-count(i);
    stpool=[stpool;ones(select(i,1),1)*i];
end
```

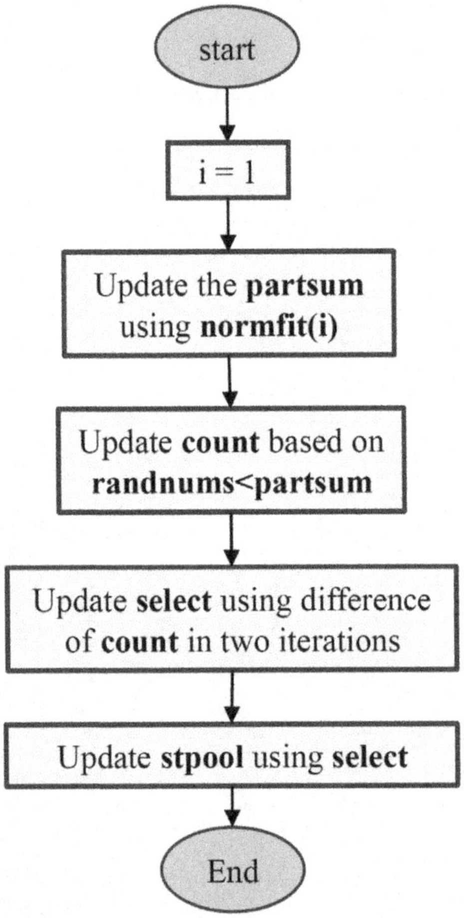

start

i = 1

Update the **partsum** using **normfit(i)**

Update **count** based on **randnums<partsum**

Update **select** using difference of **count** in two iterations

Update **stpool** using **select**

End

FIGURE 11.16 Schematic representation of Code 11.22.

Using the randperm command, the order of the values presented in the stpool changes randomly. The randperm(n) command returns a row vector containing a random permutation of the integers from 1 to "*n*" inclusive. For the numst iteration number, using a for loop and an if-then command, the aaa parameter is assigned as a 1 × numst cell so that each cell has a matrix with one row and numst columns. In the main loop and at each iteration, three parameters, including atemp, aaa, and index, are defined. Owing to the definition of these three parameters, the stpool variable is not empty, and then if-then command is then run for all iterations. The index parameter is calculated based on the position of the *i*th number in the stpool parameter. Using the calculated value for index and atemp at each iteration, the atemp and aaa (respectively) are calculated at all iterations. The schematic representation of Code 11.23 is provided in Fig. 11.17.

Code 11.23

```
1    stpool = stpool(randperm(NP));
2
3        for i=1:numst
4            atemp=zeros(1,NP);
5            aaa{i}=atemp;
6            index{i}=[];
7            if ~isempty(find(stpool == i))
8                index{i} = find(stpool == i);
9                atemp(index{i})=1;
10               aaa{i}=atemp;
11           end
12       end
```

After defining a zeros matrix with dimension *NP* by *D* as aa, its value is assigned using a for loop. In the current study, *D* is the sum of input weights (*NHN* × *InV*) and bias of hidden neurons (*NHN* × 1), where NHN is the number of hidden neurons and *InV* is the input weight. Therefore, the size of *D* is *NHN*(*InV* + 1). To assign aa, the index, cc, and *D* variables are needed (which have already been defined in the previous segments of SaE-ELM coding). The final value of all rows and columns in aa is either zero or one so that if all random numbers are lower than the crossover rate (*CR*), aa is one, and otherwise, it is zero. The final aa is considered as mui, and the transpose of the defined mui is sorted to collect all 1's in each column.

Code 11.24

```
1    aa=zeros(NP,D);
2
3     for i=1:numst
4         aa(index{i},:)=rand(length(index{i}),D) <repmat(cc(index{i},i),1,D);
5     end
6     mui=aa;
7
8     mui=sort(mui');
```

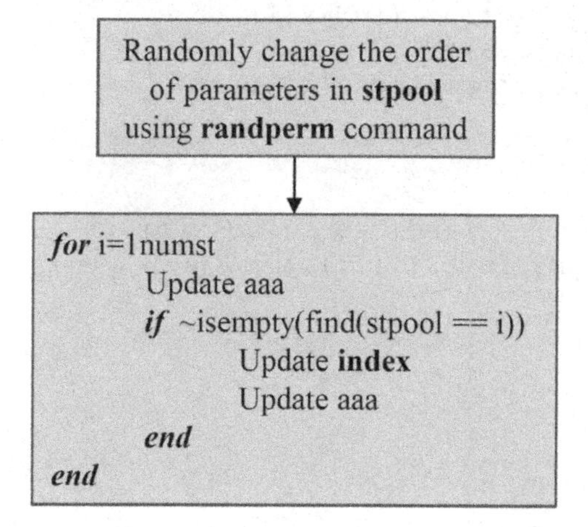

FIGURE 11.17 Schematic representation of Code 11.23.

Using the rand and ceil commands as well as the NP and D parameters, the dd parameter is assigned as a matrix with a single column and NP rows. It should be noted that the ceil(x) command rounds each element of x to the nearest integer greater than or equal to that element. Using dd, the mui is reconstructed as a $NP \times D$ matrix. Finally, using the newly reconstructed mui, the mpo is updated.

Code 11.25

```
1   dd=ceil(D*rand(NP,1));
2       for kk=1:NP
3           mui(kk,dd(kk))=1;
4       end
5       mpo = mui < 0.5;
```

In Code 11.26, for all numst values, an iterative process is performed to calculate ui based on the mathematical formulation provided in Eq. (11.7)–(11.10). Before calculation of this parameter, the m variable is considered as the length of the index{i} (calculated previously), F is calculated using the **normrnd** command and the m parameter, and finally, the F is repeated using the repmat command for all dimensions. It should be noted that the normrnd(mu,sigma) command generates random numbers from the normal distribution with mean parameter mu and standard deviation parameter sigma. To calculate the F parameter, the mu and sigma are set to 0.5 and 0.3, respectively. The schematic representation of Code 11.26 is provided in Fig. 11.18.

Code 11.26

```
1   for i=1:numst
2           F=[];
3           m=length(index{i});
4           F=normrnd(0.5,0.3,m,1);
5           F=repmat(F,1,D);
6
7           if i==1
8               ui(index{i},:) = pm3(index{i},:) + F.*(pm1(index{i},:) -
    pm2(index{i},:));
9               ui(index{i},:) = popold(index{i},:).*mpo(index{i},:) +
    ui(index{i},:).*mui(index{i},:);
10          end
11          if i==2
12              ui(index{i},:) = popold(index{i},:) + F.*(bm(index{i},:)-
    popold(index{i},:)) + F.*(pm1(index{i},:) - pm2(index{i},:) +
    pm3(index{i},:) - pm4(index{i},:));
13              ui(index{i},:) = popold(index{i},:).*mpo(index{i},:) +
    ui(index{i},:).*mui(index{i},:);
14          end
15          if i==3
16              ui(index{i},:) = pm5(index{i},:) + F.*(pm1(index{i},:) -
    pm2(index{i},:) + pm3(index{i},:) - pm4(index{i},:));
17              ui(index{i},:) = popold(index{i},:).*mpo(index{i},:) +
    ui(index{i},:).*mui(index{i},:);
18          end
19          if i==4
20              ui(index{i},:) = popold(index{i},:) + rand.*(pm5(index{i},:)-
    popold(index{i},:)) + F.*(pm1(index{i},:) - pm2(index{i},:));
21          end
22      end
```

FIGURE 11.18 Schematic representation of Code 11.26.

for i=1:numst

  if i =1

    Eq. 10.16

  end

  if i =2

    Eq. 10.17

  end

  if i =3

    Eq. 10.18

  end

  if i =4

    Eq. 10.19

  end

end

After calculation of the ui for the entire population number (*NP*), all of the ui lower than Lbound (i.e., lower bound) and greater than Ubound (i.e., upper bound) are found and replaced with random numbers generated using XRRmin and XRRmax. The schematic representation of Code 11.27 is provided in Fig. 11.19.

Code 11.27

```
1        for i=1:NP
2            outbind=find(ui(i,:) < Lbound);
3            if size(outbind,2)~=0
4
5                ui(i,outbind)=XRRmin(outbind)+(XRRmax(outbind)-
     XRRmin(outbind)).*rand(1,size(outbind,2));
6
7            end
8            outbind=find(ui(i,:) > Ubound);
9            if size(outbind,2)~=0
10
11               ui(i,outbind)=XRRmin(outbind)+(XRRmax(outbind)-
     XRRmin(outbind)).*rand(1,size(outbind,2));
12
13           end
14       end
```

In Code 11.28, the lpcount and npcount variables are predefined as two zero-matrices with a single row and numst columns to enhance the running time of the proposed code. Following this, for the entire population (i.e., *NP*), the

```
for i=1:NP

    if ui(i) < Lower bound
        Replace ui(i) with random numbers
    end

    if ui(i) > Upper bound
        Replace ui(i) with random numbers
    end

end
```

FIGURE 11.19 Schematic representation of Code 11.27.

individual ELM (which is defined as ELM_X function) is called to calculate the fitness and output weight, which are stored as tempval and OutputWeight, respectively. In this loop, the competitor cost (i.e., tempval) is checked with val, which was calculated at the initialization step. If the competitor (i.e., tempval) is better than the value in the cost array (i.e., val), the old vector (i.e., pop) is replaced with the new one for the new iteration (ui), and the cost array is updated using tempval. After that, the global best value of the DE fitness (i.e., DE_gbestval) is updated in case of success to save time. If the competitor is better than the best one ever (i.e., tempval < = DE_gbestval), the new best value (i.e., tempval) is saved in the DE_gbestval. The new best parameter vector ever found (i.e., ui) is also saved in the DE_gbest and ouputweight (i.e., the output weight calculated from ELM_X) is saved as bestweight. Moreover, if the competitor (i.e., tempval-DE_gbestval) is better than the best one ever (i.e., tolerance*DE_gbestval), the norm of OutputWeight is checked to be lower than the norm of bestweight. If the comparison condition is met (i.e., norm(OutputWeight,2) < norm(bestweight,2)), the new best value (i.e., DE_gbestval) and new best parameter (i.e., bestweight) are updated. If the competitor (i.e., tempval) is not better than the value in the cost array (i.e., **tempval > val**), the DE_gbestval, DE_gbest and bestweight are not updated. The schematic representation of Code 11.28 is provided in Fig. 11.20.

Code 11.28

```
1        lpcount=zeros(1,numst);
2        npcount=zeros(1,numst);
3
4    for i=1:NP
5            [tempval,OutputWeight] =
     ELM_X(Elm_Type,ui(i,:),P,T,VV,NumberofHiddenNeurons,ActivationFunction);
6
7            if (tempval <= val(i))
8                pop(i,:) = ui(i,:);
9                val(i)   = tempval;
10               tlpcount=zeros(1,numst);
11               for j=1:numst
12                   temp=aaa{j};
13                   tlpcount(j)=temp(i);
14                   if tlpcount(j)==1
15                       aaaa{j}=[aaaa{j};cc(i,j) iter]  ;
16                   end
17               end
18               lpcount=[lpcount;tlpcount];
19
20               if DE_gbestval-tempval>tolerance*DE_gbestval
21
22                   DE_gbestval = tempval;
23                   DE_gbest = ui(i,:);
24                   bestweight = OutputWeight;
25
26                   elseif abs(tempval-DE_gbestval)<tolerance*DE_gbestval
27                       if norm(OutputWeight,2)<norm(bestweight,2)
28                           DE_gbestval = tempval;
29                           DE_gbest = ui(i,:);
30                           bestweight = OutputWeight;
31   DE_get_flag=1;
32
33                   end
34               end
35           else
36               tnpcount=zeros(1,numst);
37               for j=1:numst
38                   temp=aaa{j};
39                   tnpcount(j)=temp(i);
40               end
41               npcount=[npcount;tnpcount];
42           end
43          break;
44
45       end
```

After running Code 11.28, the main while loop is executed for the user-defined number of generations (i.e., max_Gen). The next step is testing the performance of the best population, which is done using the syntax outlined in Code 11.29. At first, three matrices, including the number of input variables (i.e., NumberInputNeurons), the number of training samples

for i = 1:NP
  Check the cost of the competitor
  if the competitor is better than the value in the "cost array"
    ➢ Replace the old vector with the new one (for the new iteration)
    ➢ Save the value in the "cost array"
    ❑ Update the DE_gbestval
    ❑ Update the DE_gbest
    ❑ Update the best weight
  end
end

FIGURE 11.20 Schematic representation of Code 11.28.

(i.e., NumberofTrainingData) and the number of testing samples (i.e., NumberofTestingData), are recalculated. The optimized values through the training phase include the input weights matrix (with NHN rows and number of input neurons columns) and the bias of hidden neurons (with NHN rows and one column). Therefore, all of the saved values in the DE_gbest should be reshaped into NumberofHiddenNeurons rows and NumberInputNeurons + 1 columns. This is done using reshape command and saved into the variable temp_weight_bias. It should be noted that the B = reshape(A, sz) command reshapes **A** using the size vector, **sz**. For example, "reshape(A, [2,3])" reshapes A into a 2-by-3 matrix. The first NumberInputNeurons columns (from one to NumberInputNeurons) are assigned to the input weight, while the final column is assigned to the bias of the hidden neurons. By multiplying the input weight (i.e., InputWeight) matrix by the matrix of training inputs (i.e., P), the tempH variable is calculated and is updated by summation with the bias of hidden neurons (i.e., BiasMatrix). The schematic representation of Code 11.29 is provided in Fig. 11.21.

Code 11.29

```
1   Output_weight = mean(abs(OutputWeight));
2   NumberInputNeurons=size(P, 1);
3   NumberofTrainingData=size(P, 2);
4   NumberofTestingData=size(TV.P, 2);
5
6   temp_weight_bias=reshape(DE_gbest, NumberofHiddenNeurons,
    NumberInputNeurons+1);
7   InputWeight=temp_weight_bias(:, 1:NumberInputNeurons);
8   BiasofHiddenNeurons=temp_weight_bias(:,NumberInputNeurons+1);
9
10  tempH=InputWeight*P;
11  ind=ones(1,NumberofTrainingData);
12  BiasMatrix=BiasofHiddenNeurons(:,ind);
13  tempH=tempH+BiasMatrix;
14  clear BiasMatrix
```

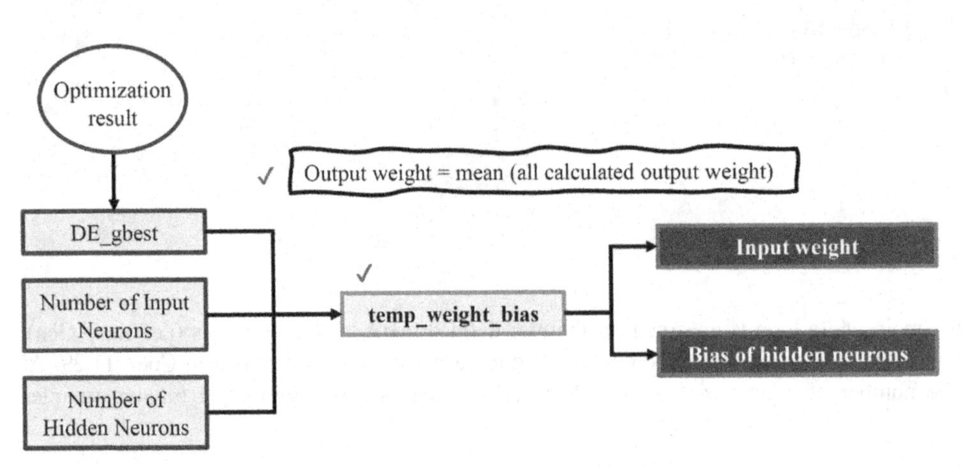

FIGURE 11.21 Schematic representation of Code 11.29.

Therefore, the matrix of the hidden neuron output (i.e., **H**) is computed using the updated tempH as well as the chosen activation function in Code 11.30.

Code 11.30

```
1  switch lower(ActivationFunction)
2      case{'s','sig','sigmoid'}
3          H = 1 ./ (1 + exp(-tempH));
4      case{'t','tanh'}
5          H = tanh(tempH);
6      case {'sin','sine'}
7          %%%%%%%% Sine
8          H = sin(tempH);
9      case {'hardlim'}
10         %%%%%%%% Hard Limit
11         H = double(hardlim(tempH));
12     case {'tribas'}
13         %%%%%%%% Triangular basis function
14         H = tribas(tempH);
15     case {'radbas'}
16         %%%%%%%% Radial basis function
17         H = radbas(tempH);
18         %%%%%%%% More activation functions can be added here
19 end
```

By multiplying H, the matrix of the output of the hidden neurons, by the bestweight (i.e., the optimized value of the output weight matrix through the training phase), the training output of the SaE-ELM is calculated. A similar process is done for the testing samples.

Code 11.31

```
1  tempH_test=InputWeight*TV.P;
2  ind=ones(1,NumberofTestingData);
3  BiasMatrix=BiasofHiddenNeurons(:,ind);
4  tempH_test=tempH_test + BiasMatrix;
5
6  switch lower(ActivationFunction)
7      case{'s','sig','sigmoid'}
8          H_test = 1 ./ (1 + exp(-tempH_test));
9      case{'t','tanh'}
10         H_test = tanh(tempH_test);
11     case {'sin','sine'}
12         H_test = sin(tempH_test);
13     case {'hardlim'}
14         H_test = double(hardlim(tempH_test));
15     case {'tribas'}
16         H_test = tribas(tempH_test);
17     case {'radbas'}
18         H_test = radbas(tempH_test);
19 End
20 TY=(H_test' * bestweight)';
```

11.1.4.5 *Saving the results of each iteration*

The testing error is computed using the RMSE at each iteration and is saved using the variable Error, as shown in Code 11.31. After completion of all iterations, the best iteration is found using the find command, with the best model performance being considered as the iteration which has the lowest Error.

Code 11.31

```
1    Error(ii)=TestingAccuracy;
2    disp(['Iteration ' num2str(ii) ': Best Cost = ' num2str(Error(ii))]);
```

For all of the iterations, the output weight (i.e., OutW), the bias of hidden neurons (i.e., BHN), and the input weight (i.e., InW) are saved so that they may be used in the event they correspond to the best iteration. The output weight and bias of hidden neurons are two matrices which contain a single column, the values of which are saved into one column for all iterations. The **InW** is a matrix which contains two dimensions initially. On the other hand, this matrix has been defined for all iterations, and should therefore be specified as a matrix with three dimensions. The first two dimensions of this matrix will contain its value, while the third dimension should contain the iteration number corresponding to the specific matrix. After finding the lowest error and its iteration number, the **InW** can be selected using the third dimension that is related to the best iteration (i.e., lowest error).

Code 11.32

```
1        OutW(:,ii) = bestweight;
2        BHI(:,ii) = BiasofHiddenNeurons;
3        InW(:,:,ii) = InputWeight;
```

11.1.5 Presentation of the training results

Once the best iteration has been found, eleven statistical indices are computed using the Statistical_Indices function, including the Pearson correlation coefficient (shown as R), the variance accounted for (VAF), RMSE, normalized root mean square error (NRMSE), mean absolute error (MAE), mean absolute percentage error (MARE), root mean squared relative error (RMSRE), mean relative error (MRE), BIAS, Correction Akaike Information Criteria (AICc) and Nash Sutcliffe Efficiency. To calculate these indices, the training and testing targets (i.e., TrainTargets and TestTargets), the training and testing outputs (i.e., TestOutputs and TrainOutputs), the number of training and testing samples (i.e., LTrain and LTest), and the number of tunned parameters through the training phase (i.e., *k*) are required. Following the execution of the Statistical_Indices function, the command window in MATLAB will display the eleven statistical indices for both the training and testing data, as shown in Fig. 11.22.

In addition to providing the quantitative results related to the best iteration highlighted in Fig. 11.22, qualitative results related to this model are also presented using five different figures for both the training and testing data.

Fig. 11.23 shows three of the qualitative figures generated from the statistical indices of the developed SaE-ELM model for the training and testing phases. The top subfigure within Fig. 11.23A and B compares the actual data values

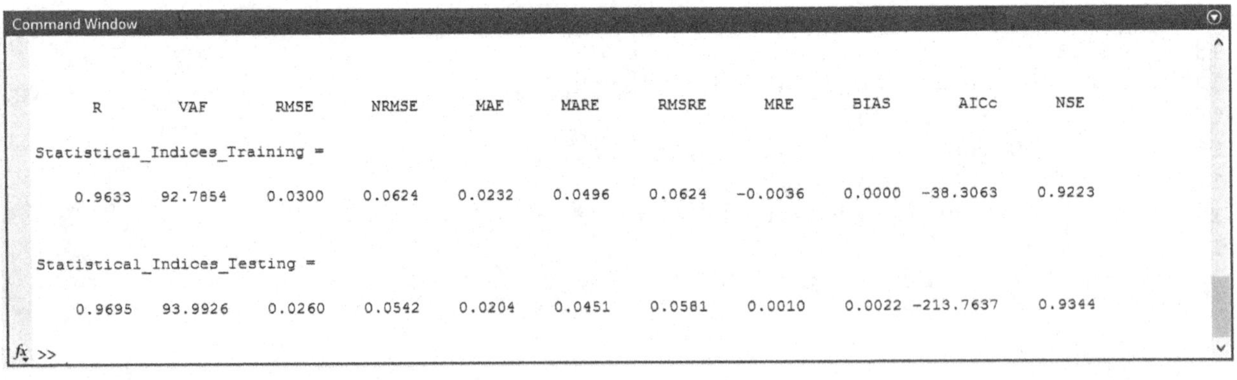

```
Command Window

      R         VAF       RMSE      NRMSE      MAE       MARE      RMSRE      MRE       BIAS       AICc        NSE

  Statistical_Indices_Training =

    0.9633   92.7854    0.0300    0.0624    0.0232    0.0496    0.0624   -0.0036    0.0000   -38.3063    0.9223

  Statistical_Indices_Testing =

    0.9695   93.9926    0.0260    0.0542    0.0204    0.0451    0.0581    0.0010    0.0022 -213.7637    0.9344

fx >>
```

FIGURE 11.22 Results of statistical indices for best iterations at training and testing phases.

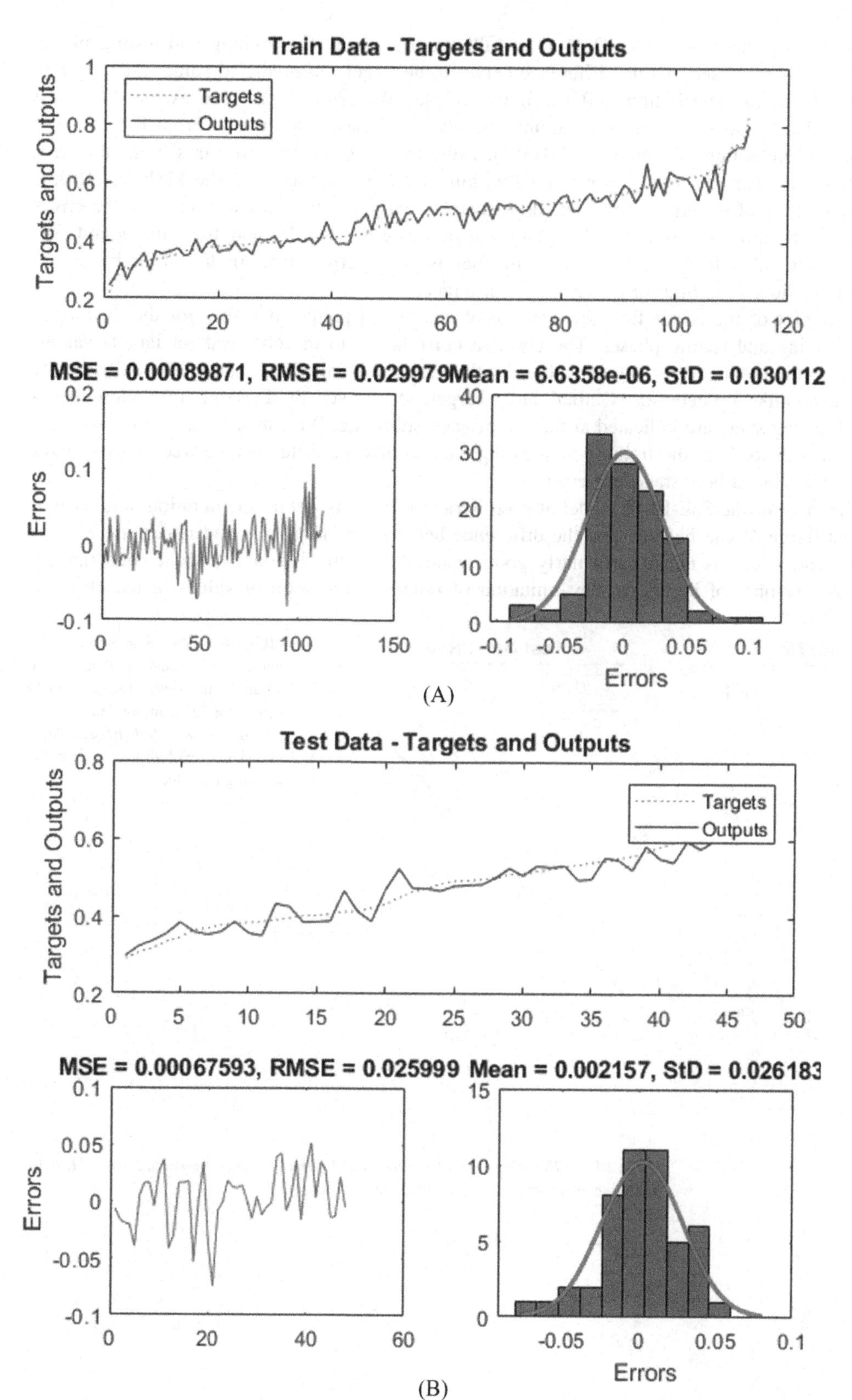

FIGURE 11.23 The quantitative results of the developed SaE-ELM mode for (A) training data and (B) testing data. *SaE-ELM*, Self-adaptive evolutionary extreme learning machine.

(i.e., Targets) and those estimated by the model (i.e., Output) for all samples during the training and testing phase, respectively. From both subfigures, we can see that the difference between the target and predicted values are very low for both the training and testing phases. The bottom left subfigures display the errors (computed as the difference between the target and output values) associated with the training and testing phases, respectively. From Fig. 11.23A, one can see that the training errors are within the range of [-0.1, 0.1], while from Fig. 11.23B, we can see that the error associated with the testing phase is lower and largely constrained within [-0.5,0.5]. Furthermore, the MSE and RMSE for each stage are also provided on top of the error graph. The bottom right subfigures present a histogram of the errors with the corresponding normal distribution (i.e., red line). We can see from this graph that for both the training and testing phases, the error follows the normal distribution. In the same manner as for the error graph, on top of the histogram graph, the mean and standard deviations (i.e., StD) of the errors are provided.

Fig. 11.24 shows the scatter plot of the correlation between the observed and predicted values for the developed SaE-ELM model during the training and testing phases. The abscissa corresponds to the observed (or target) values, while the ordinate corresponds to the predicted (or output) values by the developed SaE-ELM model. The axis title for the ordinate provides the linear relationship between "Output" and "Target." Moreover, the Pearson correlation coefficient for both the training and testing stage are indicated at the top of each subfigure. We can see that in this case, the developed SaE-ELM performed well for both the training and testing data so that the difference between each marker and the perfect correlation "Y = T" line at both stages is very low.

Fig. 11.25 demonstrates the error of the SaE-ELM model at each iteration. For this figure, the iteration number was considered to be 50. From this figure, it can be seen that the difference between the minimum and maximum error is about 25% of the minimum error, which is not a particularly good result. Therefore, it could be concluded that by considering a significantly large number of iterations, the limitation of random generation of values in the classical

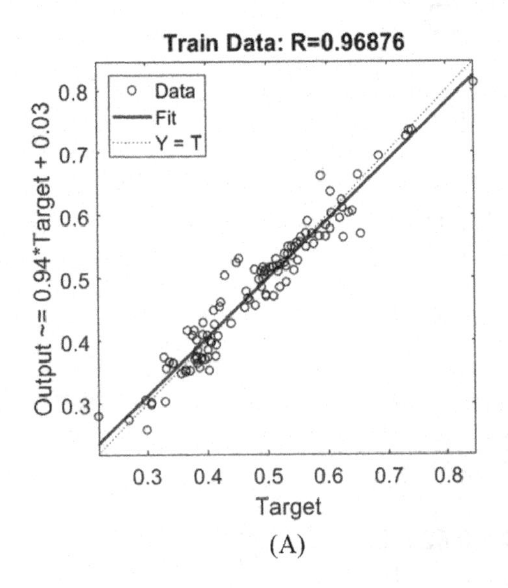

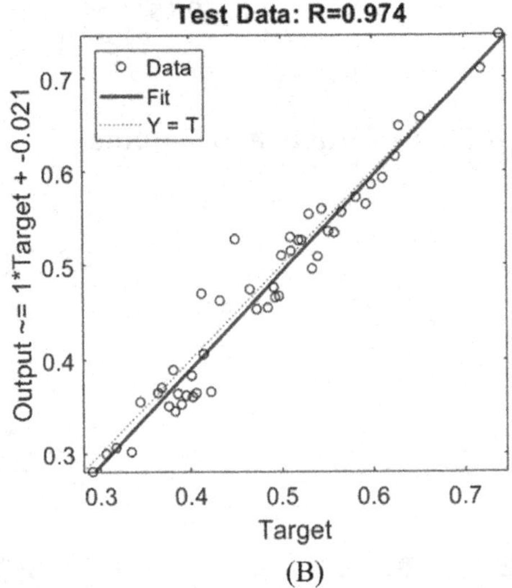

FIGURE 11.24 Scatter plot comparing the actual and estimated value of the developed SaE-ELM model for (A) training data and (B) testing data. *SaE-ELM*, Self-adaptive evolutionary extreme learning machine.

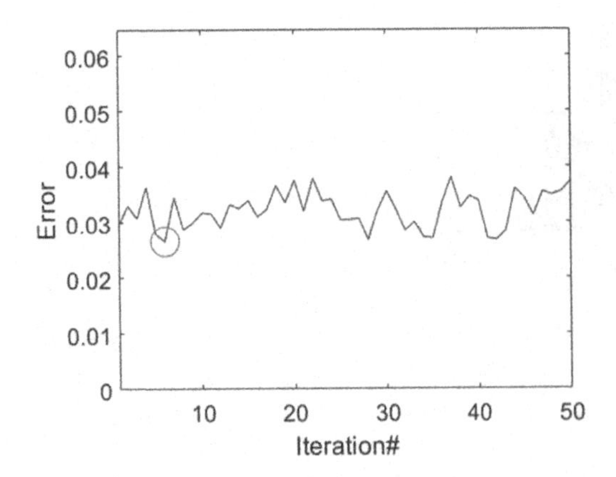

FIGURE 11.25 The error of the SaE-ELM model for each iteration. *SaE-ELM*, Self-adaptive evolutionary extreme learning machine.

ELM-based approach may be overcome. Of course, in the SaE-ELM method, an attempt has been made to overcome this problem by using the self-adapted evolutionary algorithm, with the results of Fig. 11.25 showing that the simultaneous use of both methods (optimization algorithms and considering iteration loop) has led to a better result compared to the case where they are considered individually (i.e., optimization algorithm or considering iteration loop).

11.2 Calculator for self-adaptive machine learning model

To apply the developed SaE-ELM to a given problem in future applications, a calculator is proposed for the first time. To apply this calculator, it is not required to have an in-depth understanding of the coding background of SaE-ELM in MATLAB. As with the previous model calculators, it is simply required that the user can run MATLAB and can enter the number of training samples and their respective values. A schematic representation of the dialog box used by the calculator to assign the number of input variables is provided in Fig. 11.26. As can be seen in this figure, this calculator is capable of handling one to ten input variables. Fig. 11.27A–J demonstrates a schematic representation of the dialog box as the user inputs the ten input variables into the calculator. It should be noted that the number of input variables (i.e., one to ten) could be increased according to the user's problem-specific requirements.

11.3 Sensitivity analysis of the self-adaptive machine learning model parameters

In this subsection the effect of the "generation number" and "population number" parameters on the SaE-ELM modeling performance are examined using five different example datasets. The main features of the five different datasets, including the minimum, average, standard deviation, sample size, and input number, were presented in Chapter 1.

11.3.1 Initial definition of the parameters

To evaluate the effect of the "generation number" and "population number" on the SaE-ELM model, other model parameters will be held constant and considered as the optimal values (as determined from previous chapters). Considering a constant number of hidden neurons and the orthogonality of the randomly allocated matrices (Input weights and bias of hidden neurons), the effect of generation number was studied for all examples. The optimum values of the iteration number, number of hidden neurons, and activation function type for all examples are provided in Fig. 11.28. With fixed values of these parameters, the effect of the generation number and population size parameters are examined.

11.3.2 Generation number

11.3.2.1 Optimum value of the generation number for Example 1

The examination of the impact of generation number on the statistical indices for Example 1 is shown in Fig. 11.29. For the correlation-based indices [i.e., correlation coefficient (R), Nash–Sutcliffe error (NSE), and variance accounted for (VA)], there is no direct (or indirect) relationship observed from increasing the number of generations (GN), so that the optimal GN for all three indices is found at GN = 25. However, it can be observed that the difference between the values recorded at GN = 25 and GN = 15 is very small. The difference between the lowest and highest values for the three indices is insignificant and is found to be 0.59, 1.13, and 1.48 for R, NSE, and VAF, respectively. Therefore, it is required to check the relative-based (i.e., MAE, RMSRE) and absolute-based (MAE, RMSE, NRMSE) indices. Similar to the correlation-based indices, the difference between the minimum and maximum values of each index for varying GN is also not remarkable. The relative error value for all GNs is approximately 4%.

Considering that increasing the GN leads to an increase in the modeling time, the optimal GN to be selected should take into account not only the statistical modeling performance, but also the modeling time required to arrive at the output. Although the best value of nearly every index is recorded at GN = 25, the corresponding values of the indices at GN = 15 have very small differences in value. Additionally, the modeling time at GN = 25 is equal to 1103 seconds,

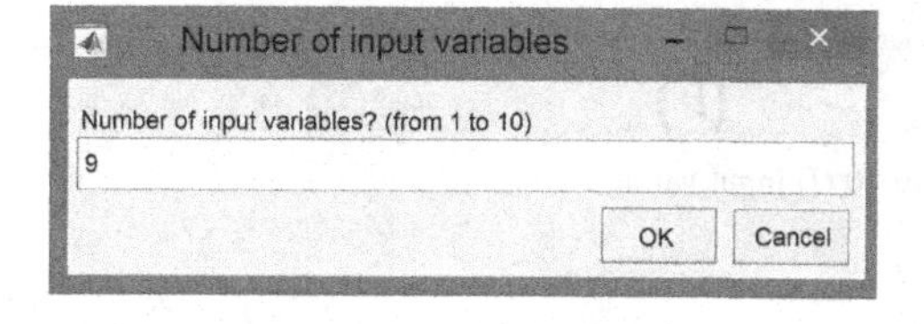

FIGURE 11.26 Dialog box used to assign the number of input variables to the calculator tool.

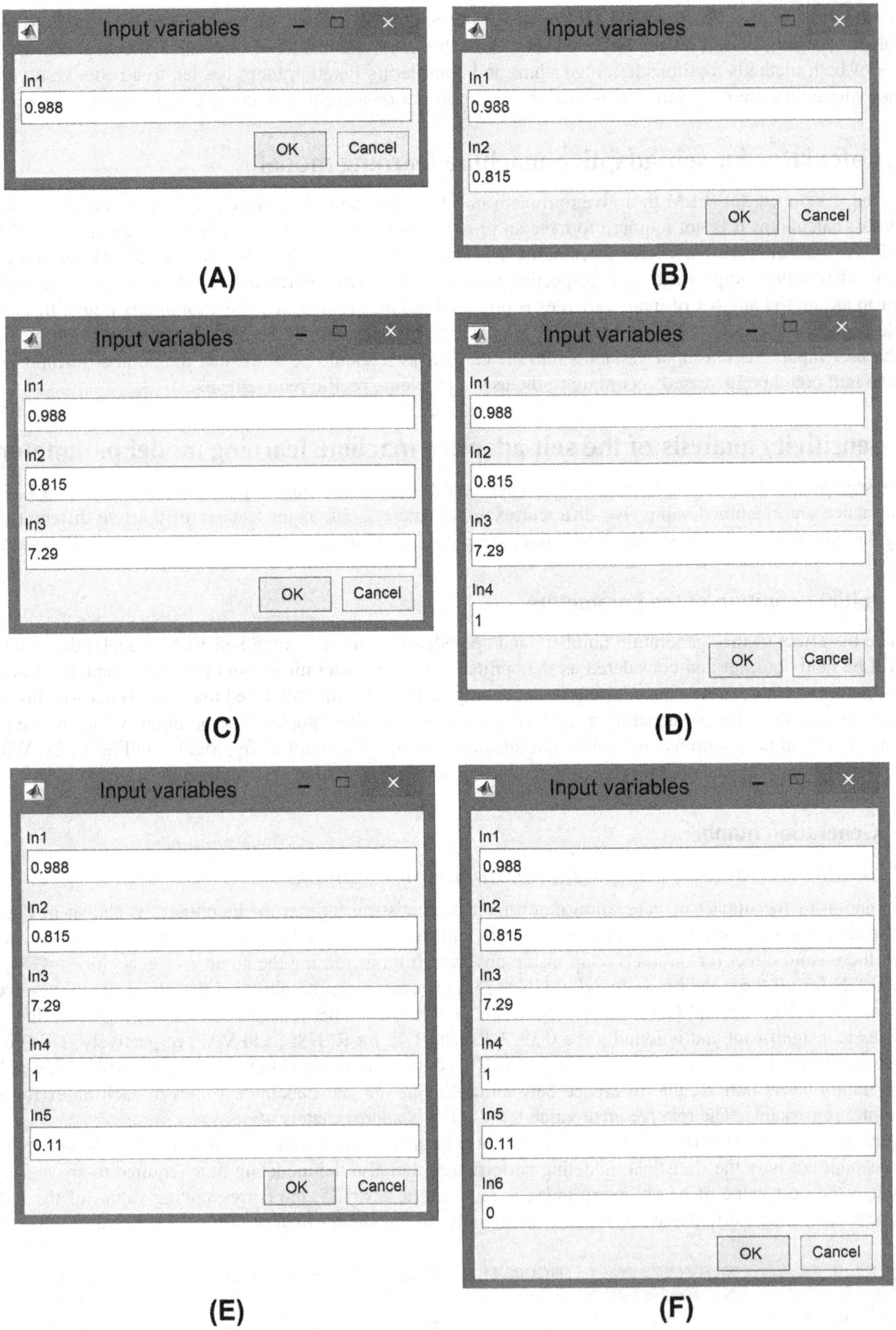

Input variables
In1
0.988
OK Cancel

(A)

Input variables
In1
0.988
In2
0.815
OK Cancel

(B)

Input variables
In1
0.988
In2
0.815
In3
7.29
OK Cancel

(C)

Input variables
In1
0.988
In2
0.815
In3
7.29
In4
1
OK Cancel

(D)

Input variables
In1
0.988
In2
0.815
In3
7.29
In4
1
In5
0.11
OK Cancel

(E)

Input variables
In1
0.988
In2
0.815
In3
7.29
In4
1
In5
0.11
In6
0
OK Cancel

(F)

FIGURE 11.27 Examples of the dialog box output used for assigning the value of one (A) to 10 (J) input variables.

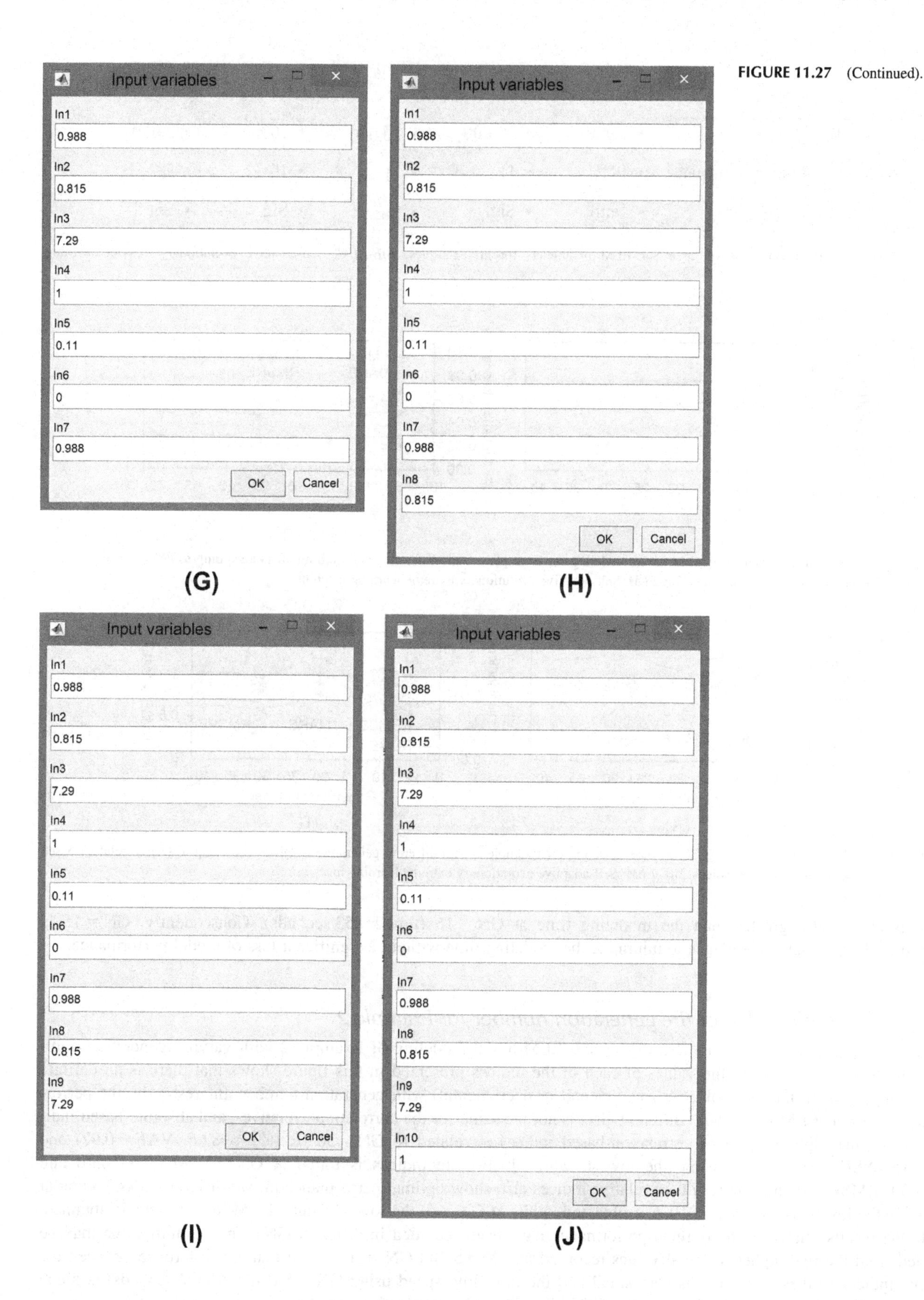

FIGURE 11.27 (Continued).

	Ex. 1	Ex. 2	Ex. 3	Ex. 4	Ex. 5
Iteration No.	• 1,000	• 10,000	• 10,000	• 500	• 2,000
Number of hidden neurons	• 12	• 11	• 11	• 16	• 5
Activation function	• Tanh	• Sig.	• Sig.	• Sig.	• Sig

FIGURE 11.28 The optimum value of three SaE-ELM parameters for all examples. *SaE-ELM*, Self-adaptive evolutionary extreme learning machine.

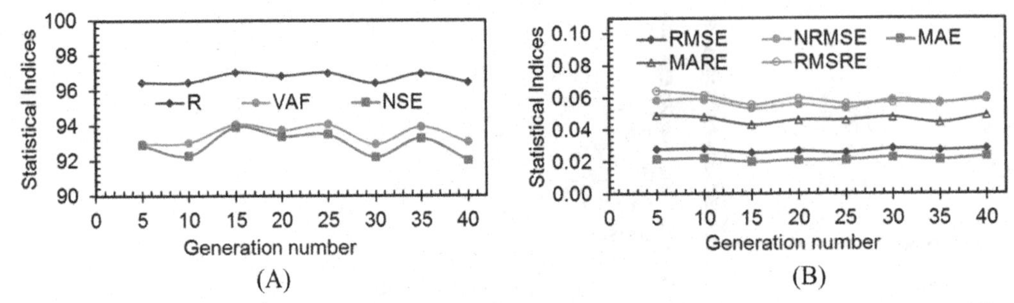

FIGURE 11.29 Performance of the SaE-ELM-based modeling for Example 1 with different generation numbers according to (A) correlation-based indices and (B) relative and absolute indices. *SaE-ELM*, Self-adaptive evolutionary extreme learning machine.

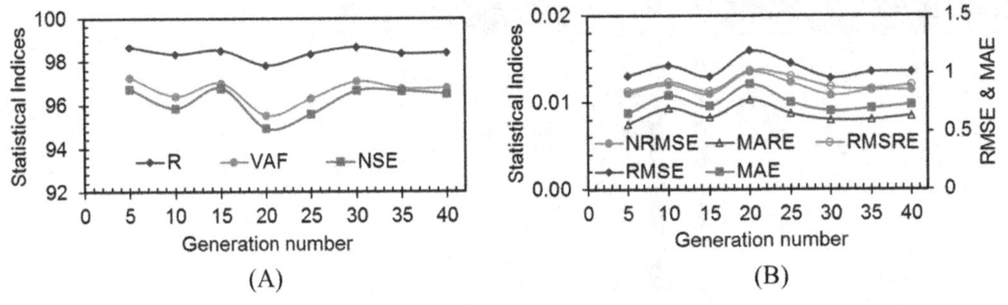

FIGURE 11.30 Performance of the SaE-ELM-based modeling for Example 2 with different generation numbers according to (A) correlation-based indices and (B) relative and absolute indices. *SaE-ELM*, Self-adaptive evolutionary extreme learning machine.

which is about 30% greater than the modeling time at GN = 15 (time = 853 seconds). Consequently, GN = 15 is selected as the optimal performance to minimize the modeling time without a significant loss of model performance.

11.3.2.2 Optimum value of the generation number for Example 2

Fig. 11.30 shows the statistical indices of the SaE-ELM-based modeling of Example 2 with varying generation numbers. The trend of changes in the values of each of the indices presented in this figure shows that there is no definite relationship between the trend of error (incremental or decremental) with generation number. Interestingly, the performance of the SaE-ELM model with different GNs is not the same for the correlation-, relative-, and absolute-based indices, so that the highest value of correlation-based indices is related to GN = 30 (R (%) = 98.66, VAF = 0.97) and GN = 15 (NSE (%) = 97.78), while the lowest value of absolute indices is found at GN = 5 (MAE = 0.656) and GN = 30 (RMSE = 0.96). The results of relative indices also show optimal performance in two different GNs, so that at GN = 30, the lowest NRMSE = 0.01086 is obtained, while at GN = 5, the lowest value of RMSRE = 0.011 is obtained. While the results related to the optimal performance have been recorded in different GNs, an optimum value may be selected from the two highest optimal values recorded at GN = 5 and GN = 30. Given that the difference between the optimal indices in these two GNs is very small and the modeling speed using GN = 5 (time = 4502 seconds) is much lower than GN = 30 (time = 18,305 seconds), GN = 5 is selected as the optimal model.

11.3.2.3 Optimum value of the generation number for Example 3

Fig. 11.31 shows statistical indices of the SaE-ELM-based modeling of Example 3 with different generation numbers. Unlike the previous two examples, the differences between the indices for different GNs are relatively large, with the difference between the lowest and highest values of R, VAF, and NSE being 6.11, 11.5, and 13.5 (respectively). In comparison, these same differences were found to be 0.6, 1.32, and 1.48 (respectively) for Example 1, and 0.85, 1.73, and 1.74, (respectively) for Example 2. However, similar to Examples 1 and 2, the best performance is not observed for the largest GN value, and was instead found for all indices GN = 15.

Increasing the GN does not necessarily lead to increasing model performance in all cases, but instead can also reduce the modeling performance, as is seen in this example. For example, GN = 35 has the second-lowest performance after GN = 5 in almost all indices. The reason for this is considered to be the large number of iterations for the SaE-ELM model using randomly generated matrices. Although SaE-ELM has tried to overcome this limitation by using SADE, it is seen in this example that the results may still be significantly impacted.

11.3.2.4 Optimum value of the generation number for Example 4

Fig. 11.32 shows the statistical indices of the SaE-ELM-based modeling of Example 4 with different generation numbers. From this figure, we can see that the behavior is similar to Examples 1 and 2 in that there is no overall trend of the model performance with GN. A noteworthy point in the modeling results presented in this figure is that the optimal model performance in correlation (R, VAF, NSE), relative (MARE, RMSRE), and absolute (MAE, NRMSE, MAE) based indices are related to different GNs, as was seen in Examples 1 and 2. Indeed, the lowest MAE, NRMSE, and RMSE are provided at GN = 10, the highest correlation-based indices are relative to the GN = 30, and the lowest MARE and RMSRE are related to the GN = 35. Similar to the previous examples, the optimum value of indices is not found at the highest value of GN, which in this case has been specified as GN = 40. The difference between the highest and lowest values of the different indices in this example is very low, so that the greatest difference in indices is for the VAF and NSE (1.65 and 1.58, respectively). To select the optimum model, we now consider the modeling time for the three GNs since the performance difference was found to be negligible. Comparing the three GNs training times, GN = 10 (time = 374 seconds) had a lower modeling time than GN = 30 (time = 803 seconds) and GN = 35 (time- = 996 seconds) and was therefore selected as the optimum value.

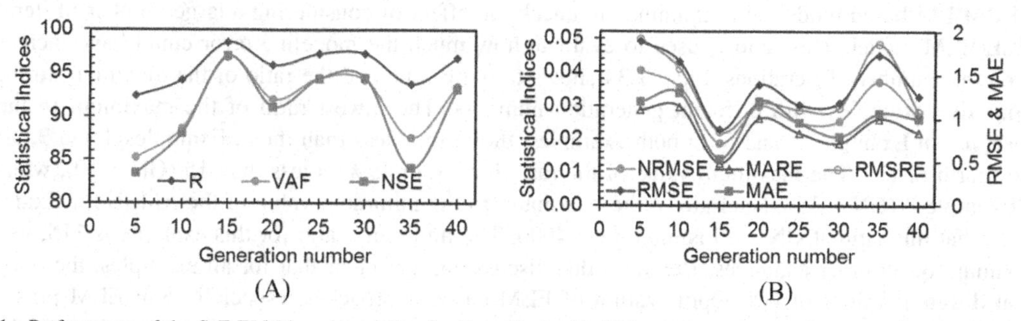

FIGURE 11.31 Performance of the SaE-ELM-based modeling for Example 3 with different generation numbers according to (A) correlation-based indices and (B) relative and absolute indices. *SaE-ELM*, Self-adaptive evolutionary extreme learning machine.

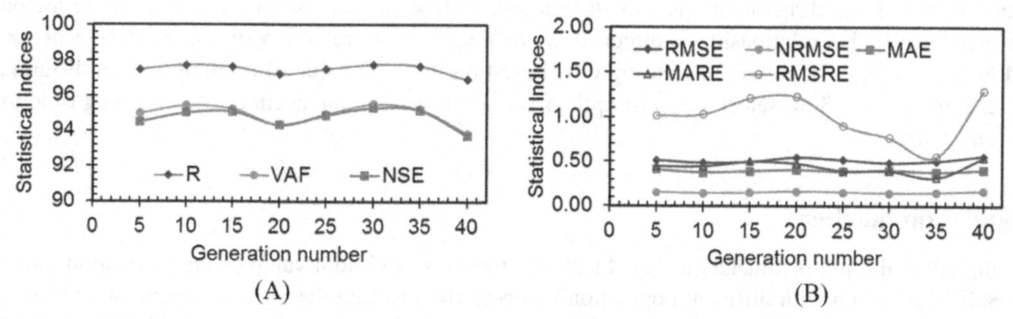

FIGURE 11.32 Performance of the SaE-ELM-based modeling for Example 4 with different generation numbers according to (A) correlation-based indices and (B) relative and absolute indices. *SaE-ELM*, Self-adaptive evolutionary extreme learning machine.

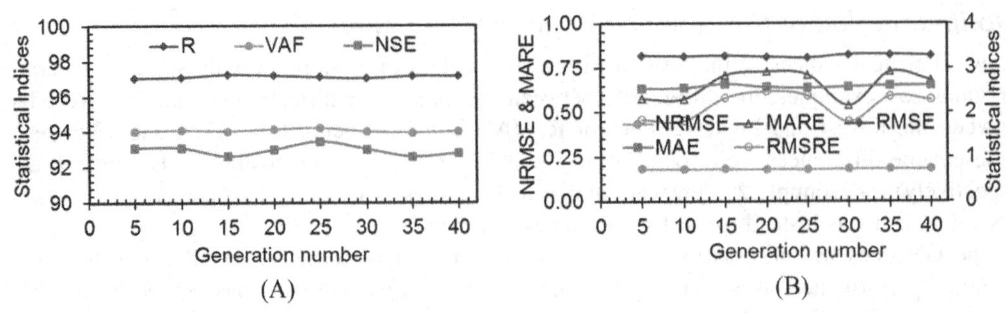

FIGURE 11.33 Performance of the SaE-ELM-based modeling for Example 5 with different generation numbers according to (A) correlation-based indices and (B) relative and absolute indices. *SaE-ELM*, Self-adaptive evolutionary extreme learning machine.

11.3.2.5 Optimum value of the generation number for Example 5

The statistical indices of the SaE-ELM-based modeling of Example 5 for different generation numbers are provided in Fig. 11.33. From the figure, it can be observed that the correlation-based R and VAF indices, as well as the absolute-based NRMSE and RMSE indices, yielded almost constant values with the difference between the lowest and highest values being 0.21, 0.25, 0.07, and 0.004, respectively. It should be understood that considering one correlation-based index such as R or VAF may be led to the misunderstanding of model performance. In this figure, the Pearson correlation coefficient demonstrates a consistently high value (>0.97) across GNs, however, the NSE value (another correlation-based index) is consistently below 0.94. The results of the absolute-based indices (i.e., NRMSE, RMSE, MAE) in all GNs present almost constant values for each index so that the difference between the minimum and maximum of each one is less than 0.1. Unlike the R, VAF, and absolute-based indices, the relative-based indices (i.e., MARE and RMSRE) demonstrated varying model performance for different GN values. In both the MARE and RMSRE, the lowest value is calculated at GN = 30 with MARE = 0.54 and RMSRE = 1.79. While a clear minimum in the MARE and RMSRE is observed at GN = 30 in Fig. 11.33, it should be understood that the difference between the highest and lowest values of these indices is only 0.04 and 0.05, respectively. For this reason, the optimal value will be selected by considering the modeling time, where $N = 5$ has the lowest training time (i.e., 1277 seconds).

After selecting the optimal GN for each example, the ratio of the maximum to the minimum RMSE and training time of the SaE-ELM-based modeling is examined to check the effect of considering a large number of iterations in the developed SaE-ELM model. This ratio is used to evaluate how much the modeling error could have increased had we not considered this number of iterations. Fig. 11.34 shows training time and the ratio of the maximum to the minimum RMSE for our five examples with different generation numbers. The lowest ratio of the maximum to the minimum RMSE is observed in Examples 1 and 5. In both examples, the ratio is less than three. Examples 4 and 3, are ranked in the third and fourth places. The maximum value of the ratio for Example 4, is less than 15 (GN = 10), while it is more than 12 for Example 3 (GN = 35). The ratio of the maximum to the minimum value of the RMSE for Example 2 is the highest one, so that this ratio at GN = 25 is more than 4000. The minimum ratio for this example is 115, which is more than the maximum of all other examples. Based on this discussion, it is clear that for all examples, the maximum ratio is recorded at different values of GN. Optimization of ELM-based approaches, especially SaE-ELM presented in this chapter, is of considerable importance so that the nonconsideration of the GN in the original version of ELM can lead to failure in the modeling of a particular problem.

The modeling times related to different GNs for all examples are also provided in Fig. 11.34. As can be seen in the figure, the training time has a direct relationship with the GN, so that an increase in GN results in an increase in training time. However, while this relationship is direct, it is not linear in nature, and with the exception of Example 3, is characterized by a decreasing slope with increasing GN. Based on the high value of training time in different examples, especially in Examples 2 and 3, to select the most optimal value of the GN, the accuracy and training time must be considered simultaneously.

11.3.3 Population number

According to the optimal input parameters in Fig. 11.28 and the most optimum value of the generation number for each example, the SaE-ELM is run with different population numbers (NP) to find the most optimum value of this parameter. It should be noted that in the previous subsection, a constant NP = 20 was used for the optimization of the GN. In this subsection, five different values are considered to find the most optimal one for each example.

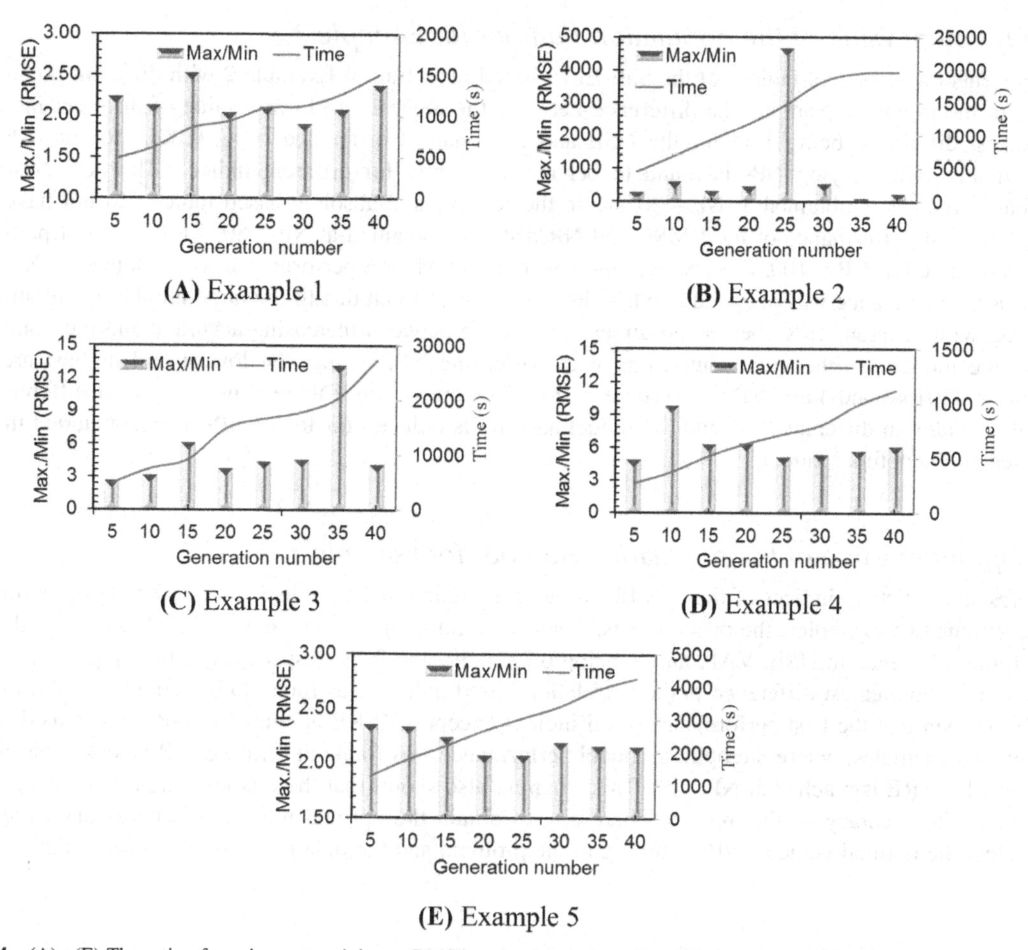

FIGURE 11.34 (A)–(E) The ratio of maximum to minimum RMSE and training time with different generation numbers for Examples 1–5. *RMSE*, Root mean square error.

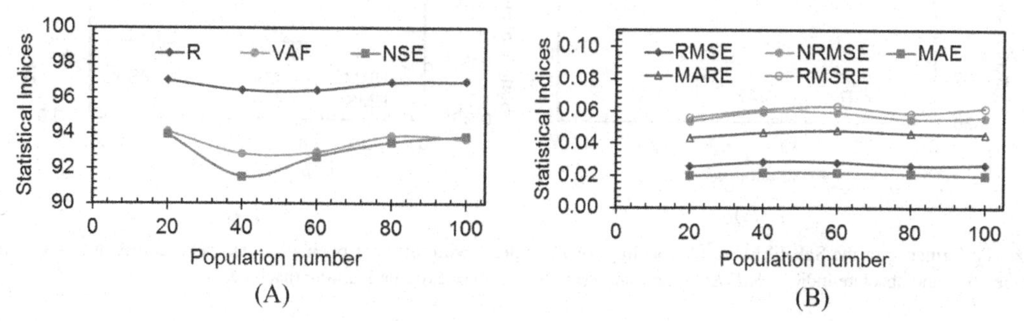

FIGURE 11.35 Performance of the SaE-ELM-based modeling for Example 1 with different population numbers according to (A) correlation-based indices and (B) relative and absolute indices. *SaE-ELM*, Self-adaptive evolutionary extreme learning machine.

11.3.3.1 Optimum value of the population number for Example 1

Fig. 11.35 presents statistical indices of the SaE-ELM-based modeling of Example 1 with different population numbers. It can be seen that the differences between the minimum and maximum values of all indices are very low, with the greatest difference of 1.25 related to VAF and the smallest difference of 0.0019 related to the MAE. The best performance of all correlation-based (i.e., R, VAF, NSE), absolute-based (i.e., RMSE, NRMSE, MAE), and relative-based (i.e., MARE, RMSRE) indices were found to occur at NP = 20. Therefore, NP = 20 is selected as the most accurate SaE-ELM-based model for Example 1. It should be noted that the training time of this model would also be the lowest when compared to models with a larger NP. Changing NP from 20 to 100 results in an increase of the training time from 853 to 1809 seconds.

11.3.3.2 Optimum value of the population number for Example 2

Fig. 11.36 presents the statistical indices of the SaE-ELM-based modeling of Example 2 with different population numbers. Similar to the previous example, the difference between the highest and lowest value of different indices is low, with the greatest difference being 1.34 for the NSE and the smallest difference being 0.0017 for the NRMSE. The model performance with varying NPs is found to behave differently for different indices. The best performance of correlation-based indices is obtained at NP = 20, while the relative and absolute-based indices do not have a singular optimal NP. The best performance of the RMSE and NRMSE was obtained at NP = 80, while the best performance of the MAE was obtained at NP = 100. For absolute indices, the best MARE performance was obtained at NP = 100, and the best RMSRE performance was obtained at NP = 40. It is observed that due to the large number of iterations in SaE-ELM modeling with different NPs, there is no direct relationship between increasing accuracy and the number of NPs. Furthermore, the modeling time greatly increases with increasing NP, so that the lowest and highest are related to NP = 20 (time = 4501 seconds) and NP = 100 (time = 26,077 seconds). Since there is no significant difference between the values of an index in different NPs, and the modeling time is reduced for lower NPs, it is concluded that NP = 20 can be selected as the optimal model.

11.3.3.3 Optimum value of the population number for Example 3

Fig. 11.37 presents statistical indices of the SaE-ELM-based modeling of Example 3 with different population numbers. Unlike the previous two examples, the difference between the minimum and maximum of each index for different NPs is large, with the difference in NSE, VAF, and R being 6.22, 6.59, and 3.18 (respectively). In comparison to the previous two examples, the greatest difference in the correlation-based indices was found to be less than 2. According to the figure, it can be seen that the best performance of all indices except RMSRE is related to NP = 80. This differs greatly from the previous examples, where the optimal model performance was found at different NP values for each statistical index. The best RMSRE is reached at NP = 20. This example also shows that there is no completely direct relationship between NP and the accuracy of the models. Therefore, to obtain the optimal model, it is necessary to test different values and select the optimal value of NP according to the problem and the obtained results of each index.

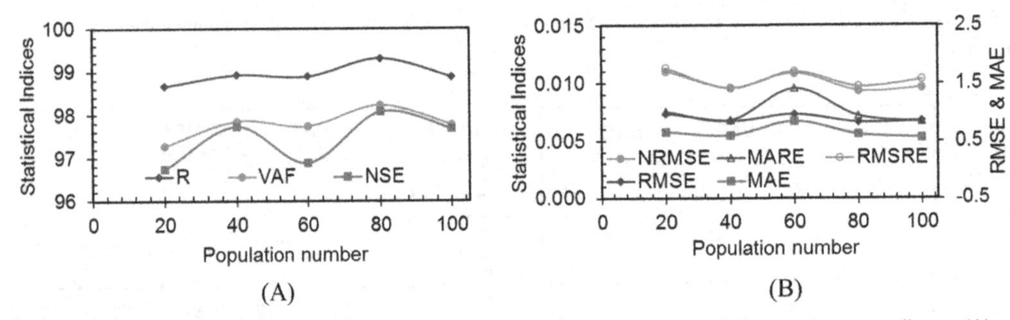

FIGURE 11.36 Performance of the SaE-ELM-based modeling for Example 2 with different population numbers according to (A) correlation-based indices and (B) relative and absolute indices. *SaE-ELM*, Self-adaptive evolutionary extreme learning machine.

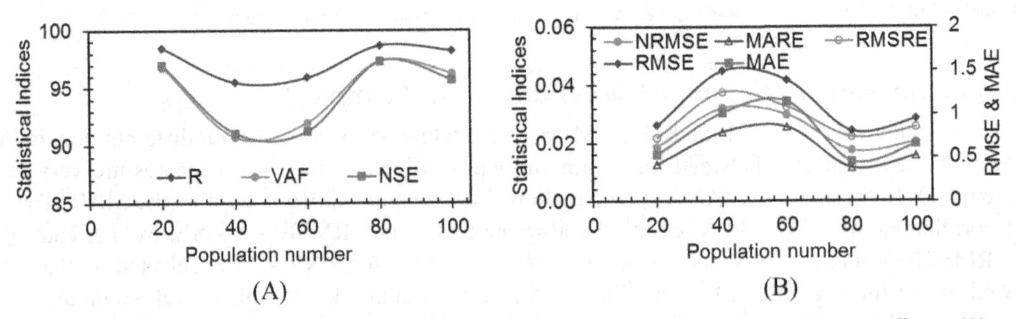

FIGURE 11.37 Performance of the SaE-ELM-based modeling for Example 3 with different population numbers according to (A) correlation-based indices and (B) relative and absolute indices. *SaE-ELM*, Self-adaptive evolutionary extreme learning machine.

11.3.3.4 Optimum value of population number for Example 4

Fig. 11.38 provides statistical indices of the SaE-ELM-based modeling of Example 4 with different population numbers. Similar to previous examples, there is no direct (or indirect) relationship between the NP and model accuracy, so that the best performance of the correlation-based indices (i.e., R, VAF, NSE) is obtained at NP = 60 while the worst performance is seen for NP = 40. It is clear that the best and worst performances are not obtained at the extremes of the examined NP range. The most significant change that is observed among the eight different indicators is in the RMSRE, so that by increasing the value of NP from 20 (RMSRE = 1.02) to 80 (RMSRE = 0.47), the value of this index is almost halved. Similar to Examples 1 and 2, and unlike Example 3, the differences between the lowest and highest values of the correlation-based indices are very low, with the greatest difference being associated with the NSE (1.45). In addition to the best correlation-based indices being obtained at NP = 60, the lowest values of the RMSE and NRMSE are also obtained at this NP. The lowest values of other indices, including MAE (absolute-based index) and the relative-based indices (i.e., MARE and RMSRE), are obtained at NP = 80. To select the most optimum conditions, the training time must also be included with the index performance. In addition to the fact that five of the eight indices (i.e., more than 60% of all indices) used in this figure are related to NP = 60, the training time in this NP (time = 523 seconds) is also less than NP = 80 (time = 657 seconds). Hence NP = 60 is selected as the optimal value for this parameter.

11.3.3.5 Optimum value of the population number for Example 5

Fig. 11.39 presents statistical indices of the SaE-ELM-based modeling of Example 5 with different population numbers. According to the R and VAF indices (two correlation-based indices), the performance of the SaE-ELM model for different NPs is identical. In contrast to this, the NSE values indicate that NP = 40 has a reduced performance compared to other NPs. When looking at the correlation-based indices, the optimal NP values based on R, VAF and NSE are obtained at NP = 40, 80, and 80, respectively. Similar to Example 3, the best value of each index is obtained at different NPs. The best relative-based indices were obtained at NP = 20, while the best RMSE and NRMSE were obtained at NP = 80. The best result for the absolute-based indices (i.e., MAE) is obtained at NP = 60.

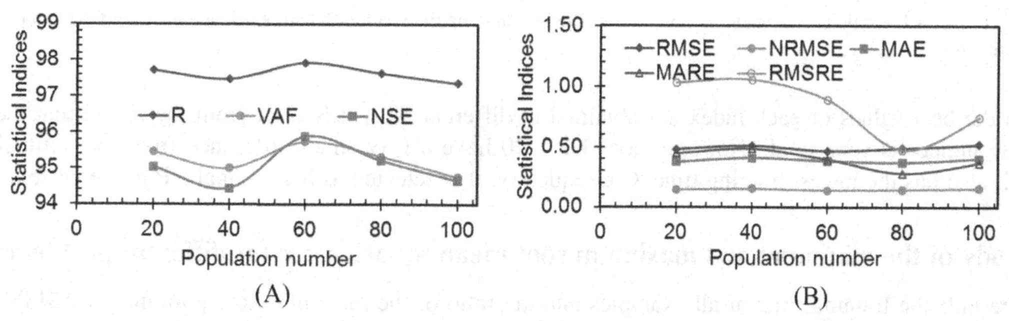

FIGURE 11.38 Performance of the SaE-ELM-based modeling for Example 4 with different population numbers according to (A) correlation-based indices and (B) relative and absolute indices. *SaE-ELM*, Self-adaptive evolutionary extreme learning machine.

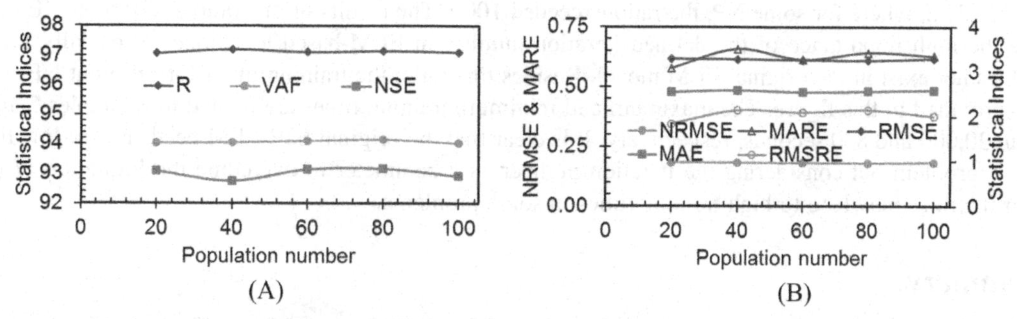

FIGURE 11.39 Performance of the SaE-ELM-based modeling for Example 5 with different population numbers according to (A) correlation-based indices and (B) relative and absolute indices. *SaE-ELM*, Self-adaptive evolutionary extreme learning machine.

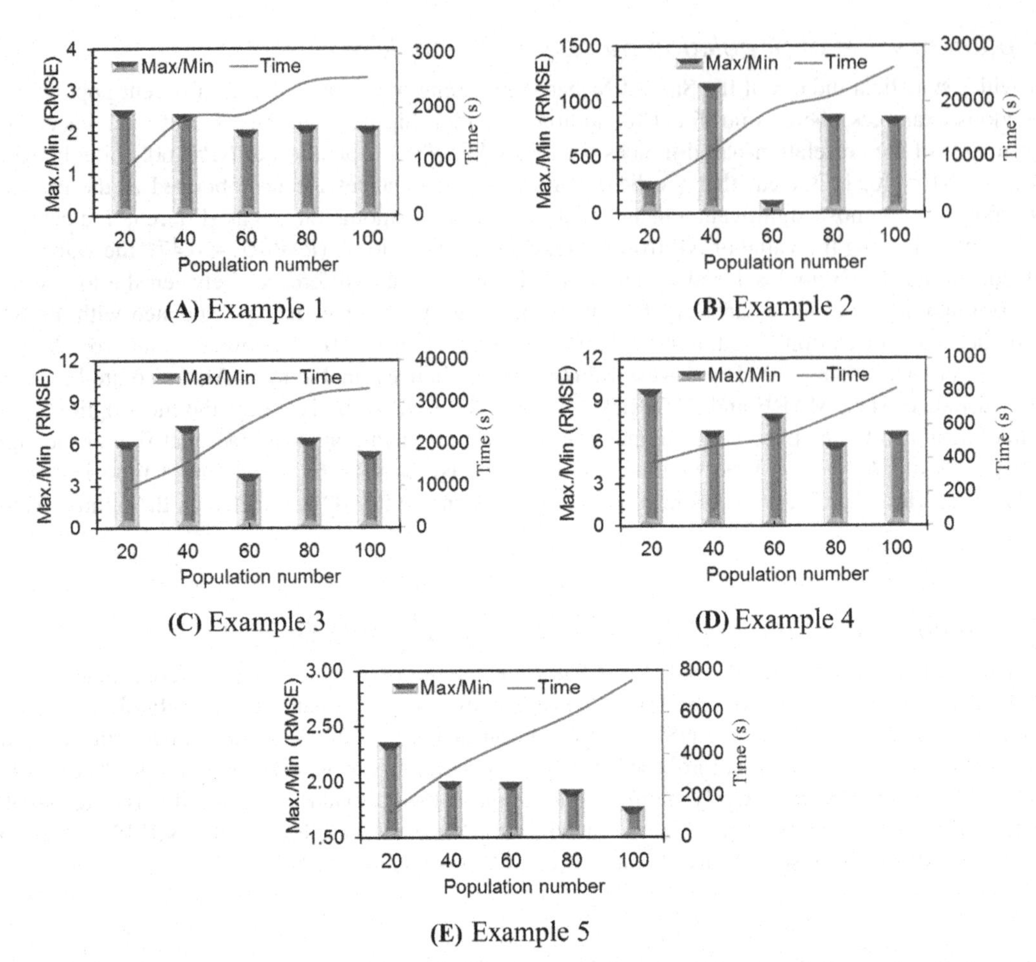

FIGURE 11.40 (A)–(E) The ratio of maximum to minimum RMSE and training time with different population numbers for Examples 1–5. *RMSE*, Root mean square error.

Although the best values of each index are obtained in different NPs, it is clear from the figure that the difference between these indices is very small. Not only does NP = 20 have a very small difference from the optimal NP for all indices, but it also has the lowest training time. Consequently, it is selected as the optimal NP in the current example.

11.3.4 Study of the minimum and maximum root mean square error for different problems

Fig. 11.40 presents the training time of all examples and the ratio of the maximum to the minimum RMSE with different population numbers. The lowest ratio of the maximum to the minimum RMSE is related to Examples 5 and 1, with the value of the ratio being less than 3. This ratio for Examples 3 and 4, was found to be less than 12, although, this value still represents a ratio that is four times as large as the values seen for Examples 1 and 5. The highest ratio is related to Example 2, where for some NP, the ratio exceeded 1000. The results of this ratio at different NPs of all examples indicate the high importance of the defined iteration number in ELM-based techniques, especially the SaE-ELM model that does not exist in the original ELM model. Besides this ratio, the training times for different NPs in all examples are also provided in this figure. The maximum and minimum training times are related to Examples 2 and 4, which are more than 20,000 and 800 seconds, respectively. It is clear that the original SaE-ELM needs a very low time to train for the desired problem but considering the iteration number as a parameter to overcome the limitation of the random generation of matrices may lead to high training times in some problems.

11.4 Summary

This chapter presents the SaE-ELM as an optimized version of the ELM that can learn self-adaptively. A detailed explanation of the mathematical formulation of the SaE-ELM is provided after providing an overview of the ELM, its limitations, and a proposed solution to overcome them in the previous chapter. Moreover, a line-by-line explanation of the coding required for this modeling approach in MATLAB was presented. In addition, a calculator is introduced to enable users to employ the developed SaE-ELM model in future applications, which were involved in the development of the model. To overcome the main limitation of the original ELM, the developed SaE-ELM applies a SADE algorithm to optimize the randomly generated matrices. The algorithm also achieves a high level of accuracy by applying a large number of iterations through an iterative training process. To apply the developed SaE-ELM, in addition to the number of hidden neurons, iteration number, and activation function type, the generation number and population number are required. A detailed sensitivity analysis was undertaken to investigate the effect of the two parameters of the differential evolution algorithm for five examples with different inputs and training samples. After reading this chapter, the reader has the ability to model any problem using the developed SaE-ELM algorithm. Based on the analysis of the five different examples in this chapter, as well as the discussion provided regarding the effect of each of the parameters on the modeling results, the reader can confidently determine the optimal modeling parameters that must be user-defined.

Appendix 11A Supporting information

Data on SaE-ELM can be found in the online version at doi:10.1016/B978-0-443-15284-9.00004-5.

References

Arabameri, A., Asadi Nalivan, O., Chandra Pal, S., Chakrabortty, R., Saha, A., Lee, S., Pradhan, B., & Tien Bui, D. (2020). Novel machine learning approaches for modelling the gully erosion susceptibility. *Remote Sensing*, *12*(17), 2833. Available from https://doi.org/10.3390/rs12172833.

Azimi, H., Shabanlou, S., Ebtehaj, I., Bonakdari, H., & Kardar, S. (2018). Closure to "combination of computational fluid dynamics, adaptive neuro-fuzzy inference system, and genetic algorithm for predicting discharge coefficient of rectangular side orifices" by Hamed Azimi, Saeid Shabanlou, Isa Ebtehaj, Hossein Bonakdari, and Saeid Kardar. *Journal of Irrigation and Drainage Engineering*, *144*(5), 07018021. Available from https://doi.org/10.1061/(ASCE)IR.1943-4774.0001294.

Azimi, H., Shiri, H., & Malta, E. R. (2021). A non-tuned machine learning method to simulate ice-seabed interaction process in clay. *Journal of Pipeline Science and Engineering*, *1*(4), 379–394. Available from https://doi.org/10.1016/j.jpse.2021.08.005.

Bonakdari, H., & Ebtehaj, I. (2021). Discussion of "time-series prediction of streamflows of Malaysian rivers using data-driven techniques" by Siraj Muhammed Pandhiani, Parveen Sihag, Ani Bin Shabri, Balraj Singh, and Quoc Bao Pham. *Journal of Irrigation and Drainage Engineering*, *147*(9), 07021014. Available from https://doi.org/10.1061/(ASCE)IR.1943-4774.0001602.

Bonakdari, H., Ebtehaj, I., Gharabaghi, B., Sharifi, A., & Mosavi, A. (2020). Prediction of discharge capacity of labyrinth weir with gene expression programming. In: *Proceedings of SAI intelligent systems conference* (pp. 202–217). Cham: Springer. https://doi.org/10.1007/978-3-030-55180-3_1.

Bonakdari, H., Gholami, A., Mosavi, A., Kazemian-Kale-Kale, A., Ebtehaj, I., & Azimi, A. H. (2020). A novel comprehensive evaluation method for estimating the bank profile shape and dimensions of stable channels using the maximum entropy principle. *Entropy*, *22*(11), 1218. Available from https://doi.org/10.3390/e22111218.

Calabrò, F., Fabiani, G., & Siettos, C. (2021). Extreme learning machine collocation for the numerical solution of elliptic PDEs with sharp gradients. *Computer Methods in Applied Mechanics and Engineering*, *387*, 114188. Available from https://doi.org/10.1016/j.cma.2021.114188.

Cao, J., Lin, Z., & Huang, G. B. (2012). Self-adaptive evolutionary extreme learning machine. *Neural processing letters*, *36*(3), 285–305. Available from https://doi.org/10.1007/s11063-012-9236-y.

Ebtehaj, I., & Bonakdari, H. (2022a). Discussion of "multivariate drought forecasting in short-and long-term horizons using MSPI and data-driven approaches" by Pouya Aghelpour, Ozgur Kisi, and Vahid Varshavian. *Journal of Hydrologic Engineering*, *27*(11), 07022007. Available from https://doi.org/10.1061/(ASCE)HE.1943-5584.0002216.

Ebtehaj, I., Bonakdari, H., & Kisi, O. (2021c). Discussion of "ANFIS modeling with ICA, BBO, TLBO, and IWO optimization algorithms and sensitivity analysis for predicting daily reference evapotranspiration" by Maryam Zeinolabedini Rezaabad, Sadegh Ghazanfari, and Maryam Salajegheh. *Journal of Hydrologic Engineering*, *26*(12), 07021006. Available from https://doi.org/10.1061/(ASCE)HE.1943-5584.0002141.

Ebtehaj, I., Bonakdari, H., Moradi, F., Gharabaghi, B., & Khozani, Z. S. (2018). An integrated framework of extreme learning machines for predicting scour at pile groups in clear water condition. *Coastal Engineering*, *135*, 1–15. Available from https://doi.org/10.1016/j.coastaleng.2017.12.012.

Ebtehaj, I., Bonakdari, H., Safari, M. J. S., Gharabaghi, B., Zaji, A. H., Madavar, H. R., & Mehr, A. D. (2020). Combination of sensitivity and uncertainty analyses for sediment transport modeling in sewer pipes. *International Journal of Sediment Research*, *35*(2), 157–170. Available from https://doi.org/10.1016/j.ijsrc.2019.08.005.

Ebtehaj, I., Bonakdari, H., Zaji, A. H., Bong, C. H., & Ab Ghani, A. (2016). Design of a new hybrid artificial neural network method based on decision trees for calculating the Froude number in rigid rectangular channels. *Journal of Hydrology and Hydromechanics*, *64*(3), 252. Available from https://doi.org/10.1515/johh-2016-0031.

Ebtehaj, I., Sammen, S. S., Sidek, L. M., Malik, A., Sihag, P., Al-Janabi, A. M. S., & Bonakdari, H. (2021). Prediction of daily water level using new hybridized GS-GMDH and ANFIS-FCM models. *Engineering Applications of Computational Fluid Mechanics*, *15*(1), 1343–1361. Available from https://doi.org/10.1080/19942060.2021.1966837.

Ebtehaj, I., Sattar, A. M., Bonakdari, H., & Zaji, A. H. (2017). Prediction of scour depth around bridge piers using self-adaptive extreme learning machine. *Journal of Hydroinformatics*, *19*(2), 207–224. Available from https://doi.org/10.2166/hydro.2016.025.

Ebtehaj, I., Soltani, K., Amiri, A., Faramarzi, M., Madramootoo, C. A., & Bonakdari, H. (2021). Prognostication of shortwave radiation using an improved no-tuned fast machine learning. *Sustainability*, *13*(14), 8009. Available from https://doi.org/10.3390/su13148009.

Grégoire, G., Fortin, J., Ebtehaj, I., & Bonakdari, H. (2022). Novel hybrid statistical learning framework coupled with random forest and grasshopper optimization algorithm to forecast pesticide use on golf courses. *Agriculture*, *12*(7), 933. Available from https://doi.org/10.3390/agriculture12070933.

Herrera, G. P., Constantino, M., Tabak, B. M., Pistori, H., Su, J. J., & Naranpanawa, A. (2019). Data on forecasting energy prices using machine learning. *Data in Brief*, *25*, 104122. Available from https://doi.org/10.1016/j.dib.2019.104122.

Lanera, C., Berchialla, P., Sharma, A., Minto, C., Gregori, D., & Baldi, I. (2019). Screening PubMed abstracts: Is class imbalance always a challenge to machine learning? *Systematic Reviews*, *8*(1), 317. Available from https://doi.org/10.1186/s13643-019-1245-8.

Maimaitiyiming, M., Sagan, V., Sidike, P., & Kwasniewski, M. T. (2019). Dual activation function-based extreme learning machine (ELM) for estimating grapevine berry yield and quality. *Remote Sensing*, *11*(7), 740. Available from https://doi.org/10.3390/rs11070740.

Mojtahedi, S. F. F., Ebtehaj, I., Hasanipanah, M., Bonakdari, H., & Amnieh, H. B. (2019). Proposing a novel hybrid intelligent model for the simulation of particle size distribution resulting from blasting. *Engineering with Computers*, *35*(1), 47–56. Available from https://doi.org/10.1007/s00366-018-0582-x.

Owolabi, T. O., & Abd Rahman, M. A. (2021). Prediction of band gap energy of doped graphitic carbon nitride using genetic algorithm-based support vector regression and extreme learning machine. *Symmetry*, *13*(3), 411. Available from https://doi.org/10.3390/sym13030411.

Ratnawati, D.E., Marjono, W., & Anam, S. (2020). Comparison of activation function on extreme learning machine (ELM) performance for classifying the active compound. In: *AIP conference proceedings* (Vol. 2264, No. 1, pp. 140001). AIP Publishing LLC. https://doi.org/10.1063/5.0023872.

Ratzinger, F., Haslacher, H., Perkmann, T., Pinzan, M., Anner, P., Makristathis, A., & Dorffner, G. (2018). Machine learning for fast identification of bacteraemia in SIRS patients treated on standard care wards: A cohort study. *Scientific Reports*, *8*(1), 12233. Available from https://doi.org/10.1038/s41598-018-30236-9.

Saha, S., Roy, J., Pradhan, B., & Hembram, T. K. (2021). Hybrid ensemble machine learning approaches for landslide susceptibility mapping using different sampling ratios at East Sikkim Himalayan, India. *Advances in Space Research*, *68*(7), 2819–2840. Available from https://doi.org/10.1016/j.asr.2021.05.018.

Samal, D., Dash, P. K., & Bisoi, R. (2022). Modified added activation function based exponential robust random vector functional link network with expanded version for nonlinear system identification. *Applied Intelligence*, *52*(5), 5657–5683. Available from https://doi.org/10.1007/s10489-021-02664-0.

Sattar, A., Ertuğrul, Ö. F., Gharabaghi, B., McBean, E. A., & Cao, J. (2019). Extreme learning machine model for water network management. *Neural Computing and Applications*, *31*(1), 157–169. Available from https://doi.org/10.1007/s00521-017-2987-7.

Suchithra, M. S., & Pai, M. L. (2020). Improving the prediction accuracy of soil nutrient classification by optimizing extreme learning machine parameters. *Information processing in Agriculture*, *7*(1), 72–82. Available from https://doi.org/10.1016/j.inpa.2019.05.003.

Thottakkara, P., Ozrazgat-Baslanti, T., Hupf, B. B., Rashidi, P., Pardalos, P., Momcilovic, P., & Bihorac, A. (2016). Application of machine learning techniques to high-dimensional clinical data to forecast postoperative complications. *PLoS One*, *11*(5), e0155705. Available from https://doi.org/10.1371/journal.pone.0155705.

Tripathi, D., Edla, D. R., Kuppili, V., & Bablani, A. (2020). Evolutionary extreme learning machine with novel activation function for credit scoring. *Engineering Applications of Artificial Intelligence*, *96*, 103980. Available from https://doi.org/10.1016/j.engappai.2020.103980.

Zeynoddin, M., Bonakdari, H., Azari, A., Ebtehaj, I., Gharabaghi, B., & Madavar, H. R. (2018). Novel hybrid linear stochastic with non-linear extreme learning machine methods for forecasting monthly rainfall a tropical climate. *Journal of Environmental Management*, *222*, 190–206. Available from https://doi.org/10.1016/j.jenvman.2018.05.072.

Index

Note: Page numbers followed by "*f*" and "*t*" refer to figures and tables, respectively.

F

G

H

I

K

L

M

N

O

Printed in the United States
by Baker & Taylor Publisher Services

Printed in the United States
by Baker & Taylor Publisher Services